HORMONES AND GROWTH FACTORS IN DEVELOPMENT AND NEOPLASIA

HORMONES AND GROWTH FACTORS IN DEVELOPMENT AND NEOPLASIA

Edited by
ROBERT B. DICKSON
and
DAVID S. SALOMON

WILEY-LISS
A JOHN WILEY & SONS, INC., PUBLICATION
New York · Chichester · Weinheim · Brisbane · Singapore · Toronto

This book is printed on acid-free paper.

While the authors, editor, and publisher believe that drug selection and dosage and the specification and usage of equipment and devices, as set forth in this book, are in accord with current recommendations and practice at the time of publication, they accept no legal responsibility for any errors or omissions, and make no warranty, express or implied, with respect to material contained herein. In view of ongoing research, equipment modifications, changes in governmental regulations and the constant flow of information relating to drug therapy, drug reactions, and the use of equipment and devices, the reader is urged to review and evaluate the information provided in the package insert or instructions for each drug, piece of equipment, or device for, among other things, any changes in the instructions or indication of dosage or usage and for added warning and precautions.

Library of Congress Cataloging-in-Publication Data

Hormones and growth factors in development and neoplasia / Robert B.
Dickson, editor ; David S. Salomon, editor.
p. cm.
Includes bibliographical references and index.
ISBN 0-471-16899-8 (cloth : alk. paper)
1. Carcinogenesis. 2. Hormones—Pathophysiology. 3. Growth factors—Pathophysiology. 4. Endocrinology, Developmental. 5. Hormones—Physiological effect. 6. Growth factors—Physiological effect. I. Dickson, Robert B. (Robert Brent), 1952– . II. Salomon, David S., 1947– .
[DNLM: 1. Genitalia—growth & development. 2. Breast—growth & development. 3. Cell Transformation, Neoplastic. 4. Neoplasms--etiology. 5. Hormones—physiology. 6. Growth Substances--physiology. 7. Receptors, Growth Factor—physiology. WJ 700 H813 1998]
RC268.5.H663 1998
616.99'4071—dc21
DNLM/DLC
for Library of Congress 97-17442

CONTENTS

PREFACE
From Worms to People: The Evolution and Devolution of Intracellular Communication ix

CONTRIBUTORS xiii

PART I GROWTH FACTORS AND STEROID HORMONES IN THE DEVELOPMENT OF INVERTEBRATES AND AMPHIBIANS 1

1 **Pattern Formation by Sequential Signaling during *C. elegans* Vulval Induction** 3
Jeffrey S. Simske and Stuart K. Kim

2 **EGF-Receptor and TGF-β-like Dpp Signaling During *Drosophila* Development** 19
Ethan Bier

3 **Ecdysone Response in *Drosophila*** 45
William Segraves

4 **Estrogen Control of *Xenopus Laevis* Egg Yolk mRNA Synthesis and Degradation** 79
David J. Shapiro, Robin E. Dodson, Maria Acena, Vincent A. DiPippo, and Hiroshi Kanamori

PART II GROWTH FACTORS AND HORMONES IN MAMMALIAN DEVELOPMENT 91

5 Growth Factors in the Mammalian Pre- and Postimplantation Embryos 93
Daniel A. Rappolee

6 Transcriptional Regulation in an *in Vitro* Model System for Mammalian Embryogenesis 115
Angie Rizzino

7 Cellular Interactions Mediated by Tyrosine Kinase Receptors During Development: Driving Forces for Growth, Motility, and Differentiation 131
Carmen Birchmeier and Walter Birchmeier

8 Developmental and Physiologic Roles of ErbB Receptors and Their Ligands in Mammals 145
Noa Ben-Baruch, Iris Alroy, and Yosef Yarden

9 The Estrogen Receptor in Mammalian Development 169
Sylvia W. Curtis and Kenneth S. Korach

10 Progesterone and Development 177
J.K. Leighton and L.L. Wei

PART III POSTNATAL DEVELOPMENT PROCESSES IN THE ADULT: REPRODUCTIVE TRACTS AND MAMMARY GLANDS 191

11 Prolactin in Development of the Mammary Gland and Reproductive Tract 193
Barbara K. Vonderhaar

12 Growth Factors as Mediators of Stromal–Epithelial Interactions in Steroid Hormone Target Organs 207
Gerald R. Cunha, Simon W. Hayward, Yun-Kit Hom, Annemarie A. Donjacour, Takeshi Kurita, Paul S. Cooke, and Dennis B. Lubahn

13 Hormones, Insulin-Like Growth Factors, and Their Binding Proteins in the Female Reproductive Tract 229
Liam J. Murphy

14 Signal Networks in the Mammary Gland: Lessons from Animal Models 239
Lothar Hennighausen

PART IV GROWTH FACTOR–HORMONAL INTERACTIONS IN TUMORIGENESIS AND MALIGNANT PROGRESSION 251

15 Sex Steroids and Cancer 253
Bradley D. Fenster, Elly-Gerald Stoica, Jianming Lu, Mark Danielsen, and Mary Beth Martin

16 The IGF-I Receptor in Normal and Abnormal Growth 269
Renato Baserga

17 Mutational Activation of Receptor Tyrosine Kinases 289
Matthias Lohmeyer and William J. Gullick

18 Estrogen, Growth Factors, and Carcinogenesis of the Reproductive Tract 311
Karen Gray, Benjamin Eitzman, Kimberly Raszmann Washburn, Tsuneo Takahashi, Yasuhiro Sakai, Glenn Merlino, and John McLachlan

19 Oncogenes, Growth Factors, and Hormones in Prostate Cancer 327
Timothy C. Thompson, Terry L. Timme, and Inder Sehgal

20 WNT and Fibroblast Growth Factor Gene Expression During Development of the Mammary Gland and Role of WNTs in Human Cancer 361
Marzieh Jonsson, Kenneth Smith, and Adrian L. Harris

21 *INT3*, a Novel-NOTCH-Related Gene in Mammary Gland Development and Neoplasia 383
Edith C. Kordon and Gilbert H. Smith

22 Tyrosine Kinases and Signal Transduction in Mouse Mammary Tumorigenesis 397
Peter M. Siegel and William J. Muller

23 The *erb*B-2 Gene in Human Cancer: Translation from Research to Application 421
Francisco J. Esteva-Lorenzo, Lakshmi Sastry, and C. Richter King

INDEX 445

PREFACE

FROM WORMS TO PEOPLE: THE EVOLUTION AND DEVOLUTION OF INTRACELLULAR COMMUNICATION

At the dawn of evolution of multicellular organisms, multiple independent body plans must have rapidly become established to fill the Precambrian seas (1). Unspecialized colonies of cells must have learned new ways of communicating with each other for improved survival and reproduction. Although we lack a window through time with which to view the biochemical and molecular bases of these critical processes, we can examine in great detail the mechanisms of cell–cell communication in the developmental and reproductive biologies of the simplest metazoans that are still extant. Although these simple organisms, such as represented by the worm *Caemorhabditis elegans,* have themselves been subject to continual selective pressures for the millions of years that separate them from humans, they provide remarkable insights into early versions of the same signaling pathways that form the bases of our own developmental and pathologic states.

C. elegans has provided a bonanza of clues to the evolution of cell–cell communication. The development of a hermaphrodite vulva in this tiny worm is initiated when a primitive gonadal cell expresses LIN-3, an ancient growth factor-like molecule. LIN-3 triggers division and differentiation in nearby vulval precursor cells. *C. elegans* is such an elegant system for these types of studies because it is so small that the fate of each of its cells is known throughout development. In addition, the study of its genetic material is very advanced and investigators can establish, by gene addition or knockout techniques, the relative importance of individual developmental signals, their cellular receptors, and their intracellular signaling molecules. Although the evolutionary origins of *C. elegans* are shrouded in the Cambrian era, LIN-3 is strikingly similar to a well-known growth factor family (epidermal growth factor,

EGF) of considerable importance to mammalian development and cancer. LIN-3 acts through a receptor termed let-23 and a signal transduction pathway involving tyrosine phosphorylation, the Ras protein, and a mitogen-activated protein (MAP) kinase molecule; again, each is strikingly similar to its mammalian counterparts. *C. elegans* thus provides us with a profound informative system for dissection of the roles of signaling molecules in development, particularly of reproductive structures.

Reproductive tissues became the focus of this book for several reasons. First, due to the importance of these structures for the propagation of the species, they have become complex in their development, involving interaction of multiple extracellular influences. Second, at least some component of the development of reproductive structures occurs postnatally. In many cases, periodic tissue growth, differentiation, and death characterize reproductive cycles. Third, and of particular concern to human society today, cancers of the reproductive tissues (breast, prostate, uterus, cervix, ovaries, testes) are extraordinarily prevalent. Thus, as we move from relatively simple developmental signaling in the reproductive system of a small worm to greater and greater complexity and greater longevity through evolution, there may be greater opportunities for signaling-associated pathologies. Reproductive cancers may be particularly important in this respect (2,3).

With the evolution of arthropods, cell–cell signaling becomes more complex than in *C. elegans,* with the need to orchestrate more different cell types through more stages and with the onset of sexual reproduction. Here our model system is *Drosophila,* the common fruit fly. Growth factor families have now expanded to encompass more members, receptors are now interactive receptor families with multiple intracellular signal transduction pathways, and certain growth factors counter the effects of other factors to finely tune developmental and homeostatic processes. Of critical importance in the overall schema of arthropod development is the need to orchestrate the behavior of distant tissues by neural, neuroendocrine, and endocrine mechanisms. A key family of endocrine effectors, the steroid hormones, typified by ecdysone, are now produced by lipid metabolic processes. This family, as we will see, plays an ever-increasing role in vertebrates (4,5).

We have several important model systems with which to study early vertebrate evolution. Lowest on the evolutionary tree is *Amphioxus,* a tiny fish with a notochord rather than a long vertebrate. A very important fish model also available to researchers is the Zebrafish. We have chosen an amphibian representative, *Xenopus,* the common frog, as a model system to highlight. Our interest here is in the development of the complex egg. Steroid hormones have now evolved to the forms important in our own reproduction: estrogens, progestins, and androgens (6).

Next, we move to the development of mammals. The common house mouse, *Mus musculus,* is the most well established model system here. The mouse has been well studied with respect to its genes, chromosomes, embryogenesis, later development, and reproductive biology. The ability of biologists to selectively inactivate or activate expression of genes in this model with tissue- and development-dependent specificity has led to remarkable insights into human development and pathology. In the reproductive tissues of the mouse, incredible complexity of interaction has developed between hormones and growth factors (2,6).

As we sequentially examine cell–cell communication from both local and systemic perspectives, it becomes clear that the two types of communication have coevolved, each interacting with the other. This interdependence of local and systemic signaling is particularly evident in the male and female reproductive tracts and in the female accessory tissue—the mammary glands. Each of these tissues requires a highly orchestrated endocrine environment in order to coordinate and control local cell–cell and tissue–tissue interactions. Each of these tissues is designed to undergo growth during sexual maturation (puberty) and cycles of growth (male reproductive tissues) or growth and death (female primary and accessory tissues) (6–8).

It is within the complex regulation of postnatal cellular proliferation and death in these male and female reproductive tissues that human cancer most often arises and most often leads to death (with only the important exception of cigarette smoking-induced lung cancer). One in eight women in the United States will contract breast cancer in her lifetime, and one in five men will contract prostate cancer. The costs of this epidemic in human suffering, economic hardship, legal deliberations, and regulatory paralysis are incalculable. It may be that this state of affairs is the result of an unfortunate interaction of greater longevity, gradual accumulation of autonomous or environmentally induced genetic defects, and a tendency for such complex regulatory systems to move toward chaos and disease (3,9,10).

REFERENCES

1. Gould SJ: Wonderful Life. New York: WW, Norton, 1989, pp 1–347.
2. Thibault C, Levasseur MC, Hunter RHF: Reproduction in Mammals and Man. Paris: Ellipses, 1993, pp 1–800.
3. Parker SL, Tong T, Bolden S, Wingo PA: Cancer Statistics 1997. Ca-Cancer J Clin 47:5–27, 1997.
4. Nilsen-Hamilton M (ed): Growth Factors and Development. San Diego, CA: Academic Press, 1990, pp 1–347.
5. Nilsen-Hamilton M (ed): Growth Factors and Signal Transduction in Development. New York: Wiley-Liss, 1994, pp 1–244.
6. Kahn SA, Stancel GM (eds): Protooncogenes and Growth Factors in Steroid Hormone Induced Growth and Differentiation. Ann Arbor, MI: CRC Press, 1994, pp 1–277.
7. Neville MC, Daniel CV (eds): The Mammary Gland. New York: Plenum Press, 1987, pp 1–625.
8. Dickson RB, Lippman ME: Growth factors in breast cancer. Endocrine Rev 16(5): 559–589, 1995.
9. Parker UIG (ed): Nuclear Hormone Receptors. New York: Academic Press, 1991, pp 1–404.
10. Waxman J (ed): Molecular Endocrinology of Cancer. Cambridge University Press, Cambridge, UK, 1996, pp 1–428.

This book was inspired by a Fogarty Symposium held at the National Institutes of Health, under the same title, in 1995.

CONTRIBUTORS

Maria Acena, Department of Biochemistry, University of Illinois, 600 S. Matthews Avenue, Urbana, IL 61801

Iris Alroy, Department of Chemical Immunology, Weizmann Institute of Science, Rehovot, Israel 76100

Renato Baserga, Jefferson Cancer Center, Thomas Jefferson University, 624 Bluemle Life Science Building, 233 South 10th Street, Philadelphia, PA 19107

Noa Ben-Baruch, Department of Chemical Immunology, Weizmann Institute of Science, Rehovot, Israel 76100

Ethan Bier, Department of Biology, University of California, San Diego, La Jolla, CA 92093-0322

Carmen Birchmeier, Max Delbrück Center for Molecular Biology, Department of Cell Biology, Robert Rossle-Strasse 10, 13125 Berlin, Germany

Walter Birchmeier, Max Delbrück Center for Molecular Biology, Department of Cell Biology, Robert Rossle-Strasse 10, 13125 Berlin, Germany

Paul S. Cooke, Department of Veterinary Biosciences, University of Illinois, Urbana, IL 61801

Gerald R. Cunha, Department of Anatomy, University of California, San Francisco, 3rd and Paranassus Avenues, San Francisco, CA 84143-0001

Sylvia W. Curtis, Laboratory of Reproductive and Developmental Toxicology, NIEHS, NIH, Post Office Box 12233, Research Triangle Park, NC 27709

Mark Danielsen, Lombardi Cancer Center, The Research Building, Georgetown University, 3970 Reservoir Rd., Washington, DC 20007

Robert B. Dickson, Lombardi Cancer Center, Georgetown University Medical Center, 3970 Reservoir Road, The Research Bldg, W417, Washington, DC 20007

Vincent A. DiPippo, Department of Biochemistry, University of Illinois, 600 S. Matthews Avenue, Urbana, IL 61801

Robin E. Dodson, Department of Biochemistry, University of Illinois, 600 S. Matthews Avenue, Urbana, IL 61801

Annemarie A. Donjacour, Department of Anatomy, University of California, San Francisco, CA 94143

Benjamin Eitzman, National Institute of Environmental Health Sciences, Research Triangle Park, NC 27709

Francisco J. Esteva-Lorenzo, Lombardi Cancer Center, The Research Building, Georgetown University, 3970 Reservoir Rd., Washington, DC 20007

Bradley D. Fenster, Lombardi Cancer Center, The Research Building, Georgetown University, 3970 Reservoir Rd., Washington, DC 20007

Karen Gray, Department of Obstetrics and Gynecology, Uniformed Services University, 4301 Jones Bridge Rd., Bethesda, MD 20814

William J. Gullick, Hammersmith Hospital, ICRF, Molecular Oncology Group, Du Cane Road, London W12 ONN, England

Adrian L. Harris, ICRF Clinical Oncology UNH, Churchill Hospital, Headington, Oxford OX3 71J England

Simon W. Hayward, Department of Anatomy, University of California, San Francisco, CA 94143

Lothar Henninghausen, Laboratory of Metabolism and Biochemistry, Building 10, Room 9N113, NIDDK, NIH, Bethesda, MD 20892

Yun Kit Hom, Department of Anatomy, University of California, San Francisco, CA 94143

Marzieh Jonsson, Latinlinjen 17, SE-224 73 Lund, Sweden

Hiroshi Kanamori, Department of Biochemistry, University of Illinois, 600 S. Matthews Avenue, Urbana, IL 61801

Stuart K. Kim, Department of Developmental Biology, Stanford University Medical School, Stanford, CA 94305

C. Richter King, Lombardi Cancer Center, The Research Building, Georgetown University, 3970 Reservoir Rd., Washington, DC 20007

Kenneth S. Korach, Laboratory of Reproductive and Developmental Toxicology, NIEHS, NIH, Post Office Box 12233, Research Triangle Park, NC 27709

Edith C. Kordon, Laboratory of Tumor Immunology and Biology, 9000 Rockville Pike, Building 10, Room 5B52, NCI, NIH, Bethesda, MD 20892

Takeshi Kurita, Department of Anatomy, University of California, San Francisco, CA 94143

J.K. Leighton, The Research Building, Lombardi Cancer Center, Georgetown University, 3790 Reservoir Rd., Washington, DC 20007

Matthias Lohmeyer, Hammersmith Hospital, ICRF, Molecular Oncology Group, Du Cane Road, London W12 ONN, England

Jianming Lu, Lombardi Cancer Center, The Research Building, Georgetown University, 3970 Reservoir Rd., Washington, DC 20007

Dennis B. Lubahn, Departments of Biochemistry and Child Health, University of Missouri, Columbia, MO 65211

MaryBeth Martin, Lombardi Cancer Center, The Research Building, Georgetown University, 3970 Reservoir Rd., Washington, DC 20007

John McLachlan, Tulane University, New Orleans, LA 70112

Glenn Merlino, National Cancer Institute, National Institutes of Health, Bethesda, MD 20892

William J. Muller, McMaster University, LSB 43b, 1280 Mainstreet West, Hamilton, Ontario L8S-4K1, Canada

Liam J. Murphy, Departments of Physiology and Internal Medicine, University of Manitoba, Room 435, 730 Williams Avenue, Winnipeg, Manitoba R3E 3JT, Canada

Daniel Rappolee, Department of Obstetrics and Gynecology, Northwestern University Medical School, Tarry Building, Room 4-725, 303 East Chicago Street, Chicago, IL 60611

Angie Rizzino, Eppley Institute for Research in Cancer and Allied Diseases, Department of Pathology and Microbiology, University of Nebraska Medical Center, Omaha, NE 68198

Yasuhiro Sakai, Department of Anatomy, School of Medicine, Kitasato University, Sagamihara 1-15-1, Kanagawa 229, Japan

David S. Salomon, LTIB-OCBD, National Cancer Institute, Bldg. 10, Room 5B43, Bethesda, MD 20892

Lakshmi Sastry, Lombardi Cancer Center, The Research Building, Georgetown University, 3970 Reservoir Rd., Washington, DC 20007

William A. Segraves, Department of Biology, Yale University, P.O. Box 6666, New Haven, CT 06511

Inder Sehgal, Scott Department of Urology, Baylor College of Medicine, 6535 Fannin Street 1035, Houston, TX 77030

David J. Shapiro, Department of Biochemistry B-4 RAL, University of Illinois, 600 S. Matthews Ave., Urbana, IL 61801

Peter M. Siegel, McMaster University, LSB 43b, 1280 Mainstreet West, Hamilton, Ontario L8S-4K1, Canada

Jeffrey S. Simske, Department of Developmental Biology, Stanford University Medical School, Stanford, CA 94305

Gilbert H. Smith, Laboratory of Tumor Immunology and Biology, 9000 Rockville Pike, Building 10, Room 5B52, NCI, NIH, Bethesda, MD 20892

Kenneth Smith, ICRF Clinical Oncology UNH, Churchill Hospital, Headington, Oxford OX3 71J, England

Elly-Gerald Stoica, Lombardi Cancer Center, The Research Building, Georgetown University, 3970 Reservoir Rd., Washington, DC 20007

Tsuneo Takahashi, Yokohama City University, School of Medicine, 3-9 Fukuura, Kanazawa-ku, Yokohama 236, Japan

Timothy C. Thompson, Scott Department of Urology, Baylor College of Medicine, 6535 Fannin Street 1035, Houston, TX 77030

Terry L. Timme, Scott Department of Urology, Baylor College of Medicine, 6535 Fannin Street 1035, Houston, TX 77030

Barbara K. Vonderhaar, Chief, Molecular and Cellular Endocrinology Section, 9000 Rockville Pike, Building 10, Room 5B56, Laboratory of Tumor Immunology and Biology, NCI, NIH, Bethesda, MD 20892

Kimberly Raszmann Washburn, National Institute of Environmental Health Sciences, Research Triangle Park, NC 27709

L.L. Wei, The Research Building, Lombardi Cancer Center, Georgetown University, 3790 Reservoir Rd., Washington, DC 20007

Yosef Yarden, Department of Chemical Immunology, Weizmann Institute of Science, Rehovot, Israel 76100

HORMONES AND GROWTH FACTORS IN DEVELOPMENT AND NEOPLASIA

PART I

GROWTH FACTORS AND STEROID HORMONES IN THE DEVELOPMENT OF INVERTEBRATES AND AMPHIBIANS

1

PATTERN FORMATION BY SEQUENTIAL SIGNALING DURING *C. elegans* VULVAL INDUCTION

JEFFREY S. SIMSKE

Department of Zoology, University of Wisconsin, Madison, WI 53706

STUART K. KIM

Department of Developmental Biology, Stanford University Medical School, Stanford, CA 94305

1.1. INTRODUCTION

Developmental patterns are often regulated and coordinated by cell signaling. For example, the number and types of fingers in our hands are determined by signals in the limb bud (1,2). Neuronal identity in the spinal cord along the dorsal/ventral axis is established by signals from the notochord (3). The pattern of veins in the wings and the types of bristles on the body of *Drosophila* are determined by cell signaling (4,5). In *Xenopus* and the chick, the entire body axis is established by signals from an organizer called the *Spemann's organizer* or *Henson's node*, respectively (6). It is fascinating how a relatively simple signal or set of signals acting early in development can result in complex and intricate developmental patterns at later stages, and how this patterning process can be so reproducible during development from individual to individual.

How do intercellular signals regulate developmental patterns? How can a signal from a localized source convey spatial information to cells at distinct locations and cause them to develop differently? Two general mechanisms have been proposed to

Hormones and Growth Factors in Development and Neoplasia, Edited by Robert B. Dickson and David S. Salomon
ISBN 0-471-16899-8

explain spatial patterning. One long-standing hypothesis is that a signal might act as a morphogen such that different concentrations along a spatial gradient could induce discrete cell types (7–10). Another hypothesis is that spatial patterning might be established by a signaling relay mechanism such that the first signal induces differentiation in the nearest cells; these cells could subsequently express a second signal that acts directly on the next set of cells, and so on.

1.2. VULVAL DEVELOPMENT

The purpose of this review is to examine the molecular mechanism of induction (morphogen vs. sequential relay) that is used to pattern a simple developmental system: the induction of the hermaphrodite vulva in *Caenorhabditis elegans.* The vulva is formed when a signal (LIN-3) from the gonadal anchor cell induces certain vulval precursor cells to divide and differentiate to produce the vulva (11). There are six vulval precursor cells (P3.p to P8.p) that lie in an anterior-posterior row along the ventral midline and comprise the ventral epidermis of the second and third larval stages (12). LIN-3 induces the vulval precursor cell that is immediately ventral to the anchor cell (P6.p) to divide three times and produce eight cells that form the inner portion of the vulva (termed the *1° cell fate*). The vulval precursor cells that are immediately anterior and posterior to the 1° cell (normally P5.p and P7.p) each produce seven progeny cells that form the outer portions of the vulva (termed the *2° cell fate*). The vulval precursor cells located far from the anchor cell (P3.p, P4.p, and P8.p) each divide once to generate two cells that fuse with the epidermis (termed the *3° cell fate*). All of the vulval precursor cells are equivalent in developmental potential since each cell can express the 1°, 2°, or 3° cell fate (13,14). Thus, the vulval precursor cells are termed an *equivalence group.* During normal development, the six vulval precursor cells (P3.p to P8.p) always express the same pattern of cell fates (3° - 3° - 2° - 1° - 2° - 3) (12).

Elegant experiments have shown that the position of the anchor cell determines which vulval precursor cells are induced to form the vulva (13–16). Thus, the anchor cell acts as a developmental organizer by inducing vulval precursor cells at varying distances to express different developmental fates. All of the vulval precursor cells express the 3° fate when the anchor cell is ablated by a laser microbeam, showing that the anchor cell induces expression of the 1° and 2° vulval cell fates (16). In *dig-1* mutants, the anchor cell is located at different positions (15). When the anchor cell is shifted anteriorly, the position of the vulva is also shifted to correspond to the new position of the anchor cell. Specifically, when the anchor cell is positioned dorsal to P5.p, P5.p expresses the 1° cell fate and the two adjacent cells (P4.p and P6.p) express the 2° cell fate.

Organizers play important roles in establishing pattern in all animals; examples include the Spemann's organizer, the zone of polarizing activity (ZPA), and the notochord (1,3,17). Each of these organizers controls pattern formation by inducing different developmental fates at varying distances in either a direct or an indirect fashion. The Spemann's organizer in the *Xenopus* embryo organizes the body axis, the

ZPA organizes digit identity in the vertebrate limb, and the notochord organizes neuronal fates along the dorsal-ventral axis in the mammalian neural tube. One advantage of studying the organization of pattern formation in *C. elegans* vulval development is that developmental fates can be studied at the single-cell level. The organizer consists of the single anchor cell, the responding tissue consists of only six vulval precursor field cells in a linear row, and there are only three discrete developmental fates (1°, 2°, or 3°). Another advantage is that genetic approaches can be used to distinguish between the different inductive mechanisms. A final point is that the LIN-3 signal functions by activating a highly conserved receptor tyrosine kinase/Ras/mitogen-activated protein (MAP) kinase signaling pathway, suggesting that the mechanisms of pattern formation using this pathway may be conserved (18).

The signal from the anchor cell is encoded by the *lin-3* gene (in *C. elegans,* gene names are italicized and protein names are given in capital letters) (19). The sequence of LIN-3 is similar to that of epidermal growth factor (EGF). In loss-of-function *lin-3* mutants, all vulval precursor cells express the uninduced 3° cell fate instead of the 1° and 2° cell fates. Below, we discuss experiments that address how the LIN-3 signal regulates vulval pattern formation. First, we review the receptor and signal transduction pathway that mediates the direct response to LIN-3. Then we review how LIN-3 might induce different vulval fates. LIN-3 may act directly as a morphogen and induce different vulval fates in a concentration-dependent manner. Alternatively, LIN-3 may initiate a signaling cascade by directly inducing P6.p to adopt the 1° cell fate and express a lateral signal, and then the lateral signal may induce the 2° cell fate in neighboring Pn.p cells.

The anchor cell signal LIN-3 induces vulval cell fates by activating a signaling pathway consisting of at least eight genes in the well-conserved receptor tyrosine kinase/Ras/MAP kinase signal transduction pathway (18,20,21). The receptor for LIN-3 is encoded by *let-23* and is a receptor tyrosine kinase similar to mammalian EGF receptors (22). A detailed summary of this pathway is beyond the scope of this review; we briefly mention some of the genes that play important roles in this signaling pathway. Signaling downstream of the receptor is transduced by SEM-5 (the *C. elegans* homolog of vertebrate GRB2 and *Drosophila* Drk) (23). GRB2 and Drk are adaptor proteins linking receptor tyrosine kinases to exchange factors for Ras (24–26). In *C. elegans,* SEM-5 presumably links the LET-23 receptor to an exchange factor for the Ras homolog LET-60 (27–29). LET-60 Ras functions by activating a protein kinase cascade consisting of LIN-45 (similar to mammalian Raf), MEK-2 (similar to mammalian MEK), and MPK-1/SUR-1 (similar to mammalian MAP kinase) (30–34).

Loss-of-function mutations in any of the genes described above cause a vulvaless phenotype that is identical to the signaling defect seen in *lin-3* mutants (35). Furthermore, constitutive activation of *let-23 receptor* or *let-60 ras* results in the formation of excess vulval tissue (29,36). Epistasis analysis of double-mutant animals containing one mutation that prevents vulval induction and another mutation that causes constitutive vulval induction has been used to order the signaling genes into a genetic pathway that is consistent with the biochemical pathway described above. These genetic data indicate that the LET-23 EGFR/LET-60 Ras signaling pathway is both necessary and sufficient to transduce the LIN-3 signal.

Two other signaling genes, *gap-1* and *ksr-1*, may act in a branch of the anchor cell signaling pathway rather than directly in the pathway. *gap-1* encodes a protein related to the Ras guanosine triphosphatase (GTPase) activating protein and acts to inhibit vulval induction, most likely by inhibiting signaling by LET-60 Ras (A. Hajnal and S. Kim, unpublished observations). *gap-1* mutations do not cause a vulval phenotype in single mutants, but they can suppress the vulvaless phenotype of weak reduction-of-function mutations in *let-60*. *ksr-1* encodes a novel protein kinase that may act to induce vulval cell fates, since loss-of-function *ksr-1* mutations can suppress the multivulva phenotype of gain-of-function *let-60 ras* mutations (37,38).

1.3. TESTS OF THE MORPHOGEN AND SEQUENTIAL SIGNALING MODELS

How does the anchor cell signal (LIN-3) induce the vulva? The key difference between the morphogen and sequential signaling models lies in the manner in which LIN-3 normally induces the presumptive 2° cells (P5.p or P7.p). According to the morphogen model, LIN-3 acts directly on the presumptive 2° cells by weakly activating the LET-23 receptor/LET-60 Ras/MPK-1 MAP kinase signaling pathway. One consequence of this model is that strong activation of the signaling pathway by LIN-3 would have to be functionally distinct from weak activation, since strong activation would result in the expression of the 1° cell fate and weak activation would result in expression of the 2° cell fate. According to the sequential signaling model, a lateral signal from the 1° cell induces the presumptive 2° cell, so that LIN-3 acts indirectly on the presumptive 2° cell by inducing the lateral signal. Below we first describe five experiments indicating that LIN-3 may initiate a signaling relay, and then we review two experiments suggesting that LIN-3 may act directly on the presumptive 2° cells. Finally, we discuss possible explanations to reconcile the differences between the experiments and propose a model for vulval induction by LIN-3.

The first experiment tests whether LIN-3 acts directly on the presumptive 2° cells (Fig. 1.1B). Genetic mosaic analyses were used to observe the consequences of loss of *let-23* activity from the presumptive 2° cells in mosaic clones (39,40). Genetic mosaic animals were identified in which *let-23*(+) activity was missing in either of the presumptive 2° cells (P5.p or P7.p) but was present in P6.p, allowing expression of the 1° cell fate. According to the morphogen hypothesis, LET-23 activity should be required in the presumptive 2° cells to transduce intermediate levels of LIN-3, so that removing *let-23* activity from these cells should cause them to express the uninduced 3° cell fate. According to the signal relay hypothesis, *let-23* activity should not be required in the presumptive 2° cells since these cells could be induced by a lateral signal from the 1° cell. In these experiments, the vast majority of cases showed the latter result; *let-23*(+) activity is not required in the presumptive 2° cells of expression of the 2° cell fate. These results suggest that a lateral signal can induce these cells to express the 2° cell fate.

The second set of experiments show that the 1° cell produces a lateral signal that induces the presumptive 2° cells to express the 2° cell fate (Fig. 1.1C). In *lin-15* multivulva mutants, each of the six vulval precursor cells expresses induced (1° or 2°)

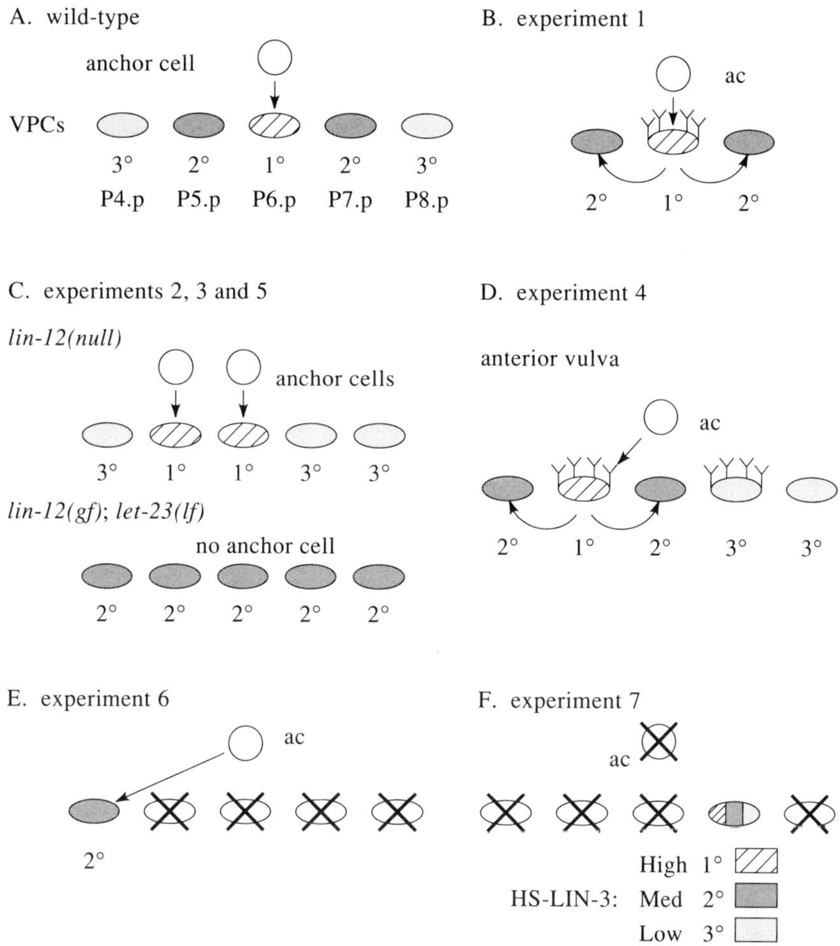

Fig. 1.1. Experimental tests of the morphogen and sequential signaling models. (A) The organization of the cells involved in vulval induction is shown. The inductive anchor cell is positioned dorsal to the vulval precursor cells (VPCs). P3.p is not shown, as occasionally it is not part of the vulval equivalence group. In each case, anterior is to the left, posterior to the right. Expression of the LET-23 receptor is indicated by Y. X indicates either genetic or laser ablation of a cell. HS-LIN-3 refers to expression of LIN-3 from the heat shock promoter. B–F refer to experiments described in the text.

cell fates (35). Sternberg noticed that the pattern of cell fates alternated between 1° and 2° along the row of six vulval precursor cells. He used laser ablation experiments to show that this alternating pattern arose from a lateral signal between the vulval precursor cells; each of the 1° cells sends a lateral signal to adjacent cells that causes them to express the 2° cell fate, thereby producing a 2° cell between each of the 1° cells (41).

Work by Greenwald and colleagues has extended our understanding of the lateral signaling pathway by showing that *lin-12* is likely to encode the lateral signaling receptor. *lin-12* encodes a protein similar to transmembrane receptor proteins such as *Drosophila* Notch and *C. elegans* GLP-1 (42–44), and LIN-12 is expressed in the vulval precursor cells (45). Loss-of-function *lin-12* mutations prevent the expression of the 2° cell fate by vulval precursor cells, whereas gain-of-function *lin-12* mutations cause vulval precursor cells to express the 2° cell fate independently of the lateral signal (42). These results demonstrate that *lin-12* activity is both necessary and sufficient for the expression of the 2° cell fate.

Do the *let-23* and *lin-12* signaling pathways act independently such that the *let-23* pathway specifies the 1° cell fate and the *lin-12* pathway specifies the 2° cell fate? Alternatively, could both signaling pathways collaborate in the same cell to specify the 2° cell fate? In the third experiment, Sternberg and Horvitz tested whether defects in one pathway might be suppressed by activation of the other pathway (46) (Fig. 1.1C). The general finding was that the two pathways act independently; the *let-23* receptor tyrosine kinase signaling pathway appears to specify the 1° cell fate, and the *lin-12* signaling pathway appears to specify the 2° cell fate. For example, loss of *lin-12* activity prevents expression of the 2° cell fate. In *lin-12(lf);let-60(gf)* double mutants, constitutive activation of *let-60* does not suppress the *lin-12(lf)* phenotype by permitting expression of the 2° cell fate. Similar results have been obtained with all double mutants involving one mutation that eliminates *lin-12* activity and another mutation that activates the *let-23* signaling pathway. Next, constitutive activation of *lin-12* causes all vulval precursor cells to express the 2° cell fate. Constitutive expression of the 2° cell fate is not prevented by a *let-23* loss-of-function mutation in *let-23(lf);lin-12(gf)* double mutants, indicating that *let-23* activity is not required for the 2° cell fate as long as the *lin-12* signaling pathway is activated. These experiments indicate that the *lin-12* signaling pathway is both necessary and sufficient to express the 2° cell fate (in the absence of *let-23* signaling) and that the *let-23* pathway is neither necessary nor sufficient to express the 2° cell fate (in the absence of *lin-12* signaling). Thus, these experiments indicate that the two pathways act independently (46). The simplest interpretation of these results is that the two pathways act in a signaling relay; the anchor cell signal activates the *let-23* pathway, which induces expression of the 1° cell fate and the lateral signal, and the lateral signal activates the *lin-12* signaling pathway, which induces expression of the 2° cell fate.

In addition to the lateral signal, the LIN-3 anchor cell signal may also be capable of directly inducing the 2° cell fate. The next set of experiments (experiments 4–7) tests whether a vulval precursor cell can express the 2° cell fate without an adjacent 1° cell (the isolated 2° cell experiment). This experiment has been done four times in different ways. The results of the first two experiments (experiments 4 and 5) indicate that vulval precursor cells rarely or never express the 2° cell fate without an adjacent 1° cell, supporting the sequential relay model for LIN-3 signaling (39,40,42). However, the results of the last two experiments (experiments 6 and 7) show that an isolated vulval precursor cell can sometimes express the 2° cell fate, suggesting that the LIN-3 signal can directly induce the 2° cell fate without an adjacent 1° cell (14,47). These last two experiments appear to support the morphogen model for LIN-3 cell fate determination.

The fourth experiment involves observing the cell fates expressed by the presumptive 2° cells (normally P5.p or P7.p) in the absence of an adjacent 1° cell (normally P6.p) (39,40) (Fig. 1.1D). In this case, a genetic mosaic approach was used to remove *let-23*(+) activity from P6.p (to prevent it from expressing the 1° cell fate) but not from the adjacent presumptive 2° cells. According to the morphogen model, the graded anchor cell signal should be sufficient to directly induce the presumptive 2° cells to express the 2° cell fate. However, none of the presumptive 2° cells were observed to express the 2° cell fate; all expressed either the 1° or 3° cell fate instead. The observation that some presumptive 2° cells expressed the 3° cell fate is particularly difficult to explain using the morphogen model. The presumptive 2° cells are genetically wild type and should be competent to receive a graded signal from the anchor cell, yet these cells often expressed the 3° uninduced cell fate, thereby indicating absence of the postulated graded another cell signal.

The patterns of cell fates expressed in these mosaic animals are interesting and informative. Generally, these mosaics expressed a "shifted vulva" in which a normal pattern of vulval cell fates (2° - 1° - 2°) was expressed except that the vulva was "shifted" either anteriorly or posteriorly from the anchor cell by one cell. We discuss the anteriorly shifted case as an example; the posteriorly shifted vulva is similar except that the vulval precursor cells expressed a mirror symmetric pattern of cell fates. First, as expected, P6.p did not express the 1° cell fate since it lacked *let-23*(+). Instead of P6.p, one of the presumptive 2° cells (P5.p for the anterior vulva) expressed the 1° cell fate. This observation suggests that LIN-3 can diffuse from the anchor cell to P5.p in these mosaic animals and cause them to express the 1° cell fate. Why don't P5.p or P7.p express the 1° cell fate in wild-type animals? One possibility is that the LIN-3 signal may not diffuse to either of these cells in wild-type animals because it is normally sequestered by binding to the LET-23 receptor on P6.p. Another possibility is that in wild-type animals the lateral signal from the 1° cell (P6.p) may normally induce P5.p and P7.p before the LIN-3 signal from the anchor cell, so that P5.p may become committed to the 2° cell fate and be unable to express the 1° cell fate in response to LIN-3.

In genetic mosaics containing an anteriorly shifted vulva, both of the vulval precursor cells adjacent to the 1° cell (P4.p and P6.p) expressed the 2° cell fate, suggesting that they were induced by a lateral signal from the 1° cell. The *let-23* signaling pathway likely did not play a role in inducing the 2° cell fate in these cells since P6.p lacked *let-23*(+) receptor activity (and could not receive LIN-3) and because P4.p was far away from the anchor cell and should not be induced by LIN-3 (P4.p is never induced in wild-type animals). These results point out a possible difference in the function of LIN-12 in *C. elegans* and Notch in *Drosophila*. Experiments in *Drosophila* suggest that Notch inhibits neighboring cells from expressing neural fates (48). In contrast, *C. elegans* LIN-12 does not merely inhibit expression of the 1° cell fate that also promotes the expression of the 2° cell fate. In the experiment described above, the lateral signal mediated by LIN-12 receptor caused vulval precursor cells to express the 2° cell fate that would otherwise be uninduced (39,40).

The remaining presumptive 2° cell (e.g., P7.p for mosaics with an anteriorly shifted vulva) expressed the 3° cell fate. According to the morphogen model, this cell should express the 2° cell fate due to induction by the graded anchor cell signal. As mentioned above, the observation that these cells expressed the 3° rather than the 2° cell fate strongly argues that the graded anchor cell signal either is not normally present or plays a relatively minor role in directly specifying the 2° cell fate.

The fifth experiment used loss-of-function mutations in the lateral signaling receptor, *lin-12* (42) (Fig. 1.1C). In *lin-12(lf)* mutants, vulval precursor cells are "isolated" since they can respond to the anchor cell signal using LET-23 but cannot signal to each other using LIN-12. In these mutants, the vulval precursor cells do not express 2° cell fates. Interestingly, there are also extra anchor cells in *lin-12(lf)* animals due to a defect in the signaling pathway between the anchor cell and the ventral uterine cell. The extra anchor cells result in the production of excess anchor cell signal. If LIN-3 could directly induce the vulval precursor cells to express the 2° cell fate, there should be some location that is an appropriate distance from the anchor cells such that the concentration of LIN-3 can cause expression of the 2° cell fate. This result is never observed. The vulval precursor cells adjacent to an anchor cell express the 1° cell fate, and all other vulval precursor cells express the 3° cell fate. This result strongly implies that the lateral signal mediated by LIN-12 is necessary for expression of the 2° cell fate.

In summary, the results of the first five experiments support a sequential relay mechanism for establishing cell fates in vulval development. The anchor cell signal induces P6.p to express the 1° cell fate, P6.p expresses the lateral signal, and then the lateral signal induces P5.p and P7.p to express the 2° cell fate. These experiments indicate that the anchor cell signal/LET-23 receptor pathway is not necessary for expression of the 2° cell fate (experiments 1 and 3), nor is it sufficient to induce the 2° cell fate in the absence of the lateral signaling pathway (experiments 2, 3, and 4). Further, the lateral signal/LIN-12 receptor pathway is both necessary to induce the 2° cell fate (experiments 2, 3, and 5) and sufficient to do so in the absence of the anchor cell signaling pathway (experiments 2, 3, and 4).

Below, we discuss two experiments that differ from the preceding experiments in that they appear to support the graded morphogen model for vulval induction. In the sixth experiment, a mutation that prevents the generation of almost all vulval precursor cells was used to produce animals containing only a single vulval precursor cell (14) (Fig. 1.1E). The single vulval precursor cell adopted a 1° cell fate if it was located near the anchor cell or a 3° cell fate if it was located far from the anchor cell. In a small number of cases, the vulval precursor cell adopted the 2° cell fate when it was at an intermediate distance from the anchor cell.

The seventh experiment is similar to the sixth except that a heat shock promoter was used to express different levels of LIN-3 using either severe or mild heat shock regimens (47) (Fig. 1.1F). The test animals had only a single vulval precursor cell since all of the other vulval precursor cells had been killed by laser ablation. In this experiment, high levels of LIN-3 generated 1° cells, lower levels of LIN-3 generated 2° cells, and absence of LIN-3 expression resulted in 3° cells. In some cases, a vulval precursor cell expressed the 2° cell fate without an adjacent primary cell. The results of both the sixth (isolated Pn.p cell) and seventh (heat shock LIN-3) experiments suggest that the 2° cell fate can be expressed without an adjacent 1° cell.

How can we reconcile the differences in the experimental results between the first set of experiments that support the sequential relay model (experiments 1–5) and the second set that support the morphogen model (experiments 6 and 7)? The simplest explanation is that in the second set of experiments, intermediate levels of LIN-3 caused the sole remaining vulval precursor cell to express the lateral signal without also causing it to differentiate as a 1° cell. By expressing the lateral signal, the vulval precursor cell might induce itself to become 2° in an autocrine fashion.

A question that remains is whether autocrine signaling normally occurs in wild-type animals; does the LIN-3 anchor cell signal directly induce the presumptive 2° cells (P5.p and P7.p) to signal themselves with the lateral signal in an autocrine fashion? So far, autocrine action of the lateral signal has been observed only following physical or genetic manipulations that disrupt normal cell–cell contacts or by expression of LIN-3 from a heat shock promoter. These experimental manipulations may generate signaling interactions that do not normally occur. Furthermore, autocrine signaling with the lateral signal was not observed in the genetic mosaics described in experiment 4, and this experiment did not perturb cell contacts or significantly alter levels of the LIN-3 anchor cell signal (Fig. 1.1D). Thus, the results of experiment 4 suggest that the lateral signal does not normally function in an autocrine fashion in worms that have not been operated on. Finally, expression of the lateral signal by P5.p and P7.p for autocrine induction of the 2° cell fate might also result in paracrine induction of adjacent vulval precursor cells (P4.p and P8.p). However, this result is not observed in wild-type animals, as P4.p and P8.p never express the 2° cell fate.

1.4. ESTABLISHMENT OF CELL FATES DURING VULVAL INDUCTION

Figure 1.2 presents a model for the signaling pathways that regulate pattern formation during vulval induction in *C. elegans*. Before signaling, the vulval precursor cells form an unpatterned linear row of cells that are capable of responding to at least two signals, the anchor cell signal and the lateral signal (Fig. 1.2A). Vulval induction begins when the LIN-3 signal from the anchor cell activates the LET-23 receptor in the nearest vulval precursor cell (normally P6.p), causing it to adopt the 1° cell fate and also express the lateral signal (Fig. 1.2B). The lateral signal from the 1° cell then activates the LIN-12 receptor in adjacent vulval precursor cells (normally P5.p and P7.p), causing them to express the 2° cell fate (Fig. 1.2C). Vulval precursor cells that receive neither anchor cell nor lateral signals adopt the uninduced 3° cell fate. In addition to the sequential signaling mechanism, it is possible that the LIN-3 signal may act in a graded fashion in some instances (Fig. 1.2D). Intermediate levels of LIN-3 may be able to directly induce presumptive 2° cells to express the lateral signal themselves. The presumptive 2° cells may then respond to their own signal in an autocrine fashion and adopt the 2° cell fate.

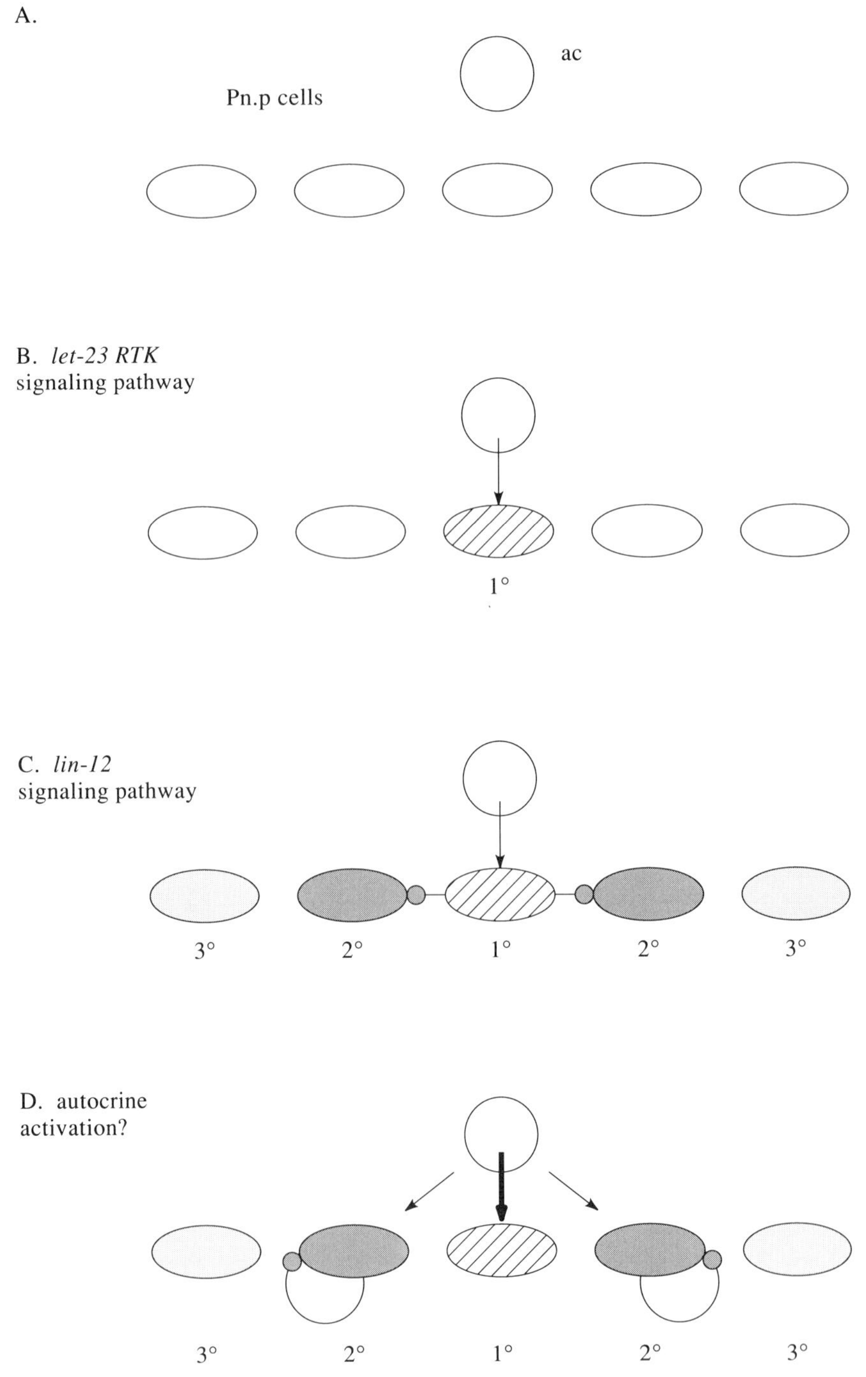
A.
ac
Pn.p cells
B. *let-23 RTK*
signaling pathway
1°
C. *lin-12*
signaling pathway
3°
2°
1°
2°
3°
D. autocrine
activation?
3°
2°
1°
2°
3°

A striking aspect of vulval development is that each cell expresses the same cell fate in different individual animals; the six vulval precursor cells (P3.p to P8.p) always express a 3° - 3° - 2° - 1° - 2° - 3° pattern of cell fates (sometimes P3.p is not a vulval precursor cell and does not express the 3° cell fate). To further illustrate this point, consider examples in which the distance between vulval precursor cells or the strength of the anchor cell signaling pathway are altered. In dumpy or long mutant worms, the distance between vulval precursor cells is either decreased (as much as 3-fold) or increased (about 1.5- to 2-fold), respectively. The changes in the relative spacing between cells would be predicted to change the concentration of LIN-3 in a diffusion gradient. Yet these mutants show normal patterns of vulval induction. Next, worms that are heterozygous for mutations in the anchor cell signaling pathway (such as *lin-3*(+) or *let-23*(+) heterozygotes) would be expected to have lower levels of anchor cell signal or a decreased response to that signal. Nevertheless, these heterozygous mutants exhibit wild-type patterns of vulval induction, suggesting that small reductions (about two-fold) in the strength of the anchor cell signal does not affect vulval induction.

An appealing aspect of the sequential signaling model proposed in Figure 1.2 is that it can account for the reproducibility of the pattern of vulval cell fates. Only a single vulval precursor cell is 1° because the anchor cell only induces the cell located immediately ventral to it (normally P6.p). Only a single anterior and a single posterior vulval precursor cell are 2° because the 1° cell only induces the two adjacent vulval precursor cells (normally P5.p and P7.p). The combined action of the two signaling pathways ensures that three vulval precursor cells centered on the anchor cell express a 2° - 1° - 2° pattern of cell fates. According to the model, signaling occurs between adjacent cells, so that changes in the relative spacing of vulval precursor cells in dumpy or long mutants do not affect intercellular signaling. Worms that are heterozygous for mutations in the *let-23* signaling pathway express normal vulval cell fates because vulval cell fates are determined by independent signals acting in an on/off fashion rather than by one graded signal. In the heterozygotes, the weakly defective *let-23* signaling pathway induces P6.p to express the 1° cell fate and secrete the lateral signal, and the lateral signal induces the adjacent vulval precursor cells to express the 2° cell fate.

Fig. 1.2. Sequential signaling during vulval induction. (A) The organization of the vulval precursor cells before the initiation of vulval induction. (B) LIN-3 from the anchor cell (ac) induces P6.p to express the 1° cell fate by activating the *let-23 receptor tyrosine kinase* signaling pathway. (C) The 1° cell subsequently expresses a lateral signal (straight with circles) that induces P5.p and P7.p to express the 2° cell fate by activating the *lin-12* signaling pathway. (D) Medium levels of anchor cell signal (LIN-3) may occasionally induce the expression of the lateral signal in presumptive 2° cells, resulting in their induction via an autocrine loop (lines with circles).

1.5. OTHER EXAMPLES OF PATTERN FORMATION BY SEQUENTIAL SIGNALING

Both morphogen gradients and sequential signaling appear to be mechanisms used to establish pattern by organizers in other animals. Elegant experiments in *Drosophila* have elucidated an example of sequential signaling in the establishment of the anterior-posterior margin in the imaginal disc (49–51). Cells near the posterior compartment border secrete one signal (hedgehog), and hedgehog induces adjacent cells to secrete another signal (decapentaplegic). The first signal (hedgehog) induces imaginal disc cells adjacent to the compartment border, and the second signal (decapentaplegic) induces cells located away from the compartment boundary. In the chick limb, many signals are involved in establishing the anterior-posterior axis, including Sonic hedgehog and FGF4 (52), and these signals may act in a sequential manner. In *Xenopus*, a molecule related to TGBβ/activin (such as Vg1) is likely involved in establishing the body axis. One view is that this molecule acts as a morphogen and can directly induce distinct fates in a concentration-dependent manner (53–55). However, a recent alternative view is that these molecules do not act directly but rather initiate a signaling relay by inducing a second signal that is responsible for establishing the fates of distal cells (56,57). In these examples of signaling relays, the relay may be initiated by distinct types of signals, such as LIN-3 growth factor, hedgehog, Sonic hedgehog, or Vg1. However, although the signals are different, the general mechanism of sequential signaling may be similar.

ACKNOWLEDGMENTS

This work was supported by grants from the National Institutes of Health, the Lucille P. Markey Charitable Trust, and the Searle Scholars Program/The Chicago Community Trust. S.K.K. was a Lucille P. Markey Scholar and a Searle Scholar.

REFERENCES

1. Saunders JW, Gasseling M: Ectodermal–mesenchymal interaction in the origin of limb symmetry. In Fleischmayer R, Billingham RE (eds): Epithelial–Mesenchymal Interaction. Baltimore: Williams & Wilkins, 1968, pp 78–970.
2. Tickle C, Summerbel D, Wolpert L: Positional signaling and specification of digits in chick limb morphogenesis. Nature 254:199–202, 1975.
3. Yamada T, Placzek M, Tanaka H, Dodd J, Jessell TM: Control of cell pattern in the developing nervous system: Polarizing activity of the floor plate and notochord. Cell 64:635–647, 1991.
4. Garcia-Bellido A, de Celis JF: Developmental genetics of the venation pattern of *Drosophila*. Annu Rev Genet 26:277–304, 1992.
5. Ingham PW, Martinez AA: Boundaries and fields in early embryos. Cell 68:221–235, 1992.

6. Stern CD, Hatada Y, Selleck MA, Storey KG: Relationships between mesoderm induction and the embryonic axes in chick and frog embryos. Development (Suppl):151–156, 1992.
7. Wolpert L: Positional information and the spatial pattern of cellular differentiation. J Theor Biol 25:1–47, 1969.
8. Stumpf HF: Mechanism by which cells estimate their location within the body. Nature 212:430–431, 1966.
9. Lawrence PA: Development and determination of hairs and bristles in the milkweed bug, *Oneopeltus fasciatus (Lygaeidae, Hemiptera)*. J Cell Sci 1:475–498, 1966.
10. Turing AM: The chemical basis of morphogenesis. Phil Trans Soc (Lond) 237:37–72, 1952.
11. Sternberg PW, Horvitz HR: Signal transduction during *C. elegans* vulval induction. Trends Genet 7:366–371, 1991.
12. Sulston JE, Horvitz HR: Post-embryonic cell lineages of the nematode, *Caenorhabditis elegans*. Dev Biol 56:110–56, 1977.
13. Sulston JE, White JG: Regulation and cell autonomy during postembryonic development of *Caenorhabditis elegans*. Dev Biol 78:577–597, 1980.
14. Sternberg PW, Horvitz HR: Pattern formation during vulval development in *C. elegans*. Cell 44:761–772, 1986.
15. Thomas JH, Stern MJ, Horvitz HR: Cell interactions coordinate the development of the *C. elegans* egg-laying system. Cell 62:1041–1052, 1990.
16. Kimble J: Alterations in cell lineage following laser ablation of cells in the somatic gonad of *Caenorhabditis elegans*. Dev Biol 87:286–300, 1981.
17. Spemann H, Mangold H: Umber Induktion von Embryonanlagen durch Implantation artfremder Organisatoren. Wilh Roux Arch EntwMech Organ 100:599–638, 1924.
18. Eisenmann DM, Kim SK: Signal transduction and cell fate specification during *Caenorhabditis elegans* vulval development. Curr Opin Genet Dev 4:508–516, 1994.
19. Hill RJ, Sternberg PW: The gene *lin-3* encodes an inductive signal for vulval development in *C. elegans*. Nature 358:470–476, 1992.
20. Duffy JB, Perrimon N: Recent advances in understanding signal transduction pathways in worms and flies. Curr Opin Cell Biol 8:231–238, 1996.
21. Perrimon N: The torso receptor protein-tyrosine kinase signaling pathway: An endless story. Cell 74, 219–22.
22. Aroian RV, Koga M, Mendel JE, Ohshima Y, Sternberg PW: The *let-23* gene necessary for *Caenorhabditis elegans* vulval induction encodes a tyrosine kinase of the EGF receptor subfamily. Nature 348:693–699, 1990.
23. Clark SG, Stern MJ, Horvitz HR: *C. elegans* cell-signalling gene *sem*-5 encodes a protein with SH2 and SH3 domains. Nature 356:340–344, 1992.
24. Lowenstein EJ, Daly RJ, Batzer AG, Li W, Margolis B, Lammers R, Ulrich A, Skolnik EY, Bar SD, Schlessinger J: The SH2 and SH3 domain-containing protein GRB2 links receptor tyrosine kinases to ras signaling. Cell 70:431–442, 1992.
25. Simon MA, Dodson GS, Rubin GM: An SH3-SH2-SH3 protein is required for p21Ras1 activation and binds to sevenless and Sos proteins in vitro. Cell 73:169–177, 1993.
26. Olivier JP, Raabe T, Henkemeyer M, Dickson B, Mbamalu G, Margolis B, Schlessinger J, Hafen E, Pawson T: A *Drosophila* SH2-SH3 adaptor protein implicated in coupling *sevenless* tyrosine kinase to an activator of Ras guanine nucleotide exchange, Sos. Cell 73:179–191, 1993.

27. Han M, Sternberg PW: *let-60*, a gene that specifies cell fates during *C. elegans* vulval induction, encodes a ras protein. Cell 63:921–931, 1990.

28. Han M, Aroian RV, Sternberg PW: The *let-60 locus* controls the switch between vulval and nonvulval cell fates in *Caenorhabditis elegans*. Genetics 126:899–913, 1990.

29. Beitel GJ, Clark SG, Horvitz HR: *Caenorhabditis elegans ras* gene *let-60* acts as a switch in the pathway of vulval induction. Nature 348:503–509, 1990.

30. Han M, Golden A, Han Y, Sternberg PW: *C. elegans lin-45 raf* gene participates in *let-60 ras*-stimulated vulval differentiation. Nature 363:133–139, 1993.

31. Wu Y, Han M: Suppression of activated Let-60 *ras* protein defines a role of *Caenorhabditis elegans* Sur-1 MAP kinase in vulval differentiation. Genes 9:742–755, 1994.

32. Lackner MR, Kornfeld K, Miller LM, Horvitz HR, Kim SK: A MAP kinase homolog, *mpk-1*, is involved in *ras*-mediated induction of vulval cell fates in *C. elegans*. Genes Dev 8:160–173, 1994.

33. Kornfeld K, Guan K-L, Horvitz HR: The *Caenorhabditis elegans* gene *mek-2* is required for vulval induction and encodes a protein similar to the protein kinase MEK. Genes Dev 9:756–768, 1995.

34. Wu Y, Han M, Guan K-L: MEK-2, a *Caenorhabditis elegans* MAP kinase, functions in Ras-mediated vulval induction and other developmental events. Genes Dev 9:742–755, 1995.

35. Ferguson EL, Sternberg PW, Horvitz HR: A genetic pathway for the specification of the vulval cell lineages of *Caenorhabditis elegans*. Nature 326:259–267, 1987.

36. Katz WS, Lesa GM, Yannoukakos D, Clandinin TR, Schelessinger J, Sternberg PW. A point mutation in the extracellular domain activates LET-23, the *Caenorhabditis elegans* epidermal growth factor receptor homolog. Mol Cell Biol 16:529–537, 1996.

37. Kornfeld K, Guan K-L, Horvitz HR: The *ksr-1* gene encodes a novel protein kinase involved in Ras-mediated signaling in *C. elegans*. Cell 83:903–913, 1995.

38. Sundaram M, Han M: The *C. elegans ksr-1* gene encodes a novel Raf-related kinase involved in Ras-mediated signal transduction. Cell 83:889–901, 1995.

39. Simske JS, Kim SK: Sequential signalling during *C. elegans* vulval induction. Nature 375:142–145, 1995.

40. Koga M, Ohshima Y: Mosaic analysis of the *let-23* gene function in vulval induction of *Caenorhabditis elegans*. Development 121:2655–2666, 1995.

41. Sternberg PW: Lateral inhibition during vulval induction in *Caenorhabditis elegans*. Nature 335:551–554, 1988.

42. Greenwald IS, Sternberg PW, Horvitz HR: The *lin-12* locus specifies cell fates in *Caenorhabditis elegans*. Cell 34:435–444, 1983.

43. Yochem J, Weston K, Greenwald I: The *Caenorhabditis elegans lin-12* gene encodes a transmembrane protein with overall similarity to *Drosophila* Notch. Nature 1988:547–550, 1988.

44. Yochem J, Greenwald I: *glp-1* and *lin-12*, genes implicated in distinct cell–cell interactions in *C. elegans*, encode similar transmembrane proteins. Cell 58:553–563, 1989.

45. Wilkinson HA, Greenwald I: Spatial and temporal patterns of *lin-12* expression during *C. elegans* hermaphrodite development. Genetics 141:513–526, 1995.

46. Sternberg PW, Horvitz HR: The combined action of two intercellular signaling pathways specifies three cell fates during vulval induction in *C. elegans*. Cell 58:679–693, 1989.

47. Katz WS, Hill RJ, Clandinin TR, Sternberg PW: Different levels of the *C. elegans* growth factor LIN-3 promote distinct vulval precursor fates. Cell 82:297–307, 1995.
48. Artavanis-Taakonas S, Matsuno K, Fortini ME: Notch signaling. Science 268:225–232, 1995.
49. Basler K, Struhl G: Compartment boundaries and the control of *Drosophila* limb pattern by hedgehog protein. Nature 368:208–214, 1994.
50. Nellen D, Burke R, Struhl G, Basler K: Direct and long-range action of a DPP morphogen gradient. Cell 85:357–368, 1996.
51. Zecca M, Basler K, Struhl G: Sequential organizing activities of engrailed, hedgehog and decapentaplegic in the *Drosophila* wing. Development 121:2265–2278, 1995.
52. Tabin C: The initiation of the limb bud: Growth factors, Hox genes, and retinoids. Cell 80:671–674, 1995.
53. Green JB, New HV, Smith JC: Responses of embryonic *Xenopus* cells to activin and FGF are separated by multiple dose thresholds and correspond to distinct axes of the mesoderm. Cell 71:731–739, 1992.
54. Gurdon JB, Harger P, Mitchell A, Lemaire P: Activin signalling and response to a morphogen gradient. Nature 371:487–492, 1994.
55. Gurdon JB, Mitchell A, Mahony D: Direct and continuous assessment by cells of their position in a morphogen gradient. Nature 376:520–521, 1995.
56. Wilson PA, Melton DA: Mesodermal patterning by an inducer gradient depends on secondary cell–cell communication. Curr Biol 4:676–686, 1994.
57. Reilly KM, Melton DA: Short-range signaling by candidate morphogens of the TGF beta family and evidence for a relay mechanism of induction. Cell 86:743–754, 1996.

2

EGF-RECEPTOR AND TGF-β-LIKE Dpp SIGNALING DURING *Drosophila* DEVELOPMENT

ETHAN BIER

Department of Biology, University of California, San Diego, La Jolla, CA 92093

2.1. INTRODUCTION

Pattern formation during development depends on cell–cell communication to establish and maintain different cell fates. Cells may communicate with their immediate neighbors through surface-bound molecules or may signal to more distant targets by secreting diffusible molecules. In this tripartite chapter, I consider the function of the epidermal growth factor-receptor (EGF-R) and transforming growth factor-β-like (TGF-β-like) decapentaplegic (Dpp) signaling pathways during development of *Drosophila* embryos and adults. In Section 2.2, I briefly describe components of the EGF-R and Dpp pathways and mechanisms by which they alter gene expression. Special emphasis is given to molecules that modulate activity levels of these pathways. In Section 2.3, I summarize the various developmental decisions governed by each of these two pathways, and in Section 2.4, I discuss in greater detail how the EGF-R and Dpp pathways interact during early neurogenesis and adult wing vein development to influence binary cell fate choices.

Hormones and Growth Factors in Development and Neoplasia, Edited by Robert B. Dickson and David S. Salomon
ISBN 0-471-16899-8

2.2. ORGANIZATION OF THE EGF-R AND Dpp SIGNALING PATHWAYS

2.2.1. The EGF-R Pathway

The *Drosophila* EGF-R is a typical receptor tyrosine kinase (RTK) and is encoded by the *Egf-r* gene. Mutations disrupting the function of the *Egf-r* gene have been recovered in genetic screens and are known as *faint little ball (flb)* or *torpedo (top* (1–3) mutants. It is generally thought that EGF-R signaling is initiated by binding of an EGF domain containing ligand to the receptor. While the mechanism of EGF activation of EGF-R remains poorly understood, the important step seems to be ligand-induced receptor dimerization (4). Once two EGF-R chains are brought into close proximity through dimerization, they cross-phosphorylate each other (autophosphorylation) on a series of tyrosine residues (5). Studies involving mammalian cells in culture have implicated several intracellular second messenger systems as mediators of EGF-R activation. Here, however, I will focus on the RAS/mitogen activated protein kinase (RAS/MAPK) pathway (Fig. 2.1) since a variety of evidence suggests that this is the predominant pathway propagating the EGF-R signal during *Drosophila* development (6, review). It should be kept in mind, however, that other potential mediators of EGF-R activation such as Corkscrew (7,8) and Daughter of Sevenless (9,10) the (JAK/STAT) pathway (11–13), phospholipase-Cγ (14,15), P13 kinase (16), and *Drosophila shc* (17) also could play roles in determining the magnitude of EGF-R activation or might contribute to defining the specificity of the EGF-R signal.

Figure 2.1 shows three putative EGF-R ligands encoded by the *spitz, gurken,* and *vein* genes. The primary products of the *spitz* (Spi) (18) and *gurken* (GRK) (19) genes are membrane-bound ligands related to vertebrate TGF-α, while *vein* encodes a protein (Vein) related to the neuregulin family of secreted ligands (20). In the case of Spitz, there is direct biochemical evidence that a secreted form of Spitz (sSpi) can bind to EGF-R and induce receptor autophosphorylation and MAPK activity in cell culture (21). In addition, it has been shown that sSpi is much more active than the unprocessed, membrane-bound form of Spitz (mSpi) (21).

Two other genes (*rhomboid* and *Star*) encode membrane proteins thought to promote EGF-R activity. The predicted *rhomboid* product (Rhomboid) is an integral membrane protein with seven hydrophobic domains (22). Rhomboid is not obviously related to known G-coupled serpentine receptors, but it shares a high degree of sequence similarity with two functionally uncharacterized *Caenorhabditis elegans* proteins identified by conceptual processing and translation of genomic DNA sequenced as part of the genome project (23). The predicted product of the *Star* gene (Star) is a type II single-pass transmembrane protein with a large amino terminal extracellular domain and a short cytoplasmic carboxy terminal tail (24). Genetic evidence discussed in Section 2.3 indicates that Rhomboid and Star collaborate to hyperactivate EGF-R signaling in various developmental contexts, but that these accessory molecules are not required for the basal receptor activity essential for epithelial cell viability.

The *rhomboid* gene is expressed in a highly localized pattern during development (22,25,26) that corresponds to the pattern of cells requiring hyperactive EGF-R

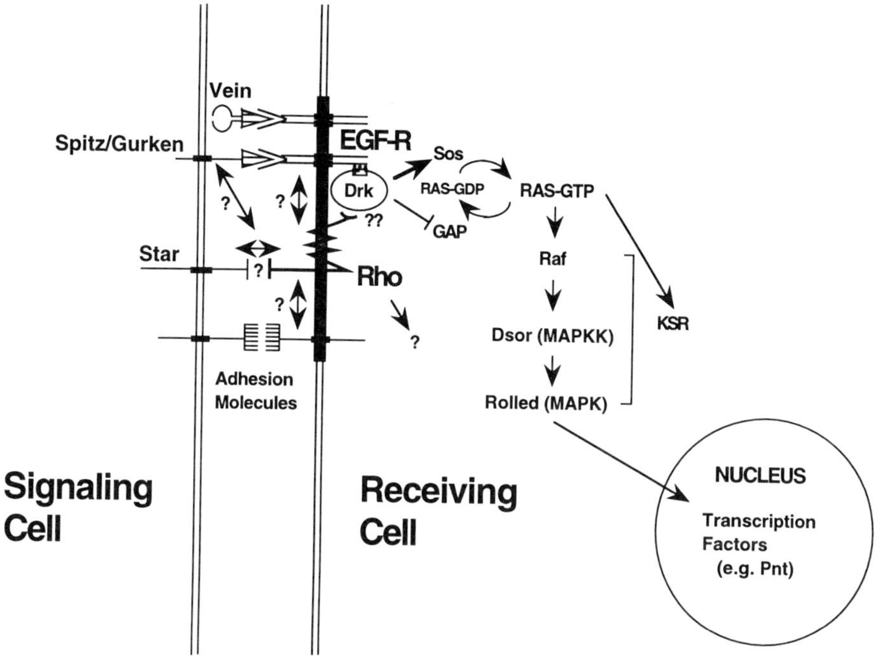

Fig. 2.1. The EGF-R/RAS signaling pathway. In the absence of Rhomboid or Star, there is a basal level of EGF-R activity sufficient to maintain cell viability. In cells containing the Rhomboid (Rho) and Star accessory proteins, EGF-R signaling is hyperactivated. The darkened band at the cell surface indicates a patch of the plasma membrane in which the Rho protein is highly concentrated. Major models for Rho hyperactivation of Spi-EGF-R signaling are as follows: Rho acts directly to increase EGF-R activity by increasing the affinity of EGF-R for Spi, by processing the membrane-bound form of Spi (mSpi) to a secreted, more active form (sSpi), by increasing the EGF-R catalytic rate on engaging ligand, or by promoting an interaction between EGF-R and downstream effectors necessary for RAS activation; alternatively, Rho functions indirectly by increasing cell–cell adhesion between communicating cells (these models are mutually exclusive). Once the EGF-R has been activated, the signal is propagated through a phosphorylation cascade of RAS → Raf → Dsor → Rolled. KSR kinase also is required for propagation of the signal from RAS to Rolled. Phosphorylated Rolled then enters the nucleus, where it phosphorylates transcription factors already present in the cell (e.g., Pnt), which either activates or inactivates them. *Key:* The primary Spitz and Gurken proteins are TGFα-like membrane-bound ligands with single EGF domains (triangles); Vein is a neuregulin-type ligand with an EGF domain (triangle) and an immunoglobulin domain (loop); Star is a type II membrane protein. EGF-R = EGF-receptor, Rho = Rhomboid protein, which has seven hydrophobic domains; Drk = downstream receptor of kinase = GRB2; Sos = nucleotide exchange factor; GAP = GTPase-activating protein; RAS-GDP = inactive form of the RAS protein; RAS-GTP = active form of the RAS protein; Raf = serine threonine kinase; Dsor = MAPK kinase, Rolled = MAPK; KSR = kinase suppressor of RAS; Pnt = Pointed ETS domain protein.

signaling (2,27–31). Localized expression of *rhomboid* plays a key role in limiting hyperactive EGF-R signaling to appropriate cells since misregulation of *rhomboid* induces ectopic EGF-R hyperactivation (see below).

The Rhomboid protein is highly concentrated in plaque-like structures at the apical cell surface (32) where signaling is thought to take place. Rhomboid plaques consist of patches of plasma membrane and underlying vesicular or reticular structures. It is possible that these membrane patches of Rhomboid define specialized subregions of the cell surface at which EGF-R signaling is potentiated.

A series of elegant genetic screens using sensitized mutant backgrounds have identified many of the genes encoding components common to all tyrosine kinase signaling pathways in *Drosophila* (33–35). Most of the proteins encoded by these genes also have been identified as mediators of tyrosine kinase signaling in mammalian cell culture systems (36, review). The heart of this intracellular signaling cascade appears to be the conversion of inactive guanosine diphosphate-bound RAS (GDP-RAS) to active guanosine triphosphate-bound RAS (GTP-RAS), which is catalyzed by a nucleotide exchange factor encoded by the *son-of-sevenless (sos)* gene (37,38). A wealth of genetic and biochemical data suggests that the Sos protein becomes bound to active tyrosine-phosphorylated RTKs through an adapter protein encoded by the *drk* gene (39,40), also known as GRB2 from mammalian studies (41). The DRK/GRB2 adapter protein has an SH2 homology domain that binds to specific phosphotyrosine residues on the activated RTK and an SH3 homology domain that binds to a string of polyproline residues on Sos (39–41). Another component associated with activated RTKs is encoded by the *gap1* gene (GAP) (42). GAP opposes the action of Sos by accelerating an intrinsic guanosine triphosphatase (GTPase) activity of RAS. EGF-R activation presumably shifts the balance between Sos and GAP activities to increase the proportion of RAS in the GTP- versus GDP-bound form.

GTP-RAS bound to the EGF–R–Drk–Sos complex then recruits the Raf serine–threonine kinase to the cell surface (43), which presumably phosphorylates and activates a MAPK-kinase encoded by the *dsor* gene (44). Phosphorylated Dsor, in turn, kinases a MAPK encoded by the *rolled* gene (Rl) (45,46). Propagation of the signal from RAS to Rl also depends on the action of the KSR kinase (47), which may organize flux through this portion of the pathway. Phosphorylated Rl then enters the nucleus, where it phosphorylates various transcription factors such as an (ETS) domain protein encoded by the *pointed* gene (48,49). It is clear that the cellular response to EGF-R/RAS pathway activation varies greatly based on when and where the pathway is active. One explanation for the specificity of EGF-R signaling is that distinct sets of MAPK-sensitive transcription factors are present in different cells. Because there is a unique complement of transcription factors in a given cell, EGF-R/RAS signaling will have distinct effects on gene expression in different cells. In addition to this cell context model, however, it should be considered that different RTKs also have been shown to associate with various subsets of proteins linking receptor activation to alternative second messenger pathways (9–17). It is possible, therefore, that these secondary pathways also contribute to defining the specificity of RTK activity.

One important form of EGF-R regulation is negative feedback following receptor activity. In mammalian systems, it has been observed that over 90% of the receptor is internalized within 5 min of ligand addition (50,51). Mutant versions of the receptor that cannot be internalized accumulate to high levels in the membrane and lead to aggressive tumor formation in mice (52). Another form of receptor inactivation is mediated by protein kinase C (PKC) phosphorylation (53). PKC phosphorylation of EGF-R reduces the affinity of the receptor for ligand.

In *Drosophila,* two forms of EGF-R-mediated negative feedback have been observed. The first mechanism is downregulation of *Egf-r* mRNA levels (54). Following receptor hyperactivation, *Egf-r* message levels drop dramatically. In addition, ectopic EGF-R hyperactivation induced by misexpression of the Rhomboid accessory protein leads to ectopic EGF-R RNA downregulation. These results suggest that EGF-R hyperactivation triggers RNA downregulation. Other mechanisms must also contribute to receptor downregulation, however, since the pattern of RNA downregulation is largely normal in *Egf-r* null mutant embryos. The second form of EGF-R inactivation is mediated by a secreted EGF antagonist encoded by the *argos* gene (AOS) (55). AOS can bind to EGF-R and competitively inhibit Spitz binding (56). Since *aos* expression is induced by EGF-R activity, this gene also mediates a negative feedback loop (56,57). These various forms of EGF-R negative feedback presumably limit the duration and domain of hyperactive EGF-R signaling, thereby permitting this pathway to be reused in a rapid succession of developmental decisions.

2.2.2. The Dpp Pathway

The *dpp* gene encodes a secreted protein (Dpp) (58) in the TGF-β superfamily most highly related to vertebrate bone morphogenetic proteins 2 and 4 (BMP-2 and BMP-4). While less is known about the mechanism of signal transduction through TGF-β receptors than through RTK pathways, rapid progress is being made in the analysis of this class of signaling pathways (Fig. 2.2). TGF-β receptors are heterodimers composed of related type I and type II serine-threonine kinases. Upon binding of homodimeric or heterodimeric ligands, the type I and type II chains associate, triggering phosphorylation of the type I chain by the type II chain (59). The phosphorylated type I chain then propagates the signal, presumably by phosphorylating cytoplasmic substrates. In *Drosophila,* the *saxophone (sax)* and *thick veins (tkv)* genes encode type I Dpp receptors (60–63). The Sax and Tkv type I receptors can complex with the single known type II Dpp receptor encoded by the *punt (put)* gene (64,65).

One important mediator of Dpp signaling is a protein encoded by a gene called *mothers against Dpp (mad)* (66–68). Since the original genetic identification of *mad* and another gene called *medea,* which are maternally supplied components of the Dpp pathway in *Drosophila* (66), a large number of related MAD proteins have been implicated in other TGF-β signaling pathways (68–71). Current data suggest that on receptor activation, receptor-associated MAD proteins are phosphorylated (perhaps directly by type I chains) and then enter the nucleus, where they bind DNA and activate transcription of target genes (72,73). MAD proteins may function to alter gene expression in concert with other transcription factors such as the

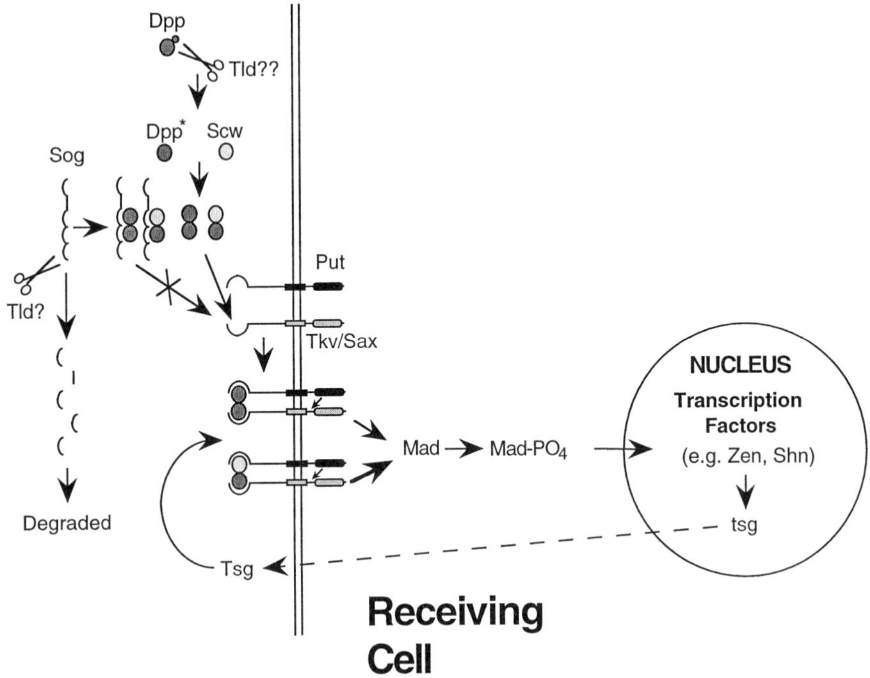

Fig. 2.2. The Dpp signaling pathway. The primary TGF-β-related Dpp product (solid circle with bud) is cleaved by the Tld protease to generate the more active Dpp* species (solid circle). Scw (gray filled circle) is another TGF-β family member. Dpp* may either homodimerize (double filled circles) or heterodimerize with Scw (filled circle plus gray circle) to form active dimeric ligands, which leads to the dimerization of heteromeric Dpp receptors composed of one of the two type I chains (Tkv or Sax) and a single known type II chain (Put). The Sog protein containing four CR repeats (linked semicircles) may bind Dpp*-Dpp* or Dpp*-Scw dimers and prevent them from activating Dpp receptors. On Dpp-receptor dimerization, the type II chain phosphorylates the type I chain, which then propagates the signal, perhaps by directly phosphorylating Mad. Phosphorylated Mad (Mad-PO_4) then enters the nucleus, where it functions in concert with other transcription factors such as Zen and Schnurri to activate gene expression. One potential Dpp target is *tsg*, which encodes a secreted molecule (Tsg) required to maintain peak Dpp activity.

Drosophila Schnurri protein (74–76), homeodomain proteins such as Zen in *Drosophila* (77) or Vox/Xvent in *Xenopus* (78–80), and winged helix proteins (81).

A variety of putative extracellular proteins have been implicated as modulators of Dpp signaling in *Drosophila*. Mutants in the dpp *screw (scw), tolloid (tld), twisted gastrulation (tsg),* and *short gastrulation (sog)* genes all display defects in patterning the dorsal region of the *Drosophila* embryo (82,83). The products of these genes are extracellular proteins that are likely to modulate Dpp activity since loss-of-function mutations in any of these genes results in phenotypes significantly less severe than those of *dpp* null mutant embryos (83,84). The *scw, tld,* and *tsg* genes promote Dpp signaling since mutants in these genes show the same types of defects as *dpp* mutants. In contrast, Sog antagonizes Dpp signaling, as revealed by the opposite phenotypes of *sog* and *dpp* mutants (see Section 2.4).

scw encodes another protein in the TGF-β superfamily (Scw) (85) most similar to the vertebrate BMP-7 protein. It has been proposed that a heterodimer of Dpp and Scw may be more effective than Dpp homodimers in activating Dpp signaling. According to this model, *scw* mutants have a less severe phenotype than *dpp* null mutants because Dpp homodimers can form in *scw* mutants and activate minimal Dpp signaling, while in *dpp* mutants no signaling takes place, thus Scw homodimers either do not form or cannot activate the Dpp receptor. The *tld* gene encodes a protein (Tld) (86) similar to vertebrate BMP-1. BMP-1 is not structurally related to TGF-β-like BMPs but copurifies with them (87). One form of BMP-1 recently has been shown to possess proteolytic activity (88). Tld has been proposed to function analogously in *Drosophila,* possibly by processing Dpp to a maximally active form. Alternatively, Tld could degrade a Dpp antagonist such as Sog. The *tsg* gene product (Tsg) (89) is a secreted protein related to a family of growth factors typified by βIG-M2 (90) that is induced in mouse cells by TGF-β1 treatment. Thus, Tsg may function in a positive feedback loop to maintain peak Dpp signaling.

sog encodes a protein (Sog) (91) with a large extracellular domain likely to be secreted from cells. A variety of genetic evidence (described in more detail in Section 1.4) suggests that Sog may diffuse over a distance of 10–15 cells to exert an inhibitory influence on Dpp signaling (91–93). Since a vertebrate homolog of Sog known as Chordin (94) has been identified in *Xenopus,* which binds with high affinity to BMP-4 (the vertebrate Dpp homolog) (95), it is likely that the inhibitory action of Sog results from the formation of a Sog-Dpp complex. This mechanism is similar to that by which follistatin binds and sequesters the TGF-β family member Activin in an inactive form (96). One difference between Sog and follistatin, however, is that Sog does not appear to block all Dpp activity (97,98). Two possible explanations for why overexpressed Sog fails to mimic the *dpp* mutant phenotype are that Sog may not bind to some active form of Dpp or Dpp may retain partial activity when bound to Sog. It is possible that other proteins function together with Sog to limit Dpp activity. For example, in *Xenopus,* another protein called Noggin has been shown to bind BMP-4 (99) and inhibit BMP-4 signaling in the neuroectoderm. As *Xenopus* Noggin can inhibit Dpp activity in *Drosophila* embryos (100), it is possible that endogenous Nogging-like activity also exists in *Drosophila.*

2.3. OVERVIEW OF EGF-R AND Dpp SIGNALING DURING *Drosophila* DEVELOPMENT

In this section, I provide a brief overview of the various requirements for EGF-R and Dpp signaling during embryonic and adult development. In the following section (Section 2.4), I focus in greater detail on the role that these two pathways play during early embryonic neurogenesis and adult wing vein development.

2.3.1. Localized Hyperactive EGF-R Signaling during Development

Mutants in genes encoding components of the EGF-R/RAS pathway have revealed at least two distinct requirements for EGF-R activity. The viability of nearly all epidermal cells depends on a low level of EGF-R and RAS pathway activity, while strong EGF-R/RAS signaling is required for a variety of cell fate decisions. This conclusion is based in part on the observation that mutants entirely lacking EGF-R or Ras1 activity exhibit widespread epidermal cell death (2,3,28,29,101,102), while epidermal cells survive in mutants with reduced activity of these genes but fail to differentiate correctly during embryonic and adult development (28–31,101,102). Mutants in one of the genes encoding a TGFα-like ligand *(spitz)* (18,27) and in genes encoding the accessory molecules Rhomboid (22,27) and Star (24,27) lend further support to a two-tier model of EGF-R activity. *rhomboid, Star,* and *spitz* mutants exhibit a set of developmental defects strikingly similar to those observed in moderate-strength *Egf-r* mutants (28,29,101) and have nearly normal cell viability.

Egf-r, spitz, Star, and genes encoding components of the RAS pathway are expressed ubiquitously during development (24,25,54,103–107), consistent with the requirement of basal EGF-R/RAS activity for general epidermal cell viability. In contrast, *rhomboid* is expressed in a highly localized pattern corresponding to cells that require hyperactive EGF-R signaling to differentiate (22,25,26). During embryonic and adult development, the correlation between *rhomboid* expression and a requirement for EGF-R activity is striking. For example, *rhomboid* is expressed in broad lateral stripes at the cellular blastoderm stage, along the ventral midline during early gastrulation, in a subset of stretch receptor organ precursor cells in the developing peripheral nervous system, and at segment boundaries in germband-retracted embryos (22), *rhomboid, spitz, Star,* and moderate *Egf-r* mutants all exhibit similar defects in the formation of these unrelated structures (18,22,24,27–31). Localized *rhomboid* expression also is required to specify cell fates during metamorphosis (25,107) and during oogenesis (26) in adult female flies.

Although the parallel between the pattern *rhomboid* expression and the array of defects in ventrolateral group mutants is compelling, *rhomboid* appears to alter cell fates in neighboring cells, as well as in cells expressing *rhomboid.* Thus, although *rhomboid* expression is confined to the midline of the embryo (22), there are defects in gene expression in adjacent ventral epidermal cells in *rhomboid* mutant embryos (31,108,109). Similarly, loss of *rhomboid* function in mutant wing clones is not strictly correlated to loss of veins (i.e., cells mutant for *rhomboid* that are next to wild-type cells occasionally make veins; reciprocally, wild-type cells that are next to *rhomboid* mutant cells may fail to form veins) (110,111). In contrast, vein loss in *Egf-r* mutant clones is strictly limited to mutant cells (110). These data suggest that Rhomboid may participate in generating a short-range signal. This signal could be an EGF-R ligand or could activate separate receptor signaling in concert with EGF-R to activate the RAS/MAPK cascade or other parallel intracellular pathways.

The restricted pattern of *rhomboid* expression is critical for limiting hyperactive EGF-R signaling to appropriate cells since misexpression of *rhomboid* leads to ectopic EGF-R hyperactivation. For example, *rhomboid* expression is restricted to vein primordia during adult wing vein development (25). As in other developmental contexts, vein-specific *rhomboid* expression is necessary since elimination of this expression leads to vein loss phenotypes. Limiting *rhomboid* expression to veins also is necessary, as misexpression of *rhomboid* in intervein regions redirects these cells to differentiate as veins (24,54).

2.3.2. Dpp Signaling during Development

The *dpp* gene derives its name from the large variety of developmental defects observed in mutants mapping to this locus. During embryogenesis, Dpp signaling is required for patterning the dorsal region of the blastoderm embryo (82,83,112) (see Section 2.4 for a detailed discussion of this function), for subdivision of the mesoderm into somatic versus visceral or cardiac components (113,114), and for induction between the visceral mesoderm and the underlying endoderm (115–117). Similar to the latter actions of Dpp in *Drosophila,* vertebrate BMP-5 may function as an inductive signal between primary germ layers (118).

The role of Dpp signaling in the induction of the secondary midgut constriction has been particularly well studied. Expression of the homeotic gene *Ubx* in the visceral mesoderm of parasegment 7 (PS7) in collaboration with the ubiquitously expressed extradenticle homeodomain protein directly activates *dpp* expression (119–121). Dpp produced in the visceral mesoderm of PS7 functions by a paracrine mechanism to induce expression of the *labial* gene in the underlying endoderm (115–117). In addition, Dpp functions through an autocrine mechanism to sustain expression of *Ubx* and *dpp* in the visceral mesoderm (68,115–117,120).

Dpp signaling also has been shown to govern developmental decisions during metamorphosis, including outgrowth of imaginal discs (122), establishment of anterior-posterior (A/P) polarity in leg and wing imaginal discs during larval development (123–128), propagation of the morphogenetic furrow in the eye imaginal disc (129,130), optic lobe development (131), promotion of vein differentiation during pupal wing development (98), and anterior follicle cell development during oogenesis (132).

The role of Dpp as a morphogen in establishing A/P positions in the wing primordium has been analyzed in detail. *dpp* expression in a narrow stripe running along the A/P compartment boundary in the center of the wing disc influences gene expression in neighboring cells in a dose-dependent fashion (126–128). For example, cells in a broad central wedge of the wing disc straddling the narrow *dpp* stripe receive sufficient Dpp signal to activate expression of a zinc finger transcription factor encoded by the *spalt* gene (133), which plays a role in patterning the central region of the wing primordium (134,135). One action of *spalt* is to define the position of the future L2 longitudinal wing vein in a narrow strip of cells adjacent to the anterior boundary of *spalt* expression (135). Lower levels of Dpp activity determine the positions of wing structures farther from the source of Dpp (126–128), consistent with the view that Dpp protein emanates from the narrow stripe of cells in which it is produced to engage Dpp receptors and impact cell fates throughout the wing primordium. Whether this reflects passive diffusion of Dpp protein or some form of facilitated transport of Dpp from cell to cell is unknown. There also is evidence that Dpp may function at different developmental stages to imprint proliferating cells with distinct identities (127). The degree to which these two mechanisms contribute to patterning the A/P axis of the wing remains to be determined.

2.4. INTERACTION OF EGF-R AND Dpp SIGNALING DURING EARLY NEUROGENESIS AND ADULT WING VEIN DEVELOPMENT

In this section, I discuss the collaboration of EGF-R and Dpp signaling during two selected stages of development. The first period is the cellular blastoderm stage, when the ectoderm is subdivided into a ventral neuroectodermal domain and a dorsal nonneural territory (Fig. 2.3a). The second developmental setting is the wing primordium during late larval and early pupal stages, when the binary vein-versus-intervein cell fate choice is made (Fig. 2.3b).

2.4.1. Subdivision of the Ectoderm Into Neural versus Nonneural Components

The EGF-R and Dpp pathways play an integral part in establishing cell fates in the lateral and dorsal regions of the blastoderm stage embryo (Fig. 2.3a). The EGF-R accessory protein Rhomboid and the Dpp antagonist Sog are produced in the lateral neuroectodermal region of the embryo, while *dpp* (136) and other genes such as *tld* (86), *zen* (77), and *tsg* (89), required for patterning the dorsal nonneural region of the embryo, are expressed only in dorsal cells. As the *rhomboid* and *sog* genes collaborate to promote neuroectodermal fates in the lateral neurogenic region of the embryo, the EGF-R and Dpp pathways function in opposition during this stage of development.

Loss of *rhomboid, spitz,* or *Star* function or partial loss of *Egf-r* function leads to the loss of ventral denticle structures in the cuticle of late-stage embryos (18,22,24,27,30). These cells derive from lateral neuroectodermal cells expressing *rhomboid* at the cellular blastoderm stage (27,122). The ventral epidermal cuticle phenotype of *rhomboid* mutants derives in part from a later function of *rhomboid* and EGF-R signaling along the ventral midline during gastrulation since midline expression of *rhomboid* driven by the GAL4-UAS system partly rescues the *rhomboid* ventral epidermal mutant phenotype (108,109). Early expression of *rhomboid* in lateral stripes also plays a role in ventral cuticle development, however, since complete loss of midline function, as observed in *single-minded (sim)* mutants, results in a less severe cuticle phenotype than that observed in *rhomboid* mutants (27). As *rhomboid* expression along the ventral midline is lost in *sim* mutants (137), these data cumulatively reveal a requirement for *rhomboid* and EGF-R function to promote epidermal cell fates in the neuroectoderm of blastoderm-stage embryos. EGF-R and Dpp signaling collaborate to restrict expression of the *msh* homeodomain protein to dorso-lateral cells (138), since *msh* expression expands into the ventral neuroectoderm in *Egf-r* mutants and into the dorsal region in *dpp* mutants.

sog collaborates with *rhomboid* to promote neural cell fates by blocking Dpp activity in the neuroectoderm. Dpp signaling must be opposed in the neuroectoderm, as it exerts two major effects that are incompatible with neural development: (1) Dpp activity suppresses expression of genes required for neural development, such as genes of the proneural achaete–scute complex (97), and (2) Dpp signaling activates expression of genes functioning to specify dorsal cell fates such as *zen*

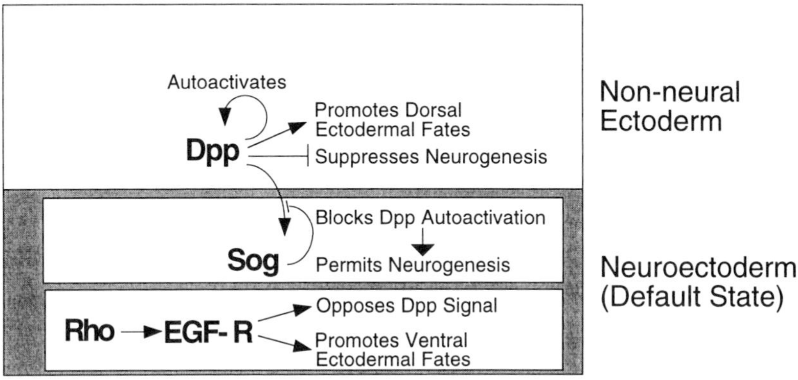

(*a*)

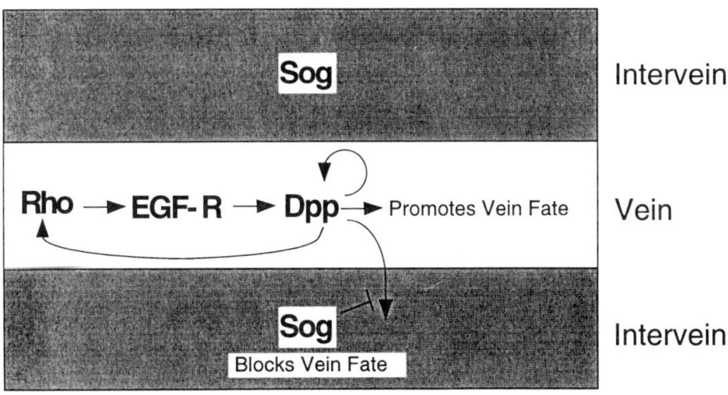

(*b*)

Fig. 2.3. (a) EGF-R and Dpp signaling in the embryonic ectoderm. (b) EGF-R and Dpp signaling in pupal wings.

(82,97), *tld* (97), and *dpp* (97) itself. This ability of Dpp to activate its own expression through an autocrine mechanism has been termed *autoactivation*. It is likely that the primary mechanism by which Sog opposes the neural suppressive activity of Dpp signaling in the neuroectoderm is to prevent *dpp* autoactivation in the neuroectoderm, which otherwise would be triggered by the low levels of Dpp diffusing into the lateral region from adjacent dorsal cells. Were the chain of Dpp diffusion and autoactivation to take place in the neuroectoderm, neural development would be suppressed since it requires significantly less Dpp signaling to suppress neurogenesis than it does to induce *dpp* expression (97).

Although *sog* is not expressed in dorsal cells, *sog* mutants exhibit defects in patterning the dorsal region of the embryo (91–93). This action at a distance presumably results from Sog diffusing from the neuroectoderm into the dorsal region to interfere with Dpp activity. A variety of evidence suggests that Dpp functions as a dose-dependent signal in the dorsal region of the embryo during this stage, with different levels of Dpp specifying distinct developmental fates (85,139,140). In this model, peak Dpp activity specifies the dorsalmost cell type (amnioserosa), while lower levels of activity specify the dorsal ectoderma. As mentioned above, small amounts of Dpp also are likely to diffuse into the lateral zone to influence patterning within the neuroectoderm. As the level of *dpp* mRNA appears to be uniform throughout the dorsal zone, it has been speculated that a posttranscriptional mechanism is responsible for establishing graded Dpp activity. Sog is likely to contribute to generating such a Dpp activity gradient. Loss of *sog* function leads to expression of early amnioserosa markers throughout nearly all dorsal cells (91). This phenotype is opposite to that of other zygotic mutants disrupting dorsal patterning such as *dpp,* in which expression of amnioserosal markers is lost rather than expanded (82,91,140). Another indication that *sog* opposes Dpp activity in the dorsal region is that there are strong gene dosage interactions between *sog* and *dpp* (91,92). For example, embryos with only one functional copy of *dpp* ($dpp^-/^+$) usually do not survive since two active copies of this gene are required to generate the amnioserosa. However, embryos with only one functional copy of *dpp* do survive at nearly wild-type frequencies if the dose of *sog* is also reduced to 50% (i.e., in $sog^-/^+$; $dpp^-/^+$ embryos) (91). This antagonistic genetic interaction suggests that Dpp activity levels in the dorsal region of the embryo result from a balance of the opposing actions of Sog and Dpp.

There is evidence that Sog functions as a morphogen to generate a Dpp activity gradient in dorsal cells (97). The term *morphogen* here refers to a molecule that is produced in one set of cells (e.g., lateral cells) but affects cell fates in other cells (e.g., dorsal cells) in a dose-dependent fashion (i.e., minimally, that low levels of the morphogen trigger a different developmental program than high levels of that molecule). The first condition presumably reflects diffusion of the morphogen (passive or active), and the second condition requires different threshold responses to that morphogen. *sog* is likely to meet these minimal criteria with respect to subdividing the dorsal region of the blastoderm embryo into the dorsalmost amnioserosa and the more ventral nonneural ectoderm. Wild-type embryos with two functional copies of *sog* express an amnioserosa marker in a strip of cells 10–12 cells wide. In embryos with only one dose of *sog,* however, this domain increases to 15–17 cells across, which is intermediate between wild-type and the full 30- to 34-cell-wide domain observed in mutants lacking *sog* activity (97). Thus, Sog produced in the neuroectoderm may diffuse over 15 cell diameters to establish a boundary separating cells exposed to low levels of Dpp activity (i.e., nonneural ectoderm cells close enough to the Sog source to have the Dpp signal attenuated by Sog) from cells receiving high levels of Dpp signal (i.e., amnioserosa cells far from the Sog source experience unopposed Dpp signaling).

2.4.2. Evolutionary Conservation of Dpp Signaling in the Neuroectoderm

The action of *sog* to antagonize the neural suppressive activity of Dpp in the neuroectoderm appears to have been highly conserved during the course of evolution. Thus, the *Xenopus* homologs of *sog* (i.e., *chordin*) and *dpp* (i.e., *BMP-4*) are expressed in complementary domains comprising the neuroectoderm and nonneural ectoderm, respectively (141). Like *sog, chordin* promotes neurogenesis (142), while BMP-4 functions analogously to Dpp by suppressing dorsal neural fates 142–148. In addition, Dpp and BMP-4 are functionally interchangeable in flies and frogs (149,150), as are Sog and Chordin (150,151). The mechanism by which Sog and Chordin oppose the action of Dpp and BMP-4 also appears to have been conserved between flies and frogs. It has been shown that BMP-4 is capable of autoactivating in early *Xenopus* embryos (79,80). Furthermore, in a zebrafish mutant called *dino,* which like *sog* mutants has unopposed BMP-4 activity in the neuroectoderm, BMP-4 expression is observed throughout both the neural and nonneural ectoderm (152). Thus, Sog and Chordin are likely to function by blocking Dpp and BMP-4 autoactivation in flies and vertebrates (97), respectively. Interference with Dpp/BMP-4 autoactivation protects the neuroectoderm from the intrinsically invasive positive feedback loop resulting from Dpp or BMP-4 diffusing into the neuroectoderm, autoactivating, and suppressing neurogenesis. Transduction of the Dpp/BMP-4 signal also has been well conserved, at least to the extent that MAD proteins are important mediators of that signal (66–71).

2.4.3. EGF-R and Dpp Signals Collaborate to Promote Adult Wing Vein Development

The adult wing is composed of two major cell types, vein and intervein cells (Fig. 2.3b). Genetic analysis of vein development has identified two clearly distinct stages in this process (153–155). During late larval development, the vein pattern is initiated independently in the primordia of both the dorsal and ventral surfaces of the wing as a series of sharp stripes (25,110). Subsequently, during pupal development, several forms of cell–cell communication contribute to the ultimate differentiation of continuous and straight veins (110,111,155). These cell–cell interactions include (1) dorsal-to-ventral ($D \rightarrow V$) signal(s) required for maintaining vein fates in cells on the ventral surface of the wing (110,155), (2) lateral inhibitory signal(s) emanating from vein primordia that limit the width of veins differentiating within broad vein competent regions (25,111,154–156), and (3) vein continuity signal(s) promoting vein formation in straight lines along the axis of vein extension (110,155). The product of these various signaling events is the formation of a series of hollow tubular veins filled with fluid that have components contributed by both the dorsal and ventral surfaces of the adult wing. Veins consist of densely packed living cells covered by a thick, darkly pigmented cuticle, and can be easily distinguished from intervein cells which are tightly connected through the extracellular matrix to cells on the opposite wing surface and have a thin, lightly pigmented cuticle with sparsely packed trichome hairs. Another difference between these two cell types is that vein cells survive into adulthood, whereas intervein cells die on eclosion.

Hyperactive EGF-R signaling is required throughout wing vein development (155). As in other developmental contexts, EGF-R hyperactivation depends on localized expression of *rhomboid* (25,107). *rhomboid* expression is initiated in a series of stripes corresponding to longitudinal vein primordia in mid-third instar larval wing discs and continues to be expressed in differentiating vein cells during pupal development (25). In contrast to *rhomboid,* other known components of the EGF-R signaling pathway are expressed ubiquitously during vein development (54, M.A. Sturtevant and E. Bier, unpublished results). The localized pattern of *rhomboid* expression is required for generating the normal vein pattern since lack of *rhomboid* function leads to vein loss phenotypes, while misexpression of *rhomboid* induces formation of ectopic veins (25,107).

Dpp signaling also is required for vein development. During early pupal stages, *dpp* expression is initiated in vein primordia and continues to be expressed in veins during the period of cell–cell interactions (98). *dpp* expression is likely to be initiated by EGF-R signaling because *dpp* is not expressed in cells lacking *rhomboid* expression and ectopic *rhomboid* expression rapidly induces *dpp* expression (98). Consistent with this view, wild-type expression *rhomboid* in vein primordia precedes that of *dpp* by several hours (98). (Fig. 2.3b). Dpp, like *rhomboid,* promotes vein formation. Lack of *dpp* expression in vein primordia leads to vein-loss phenotypes (98,157), whereas ectopic expression of *dpp* during early pupal stages generates extra veins (98). During the same period in which *dpp* is expressed in vein primordia, *sog* is expressed in complementary intervein cells and functions to oppose vein formation (98). Dosage-sensitive genetic interactions between *dpp* and *sog* mutations confirm that these two genes function antagonistically during vein development (98). As during early neurogenesis, Sog blocks a Dpp autoactivation loop since ectopic expression of *sog* can eliminate *dpp* expression without affecting expression of other vein markers such as *rhomboid* (98).

Among the various forms of cell–cell communication that have been characterized during pupal development, it seems most likely that Dpp signaling is involved in establishing vein continuity. This view is suggested by a phenotype resulting from *sog* misexpression in which the posterior cross-vein is detached from the longitudinal veins (98). The floating cross-vein phenotype is similar to a known mutant called *detached,* which has tentatively been interpreted as a vein continuity mutant (155). Further support for interpretation of the detached vein phenotype as disruption of vein continuity is that a *dpp* vein-loss mutant phenotype is exacerbated in a *detached* mutant background (98). The combination of Dpp diffusion and autoactivation is ideally suited to mediating a vein continuity function. Thus, Dpp may diffuse along the length of the vein (where *sog* is absent) and autoactivate to ensure that all cells along the length of the vein are committed to the same fate. Continuity is an essential property of veins, as any interruption destroys vein function.

2.5. CONCLUSIONS

The EGF-R and Dpp signaling pathways influence a large number of cell fate decisions during *Drosophila* embryonic and adult development. In some cases, both of these pathways are involved in a given cell fate choice, while in others, only one pathway is involved. In this review, two cases of opposite interactions between the EGF-R and Dpp pathways have been discussed in detail (i.e., in the embryonic neuroectoderm and pupal wing; see Fig. 2.3). In the early embryonic ectoderm, EGF-R and Dpp signaling act in opposition to promote or suppress neuroectodermal fates, respectively. In contrast, in the pupal wing, these two pathways function synergistically to create a positive feedback cycle promoting vein development. Thus, there does not appear to be any consistent relationship between EGF-R and Dpp signaling during development. The interaction between EGF and TGF-β signaling also appears to vary in different vertebrate contexts. For example, TGF-β signaling has been shown to oppose RAS activity in driving early events in tumor formation (158), whereas the EGF and TGF-β pathways appear to act synergistically to promote invasive tumor growth at a later stage of tumorigenesis (159). The simplest explanation for the seemingly independent action of EGF-R and TGF-β signaling systems is that these two pathways are not coupled at the level of signal reception or signal transduction. Rather, they are likely to independently activate or repress the activity of specific transcription factors present in particular cell types. In cases where these two pathways promote expression of the same or allied target genes they will function synergistically, while in other cases where they either activate conflicting genetic programs or exert opposing influences on the same set of target genes, these pathways will act antagonistically. This model also is consistent with the existence of situations in which only one of the two pathways is involved in a given developmental process.

The fact that coherent modules of proteins comprising the EGF-R and Dpp pathways have been highly conserved during evolution has two repercussions. The first consequence is a practical one. Since the component organization of these pathways is similar in vertebrates and invertebrates, it is possible to exploit powerful model genetic systems offered by organisms such as *Drosophila* to gain important detailed insights into the corresponding action of vertebrate systems. Because the EGF-R and TGF-β signaling pathways play important roles in cancer and other disease conditions, the strong relationship between vertebrate and invertebrate signaling systems is of significant medical relevance. The second implication of the evolutionary conservation of these signaling pathways is that all metazoan animals descend from a single highly evolved common ancestor. The most recent common ancestor between vertebrates and invertebrates not only had already evolved the basic organization of important signaling pathways but had embedded these pathways so firmly in basic developmental programs that these complex regulatory relationships have remained immutable over 500 million years of evolution. This strongly suggests that among the variety of organisms with diverse body plans coexisting during the pre-Cambrian period, only one species survived to give rise to all modern metazoa.

REFERENCES

1. Livneh E, Glazer L, Segal D, Schlessinger J, Shilo BZ: The *Drosophila* EGF receptor gene homolog: Conservation of both hormone binding and kinase domains. Cell 40:599–607, 1985.
2. Price JV, Clifford RJ, Schupbach T: The maternal ventralizing locus *torpedo* is allelic to *faint little ball,* an embryonic lethal, and encodes the *Drosophila* EGF receptor homolog. Cell 56:1085–1092, 1989.
3. Schejter ED, Shilo BZ: The *Drosophila* EGF receptor homolog (DER) gene is allelic to *faint little ball,* a locus essential for embryonic development. Cell 56:1093–1104, 1989.
4. Sherrill JM, Kyte J: Activation of epidermal growth factor receptor by epidermal growth factor. Biochemistry 35:5705–5718, 1996.
5. Qian X, LeVea CM, Freeman JK, Dougall WC, Greene MI: Heterodimerization of epidermal growth factor and wild-type or kinase-deficient Neu: A mechanism of interreceptor kinase activation and transphosphorylation. Proc Natl Acad Sci USA 91:1500–1504, 1994.
6. Perrimon N: Signalling pathways initiated by receptor protein tyrosine kinases in *Drosophila.* Curr Opin Cell Biol 6:260–266, 1994.
7. Perkins LA, Larsen I, Perrimon N: *corkscrew* encodes a putative protein tyrosine phosphatase that functions to transduce the terminal signal from the receptor tyrosine kinase torso. Cell 70:225–236, 1992.
8. Feng GS, Hui CC, Pawson T: SH2-containing phosphotyrosine phosphatase as a target of protein-tyrosine kinases. Science 259:1607–1611, 1993.
9. Herbst R, Carroll PM, Allard JD, Schilling J, Raabe T, Simon MA: Daughter of sevenless is a substrate of the phosphotyrosine phosphatase Corkscrew and functions during sevenless signaling. Cell 85:899–909, 1996.
10. Raabe T, Riesgo-Escovar J, Liu X, Bausenwein BS, Deak P, Maroy P, Hafen E: DOS, a novel pleckstrin homology domain-containing protein required for signal transduction between sevenless and Ras1 in *Drosophila.* Cell 85:911–920, 1996.
11. Binari R, Perrimon N: Stripe-specific regulation of pair-rule genes by *hopscotch,* a putative Jak family tyrosine kinase in *Drosophila.* Genes Dev 8:300–312, 1994.
12. Yan R, Small S, Desplan C, Dearolf CR, Darnell JE Jr: Identification of a Stat gene that functions in *Drosophila* development. Cell 84:421–430, 1996.
13. Hou XS, Melnick MB, Perrimon N: *marelle* acts downstream of the *Drosophila* HOP/JAK kinase and encodes a protein similar to the mammalian STATs. Cell 84:411–419, 1996.
14. Noh DY, Shin SH, Rhee SG: Phosphoinositide-specific phospholipase C and mitogenic signaling. Biochim Biophys Acta 1242:99–113, 1995.
15. Cleghon V, Gayko U, Copeland TD, Perkins LA, Perrimon N, Morrison DK: *Drosophila* terminal structure development is regulated by the compensatory activities of positive and negative phosphotyrosine signaling sites on the Torso RTK. Genes Dev 10:556–577, 1996.
16. Akimoto K, Takahashi R, Moriya S, Nishioka N, Takayanagi J, Kimura K, Fukui Y, Osada SI, Mizuno K, Hirai Si, Kazaluskas A, Ohno S: EGF or PDGF receptors activate atypical PKClambda through phosphatidylinositol 3-kinase. EMBO J 15:788–798, 1996.
17. Lai KM, Olivier JP, Gish GD, Henkemyer M, McGlade J, Pawson T: A *Drosophila shc*

gene product is implicated in signaling by the DER receptor tyrosine kinase. Mol Cell Biol 15:4810–4818, 1995.

18. Rutledge B, Zhang K, Bier E, Jan YN, Perrimon N: The *Drosophila spitz* gene encodes a putative EGF-like growth factor involved in dorsal-ventral axis formation and neurogenesis. Genes Dev 6:1503–1517, 1992.
19. Neuman-Silberberg FS, Schupbach T: The *Drosophila* dorsoventral patterning gene *gurken* produces a dorsally localized RNA and encodes a TGF alpha-like protein. Cell 75:165–174, 1993.
20. Schnepp B, Grumbling G, Donaldson T, Simcox A: *vein* is a novel component in the *Drosophila* epidermal growth factor receptor pathway with similarity to the neuregulins. Genes Dev 10:2302–2313, 1996.
21. Schweitzer R, Shaharabany M, Seger R, Shilo BZ: Secreted Spitz triggers the DER signaling pathway and is a limiting component in embryonic ventral ectoderm determination. Genes Dev 9:1518–1529, 1995.
22. Bier E, Jan LY, Jan YN: *rhomboid*, a gene required for dorsoventral axis establishment and for peripheral nervous system development in *Drosophila melanogaster*. Genes Dev 4:190–203, 1990.
23. Wilson R, Ainscough R, Anderson K, Baynes C, Berks M, Bonfield J, Burton J, Connell M, Copsey T, Cooper J, et al: 2.2 Mb of contiguous nucleotide sequence from chromosome III of *C. elegans*. Nature 368:32–38, 1994.
24. Kolodkin AL, Pickup AT, Lin DM, Goodman CS, Banerjee U: Characterization of *Star* and its interactions with *sevenless* and *Egf-r* during photoreceptor cell development in *Drosophila*. Development 120:1731–1745, 1994.
25. Sturtevant MA, Roark M, Bier E: The *Drosophila rhomboid* gene mediates the localized formation of wing veins and interacts genetically with components of the EGF-R signaling pathway. Genes Dev 7:961–973, 1993.
26. Ruohola-Baker H, Grell E, Chou TB, Baker D, Jan LY, Jan YN: Spatially localized *rhomboid* is required for establishment of the dorsal-ventral axis in *Drosophila* oogenesis. Cell 73:953–965, 1993.
27. Mayer U, Nüsslein-Volhard C: A group of genes required for pattern formation in the ventral ectoderm of the *Drosophila* embryo. Genes Dev 2:1496–1511, 1988.
28. Clifford RJ, Schüpbach T: The *torpedo* (DER) receptor tyrosine kinase is required at multiple times during *Drosophila* embryogenesis. Development 115:853–872, 1992.
29. Raz E, Shilo BZ: Dissection of the *faint little ball* (FLB) phenotype—determination of the development of the *Drosophila* central nervous system by early interactions in the ectoderm. Development 114:113–123, 1992.
30. Raz E, Shilo BZ: Establishment of ventral cell fates in the *Drosophila* ectoderm requires DER, the EGF receptor homolog. Genes Dev 7:1937–1948, 1993.
31. Kim SE, Crews ST: Influence of *Drosophila* ventral epidermal development by the CNS midline cells and *spitz* class genes. Development 118:893–901, 1993.
32. Sturtevant MA, Roark M, O'Neill JW, Biehs B, Colley N, Bier E: The *Drosophila* Rhomboid protein is concentrated in patches at the apical cell surface. Dev Biol 174:298–309, 1996.
33. Simon MA, Bowtell D, Dodson G, Laverty TR, Rubin GM: Ras1 and a putative guanine nucleotide exchange factor perform crucial steps in signaling by the Sevenless protein tyrosine kinase. Cell 67:701–716, 1991.

34. Dickson BJ, van der Straten A, Dominguez M, Hafen E: Mutations modulating Raf signaling in *Drosophila* eye development. Genetics 142:163–171, 1996.

35. Karim FD, Chang HC, Therrien M, Wassarman DA, Laverty T, Rubin GM: A screen for genes that function downstream of Ras1 during *Drosophila* eye development. Genetics 143:315–329, 1996.

36. Margolis B, Skolnik EY: Activation of Ras by receptor tyrosine kinases. J Am Soc Nephrol 5:1288–1299, 1994.

37. Rogge RD, Karlovich CA, Banerjee U: Genetic dissection of a neurodevelopmental pathway: *son of sevenless* functions downstream of the sevenless and EGF receptor tyrosine kinases. Cell 64:39–48, 1991.

38. Bonfini L, Karlovich CA, Dasgupta C, Banerjee U: The Son of Sevenless gene product: A putative activator of Ras. Science 255:603–606, 1992.

39. Simon MA, Dodson GS, Rubin GM: An SH3-SH2-SH3 protein is required for p21Ras1 activation and binds to sevenless and Sos proteins in vitro. Cell 73:169–177, 1993.

40. Olivier JP, Raabe T, Henkemeyer M, Dickson B, Mbamalu G, Margolis B, Schlessinger J, Hafen E, Pawson T: A *Drosophila* SH2-SH3 adaptor protein implicated in coupling the sevenless tyrosine kinase to an activator of Ras guanine nucleotide exchange, Sos. Cell 73:179–191, 1993.

41. Lowenstein EJ, Daly RJ, Batzer AG, Li W, Margolis B, Lammers R, Ullrich A, Skolnik EY, Bar-Sagi D, Schlessinger J: The SH2 and SH3 domain-containing protein GRB2 links receptor tyrosine kinases to ras signaling. Cell 70:431–442, 1992.

42. Gaul U, Mardon G, Rubin GM: A putative Ras GTPase activating protein acts as a negative regulator of signaling by the Sevenless receptor tyrosine kinase. Cell 68:1007–1019, 1992.

43. Dickson B, Sprenger F, Morrison D, Hafen E: Raf functions downstream of Ras1 in the Sevenless signal transduction pathway. Nature 360:600–603, 1992.

44. Tsuda L, Inoue YH, Yoo MA, Mizuno M, Hata M, Lim YM, Adachi-Yamada T, Ryo H, Masamune Y, Nishida Y: A protein kinase similar to MAP kinase activator acts downstream of the Raf kinase in *Drosophila*. Cell 72:407–414, 1993.

45. Brunner D, Oellers N, Szabad J, Biggs WH, Zipursky SL, Hafen E: A gain-of-function mutation in *Drosophila* MAP kinase activates multiple receptor tyrosine kinase signaling pathways. Cell 76:875–888, 1994.

46. Biggs WH, Zavitz KH, Dickson B, van der Straten A, Brunner D, Hafen E, Zipursky SL: The *Drosophila rolled* locus encodes a MAP kinase required in the sevenless signal transduction pathway. EMBO J 13:1628–1635, 1994.

47. Therrien M, Chang HC, Solomon NM, Karim FD, Wassarman DA, Rubin GM: KSR, a novel protein kinase required for RAS signal transduction. Cell 83:879–888, 1995.

48. O'Neill EM, Rebay I, Tijan R, Rubin GM: The activities of two Ets-related transcription factors required for *Drosophila* eye development are modulated by the Ras/MAPK pathway. Cell 78:137–147, 1994.

49. Brunner D, Ducker K, Oellers N, Hafen E, Scholz H, Klambt C: The ETS domain protein pointed-P2 is a target of MAP kinase in the sevenless signal transduction pathway. Nature 370:386–389, 1994.

50. Haigler H, Ash JF, Singer SJ, Cohen S: Visualization by fluorescence of the binding and internalization of epidermal growth factor in human carcinoma cells A-431. Proc Natl Acad Sci USA 75:3317–3321, 1978.

51. Lund KA, Opresko LK, Starbuck C, Walsh BJ, Wiley HS: Quantitative analysis of the endocytic system involved in hormone-induced receptor internalization. J Biol Chem 265:15723–15731, 1990.
52. Wiley HS, Herbst JJ, Walsh BJ, Lauffenburger DA, Rosenfeld MG, Gill GN: The role of tyrosine kinase activity in endocytosis, compartmentation, and downregulation of the epidermal growth factor receptor. J Biol Chem 266:11083–11094, 1990.
53. Welsh JB, Gill GN, Rosenfeld MG, Wells A: A negative feedback loop attenuates EGF-induced morphological changes. J Cell Biol 114:533–543, 1991.
54. Sturtevant MA, O'Neill J, Bier E: Down-regulation of *Drosophila Egf-r* mRNA following hyperactivated receptor signaling. Development 120:2593–2600, 1994.
55. Freeman M, Klambt C, Goodman CS, Rubin GM: The *argos* gene encodes a diffusible factor that regulates cell fate decisions in the *Drosophila* eye. Cell 69:963–975, 1992.
56. Schweitzer R, Howes R, Smith R, Shilo BZ, Freeman M: Inhibition of *Drosophila* EGF receptor activation by the secreted protein Argos. Nature 376:699–702, 1995.
57. Golembo M, Schweitzer R, Freeman M, Shilo BZ: argos transcription is induced by the *Drosophila* EGF receptor pathway to form an inhibitory feedback loop. Development 122:223–230, 1996.
58. Padgett RW, St Johnson RD, Gelbart WM: A transcript from a *Drosophila* pattern gene predicts a protein homologous to the transforming growth factor-β family. Nature 325:81–84, 1987.
59. Wrana JL, Attisano L, Wieser R, Ventura F, Massague J: Mechanism of activation of the TGF-beta receptor. Nature 370:341–347.
60. Xie T, Finelli AL, Padgett RW: The *Drosophila saxophone* gene: A serine threonine kinase receptor of the TGF-beta superfamily. Science 263:1756–1759, 1994.
61. Nellen D, Affolter M, Basler K: Receptor serine/threonine kinases implicated in the control of *Drosophila* body pattern by *decapentaplegic*. Cell 78:225–237, 1994.
62. Penton A, Chen Y, Staehling-Hampton K, Wrana JL, Attisano L, Szidonya J, Cassill JA, Massagué J, Hoffmann FM: Identification of two bone morphogenetic protein type I receptors in *Drosophila* and evidence that *brk25D* is a decapentaplegic receptor. Cell 78:239–250, 1994.
63. Brummel TJ, Twombly V, Marqués G, Wrana JL, Newfeld SJ, Attisano L, Massagué J, O'Connor MB, Gelbart WM: Characterization and relationship of Dpp receptors encoded by the *saxophone* and *thick veins* genes in *Drosophila*. Cell 78:251–261, 1994.
64. Letsou A, Arora K, Wrana JL, Simin K, Twombly V, Jamal J, Staehling-Hampton K, Hoffmann FM, Gelbart WM, Massagué J, O'Connor MB: *Drosophila* Dpp signaling is mediated by the *punt* gene product: A dual ligand-binding type II receptor of the TGF receptor family. Cell 80:899–908, 1995.
65. Ruberte E, Marty T, Nellen D, Affolter M, Basler K: An absolute requirement for both the type II and the type I receptors, punt and thick veins of Dpp signaling in vivo. Cell 80:889–898, 1995.
66. Raftery LA, Twombly V, Wharton K, Gelbart WM: Genetic screens to identify elements of the decapentaplegic signaling pathway in *Drosophila*. Genetics 139:241–254, 1995.
67. Sekelsky JJ, Newfeld SJ, Raftery LA, Chartoff EH, Gelbart WM: Genetic characterization and cloning of *mothers against Dpp*, a gene required for *decapentaplegic* function in *Drosophila melanogaster*. Genetics 139:347–358, 1995.

68. Newfeld SJ, Chartoff EH, Graff JM, Melton DA, Gelbart WM: *mothers against Dpp* encodes a conserved cytoplasmic protein required in DPP/TGF-beta responsive cells. Development 122:2099–2108, 1996.

69. Hoodless PA, Haerry T, Abdollah S, Stapleton M, O'Connor MB, Attisano L, Wrana JL: MADR1, a MAD-related protein that functions in BMP2 signaling pathways. Cell 85:489–500, 1996.

70. Graff JM, Bansal A, Melton DA: *Xenopus* Mad proteins transduce distinct subsets of signals for the TGF beta superfamily. Cell 85:479–487, 1996.

71. Thomsen GH: *Xenopus mothers against decapentaplegic* is an embryonic ventralizing agent that acts downstream of the BMP-2/4 receptor. Development 122:2359–2366, 1996.

72. Zhang Y, Feng X, We R, Derynck R: Receptor-associated Mad homologues synergize as effectors of the TGF-beta response. Nature 383:168–172, 1996.

73. Liu F, Hata A, Baker JC, Doody J, Carcamo J, Harland RM, Massague J: A human Mad protein acting as a BMP-regulated transcriptional activator. Nature 381:620–623, 1996.

74. Arora K, Dai H, Kazuko SG, Jamal J, O'Connor MB, Letsou A, Warrior R: The *Drosophila schnurri* gene acts in the Dpp/TGFβ signaling pathway and encodes a transcription factor homologous to the human MBP family. Cell 81:781–790.

75. Grieder NC, Nellen D, Burke R, Basler K, Affolter M: *schnurri* is required for *Drosophila* Dpp signaling and encodes a zinc finger protein similar to the mammalian transcription factor PDRII-BF1. Cell 81:791–800, 1995.

76. Staehling-Hampton K, Laughon AS, Hoffmann FM: A *Drosophila* protein related to the human zinc finger transcription factor PRDII/MBPI/HIV-EP1 is required for Dpp signaling. Development 121:3393–3403, 1995.

77. Doyle HJ, Harding K, Hoey T, Levine M: Transcripts encoded by a homeobox gene are restricted to dorsal tissues of *Drosophila* embryos. Nature 323:76–79, 1986.

78. Gawantka V, Delius H, Hirschfeld K, Blumenstock C, Niehrs C: Antagonizing the Spermann organizer—role of the homeobox gene *Xvent-1*. EMBO J 14:6268–6279, 1995.

79. Schmidt JE, Vondassow G, Kimelman D: Regulation of dorsal-ventral patterning—the ventralizing effects of the novel *Xenopus* homeobox gene *vox*. Development 122:1711–1721, 1996.

80. Onichtchouk D, Gawantka V, Dosch R, Delius H, Hirschfeld K, Blumenstock C, Niehrs C: The *Xvent-2* homeobox gene is part of the Bmp-4 signalling pathway controlling dorsoventral patterning of *Xenopus* mesoderm. Development 122:3045–3053, 1996.

81. Chen X, Rubock MJ, Whitman M: A transcriptional partner for Mad proteins in Tgf-β signalling. Nature 383:691–696, 1996.

82. Ray RP, Arora K, Nüsslein-Volhard C, Gelbart WM: The control of cell fate along the dorsal-ventral axis of the *Drosophila* embryo. Development 113:35–54, 1991.

83. Arora K, Nüsslein-Volhard C: Altered mitotic domains reveal fate map changes in *Drosophila* embryos mutant for zygotic dorsoventral patterning genes. Development 114:1003–1024, 1992.

84. Wharton KA, Ray RP, Gelbart WM: An activity gradient of *decapentaplegic* is necessary for the specification of dorsal pattern elements in the *Drosophila* embryo. Development 117:807–822, 1993.

85. Arora K, Levine M, O'Connor MB: The *screw* gene encodes a ubiquitously expressed member of the TGF-β family required for specification of dorsal cell fates in the *Drosophila* embryo. Genes Dev 8:2588–2601, 1994.

86. Shimell MJ, Ferguson EL, Childs SR, O'Connor MB: The *Drosophila* dorsal-ventral patterning gene *tolloid* is related to human bone morphogenetic protein 1. Cell 67:469–481, 1991.

87. Wozney JM, Rosen V, Celeste AJ, Mitsock LM, Whitters MJ, Kriz RW, Hewick RM, Wang EA: Novel regulators of bone formation: Molecular clones and activities. Science 242:1528–1534, 1988.

88. Kessler E, Takahara K, Biniaminov L, Brusel M, Greenspan DS: Bone morphogenetic protein-1: The type I procollagen C-proteinase. Science 271:360–362, 1996.

89. Mason ED, Konrad KD, Webb CD, Marsh LJ: Dorsal midline fate in *Drosophila* embryos requires *twisted gastrulation*, a gene encoding a secreted protein related to human connective tissue growth factor. Genes Dev 8:1489–1501, 1994.

90. Brunner A, Chinn J, Neubauer M, Purchio AF: Identification of a gene family regulated by transforming growth factor-β. DNA Cell Biol 10:293–300, 1991.

91. François V, Solloway M, O'Neill JW, Emery J, Bier E: Dorsal-ventral patterning of the *Drosophila* embryo depends on a putative negative growth factor encoded by the *short gastrulation* gene. Genes Dev 8:2602–2616, 1994.

92. Ferguson EL, Anderson KV: Localized enhancement and repression of the activity of the TGF-β family member, *decapentaplegic*, is necessary for dorsal-ventral pattern formation in the *Drosophila* embryo. Development 114:583–597, 1992.

93. Zusman SB, Sweeton D, Wieschaus EF. *short gastrulation*, a mutation causing delays in stage specific cell shape changes during gastrulation in *Drosophila melanogaster*. Dev Biol 129:417–427, 1988.

94. Sasai Y, Lu B, Steinbeisser H, Geissert D, Gont L, De Robertis EM: *Xenopus chordin*. A novel dorsalizing factor activated by organizer specific homeobox genes. Cell 79:779–790, 1994.

95. Picolla S, Sasai Y, Lu B, de Robertis EM: Dorsoventral patterning in *Xenopus* inhibition of ventral signals by direct binding of chordin to BMP-4. Cell 86:589–598, 1996.

96. Nakamura T, Takio K, Eto Y, Shibai H, Titani K, Sugino H: Activin-binding protein from rat ovary is follistatin. Science 247:836–838, 1990.

97. Biehs B, François V, Bier E: The *Drosophila short gastrulation* gene prevents Dpp signaling from autoactivating and suppressing neurogenesis in the neuroectoderm. Genes Dev 10:2922–2934, 1996.

98. Yu K, Sturtevant MA, François V, Padgett R, Blackman R, Bier E: The *Drosophila decapentaplegic* and *short gastrulation* genes function antagonistically during adult wing vein development. Development 122:4033–4044, 1996.

99. Zimmerman LB, De Jesus-Escobar JM, Harland RM: The Spemann organizer signal noggin binds and inactivates bone morphogenetic protein 4. Cell 86:599–606, 1996.

100. Holley SA, Neul JL, Attisano L, Wrana JL, Sasai Y, O'Connor MB, De Robertis EM, Ferguson EL: The *Xenopus* dorsalizing factor noggin ventralizes *Drosophila* embryos by preventing Dpp from activating its receptor. Cell 86:607–617, 1996.

101. Díaz-Benjumea FJ, García-Bellido A: Behavior of cell mutant for an EGF *receptor* homologue of *Drosophila* in genetic mosaics. Proc R Soc Lond B 242:36–44, 1990.

102. Díaz-Benjumea FJ, Hafen E: The *sevenless* signalling cassette mediates *Drosophila* EGF receptor function during epidermal development. Development 120:569–578, 1994.

103. Kammermeyer KL, Wadsworth SC: Expression of *Drosophila* epidermal growth factor homologue in mitotic cell populations. Development 100:201–210, 1987.

104. Katzen AL, Kornberg T, Bishop JM: Expression during *Drosophila* development of DER, a gene related to *erbB-1* and *neu:* Correlations with mutant phenotypes. Dev Biol 145:287–301, 1991.

105. Zak NB, Wides RJ, Schehjter ED, Raz E, Shilo BZ: Localization of the DER/flb protein in embryos: Implications on the *faint little ball* lethal phenotype. Development 109:865–874, 1990.

106. Zak NB, Shilo BZ: Localization of DER and the pattern of cell divisions in wild-type and Ellipse eye imaginal discs. Dev Biol 149:448–456, 1992.

107. Noll R, Sturtevant MA, Gollapudi R, Bier E: New functions of the *Drosophila rhomboid* gene during embryonic and adult development are revealed by a novel genetic method, enhancer piracy. Development 120:2329–2338, 1994.

108. Xiao H, Hrdlicka LA, Nambu JR: Alternate functions of the *single-minded* and *rhomboid* genes in development of the *Drosophila* ventral neuroectoderm. Mech Dev 58:65–74, 1996.

109. Golembo M, Raz E, Shilo BZ: The *Drosophila* embryonic midline is the site of Spitz processing, and indicates activation of the EGF receptor in the ventral ectoderm. Development 122:3363–3370, 1996.

110. García-Bellido A: Inductive mechanisms in the process of wing vein formation in *Drosophila*. Wilhelm Roux's Arch Dev Biol 182:93–106, 1977.

111. García-Bellido A, De Delis JF: Developmental genetics of the venation pattern of *Drosophila*. Annu Rev Genet 26:277–304, 1992.

112. Irish VF, Gelbart WM: The *decapentaplegic* gene is required for dorsal-ventral patterning of the *Drosophila* embryo. Genes Dev 1:868–879, 1987.

113. Staehling-Hampton K, Hoffmann FM, Baylies MK, Rushton E, Bate M: *dpp* induces mesodermal gene expression in *Drosophila*. Nature 372:783–786, 1994.

114. Frasch M: Induction of visceral and cardiac mesoderm by ectodermal Dpp in the early *Drosophila* embryo. Nature 374:464–467, 1995.

115. Panganiban GE, Reuter R, Scott MP, Hoffmann FM: A *Drosophila* growth factor homolog, *decapentaplegic*, regulates homeotic gene expression within and across germ layers during midgut morphogenesis. Development 110:1041–1050, 1990.

116. Hursh DA, Padgett RW, Gelbart WM: Cross regulation of *decapentaplegic* and *Ultrabithorax* transcription in the visceral mesoderm of *Drosophila*. Development 117:1211–1222, 1993.

117. Thuringer F, Cohen SM, Bienz M: Dissection of an indirect autoregulatory response of a homeotic *Drosophila* gene. EMBO J 12:2419–2430, 1993.

118. Kingsley DM, Bland AE, Grubber JM, Marker PC, Russell LB, Copeland NG, Jenkins NA: The mouse *short ear* skeletal morphogenesis locus is associated with defects in a bone morphogenetic member of the TGF beta superfamily. Cell 71:399–410, 1992.

119. Reuter R, Panganiban GE, Hoffmann FM, Scott MP: Homeotic genes regulate the spatial expression of putative growth factors in the visceral mesoderm of *Drosophila* embryos. Development 110:1031–1040, 1990.

120. Manak JR, Mathies LD, Scott MP: Regulation of a *decapentaplegic* midgut enhancer by homeotic proteins. Development 120:3605–3619, 1994.

121. Capovilla M, Brandt M, Botas J: Direct regulation of *decapentaplegic* by *Ultrabithorax* and its role in *Drosophila* midgut morphogenesis. Cell 76:461–475, 1994.

122. Campbell G, Weaver T, Tomlinson A: Axis specification in the developing *Drosophila* appendage: The role of *wingless*, *decapentaplegic*, and the homeobox gene *aristaless*. Cell 74:1113–1123, 1993.

123. Basler K, Struhl G: Compartment boundaries and the control of *Drosophila* limb pattern by the Hedgehog protein. Nature 368:208–214, 1994.

124. Zecca M, Basler K, Struhl G: Sequential organizing activities of *engrailed*, *hedgehog*, and *decapentaplegic* in the *Drosophila* wing. Development 121:2265–2278, 1995.

125. Tabata T, Kornberg TB: *hedgehog* is a signalling protein with a key role in patterning *Drosophila* imaginal discs. Cell 76:89–102, 1994.

126. Nellen D, Burke R, Struhl G, Basler K: Direct and long range action of a Dpp morphogen gradient. Cell 85:357–368, 1996.

127. Lecuit T, Brook WJ, Ng M, Calleja M, Sun H, Cohen SM: Two distinct mechanisms for long range patterning by Decapentaplegic in the *Drosophila* wing. Nature 381:387–393, 1996.

128. Singer MA, Penton A, Twombly V, Hoffmann FM, Gelbart WM: Signaling through both type I Dpp receptors is required for anterior-posterior patterning of the entire *Drosophila* wing. Development 122:3707–3718, 1997.

129. Heberlein U, Wolff T, Rubin GM: The TGF beta homolog *dpp* and the segment polarity gene *hedgehog* are required for propagation of a morphogenetic wave in the *Drosophila* retina. Cell 75:913–926, 1993.

130. Ma CY, Beachy PA, Moses K: The segment polarity gene *hedgehog* is required for progression of the morphogenetic furrow in the developing *Drosophila* eye. Cell 75:927–938, 1993.

131. Kaphingst K, Kunes S: Pattern formation in the visual centers of the *Drosophila* brain: *wingless* acts via *decapentaplegic* to specify the dorsoventral axis. Cell 78:437–448, 1994.

132. Twombly V, Blackman RK, Jin H, Graff JM, Padgett RW, Gelbart WM: The *decapentaplegic* signalling pathway is essential for *Drosophila* oogenesis. Development 122:1555–1565, 1996.

133. Kühnlein RP, Frommer G, Friedich M, Gonazález-Gaitán M, Weber A, Wagner-Bernholz JF, Gehring WJ, Jäckle H, Schuh R: *spalt* encodes an evolutionarily conserved zinc finger protein of novel structure which provides homeotic gene function in the head and tail region of the *Drosophila* embryo. EMBO J 13:168–179, 1994.

134. de Celis JF, Barrio R, Kafatos FC: A gene complex acting downstream of *dpp* in *Drosophila* wing morphogenesis. Nature 381:421–424, 1996.

135. Sturtevant MA, Marin E, Bier E: The *spalt* gene links the A/P compartment boundary to a linear adult structure in the *Drosophila* wing. Development 124:21–32, 1997.

136. St Johnson RD, Gelbart WM: Decapentaplegic transcripts are localized along the dorsal-ventral axis of the *Drosophila* embryo. EMBO J 6:2785–2791, 1987.

137. Nambu JR, Franks RG, Hu S, Crews ST: The *single-minded* gene of *Drosophila* is required for the expression of genes important for the development of CNS midline cells. Cell 63:63–75, 1990.

138. Dalessio M, Frasch M: *msh* may play a conserved role in dorsoventral patterning of the neuroectoderm and mesoderm. Mech Dev 58:217–231, 1996.

139. Ferguson EL, Anderson KV: Decapentaplegic acts as a morphogen to organize dorsal-ventral pattern in the *Drosophila* embryo. Cell 71:451–461, 1992.

140. Rusch J, Levine M: Regulation of a *dpp* target gene in the *Drosophila* embryo. Development 124:303–311, 1997.

141. François V, Bier E: The *Xenopus chordin* and the *Drosophila short gastrulation* genes encode homologous proteins functioning in dorsal-ventral axis formation. Cell 80:19–20, 1995.

142. Sasai Y, Lu B, Steinbeisser H, De Robertis EM: Regulation of neural induction by the Chd and BMP-4 antagonistic patterning signals in *Xenopus*. Nature 376:333–336, 1995.

143. Jones CM, Lyons KM, Lapan PM, Wright CVE, Hogan BLM: DVR-4 (bone morphogenetic protein-4) as a posterior ventralizing factor in *Xenopus* mesoderm induction. Development 115:639–647, 1992.

144. Maeno M, Ong RC, Suzuki A, Ueno N, Kung HK: A truncated bone morphogenetic protein receptor alters the fate of ventral mesoderm to dorsal mesoderm—roles of animal tissue in the development of ventral mesoderm. Proc Natl Acad Sci USA 91:10260–10264, 1994.

145. Fainsod A, Steinbeisser H, De Robertis EM: On the function of BMP-4 in patterning the marginal zone the *Xenopus* embryo. EMBO J 13:5015–5025, 1994.

146. Graff JM, Theis RS, Song JJ, Celeste AJ, Melton DA: Studies with a BMP receptor suggest that ventral mesoderm-inducing signals override dorsal signals in vivo. Cell 79:169–179, 1994.

147. Schmidt J, Suzuki A, Ueno N, Kimelman D: Localized BMP-4 mediates dorsal-ventral patterning in the early *Xenopus* embryo. Dev Biol 169:37–50, 1995.

148. Wilson PA, Hemmati-Brivanlou A: Induction of epidermis and inhibition of neural fate by BMP-4. Nature 376:331–333, 1995.

149. Padgett RW, Wozney JM, Gelbart WM: Human BMP sequences can confer normal dorsal-ventral patterning in the *Drosophila* embryo. Proc Natl Acad Sci USA 90: 2905–2909, 1993.

150. Holley SA, Jackson PD, Sasai Y, Lu B, De Robertis EM, Hoffmann FM, Ferguson EL: A conserved system for dorsal-ventral patterning in insects and vertebrates involving *sog* and *chordin*. Nature 376:249–253, 1995.

151. Schmidt J, François V, Bier E, Kimelman D: The *Drosophila short gastrulation* gene induces an ectopic axis in *Xenopus*: Evidence for conserved mechanisms of dorsal-ventral patterning. Development 121:4319–4328, 1995.

152. Hammerschmidt M, Serbedzija GN, McMahon A: Genetic analysis of dorsoventral pattern formation in the zebrafish: Requirement for of a BMP-like ventralizing activity and its dorsal repressor. Genes Dev 10:2452–2461, 1996.

153. Waddington CH: The genetic control of wing development in *Drosophila*. J Genet 41:75–139, 1940.

154. Díaz-Benjumea F, García-Bellido A: Genetic analysis of the wing vein pattern of *Drosophila*. Wilhelm Roux's Arch Dev Biol 198:336–354, 1990.

155. Sturtevant MA, Bier E: Analysis of the genetic hierarchy guiding wing vein formation in *Drosophila*. Development 121:785–801, 1995.

156. Shellenberger DL, Mohler JD: Temperature-sensitive periods and autonomy of pleiotropic effects of *l*(1)N^{ts1}, a conditional *Notch* lethal in *Drosophila*. Dev Biol 62:423–446, 1978.

157. Segal D, Gelbart WM: Shortvein, a new component of the *decapentaplegic* gene complex in *Drosophila melanogaster*. Genetics 109:119–143, 1985.

158. Glick AB, Lee MM, Darwiche N, Kulkarni AB, Karlsson S, Yuspa SH: Targeted deletion of the TGF-beta 1 gene causes rapid progression to squamous cell carcinoma. Genes Dev 8:2429–2440, 1994.

159. Oft M, Peli J, Rudaz C, Schwarz H, Beug H, Reichmann E: TGF-β1 and Ha-Ras collaborate in modulating the phenotypic plasticity and invasiveness of epithelial tumor cells. Genes Dev 10:2462–2477, 1996.

3

ECDYSONE RESPONSE IN *Drosophila*

WILLIAM SEGRAVES

Department of Biology, Yale University, New Haven, CT 06511

3.1. INTRODUCTION

In 1917, a young Polish physiologist named Stefan Kopeć made a remarkable discovery (1). When full-grown gypsy moth caterpillars were ligated, isolating the abdomen from the thorax, the anterior half pupated while the posterior half remained larval. This finding gave rise to a remarkable series of classical studies showing that the temporal progression of insect development was controlled by circulating hormones (2). Berthold had shown, several decades earlier, that circulating secretions of the testis were responsible for the sexual differentiation of male birds (3). Today, we know that these and many other developmental events in vertebrates and invertebrates are under the proximate control of steroid hormones, and that the characteristic nuclear receptors and molecular mechanisms through which these hormones act are highly conserved throughout the animal kingdom.[1] I will focus here on the only known steroid hormones in invertebrates, the ecdysteroid hormones best known for their roles in molting and metamorphosis. However, the existence of other evolutionarily conserved steroid receptor homologs suggests an ancient diversity to the nuclear hormone receptor motif and the possibility that there may be other, as yet undiscovered, steroid or steroid-like insect hormones.

[1]While some plants and fungi have abundant steroids and, in some cases, show evidence of a steroid response, there is no evidence to indicate that the nuclear hormone receptor motif and associated mechanisms of the steroid response are operative in these systems.

Hormones and Growth Factors in Development and Neoplasia, Edited by Robert B. Dickson and David S. Salomon
ISBN 0-471-16899-8

3.2. PHYSIOLOGIC FUNCTIONS OF ECDYSONE[2]

While Kopeć's initial studies focused on the brain (the source of a key ecdysiotropic hormone), it soon became clear that the actual source of the molting hormone was a closely associated gland called the *prothoracic gland*. Using the development of ligated larvae as a bioassay, a substantial purification of the steroid molting hormone, consisting of α-ecdysone and β-ecdysone (20-hydroxyecdysone [20-OH ecdysone]; Fig. 3.1a) (5), was obtained, opening the door to structural analysis of the hormone (6,7) and to the use of purified hormones in ecdysone research. In the years since the initial identification of ecdysone, it has become clear that this hormone is responsible for bringing about a variety of molecular and cellular changes associated with larval molting and metamorphosis in insects, as well as playing a key role in regulation of reproductive function. Recent results suggest that ecdysone may also play previously unexpected roles in embryonic development.

Even in comparison to vertebrate steroid and thyroid hormones, ecdysone has a broad range of functions. It is intriguing that functions carried out by several different steroid hormones and thyroid hormone in vertebrates, such as the regulation of skeletal growth and maturation, sexual maturation and reproductive function, and amphibian metamorphosis, are in insects apparently carried out by one hormone, ecdysone. The remarkable diversity of ecdysone functions makes the problems of tissue and stage specificity of ecdysone response particularly compelling. Consistent with its unique and central role in insects, there is no true vertebrate homolog of ecdysone or of the ecdysone receptor. Nonetheless, the structure of the ecdysone receptor (see below) strongly suggests a link to the subfamily of nuclear hormone receptors, which includes vitamin D and nonsteroidal small lipophilic molecules such as thyroid hormone and retinoic acid.

3.3. GENE REGULATION BY ECDYSONE

3.3.1. Larval Polytene Chromosome Puffing

One of the experimental advantages of *Drosophila* and other Diptera is the presence of polytene chromosomes, which result from DNA replication without cell division in growing larval cells. The resulting giant chromosomes, with 1000 or more copies in register, have provided cytogenetic tools facilitating the analysis of chromosome structure and gene response at the light microscopic levels.

Puff Induction Using Purified Hormone Just a few years after the demonstration that decondensed regions called *polytene chromosome puffs* were sites of active transcription (8,9), the purification of ecdysone made it possible for the first time to analyze the genetic effects of ecdysone under controlled experimental conditions. In these first experiments, Clever and Karlson (10) analyzed polytene chromosome

[2]In keeping with common usage, I will use *ecdysone* to denote compounds with the appropriate biologic activity (4) and refer to specific compounds by name as indicated. The major biologically active ecdysone in *Drosophila* is 20-OH ecdysone (also known as *β-ecdysone* or *ecdysterone*). Where designation of the specific chemical ecdysone is intended, I will use the original term *α-ecdysone* for clarity.

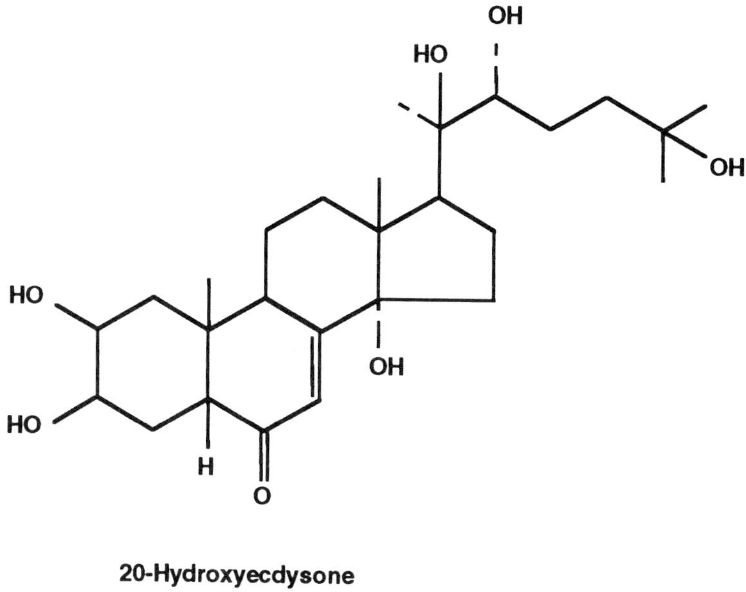

(*a*)

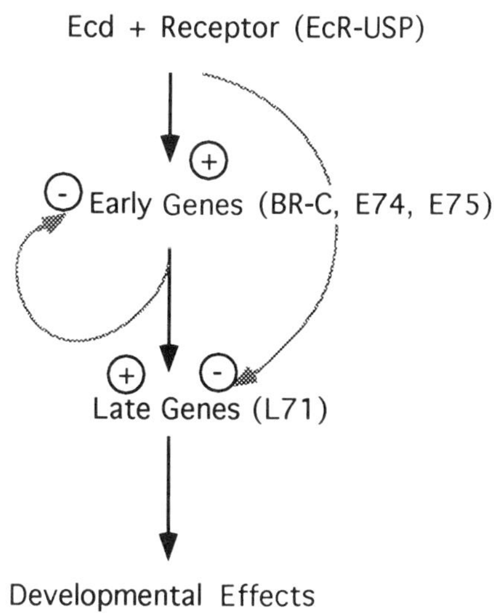

(*b*)

Fig. 3.1. (a) Structure of 20-hydroxyecdysone, the major biologically active ecdysone of *Drosophila*. (b) Clever-Ashburner model of ecdysone-regulated hierarchies in the larval salivary gland. Clever initially described the hierarchical expression of early and late ecdysone-responsive genes. Ashburner et al. further proposed (1) the direct activation of the early genes by the ecdysone receptor complex; (2) the autoregulatory inactivation of early gene expression following the initial response; and (3) the repression of late gene activity by the ecdysone receptor complex. This established the formal competition between the ecdysone receptor complex and early ecdysone-inducible proteins. Parallel regulatory hierarchies are known or presumed to control ecdysone-regulated developmental events in other tissues and stages. Products of a given hierarchical response play an important role in establishing competence for a subsequent ecdysone response.

puffing patterns in *Chironomus* larvae that had first been injected with purified ecdysone. Their remarkable results on the induction of specific polytene chromosome puffs provided the first evidence that ecdysone acted by altering patterns of gene activity and further provided evidence for a regulatory hierarchy of ecdysone response. Clever and coworkers significantly extended these studies over the next several years, first demonstrating a characteristic sequence in the appearance of puffs following ecdysone administration and then showing that the early portion of the puffing sequence, but not the latter portion, could proceed in the absence of protein synthesis (11). These findings gave birth to the concept of the early and late steroid response analogous to the regulatory hierarchies controlling expression of bacterial enzymes (12), with products of the early-responsive loci acting to turn on the later response (Fig. 3.1b).

Experimental Manipulation of the Puffing Response in Drosophila Simultaneously, Becker, Berendes, and others had been examining the process of polytene chromosome puffing in *Drosophila* and characterizing the precise patterns of puffing at the end of larval development under the control of the ring gland (13,14). What followed can only properly be described by Michael Ashburner, and I regret that I cannot refer the reader directly to Michael's humble, enlightening, and entertaining historical perspective presented at the XIIIth Ecdysone Workshop in Barcelona during the summer of 1996. Through a combination of hard work, insight, and luck, Ashburner and colleagues succeeded in carrying out a series of experiments that would lay the foundation for our current understanding of ecdysone-triggered regulatory hierarchies. First came the discovery that puffing could be induced in *Drosophila* salivary glands cultured *in vitro,* and that this allowed a substantially faithful replication of *in vivo* puffing patterns at the end of larval development when ecdysone represses the intermolt genes expressed in the mid third instar larva and stimulates the characteristic late larval puffing pattern. This opened the door to more sophisticated manipulation of experimental conditions, including ecdysone withdrawal and restimulation, inhibition of protein synthesis, and fine control of hormone levels. What emerged from these studies was a simple, elegant model (Fig. 3.1b) that has the unusual distinction of having withstood the test of time (15). While the basic principle of the model was similar to that proposed by Clever, Ashburner and colleagues extended it considerably by demonstrating (1) the nature of all-or-none response of the late genes following a threshold level of early gene expression, (2) the autoregulatory nature of early puff regression, and (3) the inhibitory effect of the ecdysone receptor on the late puffs. Based on preliminary studies indicating chromatin-associated ecdysone binding sites (16), Ashburner et al. went on to propose a direct role for the ecdysone–receptor complex in the regulation of both early and late gene expression (15).

Over the next several years, genetic evidence confirmed the role of early puff genes in regulating these hierarchies, and the development of molecular tools allowed further extension of this model. Remarkably, all features of the model remain intact, the most significant additions being other classes of genes, including early-late genes and midprepupal genes, which appear to play important roles as well.

3.3.2. Ecdysone Response in Other Tissues and Stages

Molting and The Metamorphic Ecdysone Response In some areas of ecdysone research, it is not uncommon to speak in general terms of the metamorphic ecdysone response, particularly in considering stage-specific factors underlying the differences between the metamorphic and larval molting responses to ecdysone. However, a detailed analysis of the ecdysone response must recognize that the programs leading up to and through metamorphosis likely consist of at least four interrelated ecdysone responses: a mid third instar response that may trigger wandering and likely establishes competence for the subsequent responses (17,18); the late larval program responsible for triggering pupariation; the related but distinct late prepupal program responsible for triggering pupation; and a midpupal program or programs controlling the latter stages of metamorphosis. Genetic and molecular studies on the effects of mutations or overexpression of ecdysone-responsive transcription factors suggest a complex and essential interrelationship between these programs, with many of the factors used reiteratively. It remains to be seen whether there is similar complexity to the programs controlling other stage-specific ecdysone responses.

The differences between larval molting and metamorphosis dramatically illustrate the problem of stage specificity of the ecdysone response. While the larval molts result in a relatively modest upgrade of existing machinery, metamorphosis is without a doubt the most dramatic developmental event know to man or insect (19). Direct evidence in *Drosophila* is lacking, but work on other insects suggests that this developmental stage specificity is controlled by a second hormone, juvenile hormone (JH), which acts during larval life formally to repress commitment to the metamorphic ecdysone response. The tissue specificity of the ecdysone response is also demonstrated dramatically at the initiation of metamorphosis. Most of the larval tissues, including those of the nervous system, will be largely retained but extensively remodeled. The imaginal discs that give rise to the cuticular structures of the adult head and thorax will begin their terminal differentiation. Other clusters of imaginal cells in the abdomen and internal organs will undergo extensive cell division in preparation for generation of the adult organs and abdominal integument.

Ecdysone-response hierarchies have been characterized predominantly in polytene tissues of the third instar larva, but abundant evidence indicates that similar responses are occurring in other tissues and stages as well. Clearly, one of the most compelling and interesting questions in ecdysone research concerns the mechanisms through which parallel regulatory hierarchies in other tissues and stages are controlled and how they, in turn, control different developmental fates.

Reproductive Ecdysone Response While the largest proportion of ecdysone research has been focused on the roles of ecdysone in molting and metamorphosis, there are clear indications that ecdysone is also involved in other major transitions in the life of the fly, including reproduction and early development.

As in mammalian systems, the control of reproductive function in insects is regulated by hormonal factors governing maturation and functioning of the reproductive system. Several lines of evidence indicate that both ecdysone and JH play critical roles in the regulation of oogenesis and vitellogenesis. The role of ecdysone is most elegantly demonstrated in the anautogenous mosquitoes, which must take a blood meal before initiating female reproductive function. In these mosquitoes, a blood meal triggers neuroendocrine stimulation of ovarian ecdysone production, which in turn stimulates key steps of oogenesis and vitellogenesis (20). While other insects, such as *Drosophila*, in which oogenesis is asynchronous, do not appear to maintain a comparable global hormonal control over reproduction, evidence suggests that ecdysone is critical to reproduction here as well. Several known mutations affecting ecdysone production or ecdysone response display female sterility. The temperature-sensitive *ecdysoneless-1* mutation displays reversible, temperature-dependent female sterility (21), and mutations in the ecdysone receptor subunit gene *ultraspiracle* (22) and in the ecdysone-responsive *Broad-Complex* gene (23) also result in female sterility. Ecdysone is known to stimulate yolk protein production (24), but other specific roles of ecdysone in reproduction have not yet been defined.

Embryonic Ecdysone Response While ecdysone appears to play a role in embryonic cuticle deposition in at least some insects, little is known of the roles of ecdysone in embryonic development in *Drosophila*. Ecdysone levels reach a peak midway through embryonic development (25), potentially as a result of the release of maternal ecdysone conjugates (26). Apparently in response to this ecdysone pulse, there is significant induction of a number of ecdysone-induced factors initially characterized in studies of the late larval ecdysone response (27–29). While this embryonic ecdysone pulse may be involved in the initiation of cuticle formation, recent genetic analyses suggest that there may be additional roles of ecdysone in the specification of spatial identities or the control of morphogenetic events within the embryo (see below).

3.3.3. The Ecdysone Receptor

Preliminary Characterization Based on the preliminary characterization of a nuclear ecdysone binding protein (16), Ashburner et al. speculated that the action of ecdysone in regulating early and late gene expression would be mediated by a receptor that would directly act on target genes (15). Within a few years, support for this hypothesis emerged from studies showing that 20-OH could be localized to the sites of ecdysone regulated puffs (30). Simultaneously, more detailed biochemical studies had begun to characterize ecdysone binding proteins that were almost certainly ecdysone receptors. These proteins displayed the proper affinities for ecdysone analogs relative to their biologic activities, and ecdysone-insensitive cell lines could be shown to be lacking in these putative receptors (31–34). Unfortunately, the low abundance of these receptors and perhaps the requirement for two nonidentical subunits (see below) hindered efforts to purify the receptors and characterize them in more detail. By the time a substantial purification of the receptor had been completed (35), parallel molecular studies had allowed the cloning and initial characterization of the ecdysone receptor (see below).

The Nuclear Hormone Receptor Superfamily The cloning of vertebrate glucocorticoid and estrogen receptors opened the door to the molecular genetic characterization of steroid receptors and related receptors in vertebrate and invertebrate systems. Molecular characterization of these receptors and related proteins dramatically altered the face of steroid and other small lipophilic hormone research by defining a family of proteins known as the *nuclear hormone receptor superfamily* (36). This family is now known to include more than 50 distinct members in vertebrates and approximately one third that number in *Drosophila*. The discovery of nearly 100 putative nuclear hormone receptors in conjunction with the *Caenorhabditis elegans* genome project (A. Sluder, personal communication) suggests that many more members may still be found in both vertebrates and insects.

Identification of the Ecdysone Receptor Gene EcR Characterization of the nuclear hormone receptor superfamily and identification of receptor homologs, including the ecdysone-inducible *Drosophila* E75 protein (see below), led to the straightforward prediction that the ecdysone receptor would be a member of this family and that it might be isolated on this basis. Using *E75* as a probe, two new members of the nuclear hormone receptor superfamily in *Drosophila* were identified: *EcR* (initially known as *DHR23*) and *DHR3* (27,37). At around the same time and later, genes encoding a wide array of other steroid receptor-like proteins were identified using molecular probes or following sequence analysis of genes identified based on their developmental function. These studies have been reviewed elsewhere (38–40), and I shall not describe the other receptor-like proteins here, except insofar as they relate to the ecdysone response (Fig. 3.2).

Gene	Cytology	Ecd Reg	Homolog	DNA/HBD
EcR	42A	early	FXR	81/34
USP	2C	const.	RXR	86/49
BR-C	2B	early	Zn finger	---
E74	74EF	early	ETS	---
E75	75B	early	Rev-Erb	79/35
DHR3	46F	early-late	RZR	75/35
E78	78C	early-late	Rev-Erb	72/35
Ftz-FI	75C	mid pp	SF-I	88/33
DHR38	38E	ind?	NURR1	91/60
DHR39	39C	early-late	SF-1	62/29
DHR78	78D	ind	TR2	74/32
DHR96	96B	ind	VDR	64/28
SVP	87B	n.d.	COUP	94/93

Fig. 3.2. Transcription factors in the ecdysone response. For each of the transcription factors described in the text, the cytogenetic location and ecdysone responsiveness are indicated. For nuclear hormone receptors, the most closely related vertebrate receptor is listed. For BR-C and E74, the type of DNA-binding domain is indicated. In the final column, identity within DNA- and hormone-binding domains (HBDs) in comparison to the most closely related vertebrate nuclear hormone receptor is indicated.

Several lines of evidence indicated that *DHR23*, one of the genes identified using an *E75* probe, encoded a key component of the ecdysone receptor, and this gene thus took on the name *EcR* (41). Transfection of tissue culture cells with an *EcR* cDNA resulted in a substantial increase in specific binding of a radiolabeled ecdysone analog, antibodies directed against EcR were capable of precipitating the native ecdysone receptor of tissue culture cells, and EcR could be shown to be bound to the ecdysone response element of the hsp27 gene (42). Finally, *EcR* could be shown to rescue the ecdysone resistance of cell lines selected for reduced ecdysone responsiveness. Nonetheless, it soon became clear that *EcR* was not sufficient to confer ecdysone responsiveness to a heterologous system (43,44). This was significant not just to completing the characterization of *EcR*, but also to its eagerly anticipated utilization as an exogenous inducible regulator of gene expression in other organisms, including mammals.

USP and the Requirement for Heterodimer Formation The molecular factors underlying the insufficiency of EcR became clear as a result of parallel studies demonstrating that DNA binding and function of vertebrate retinoic acid, thyroid hormone, and vitamin D receptors depended on a second factor, RXR, a nuclear hormone receptor responsive to retinoids including 9-*cis* retinoic acid (45–51). Several groups had identified a homologous gene in *Drosophila* (52–54), and this was shown to be the site of previously characterized *ultraspiracle (USP)* mutations resulting in lethality associated with an abortive first larval molt (22,55). The high level of structural similarity between USP and RXR was reflected at the functional level, and USP was shown to substitute for RXR in allowing DNA binding of vertebrate RXR partners (44,56). This naturally led to the search for USP partners in *Drosophila*. The combination of USP and EcR led to striking results; together they displayed strong selective binding to ecdysone response elements, and the simultaneous addition of *EcR* and *usp* conferred ecdysone responsiveness to mammalian cells (44). Parallel studies addressing differences in ecdysone responsiveness of various cell types transfected with *EcR* alone showed that the requirement of EcR for a partner can also be met by RXR (57). Several lines of evidence suggest that the Ecr-USP heteromer, most likely a heterodimer, is the naturally occurring ecdysone receptor; ecdysone receptors from tissue culture cells and embryos react with both anti-EcR and anti-USP antibodies, and EcR and USP are colocalized to polytene chromosome puff sites including early and late ecdysone-responsive puffs (58). Many *usp* phenotypes, including interruption of larval molting and pupation, are consistent with the proposed role of *usp* as an obligatory ecdysone receptor subunit (22,55,59). Other phenotypes, such as an embryonic posterior abdominal defect and disruptions in eye development, are less clearly related. Most enigmatic, however, is the demonstration that homozygous *usp* clones, generated by mitotic recombination, can give rise to adult head, thoracic, and abdominal tissue; it seems highly unlikely that the completion of metamorphosis by these tissues could proceed in the absence of functional ecdysonc receptors. It is not yet clear whether this finding signifies the existence of a USP-independent class of ecdysone receptors or whether it might potentially be explained by the persistence of functional USP protein in the genetically *usp* clones (22).

Despite the high degree of structural and functional conservation of the heterodimer motif, there is one significant difference between the vertebrate and *Drosophila* receptors. Unlike the situation in the vertebrate receptor heterodimers, USP is required for hormone binding as well as DNA binding, raising difficult questions as to which of these proteins actually binds ecdysone (58). Low levels of hormone binding have been reported in preparations of *in vitro*-translated EcR (58), but it seems plausible that this could be the result of RXR contamination. Further support for preliminary findings indicating ecdysone binding by bacterially expressed EcR (60) would carry considerably more weight. If indeed EcR alone can selectively bind to ecdysone, this suggests that USP may allosterically affect hormone binding. Nonetheless, it remains possible that the USP hormone binding domain directly contributes to the ecdysone-binding site. Remarkably, preliminary results suggest, however, that the USP hormone binding domain may alternatively serve as an independent, low-affinity ecdysone binding site (X. Hu and P. Cherbas, personal communication). Whether ecdysone is the natural ligand for this site remains to be determined. However, the possibility that USP may also be responding to a ligand provides interesting possibilities for broadening the dynamic range of ecdysone response or for integrating responses to ecdysone and another hormone directly at the level of a heterodimeric bifunctional receptor.

From a technical standpoint, the requirement for a heterodimer partner initially hindered the widely anticipated use of EcR as an inducible exogenous regulator of mammalian gene expression. In addition to necessitating a two-component system, the evolutionary conservation of nuclear hormone receptor DNA binding allowed for the possibility that endogenous and introduced elements might be promiscuously recognized by ecdysone receptor heterodimers and endogenous receptors with similar DNA binding specificities. No et al. (61) have designed a clever method to circumvent these problems. By engineering a modified EcR with the DNA binding specificity of the glucocorticoid receptor, they have directed EcR-dependent expression to a novel hybrid response element containing one glucocorticoid response element half site and one natural ecdysone response element half site (Fig. 3.3). This novel element is strongly activated by ecdysone in cultured cells and transgenic mice carrying the modified *EcR* along with *RXR*. This thus provides for a strongly inducible system in which an exogenous stimulus with few if any known side effects can rapidly and reversibly stimulate gene expression in mammalian cells or organisms. Clearly, this system offers significant potential for the experimental or therapeutic regulation of gene expression.

3.3.4. Ecdysone-Responsive Genes

The discussion of ecdysone-responsive genes will be divided based on the Ashburner et al. scheme for salivary gland puffing. Nonetheless, it should be borne in mind that, within each class and even among the transcription units of a single

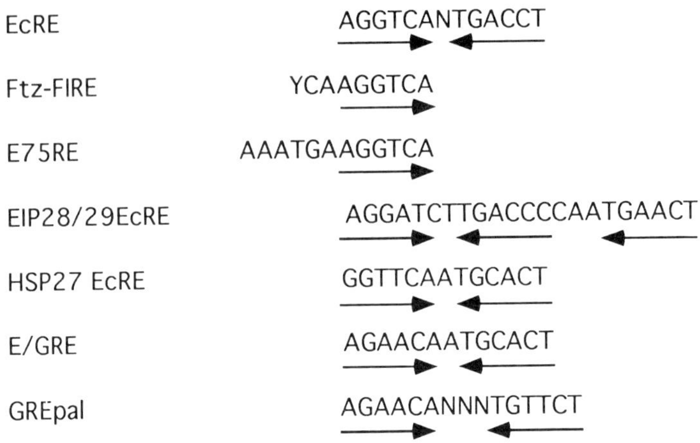

Fig. 3.3. Response elements recognized by the ecdysone receptor and ecdysone-regulated transcription factors. A canonical perfect palindromic EcRE is shown. Most natural EcREs are related to the palindromic consensus RGG/TTCAnTGAC/ACY, although some EcREs appear to consist of direct repeats. The Ftz-FI and E75 target sites consist of one AGGTCA half-site with specific flanking sequences. The EIP28/29 EcRE contains a degenerate inverted repeat EcRE, as well as DR3 that may be a site for competitive binding by SVP. the E/GRE element was created by altering one half-site of the hsp27 EcRE to match a palindromic glucocorticoid response element (GREpal). This element is recognized by a hybrid ecdysone receptor in which EcR has been modified to have the DNA- binding specificity of the glucocorticoid receptor.

gene, there are significant differences in temporal expression profiles. These differences may depend in part on differential hormone sensitivity (62), but they vary substantially from tissue to tissue (27,63). The extent to which these differences are essential or to which they may underlie the tissue-specific response at the developmental level is as yet undetermined, and I will focus on the broader aspects of the expression and function of ecdysone-inducible genes while leaving the details of the temporal patterns of response to recent comprehensive reviews (18,64).

The Classical Earlies Initial characterization of late larval polytene chromosome puffing described a small number of puffs that were induced by ecdysone within minutes of hormone exposure (65). The most prominent of these were the early puffs at 2B, 23E, 63F, 74EF, and 75B. After a few hours of hormone exposure, the early puffs regress as a large, heterogeneous set of late puffs begins to appear. Consistent with the idea that the early puff genes are directly induced by the ecdysone–receptor complex, protein synthesis inhibitors do not block their induction (with the exception of 23E). The regression of the early puffs, in contrast, is blocked by inhibitors of protein synthesis, consistent with the idea that the early puffs are repressed by an ecdysone-inducible factor or factors (66). Detailed analysis of the late prepupal ecdysone response revealed a substantial reiteration of the late larval ecdysone response, with the addition of a sixth prominent early puff at

93F (67). Products of these early ecdysone-responsive genes were considered candidates for the regulatory factors controlling ecdysone-responsive hierarchies; this hypothesis has been borne out by subsequent molecular and genetic studies.

The first direct evidence of a regulatory role for the early ecdysone-responsive puff genes came from work on the 2B5 puff region. A screen for nonpupariating mutations on the X chromosome identified lethal mutations mapping to the 2B5 puff (68–70). A large number of previously and subsequently identified mutations were found to be allelic to these mutations, and this gene is now known as the *Broad-Complex (BR-C)*, after the first identified *BR-C* mutation, *broad* (71). While loss of all function at *BR-C* results in the nonpurariating phenotype, *BR-C* encompasses at least three other genetically separable functions: *br (broad), rbp (reduced bristles on palpus),* and *l(1)2Bc* (70,72). These *BR-C* mutations result in alterations in polytene chromosome puffing (73) and in the expression of ecdysone-responsive genes (74,75), as well as in tissue-specific disruptions of metamorphosis (72,76).

While *BR-C* and other classical earlies resisted initial attempts at isolation via differential screens (77,78), the increasing availability of cytogenetically mapped probes and the advent of chromosomal walking allowed the cloning of *BR-C* on the basis of its chromosomal map position (79). Detailed molecular description of *BR-C* revealed that the genetic complexity of *BR-C* is more than adequately accounted for at the molecular level. At least three independent 5′ untranslated region-containing exons are spliced to a core region and then to a set of four alternative exons encoding distinct, isoform-specific TFIIIA-like zinc fingers (80,81). Molecular genetic analysis and analysis of the BR-C proteins indicates that there is a correlation between the genetically defined *BR-C* functions and the protein isoforms Z1, Z2, Z3, and Z4 defined by differential utilization of the alternative zinc finger encoding exons; Z1 appears to correspond to the *rbp* function, Z2 to the *br* function and Z3 to the *2Bc* function. The tissue specificity of *BR-C* isoform expression (81,82) suggests that differential *BR-C* isoform expression may be one mechanism underlying the tissue specificity of ecdysone response (see "Ecdysone Receptor and Early Gene Isoforms" in Section 3.3.5).

Only a few years after the identification of nonpurpariating *BR-C* mutations, 74EF and 75B were also linked genetically to the control of the ecdysone-triggered regulatory hierarchies by the studies of Walker and Ashburner (83), which showed that duplications and deficiencies for a region including these puffs had effects on the kinetics of early and late gene response. These genes as well were then cloned on the basis of cytogenetic criteria.

The gene from the 74EF puff consists of two overlapping transcription units giving rise to *E74A* and *E74B* products (28). Consistent with the proposed regulatory role, the E74 proteins are transcription factors, with an ETS-related (84) DNA binding domain shared by the two E74 proteins. Immunocytochemistry using anti-E74 antibodies indicates that the E74 proteins can be mapped on polytene chromosomes to a large number of sites, including early and late puffs (85). These findings can be extended to the molecular level, as E74 binding sites can be mapped to the *E74* gene (85) as well as to the regulatory region of a putative late gene target (86). Genetic analysis of *E74* suggests that *E74B* and *E74A* are required for genetically separable functions in pupariation and pupation, respectively (87). Although *E74* is expressed at earlier ecdysone pulses, including during embryogenesis, neither *E74* transcription unit is required prior to the end of larval life. Studies on the effects of *E74* mutations on late gene expression are consistent with the idea that *E74* plays a significant role in the induction of late gene expression (88).

Since *BR-C* and *E74* mutations alter the expression of many of the same genes, Fletcher and Thummel (88) asked whether they might show signs of genetic interaction. Consistent with the idea that *BR-C* and *E74* cooperate in the regulation of target genes, *BR-C* and *E74* mutations show synergistic effects on the expression of intermolt, early, and late gene targets. Less expected was the finding that combinations of *E74* and *BR-C* alleles result in earlier lethality than either class of mutation alone. This may suggest some degree of functional redundancy in their regulation of target genes acting in earlier developmental stages. Alternatively, it remains possible that functional redundancy between *E74A* and *E74B* masks an earlier requirement for *E74* that would be revealed by a common region null mutation. In this case, dependence of proper *E74* expression on *BR-C* might underlie the failure of reduced levels of *E74* products to substitute for one another.

Much like *BR-C* and *E74*, the *E75* gene within the 75B puff consists of a set of overlapping transcription units, in this case three transcription units spanning 100 kb, encoding a family of transcription factors (89). The E75A, E75B, and E75C proteins belong to the nuclear hormone receptor superfamily and contain regions of DNA and hormone binding domain conservation that are characteristic of this family. E75 proteins are also localized to the sites of early and late puff genes (90), and molecular studies suggest that E75A, which represses its own expression, may be responsible at least in part for the regression of early puffs at the onset of the late response (W. Feng, W. Pierceall, E. Wiellette, J-Y. Leu, Z. Zhang, W. Segraves, in preparation). The E75A and E75C proteins each contain complete DNA and hormone binding domains and differ only in their amino terminal A/B domain. By analogy to studies on N-terminal variations in vertebrate receptors (91), this suggests that they may have different activation domains and consequently different capacities to regulate specific target genes. In contrast, the B protein is missing the first finger of the DNA binding domain, an obligatory contributor to DNA binding by steroid receptors (92,93). Genetic analysis of *E75* indicates that *E75A* and *E75C* have genetically separable functions in the control of embryonic and larval development, molting, and metamorphosis (94). Unlike *BR-C* or *E74*, *E75A* is required for all developmental ecdysone responses, including embryonic development, larval molting, and metamorphosis. *E75C* is not needed for embryonic, larval, and early pupal development but is required for late pupal development and also for adult fertility. Despite the evolutionary conservation of *E75B* over the approximately 300 million years separating Diptera and Lepidoptera (95), *E75B* is not absolutely required, and deletion of the *E75B*-specific region has only modest effects on survival and fertility (J. Cole and W. Segraves, unpublished observations).

The large range of developmental requirements for *E75* has prompted us to investigate the role of *E75* in ecdysone-regulated developmental processes that have not previously been studied in detail. Of greatest novelty and potential significance are the potential roles of *E75* in the integration of ecdysone and JH response (see "Juvenile Hormone Modulation of the Ecdysone Response" in Section 3.3.5) and in the control of ecdysone-triggered hierarchies in embryonic development and reproduction.

The embryonic lethality of strong *E75* mutations has prompted us to consider potential roles for ecdysone in embryonic development. Of particular interest is the failure of the process of head involution, which normally gives rise to the internalized mouthparts characteristic of higher diptera. Consistent with the idea that the *E75* defects are ecdysone related is the finding that similar phenotypes are manifested by the offspring of mothers carrying the *dre-4* ecdysone-deficient mutation (T. Sliter, personal communication). Segraves et al. (94) have proposed two models to explain the role of *E75* in embryonic head involution, one involving the interface between spatial and temporal regulatory hierarchies and the other involving a more direct role for ecdysone in stimulating head involution. The complex interplay between spatial patterning in the head and head involution does not permit an unambiguous interpretation of the primary cause of the defect; mutations such as *giant* that lead to loss of head structures show head involution defects (96), but blockage of head involution by other means conversely leads to the inability to recognize certain head structures (97). If the primary defect is in the specification of pattern, the *E75* gene may be the convergence point for the specification of spatial identity by pattern formation genes and the realization of that pattern under temporal control by ecdysone. Consistent with a role for *E75* in specifying spatial pattern is the finding of denticle belt defects specific to the second thoracic segment of *E75* mutant embryos (94) and the demonstration that *E75* mutations cause alterations in the spatial patterning of embryonic gut constrictions (98). Alternatively, the primary defect may be in head involution per se, a possibility suggested by the intriguing correlation between the embryonic ecdysone pulse and the initiation of head involution. It has been proposed that the process of head involution requires substantial cell division along the dorsal midline, and the presence of minor dorsal midline defects in *E75* mutants suggests that a failure of this process could be linked to the observed head involution defects.

Similarly, the expression of *E75* in the *Drosophila* ovary and the sterility of certain *E75* mutants have prompted us to investigate the potential role of *E75* in the reproduction of *Drosophila* and other insects. The yellow fever mosquito *Aedes aegypti* is an important disease vector and an excellent model for understanding the hormonal control of reproductive function. Anautogenous mosquitoes such as *A. aegypti* require a blood meal to initiate oogenesis and vitellogenesis. Mechanical and chemical signals associated with the blood meal stimulate ovarian ecdysone production and consequent initiation of oogenesis and vitellogenesis under the influence of ecdysone. Synchronous oogenesis under ecdysone-inducible control offers the opportunity to explore reproductive ecdysone-triggered regulatory hierarchies in much the same way that the salivary gland response has been investigated. Raikhel and colleagues have shown *EcR* and *usp* expression in the ovary and fat body of vitellogenic females (99,100). Interestingly, unlike in *Drosophila,* there are two distinct USP isoforms, one expressed predominantly in the fat body and the other in the ovary (100). Consistent with the idea that ecdysone-triggered regulatory hierarchies similar to the *Drosophila* salivary gland hierarchies are operating in the control of mosquito reproduction, *E75* is expressed in the ovary and fat body in response to ecdysone, with expression corresponding to the peak of vitellogenesis (W. Pierceall, A. Biran, A. Raikhel, and W. Segraves, submitted). Interestingly, while there are significant differences in the global organization of mosquito and *Drosophila* oogenesis, the expression pattern of *E75* in the *Drosophila* ovary suggests that ecdysone-regulated genes such as *E75* could play key roles in the progression of *Drosophila* oogenesis as well (M. Buszczak and W. Segraves, unpublished observations).

Other Early Ecdysone-Responsive Genes Molecular characterization of the early puff genes at 63F and 93F indicates that these puffs differ in several respects from the previously described early puffs. Most notably, they show much greater tissue specificity than the *BR-C, E74,* or *E75* genes that appear to be induced by ecdysone in all tissues of the late third instar larva.

The 63F puff contains two nonoverlapping, ecdysone-inducible transcription units, *E63-1* and *E63-2,* rather than the overlapping transcription units characteristic of the others (101). *E63-1* appears not to encode a transcription factor, but rather encodes a calmodulin-like calcium binding protein, while *E63-2* appears not to have any large open reading frames. The expression of *E63-1* in response to ecdysone is in close correlation with the 63F puff, which is expressed in salivary glands but not in the fat body. *E63-1* is also expressed at moderate levels in the gut, but is not ecdysone-regulated, and at low levels in other tissues of the third instar.

The *E93* gene responsible for the late prepupal early puff at 93F also displays tissue specificity in its ecdysone inducibility (102). In salivary glands, it is expressed in response to ecdysone in correlation with the 93F late prepupal puff. However, it is not expressed during the late larval-prepupal period in imaginal discs and shows ecdysone-independent patterns of expression in other larval tissues. While the E93 protein is not significantly related to any known protein, preliminary results suggest that it is a nuclear protein that is a potential transcription factor (102).

The tissue specificity of *E63* and *E93* makes them good candidates for the type of regulatory early genes that might control tissue-specific late responses underlying diverse developmental ecdysone responses. E63, while not a transcription factor, a could clearly play a regulatory role via its presumed function as a calcium calmodulin, potentially serving to integrate Ca^{2+}-dependent and ecdysone-dependent signaling pathways. If E93 also proves to be a transcription factor or other regulatory protein, it could play a role in directing a prepupal stage-specific response to tissues such as the salivary gland. Baehrecke and Thummel have proposed that *E93* may be involved in directing the histolysis of the salivary gland; as a late prepupal gene, it is induced only under conditions that lead to histolysis, and it is induced just prior to initiation of programmed cell death in the salivary gland (102).

Many other early ecdysone-responsive genes have now been characterized at the molecular level. These include a number of early ecdysone-responsive genes that are likely to act directly as "effector" genes functioning in parallel with the regulatory cascades defined above. These have been reviewed recently by Richards (18) and will not be discussed extensively here. However, it is worth nothing that, like *E63* and *E93,* many of these genes display profound tissue and stage specificity. Examples include the *Fbp-1* gene, which is expressed in the fat body (103); the *EIP28/29* gene, which is ecdysone induced, ecdysone repressed or ecdysone insensitive, depending on the target tissue (104); and the imaginal disc-specific *IMP-E1* gene believed to be involved in imaginal disc morphogenesis (29). For many years, models for understanding tissue and stage specificity have tended to focus on the ways in which responsive programs, and in particular the late response, must be different, but detailed puffing studies and more recent studies such as these provide clear evidence that some of the tissue- and stage-specific differences are preprogrammed into the initial ecdysone response (18,105). This programming may well depend on the prior expression of ecdysone-inducible, tissue- or stage-specific determinants during the previous ecdysone response, but at some level it must reflect innate tissue- or stage-specific competence based on restrictive or permissive interactions between the ecdysone receptor and other tissue- or stage-specific factors.

The Early Lates Among the initially characterized ecdysone-responsive puffs were a small number of cycloheximide-sensitive puffs that arise after the early puffs but before other late puffs. These so-called early-late puffs could also be distinguished from the late puffs by their requirement for continuous exposure to ecdysone (15,83). Three representative of this class—*DHR3* (37), *DHR39* (106,107), and *E78* (108)—have been cloned and characterized. Interestingly, each of these is also a member of the nuclear hormone receptor superfamily. Characterization of the ecdysone response of each of these at the RNA level has elucidated the features underlying the early-late response (109). Each can be induced by ecdysone even in the presence of cycloheximide, but they attain peak levels of expression only after several hours in the presence of ecdysone without cycloheximide. Thus, it appears that they are activated directly by ecdysone, like the early genes, but are further activated, rather than repressed, as the early regulatory products are synthesized. All three of these genes are expressed in conjunction with the ecdysone response at the end of the third instar, but the function of these genes is not yet fully understood. The *E78* gene is nonessential, but *E78* is expressed at high levels during the late third instar and *E78* mutations result in alterations in ecdysone-induced puffing (110). Remarkably, the E78B product likely involved in the regulation of puffing lacks the entirety of the DNA binding domain, while the E78A product, which contains a complete DNA binding domain, is not expressed until midpupal development. It is intriguing to speculate that the survival of *E78* mutations may represent some functional redundancy. A good candidate for a protein with overlapping function is E75B, which is expressed at approximately the same time, lacks a full DNA binding domain, and is also nonessential. Genetic analysis of *DHR3* function in the third instar larva has been complicated by the embryonic lethality of *DHR3* mutations (M. Bender and G. Carney, personal communication), but recent findings suggest that *DHR3* may be involved in the repression of early genes and the induction of the midprepupal gene *Ftz-FI* (111,112).

Late Ecdysone-Responsive Genes Studies on tissue and stage specificity of puffing suggest that the large and heterogeneous late response likely underlies much of the developmental specificity of the ecdysone response. However, in comparison to the extensive characterization of the early ecdysone response, relatively little is known about specific genes involved in the late ecdysone response. Clearly, a detailed understanding of late gene expression and the specific developmental functions of late genes will be essential to a complete understanding of genetic regulatory hierarchies in the ecdysone response.

The best-characterized late ecdysone responsive puff is the 71E late puff. This puff contains a cluster of coordinately expressed late genes (113,114). Analysis of the 71E puff region suggests that these genes evolved by duplication of genes encoding a primordial pair of small basic proteins with a conserved cysteine backbone. These proteins appear to be related to vertebrate defensins and have been proposed to carry out an antimicrobial function during metamorphosis (114). Molecular and genetic data suggest that the 71E late puff is turned on in response to expression of *E74* early gene products. Gene expression in the 71E late puff is altered in *E74* mutants (115), and E74A binding sites found upstream of the *L71-6* gene are essential for appropriate expression of *L71-6* (86). In conjunction with the localization of E74 to a large number of late puffs (85), these results suggest that *E74* may generally be involved in the induction of late gene expression.

Other late genes, active in imaginal discs, have been identified in a screen for imaginal-disc specific proteins expressed on membrane-bound polysomes (116,117). Like the salivary gland late responsive puffs, *IMP-L1* is not induced until 6 hr after ecdysone exposure, and its expression requires protein synthesis. In conjunction with studies showing rapid cycloheximide-insensitive induction of the *IMP-E1* gene, this shows that regulatory hierarchies like those found in the salivary gland are also active in imaginal discs (29). As predicted based on the methods used to isolate these genes, *IMP-E1*, *IMP-L1*, and *IMP-L2* all appear to encode secreted or membrane-associated proteins and are proposed to be involved in regulating key steps of ecdysone-regulated imaginal disc morphogenesis (116–118). Thus, it is likely that other factors, potentially the broadly expressed classical earlies, are involved in regulating the onset of late gene expression in imaginal discs.

The classical late gene response occurs under conditions of continuous hormone exposure. However, puffing, developmental, and molecular studies make it clear that proper progression of the prepupal ecdysone response in salivary glands and imaginal discs depends on hormone withdrawal following the initial ecdysone pulse (119,120). Studies on the *Ftz-FI* gene demonstrate the significance of the expression of genes in this midprepual class for establishment of competence for the late prepupal ecdysone response.

The nuclear hormone receptor FTZ-FI was initially identified based on its binding to regulatory elements for the embryonic patterning gene *fushi tarazu* (121,122) and was subsequently found to have an isoform, FTZ-FIβ, expressed in the midprepupal period (123). Expression studies and *in situ* mapping demonstrated that the *Ftz-FI* gene is in fact responsible for the midprepupal 75CD puff. Localization of the FTZ-FI protein to thc sites of the late prepupal earlies suggested that it might be involved in regulating the late prepupal ecdysone response. This potential stage-specific role of *Ftz-FI* will be considered in "*Ftz-FI* Regulation of the Late Prepupal Ecdysone Response" in Section 3.3.5.

3.3.5. Control of the Tissue- and Stage-Specific Ecdysone Response

Juvenile Hormone Modulation of the Ecdysone Response The most significant known modulator of developmental ecdysone response is juvenile hormone (JH). That an endocrine factor might be responsible for maintaining the juvenile state was initially suggested by studies indicating that a circulating factor from penultimate stage *Rhodnius* larvae could inhibit metamorphosis of a final instar larva (124). Chemical identification of the hormonal factor, JH, responsible for maintaining the larval state in this and other insects indicated that it is a sesquiterpenoid hormone (125). Given its significance, it is remarkable that so little is known about the mechanisms through which JH functions. The difficulty of defining a direct JH response pathway and the biochemical intractability of the JH receptor have severely restricted progress at the mechanistic level. Nevertheless, a preliminary picture of the mechanisms through which JH may function is beginning to emerge.

Several lines of evidence suggest that JH modifies, rather than inhibits, the action of ecdysone (126). Circumstantial evidence that JH does not simply block the early ecdysone response dates at least to the early experiments of Clever showing that the polytene chromosomes of early instar *Chironomus* larvae showed the same early puffs (127). This suggested that JH, which is presumably present at that time, acts neither by antagonizing ecdysone function nor by simply altering the identity of the early responders. Fifteen years later, Richards attempted to extend the utility of the *Drosophila melanogaster* puffing experiments to an understanding of JH action (128). As might have been predicted based on Clever's observations, JH appeared not to block the early puffs directly, but did seem to block the reacquisition of competence to respond to ecdysone, an obligatory prerequisite to progression to the late prepupal puffing response. As molecular probes for the ecdysone response became available, it was possible to show more directly that JH does not block the induction of early ecdysone-inducible transcripts (27,129). That JH might directly modulate ecdysone regulation of a subset of target genes has been suggested by a report that JH blocks ecdysone induction from certain ecdysone receptor responsive element (EcRE)-containing constructs in transfected tissue culture cells (130). The possibility that JH might directly modulate the ecdysone response at certain EcREs is intriguing, particularly with regard to the idea that JH might act directly on the ecdysone through a potential ligand binding moiety of USP. However, further support for this model has not been forthcoming.

Recent results in the tobacco hornworm *Manduca sexta,* where the JH response is perhaps best understood, suggest an alternative mechanism of JH action. Like *Drosophila, Manduca* undergoes a complete metamorphosis following a larval development consisting of several instars punctuated by molts. Studies on the effects of surgical removal or transplantation of endocrine organs or exogenous hormone treatments have elegantly demonstrated that the stage specificity of the ecdysone response is determined by the presence or absence of JH (126). Cloning and characterization of ecdysone-responsive genes from *Manduca* shows that many aspects of the ecdysone-responsive hierarchies are highly conserved. EcR and USP are present at the time of the response (131,132) and have been shown in other Lepidoptera to form functional heteromeric EcREs (133). *E75 A* and *B* are expressed at the time of the initiation of metamorphosis and are shown to be primary ecdysone responsive (134). As in *Drosophila, DHR3* behaves as an early-late (135).

Analysis of JH effects on the expression of these genes suggests that JH may work not by directly inhibiting the ecdysone response but rather by stimulating the expression of *E75* (134,136). When tissues are incubated with ecdysone in the presence of JH, *E75* expression increases to higher levels and does not fully regress after attaining peak levels. *E75*, as a repressor of the early gene response, could thus act as a selective attenuator of the ecdysone response. A potential target of *E75* is the *BR-C*, which is underexpressed in JH-treated tissue (136a) and which is known to be required for the metamorphic ecdysone response program in *Drosophila* (see above). These findings suggest a potential basis for the premature pupariation of larval lethal *E75* mutants (94), which would thus be incapable of JH modulation of the ecdysone response via this mechanism.

The specific mechanisms through which JH might regulate *E75* expression remain to be determined. Much effort has gone into the identification of a JH receptor, but this problem has not yet yielded to biochemically based approaches. It was also hoped that the nuclear hormone receptor superfamily might include a JH receptor. This hypothesis seems particularly plausible in light of the structural similarities between JH and retinoic acid, the low but significant activity of some synthetic retinoids in JH bioassays (137), and the recent demonstration that JH and JH analog activate vertebrate farnesoid (138) and retinoid X (139) receptors, respectively. However, it has not yet been possible to demonstrate that any known insect orphan receptor is capable of responding to JH. It is hoped that molecular and genetic tools derived from studies on *E75* will provide the means to identify additional components of the JH response pathway including the JH receptor.

Ecdysone Receptor and Early Gene Isoforms The stage and tissue specificity of the ecdysone response could potentially be controlled at any of several different levels: at the level of quantitative alterations in expression of responsive genes, by JH, for instance, as suggested above; at the level of the receptor, with different forms of the receptor acting at different tissues or stages; or at the level of differential expression of different combinations of ecdysone-responsive genes or their isoforms. Preliminary findings suggest that each of these factors cooperates in the attainment of an appropriate developmental ecdysone response.

Consistent with the idea that different types of ecdysone receptors might carry out different types of ecdysone response, the EcR protein of *Drosophila* can be detected in three isoforms known as A, B1, and B2 (140). These differ in their amino terminal A/B domains, suggesting that they might have differential target gene regulatory capabilities. The tissue and stage specificity of the distribution of these proteins suggests that they might mediate different types of ecdysone responses. Most interesting are the differences between the patterns of EcRA and EcRB expression in various tissues of the late instar and in the metamorphosing nervous system (140,141). Different classes of responding tissue display stereotypic patterns of EcR isoform expression, with imaginal discs showing mostly EcRA, larval tissues showing predominantly EcRB1, and the histoblasts and other nests of imaginal cells showing exclusively EcRB1. Consistent with the idea that specific EcR isoforms would be required for metamorphosis of specific tissue classes, imaginal discs but not larval tissues initiate metamorphosis normally in *EcRB1* mutants (M. Bender, W. Talbot, and D. Hogness, unpublished, cited in ref. 140). Isoform-specific EcR expression is also correlated with a differential ecdysone response within the nervous system. Neurons that express high levels of EcRA display rapid ecdysone-suppressible degeneration following the decline in ecdysone titer at emergence of the adult (141). These findings strongly suggest that differential expression of EcR isoforms may play a role in specifying different types of ecdysone response.

Similarly, as described above, each of the characterized classical early genes also displays a number of different isoforms. The different E74 and E75 isoforms are expressed in distinct stage and tissue-specific patterns (63) and genetic analysis of isoform-specific mutations make it clear that there are specific roles for the E74A, E74B, E75A and E75C proteins (87,94). These findings suggest that differential expression of E74 and E75 isoforms could also be a mechanism underlying tissue and stage specificity of ecdysone response.

Differential expression of BR-C isoforms is to date the best candidate mechanism for determining the tissue specificity of the ecdysone response. *BR-C* mutations in the *br*, *rbp*, and *2Bc* classes clearly have distinct effects on specific tissues and target genes, and there is some correlation between the expression pattern of the corresponding isoforms and the affected tissues. Nonetheless, it is becoming increasingly clear that there is no one-to-one correlation between BR-C isoform expression and specific patterns of ecdysone response. This should perhaps not be surprising in light of data indicating that the expression patterns and DNA binding specificities of various isoforms show considerable overlap, and that target promoters may contain binding sites for more than one BR-C isoform (82,142). In light of the conservation of DNA binding domains and partially overlapping expression patterns displayed by EcR and early protein isoforms, it seems likely that there is some degree of functional redundancy between the different isoforms of these proteins as well.

Ftz-FI Regulation of the Late Prepupal Ecdysone Response The late larval and late prepupal ecdysone responses are substantially similar, with the exception of the early induction of the 93F puff in the late prepupal ecdysone response. The regulation of the *E93* gene located in this puff thus provides a useful model for understanding the basis of the stage-specific ecdysone response. A strong clue to the mechanism of *E93* regulation came from the demonstration that *Ftz-FI* was expressed in the midprepupal period and that the FTZ-FI protein is localized to the site of the 93F puff (123). Woodard et al. tested the hypothesis that *Ftz-FI* might be responsible for specifying the late prepupal-specific response of *E93* by looking at the effects of premature *Ftz-FI* expression under heart shock control (143). Consistent with a critical role for *Ftz-FI* in controlling the late prepupal response of *E93*, *Ftz-FI* overexpression allows induction of *E93* during the late larval period. Interestingly, *Ftz-FI* also greatly enhances the expression of *E74* and *E75*. Since *Ftz-FI* expression occurs in response to a fall in the ecdysone titer during the midprepupal period of reacquisition of competence to respond to ecdysone, FTZ-FI thus becomes a strong candidate for the factor responsible for setting the stage for the late prepupal ecdysone response. Given the proposed role of *Ftz-FI* and its target *E93* in determining the stage specificity of the late prepupal ecdysone response, it should be particularly interesting to see whether *Ftz-FI* overexpression is sufficient to alter the timing of phenotypically observable developmental processes such as salivary gland histolysis.

3.3.6. Interactions Between the Ecdysone Receptor and Other Nuclear Hormone Receptors

By now the reader will have noticed the preponderance of steroid receptor-like molecules now implicated in ecdysone-responsive regulatory hierarchies. This finding has allowed for and likely fueled the proliferation of models incorporating possible interactions between EcR, USP, and other ecdysone-regulated nuclear hormone receptors. These models include those in which proteins with similar DNA binding properties might compete with the ecdysone receptor or with one another, models in which various combinations of receptors form different sets of heterodimers, and a third model highlighting the role of ecdysone in coordinating putative evolutionarily ancient developmental pathways controlled by these other receptors.

The DNA-Binding Competition Model In the Ashburner model (15), there is formal competition between the early proteins and the ecdysone-receptor complex for the regulation of both early and late gene expression. The ecdysone-receptor complex induces the earlies and represses the lates, while the early proteins repress the earlies and induce the lates. The finding of significant similarity between the DNA binding domains of E75 and EcR, particularly within the DNA binding domain P region known to be involved in specifying response element sequence recognition (144–146), raised the possibility that this formal competition might be a reflection of direct competition at the level of DNA binding. The finding of other ecdysone-regulated nuclear hormone receptors has intensified interest in this possibility.

We have tested this hypothesis by looking at the regulation of the *E75A* promoter presumed to be a target for ecdysone induction and autoregulatory repression (W. Feng, W. Pierceall, E. Wiellette, J-Y. Leu, Z. Zhang, W. Segraves, in preparation). Mapping of response elements for the ecdysone receptor and the E75A protein demonstrates that these factors regulate *E75A* through response elements that are separated from one another by kilobases and that are functionally independent within the context of the *E75A* promoter or heterologous promoters. Fine mapping and characterization of the response elements shows that this distinction is made at the level of DNA binding. Whereas the ecdysone receptor binds to a palindromic or direct repeat response element (147–150), the E75A protein appears to bind as a monomer to a sequence consisting of a single core half-site, with adjacent sequences conferring additional specificity. This sequence is nearly identical to the DNA binding site of its mammalian homolog rev-ErbA (151). Characterization of the DNA binding domains of other receptors shows that relatively few of these receptors have specificities closely resembling that of the ecdysone receptor and that most of them have distinct specificities. Thus, it appears that DNA binding competition is not the only mechanism and may not even be the major mechanism through which these proteins might coregulate common sets of targets.

The sequence similarities between elements recognized by various receptors do, however, allow for the possibility of a variety of promiscuously recognized or overlapping compound response elements that could be recognized by more than one receptor. Preliminary evidence suggests the existence of overlapping DNA binding sites for E75 and the ecdysone receptor in a putative target gene (M. Buszczak, W. Pierceall, and W.A. Segraves, unpublished), and DHR78, DHR96, and SVP (seven-up), the *Drosophila* homolog of COUP-TF (152), appear to compete with EcR-USP for binding to certain ecdysone response elements (153–155). These findings suggest the possibility that competitive interactions between these factors and the ecdysone receptor could play a role in modulating the ecdysone response. Further potential for competition at the level of target sites exists within each of the different classes of monomer binding receptors, which likely have overlapping specificities and the potential to compete with one another as well (156).

The Heterodimer Formation Model The demonstration that functional ecdysone receptors are EcR-USP heterodimers, when considered in the context of the promiscuity of interactions displayed by the vertebrate RXR, led to speculation that EcR and USP might have other partners in *Drosophila*. Most intriguing in the context of ecdysone-responsive hierarchies was the possibility that ecdysone-induced factors might combine with EcR to form different types of receptor heterodimers that might mediate different tissue- or stage-specific ecdysone responses (157). There has been little direct evidence to suggest that EcR or USP heterodimerize with other proteins to form functional ecdysone receptors, but the search for other partners has identified other potentially significant protein–protein interactions that may result in competition for heterodimer partners. EcR interaction with a protein other than USP is seen most clearly in yeast two-hybrid tests showing interaction between EcR and SVP (154). Consistent with the idea that SVP can inhibit the ecdysone response by interacting with EcR is the fact that SVP overexpression can inhibit the activity of ecdysone response elements, including those that are not direct targets for SVP DNA binding. SVP overexpression also appears to inhibit the ecdysone response in third instar larvae, where overexpression leads to lethality characteristic of a deficient ecdysone response. Recent findings suggest that DHR3 may also be capable of inhibiting the ecdysone response through direct interaction with EcR (112).

USP can also be shown to interact with at least one other factor, the DHR38 nuclear hormone receptor of the NGFI-B class. DHR38 interacts strongly with USP in yeast two-hybrid assays, as well as in solution, and inhibits the ecdysone response in transfected tissue culture cells (158). While the biologic significance of the EcR–SVP and USP–DHR38 interactions is not yet known, these studies show that there is significant potential for modulation of receptor function at the level of protein–protein interactions. Such interactions also provide an intriguing alternative to the interpretation of USP phenotypes not obviously related to the ecdysone response; perhaps one role of USP is to interfere with the functions of other factors such as DHR38.

The limited number of heterodimeric interactions discovered thus far is likely related to the fact that most nuclear hormone receptors in *Drosophila* appear to bind to DNA primarily as monomers. It remains to be seen how many of the other orphan receptors that appear to bind to DNA as monomers might also interact with one another or with ecdysone receptor subunits on as yet untested response elements or under other appropriate conditions. While E75 appears to recognize DNA as a monomer and has not been shown to form heterodimers, the evolutionary conservation of the E75B protein, which lacks much of the DNA binding domain, suggests that it is likely capable of some type of protein–protein interaction. Recent findings suggest that the DHR3 protein is a potential target of E75B (112).

The Endocrine Integration Model Evidence suggesting that many of the nuclear hormone receptors carry out distinct functions through independent regulatory sequences has led to a reexamination of the assumptions underlying the proposal that nuclear hormone receptors in ecdysone-responsive hierarchies function primarily to modulate ecdysone receptor function. This hypothesis derives at least in part from the idea that the similarity between these molecules reflects evolutionary duplication and divergence of primordial ecdysone receptor genes to allow further refinement and diversification of this response. If this were the case, one would expect that the *Drosophila* receptors would show evidence of relatively recent common ancestry. The evidence, on the contrary, suggests the opposite conclusion. Comparison of *Drosophila* and vertebrate steroid receptors shows that a substantial majority of *Drosophila* receptors are far more closely related to specific vertebrate receptors than to other *Drosophila* receptors (Fig. 3.2), suggesting that the common ancestors of vertebrates and insects already had a well-differentiated set of nuclear hormone receptors. The evolutionary history that follows is therefore most likely not one of duplication and divergence, but rather of the coopting of developmentally important receptor genes into the ecdysone-dependent regulatory hierarchies that serve to coordinate arthropod development. If some of these proteins are receptors, as is suggested by their structure, then the actions of their ligands would be dependent on and coordinated by ecdysone. A precedent for such an interaction is readily seen in the case of the ecdysone–JH interaction. It is common to regard JH as a modulator of the ecdysone response, as the above discussion of JH action reflects. Nonetheless, the majority of data on the JH response allows for an alternative perspective that becomes clear if one proposes that the JH receptor might be found among the ecdysone-responsive nuclear hormone receptors described above; perhaps ecdysone is needed to induce the JH receptor and consequently the competence to respond to JH. Whether this is true for JH remains to be seen, but it seems likely that one or more of the ecdysone-regulated nuclear hormones receptors might modulate the response to another, as yet unknown, ligand.

3.3.7. Perspectives on Ecdysone Research

Steroid-Regulated Hierarchies The identification of additional genes in ecdysone-triggered regulatory hierarchies is allowing the continual refinement and expansion of models for the regulation of the ecdysone response. Of particular importance to understanding the developmental roles of these hierarchies will be the definition of late ecdysone-responsive genes involved in specific regulatory hierarchies and developmental processes. With the addition of these genes to the story, it should be possible to understand the entire process of selected ecdysone responses from the initial hormonal signal to its developmental consequences.

In understanding the functional interrelationships between factors in these hierarchies, it will also be important to begin to define the significance of temporal and quantitative variations in the expression of specific regulatory factors. Studies on the expression of early responsive genes, in particular, have demonstrated remarkably diverse and precise patterns of expression that are altered by mutations in regulatory genes. It remains to be seen whether relatively subtle variations in the timing of expression are responsible for the mutant phenotypes or whether more significant undescribed qualitative variations are of tantamount importance. Of particular interest to an understanding of the significance of specific temporal patterns will be experimental testing of the effects of the misexpression of genes within these regulatory hierarchies. If the precise timing of specific isoform expression is indeed central to the proper elaboration of these regulatory hierarchies, it will be interesting to see how flexibility is achieved in the face of genetic or epigenetic disruption of the normal temporal pattern of events, and how these patterns are altered in insects with temporally or more fundamentally distinct developmental programs of ecdysone response.

Questions relating to tissue and stage specificity of the ecdysone response will remain at the forefront as models describing the molecular mechanisms are tested and refined. Of particular importance will be the elucidation of the JH response pathway. While the direct biochemical approach based on JH binding has not been fruitful in the past, the development of molecular tools including JH-responsive genes (134,159,160) should provide the necessary tools for working backward through the JH response pathway. Mutations altering the integration of JH and the ecdysone response (94) offer opportunities for a genetic approach to identifying components of the JH response.

At the level of transcription factors controlling differential response, Ftz-FI and various receptor isoforms appear to correlate with and potentially control specific responses. In understanding the role of different isoforms in specifying distinct responses, it will then be of primary importance to determine which target genes are selectively regulated by the different isoforms. A distinctive strength of the *Drosophila* system is its ability to identify genetic interactions through suppressor/enhancer screens. Such approaches should make a detailed understanding of the genetic mechanisms underlying the differential hormone response an attainable goal.

Neuroendocrine Control of Ecdysone Production A relatively unexplored area in the field of ecdysone response concerns a set of fascinating questions underlying activation of this system at the neuroendocrine level. As more of the developmental and genetic tools of *Drosophila* are directed to this system, and to the integration of physiologic studies from other insects, it should be possible to gain a detailed understanding of the mechanisms through which ecdysone release is initiated in response to specific developmental and environmental cues. At present, however, little is known of the mechanisms that stimulate the production of the ecdysiotropin prothoracicotropic hormone (PTTH) and the mechanisms through which PTTH stimulates ecdysone production. The recent identification of a *Drosophila* adrenodoxin reductase gene (M. Freeman and J. Carlson, personal communication), which may be regulated in response to the PTTH regulatory network, provides new genetic and molecular tools that may aid in the elucidation of PTTH production and response.

Ecdysone in Reproduction and Early Development Finally, it is somewhat striking that more is not known about the reproductive ecdysone response. Despite longstanding evidence that ecdysone plays a role in *Drosophila* reproduction, little is known about the specific roles of ecdysone or the ways in which it might control the progression of oogenesis and vitellogenesis. Evidence of a regulatory role of ecdysone in the reproduction of other insects suggests that this may be an important control point in insect reproduction. While the specific mechanisms through which nutritive and other environmental factors influence progression of *Drosophila* oogenesis remain to be elucidated, it is intriguing to speculate that, as in the mosquito, progression may be regulated via neuroendocrine stimulation of ovarian ecdysone production. Similarly, greater understanding is needed concerning the possible roles of ecdysone and other hormones in the regulation of other stages during development, including embryogenesis. These responses will serve as excellent models for reproductive and early developmental roles of steroid hormones. Furthermore, a detailed understanding of reproductive and developmental ecdysone responses should lead to the development of new bioactive agents, including new ecdysone analogs or ligands for ecdysone-responsive orphan receptors, as well as to a better understanding of the vulnerability of desirable insects and other invertebrates to environmental agents that might interfere with their reproduction and development.

ACKNOWLEDGMENTS

I thank Peter Cherbas, Carl Thummel, David Hogness, Michael Bender, Ann Sluder, Tim Sliter, Marc Freeman, and John Carlson for communication of results prior to publication; Alex Raikhel, Peter Cherbas, Ken Burtis, Marc Freeman, and members of my laboratory for stimulating discussions; and Bill Pierceall and Mike Buszczak for critical comments on the manuscript.

REFERENCES

1. Kopeć S: Experiments on metamorphosis of insects. Bull Int Acad Sci Cracovie (B) 57–60, 1917.
2. Wigglesworth V: Insect Hormones. Edinburgh: Oliver and Boyd, 1970.
3. Berthold AA: Transplantation der Hoden. Arch Anat Physiol Wiss Med 16:42–46, 1849.
4. Cherbas L, Fristrom JW, O'Connor JD. *Biosynthesis, Metabolism and Mode of Action of Invertebrate Hormones*. Berlin: Springer-Verlag, 1984, pp 305–322.
5. Butenandt A, Karlson P: Isolation of moulting hormone, ecdysone. Z Naturf 9b:389–391, 1954.
6. Huber F, Hoppe W: Die kristall-und molekulstrukturanalyse des insecten verpuppungshormons ecdyson mit der automatisierten faltmolkulmethode. Chem Ber 98:2403–2424, 1965.
7. Karlson P: The chemistry of insect hormones and insect pheromones. Pure Appl Chem 14:75–87, 1967.
8. Beermann W: Nuclear differentiation and functional morphology of chromosomes. Cold Spring Harbor Symp Quant Biol 21:217–232, 1956.
9. Pelling G: Chromosomal synthesis of ribonucleic acid as shown by incorporation of uridine labelled with tritium. Nature 184:655–656, 1959.
10. Clever U, Karlson P: Induktion von puff-veranderungen in den speicheldrusenchromomen von *Chironomus tentans* durch ecdyson. Exp Cell Res 20:623–626, 1960.
11. Clever U, Romball CG: RNA and protein synthesis in the cellular response to a hormone, ecdysone. Proc Natl Acad Sci USA 56:1470–1476, 1966.
12. Jacob F, Monod J: On the regulation of gene activity. Cold Spring Harbor Symp Quant Biol 26:193–211, 1961.
13. Becker HJ: Die Puffs der Speicheldrusenchromosomen von *Drosophila melanogaster*. II. Die Auslosung der Puffbildung, ihre Spezifitat und ihr Beziehung zur Funktion der Ringdruse. Chromosoma 13:341, 1962.
14. Berendes HD: The hormone ecdysone as effector of specific changes in the pattern of gene activities of *Drosophila hydei*. Chromosoma 22:274–293, 1967.
15. Ashburner M, Chihara C, Meltzer P, Richards G: Temporal control of puffing activity in polytene chromosomes. Cold Spring Harbor Symp Quant Biol 38:655–662, 1974.
16. Emmerich H: Ecdysone binding proteins in nuclei and chromatin of *Drosophila* salivary glands. Gen Comp Endocrinol 19:543, 1972.
17. Andres AJ, Fletcher JC, Karim FD, Thummel CS: Molecular analysis of the initiation of insect metamorphosis: A comparative study of *Drosophila* ecdysteroid-regulated transcription. Dev Biol 160:388–404, 1993.
18. Richards G: The ecdysone regulatory cascades in *Drosophila*. Adv Dev Biochem, in press.
19. Kafka F: Metamorphosis. New York: Vanguard Press, 1946.

20. Clements AN: The Biology of Mosquitoes, Vol 1. London: Chapman and Hall, 1992, chap 21.
21. Garen AL, Kauvar L. Lepesant J-A: Roles of ecdysone in *Drosophila* development. Proc Natl Acad Sci USA 74:5099–5103, 1977.
22. Oro AE, McKeown M, Evans RM: The *Drosophila* retinoid X receptor homolog *ultraspiracle* functions in both female reproduction and eye morphogenesis. Development 115:449–462, 1992.
23. Orr WC, Galanopoulos VK, Romano CP, Kafatos FC: A female sterile screen of the *Drosophila melanogaster* X chromosome using hybrid dysgenesis: Identification and characterization of egg morphology mutants. Genetics 122:847–858, 1989.
24. Bownes M, Blair M, Kozma R, Dempster M: 20-Hydroxyecdysone stimulates tissue-specific yolk protein gene expression in both male and female *Drosophila*. Embyrol Exp Morphol 78:249–268, 1983.
25. Kraminsky GP, Clark WC, Estelle MA, Gietz RD, Sage BA, O'Connor JD: Induction of translatable mRNA for dopa decarboxylase in *Drosophila*: An early response to ecdysone. Proc Natl Acad Sci USA 77:4175–4179, 1980.
26. Bownes M, Shirras A, Blair M, Collins J, Coulson A: Evidence that insect embryogenesis is regulated by ecdysteroids released from yolk proteins. Proc Natl Acad Sci USA 85:1554–1557, 1988.
27. Segraves WA: Molecular and genetic analysis of the E75 ecdysone-responsive gene of *Drosophila melanogaster*. PhD thesis, Stanford University, 1988.
28. Burtis KC, Thummel CS, Jones CW, Karim FD, Hogness DS: The *Drosophila* 74EF puff contains *E74*, a complex ecdysone-inducible gene that encodes two *ets*-related proteins. Cell 61:85–99, 1990.
29. Natzle J: Temporal regulation of *Drosophila* imaginal disc morphogenesis: A hierarchy of primary and secondary 20-hydroxyecdysone-responsive loci. Dev Biol 155:516–532, 1993.
30. Gronemeyer H, Pongs O: Localization of ecdysterone on polytene chromosomes of *Drosophila melanogaster*. Proc Natl Acad Sci USA 77:2108–2112, 1980.
31. Maroy P, Dennis R, Beckers C, Sage B, O'Connor JD: Demonstration of an ecdysteroid receptor in a cultured cell line of *Drosophila melanogaster*. Proc Natl Acad Sci USA 75:6035–6038, 1978.
32. Yund MA, King DS, Fristrom JW: Ecdysteroid receptors in imaginal discs of *Drosophila melanogaster*. Proc Natl Acad Sci USA 75:6039–6043, 1978.
33. Ireland RC, Berger E, Sirotkin K, Yund MA, Osterbur D, Fristrom J: Ecdysterone induces the transcription of four heat-shock genes in *Drosophila* S3 cells and imaginal discs. Dev Biol 93:498–507, 1982.
34. Stevens B, O'Connor JD: The acquisition of resistance to ecdysteroids in cultured *Drosophila* cells. Dev Biol 94:176–182, 1982.
35. Luo Y, Amin J, Voellmy R: Ecdysterone receptor is a sequence-specific transcription factor involved in the developmental regulation of heat shock genes. Mol Cell Biol 11:3660–3675, 1991.
36. Mangelsdorf DJ, Thummel C, Beato M, Herrlich P, Schutz G, Umesono K, Blumberg B, Kastner P, Mark M, Chambon P, Evans RM: The nuclear receptor superfamily: The second decade. Cell 83:835–839, 1995.

37. Koelle MR, Segraves WA, Hogness DS: *DHR3*: A *Drosophila* steroid receptor homolog. Proc Natl Acad Sci USA 89:6167–6171, 1992.

38. Segraves WA: Steroid receptors and orphan receptors in *Drosophila* development. Semin Cell Biol 5:105–113, 1994.

39. Henrich VC, Brown NE: Insect nuclear receptors: A developmental and comparative perspective. Insect Biochem Mol Biol 25:881–897, 1995.

40. Thummel CS: From embryogenesis to metamorphosis: The regulation and function of *Drosophila* nuclear receptor superfamily members. Cell 83:871–877, 1995.

41. Koelle MR, Talbot WS, Segraves WA, Bender MT, Cherbas P, Hogness DS: The *Drosophila* EcR gene encodes an ecdysone receptor, a new member of the steroid receptor superfamily. Cell 67:59–77, 1991.

42. Riddihough G, Pelham HRB: An ecdysone response element in the *Drosophila* hsp27 promoter. EMBO J 6:3729–3734, 1987.

43. Koelle MR: Molecular analysis of the *Drosophila* ecdysone receptor complex. PhD thesis, Stanford University, 1992.

44. Yao T-P, Segraves WA, Oro AE, McKeown M, Evans RM: *Drosophila* ultraspiracle modulates ecdysone receptor function via heterodimer formation. Cell 71:63–72, 1992.

45. Mangelsdorf DJ, Ong ES, Dyck JA, Evans RM: Nuclear receptor that identifies a novel retinoic acid response pathway. Nature 345:224–229, 1990.

46. Yu VC, Delsert C, Anderen B, Holloway JM, Devary O, Naar AM, Kim SY, Boutin J-M, Glass CK, Rosenfeld MG: RXRβ: a coregulator that enhances binding of retinoic acid, thyroid hormone and vitamin D receptors to their cognate response elements. Cell 67:1251–1266, 1991.

47. Bugge TH, Phol J, Lonnoy O, Stunnenberg HG: RXRα, a promiscuous partner of retinoic acid and thyroid hormone receptors. EMBO J 11:1409–1418, 1992.

48. Kliewer SA, Umesono K, Mangelsdorf DJ, Evans RM: The retinoid X receptor interacts directly with nuclear receptors involved in retinoic acid, thyroid hormone, and vitamin D_3 signalling. Nature 355:446–449, 1992.

49. Leid M, Kastner P, Lyons R, Nakshatri H, Saunders M, Zacharewski T, Chen J-Y, Staub A, Garnier J-M, Mader S, Chambon P: Purification, cloning, and RXR identity of HeLa cell factor with which RAR or TR heterodimerizes to bind target sequences efficiently. Cell 68:377–395, 1992.

50. Marks MS, Hallenbeck PL, Nagata T, Segars JH, Appella E, Nikodem VM, Ozato K: H-2RIIBP (RXRβ) heterodimerization provides a mechanism for combinatorial diversity in the regulation of retinoic acid and thyroid hormone responsive genes. EMBO J 4:1419–1435, 1992.

51. Zhang X-K, Hoffmann B, Tran PB-V, Graupner G, Pfahl M: Retinoid X receptor is an auxiliary protein for thyroid hormone and retinoic acid receptors. Nature 355:441–446, 1992.

52. Henrich VC, Sliter TJ, Lubahn DB, MacIntyre A, Gilbert LI: A steroid/thyroid hormone receptor superfamily member in *Drosophila melanogaster* that shares extensive sequence homology with a mammalian homologue. Nucl Acids Res 18:4143–4148, 1990.

53. Oro AE, McKeown M, Evans RM: Relationship between the product of the *Drosophila ultraspiracle* locus and vertebrate retinoid X receptor. Nature 347:298–301, 1990.

54. Shea MJ, King DL, Conboy MJ, Mariani BD, Kafatos FC: Proteins that bind to *Drosophila* chorion cis-regulatory elements: A new C2H2 zinc finger protein and a C2C2 steroid receptor-like component. Genes Dev 4:1128–1140, 1990.
55. Perrimon N, Engstrom L, Mahowald AP: Developmental genetics of the 2C-D region of the *Drosophila* X chromosome. Genetics 111:23–41, 1985.
56. Yang N, Schule R, Mangelsdorf DJ, Evans RM: Characterization of DNA binding and retinoic acid binding properties of the retinoic acid receptor. Proc Natl Acad Sci USA 88:3559–3563, 1991.
57. Thomas HE, Stunnenberg HG, Steward AF: Heterodimerization of the *Drosophila* ecdysone receptor with retinoid X receptor and ultraspiracle. Nature 362:471–475, 1993.
58. Yao T-P, Forman BM, Jiang Z, Cherbas L, Chen J-D, McKeown M, Cherbas P, Evans RM: Functional ecdysone receptor is the product of *EcR* and *ultraspiracle* genes. Nature 366:476–479, 1993.
59. Henrich VC, Szekely AA, Kim SJ, Brown NE, Antoiniewski C, Hayden MA, Lepesant J-A, Gilbert LI: Expression and function of the *ultraspiracle* (*usp*) gene during development of *Drosophila melanogaster*. Dev Biol 165:38–52, 1994.
60. Elke C, Vogtli M, Rauch P, Spindler-Barth M, Lezzi M: Expression of EcR and USP in *E. coli*: purification and functional studies. Arch Insect Biochem Physiol 35:59–69, 1997.
61. No D, Yao T-P, Evans RM: Ecdysone-inducible gene expression in mammalian cells and transgenic mice. Proc Natl Acad Sci USA 93:3346–3351, 1996.
62. Karim F, Thummel CS: Temporal coordination of regulatory gene expression by the steroid hormone ecdysone. EMBO J 11:4083–4093, 1992.
63. Huet F, Ruiz C, Richards G: Puffs and PCR: The in vivo dynamics of early gene expression during ecdysone responses in *Drosophila*. Development 118:613–627, 1993.
64. Thummel CS: Flies on steroids—*Drosophila* metamorphosis and the mechanisms of steroid hormone action. Trends in Genetics 12:306–310, 1996.
65. Ashburner M: Patterns of puffing activity in the salivary gland chromosomes of *Drosophila*. VI. Induction of ecdysone in salivary glands of *D. melanogaster* cultured *in vitro*. Chromosoma 38:255–281, 1972.
66. Ashburner M: Sequential gene activation by ecdysone in polytene chromosomes of *Drosophila* melanogaster. II. The effects of inhibitors of protein synthesis. Dev Biol 39:141–157, 1974.
67. Richards G: Sequential gene activation by ecdysone in polytene chromosomes of *Drosophila melanogaster*. V. The late prepupal puffs. Dev Biol 54:264–275, 1976.
68. Stewart MC, Murphy C, Fristrom JW: The recovery and preliminary characterization of X chromosome mutants affecting imaginal discs of *Drosophila melanogaster*. Dev Biol 27:71–83, 1972.
69. Kiss I, Bencze G, Fodor A, Szabad J, Fristrom J: Prepupal larval mosaics in *Drosophila melanogaster*. Nature 262:136–138, 1976.
70. Belyaeva ES, Aizenzon MG, Semeshin VF, Kiss I, Koezka K, Baritcheva EM, Gorelova TD, Zhimulev IF: Cytogenetic analysis of the 2B3-4 to 2B11 region of the X chromosome of *Drosophila melanogaster*. I. Cytology of the region and mutant complementation groups. Chromosoma 81:281–306, 1980.
71. Bridges C: Non-disjunction as proof of the chromosome theory of heredity. Genetics 1:107–163, 1916.

72. Kiss I, Beaton AH, Tardiff J, Fristrom D, Fristrom JW: Interaction and developmental effects of mutations in the *Broad-Complex* of *Drosophila melanogaster.* Genetics 118:247–259, 1988.
73. Belyaeva ES, Vlassova IE, Biyasheva ZM, Kakpakov VT, Richards G, Zhimulev IF: Cytogenetic analysis of the 2B3-4 to 2B11 region of the X-chromosome of *Drosophila melanogaster*. II. Changes in 20-OH ecdysone puffing caused by genetic defects of puff 2B5. Chromosoma 84:207–219, 1982.
74. Guay P, Guild GM: The ecdysone-induced puffing cascade in *Drosophila* salivary glands: A *Broad-Complex* early gene regulates intermolt and late gene transcription. Genetics 129:169–175, 1991.
75. Karim FD, Guild GM, Thummel CS: The *Drosophila Broad-Complex* plays a key role in controlling ecdysone-regulated gene expression at the onset of metamorphosis. Development 118:977–988, 1993.
76. Restifo L, White K: Mutations in a steroid hormone-regulated gene disrupt the metamorphosis of internal tissues in *Drosophila*; salivary glands, muscle and gut. Roux's Arch Dev Biol 201:221–234, 1992.
77. Savakis C, Demetri G, Cherbas P: Ecdysteroid-inducible polypeptides in a *Drosophila* cell line. Cell 22:665–674, 1980.
78. Wolfner M: Ecdysone-responsive genes of the salivary gland of *Drosophila melanogaster*. PhD thesis, Stanford University, 1980.
79. Chao AT, Guild GM: Molecular analysis of the ecdysterone-inducible 2B5 "early" puff in *Drosophila melanogaster*. EMBO J 5:143–150, 1986.
80. DiBello PR, Withers DA, Bayer CA, Fristrom JW, Guild GM: The *Drosophila Broad-Complex* encodes a family of related, zinc finger-containing proteins. Genetics 129:385–397, 1992.
81. Bayer CA, Holley B, Fristrom JW: A switch in *Broad-Complex* zinc-finger isoform expression is regulated posttranscriptionally during the metamorphosis of *Drosophila* imaginal discs. Dev Biol 177:1–14, 1996.
82. Emery IF, Bedian V, Guild GM: Differential expression of *Broad-Complex* transcription factors may forecast tissue-specific developmental fates during *Drosophila* metamorphosis. Development 120:3275–3287, 1994.
83. Walker V, Ashburner M: The control of ecdysterone-regulated puffs in *Drosophila* salivary glands. Cell 26:269–277, 1981.
84. Karim FD, Urness LD, Thummel CS, Klemsz MJ, McKercher SR, Celada A, Van BC, Maki RA, Gunther CV, Nye JA, Graves BJ: The ETS-domain: A new DNA-binding motif that recognizes a purine-rich core DNA sequence. Genes Dev 4:1451–1453, 1990.
85. Urness LD, Thummel CS: Molecular interactions within the ecdysone regulatory hierarchy: DNA binding properties of the *Drosophila* ecdysone-inducible E74A protein. Cell 63:47–61, 1990.
86. Urness LD, Thummel CT: Molecular analysis of a steroid-induced regulatory hierarchy: The *Drosophila* E74A protein directly regulates L71-6 transcription. EMBO J 14:6239–6246, 1995.
87. Fletcher JC, Burtis KC, Hogness DS, Thummel CS: The *Drosophila* E74 gene is required for metamorphosis and plays a role in the polytene chromosome puffing response to ecdysone. Development 121:1455–1465, 1995.

88. Fletcher JC, Thummel CS: The ecdysone-inducible *Broad-Complex* and E74 genes interact to regulate target gene transcription and *Drosophila* metamorphosis. Genetics. 141:1025–1035, 1995.
89. Segraves WA, Hogness DS: The *E75* ecdysone-inducible gene responsible for the 75B early puff in *Drosophila* encodes two new members of the steroid receptor superfamily. Genes Dev 4:204–219, 1990.
90. Hill RJ, Segraves WA, Choi D, Underwood PA, Macavoy E: The reaction with polytene chromosomes of antibodies raised against *Drosophila* E75A protein. Insect Biochem Mol Biol 23:99–104, 1992.
91. Tora L, Gronemeyer H, Turcotte B, Gaub M-P, Chambon P: The N-terminal region of the chicken progesterone receptor specifies target gene activation. Nature 333:185–188, 1988.
92. Hard T, Kellenbach E, Boelens R, Maler BA, Dahlman K, Freedman LP, Carlstedt-Duke J, Yamamoto KR, Gustafsson J, Kaptein R: Solution structure of the glucocorticoid receptor DNA-binding domain. Science 249:157–160, 1990.
93. Schwabe JWR, Neuhaus D, Rhodes D: Solution structure of the DNA-binding domain of the oestrogen receptor. Nature 348:458–461, 1990.
94. Segraves W, Jenik P, Fichtenberg C, Hughes C: Genetic analysis of the 75B puff of *Drosophila melanogaster*. Submitted.
95. Segraves WA, Woldin CW: The *E75* gene of *Manduca sexta* and comparison with its *Drosophila* homolog. Insect Biochem Mol Biol 23:91–97, 1993.
96. Finkelstein R, Perrimon N: The molecular genetics of head development in *Drosophila melanogaster*. Development 112:899–912, 1991.
97. White K, Grether ME, Abrams JM, Young L, Farrell K, Steller H: Genetic control of programmed cell death in *Drosophila*. Science 264:677–683, 1994.
98. Bilder D, Scott MP: Genomic regions required for morphogenesis of the *Drosophila* embryonic midgut. Genetics 141:1087–1100, 1995.
99. Cho W-L, Kapitskaya MZ, Raikhel AS: Mosquito ecdysteroid receptor: Analysis of the cDNA and expression during vitellogenesis. Insect Biochem Mol Biol 25:19–27, 1995.
100. Kapitskaya M, Wang S, Cress DE, Dhadhialla TS, Raikhel AS: The mosquito ultraspiracle homolog, a partner of ecdysteroid receptor heterodimer: Cloning and characterization of isoforms expressed during vitellogenesis. Mol Cell Endocrinol 121:119–132, 1996.
101. Andres AJ, Thummel CS: The *Drosophila* 63F early puff contains E63-1, an ecdysone-inducible gene that encodes a novel Ca2+-binding protein. Development 121:2667–2769, 1995.
102. Baehrecke EH, Thummel CS: The *Drosophila* E93 gene from the 93F early puff displays stage- and tissue-specific regulation by 20-hydroxyecdysone. Dev Biol 171:85–97, 1995.
103. Lepesant JA, Levine M, Garen A, Lepesant-Kejzlarova J, Rat L, Somme-Martin G: Developmentally regulated gene expression in *Drosophila* larval fat bodies. J Mol Appl Genet 1:371–383, 1982.
104. Andres AJ, Cherbas P: Tissue-specific ecdysone responses: Regulation of the *Drosophila* genes Eip28/29 and Eip40 during larval development. Development 116:865–876, 1992.
105. Richards G: Sequential gene activation by ecdysone in polytene chromosomes of

Drosophila melanogaster. VII. Tissue specific puffing. Wilhelm Roux's Arch Dev Biol 191:103–111, 1982.

106. Ayer S, Benyajati CS: The binding site of a steroid hormone receptor-like protein within the *Drosophila Adh* adult enhancer is required for high levels of tissue-specific alcohol dehydrogenase expression. Mol Cell Biol 12:661–673, 1992.
107. Ohno CK, Petkovich M: FTZ-FIB, a novel member of the *Drosophila* nuclear receptor family. Mech Dev 40:13–24, 1992.
108. Stone B, Thummel C: The *Drosophila* 78C early late puff contains E78, an ecdysone-inducible gene that encodes a novel member of the steroid receptor superfamily. Cell 75: 307–320, 1993.
109. Huet F, Ruiz C, Richards G: Sequential gene activation by ecdysone in *Drosophila melanogaster:* The hierarchical equivalence of early and early late genes. Development 121:1195–1204, 1995.
110. Russell SRH, Heimbeck G, Goddard CM, Carpenter ATC, Ashburner M: The *Drosophila* Eip78C gene is not vital but has a role in regulating chromosome puffs. Genetics 144:159–170, 1996.
111. Lam GT, Jiang C, Thummel CS: Coordination of larval and prepupal gene expression by the DHR3 orphan receptor during *Drosophila* metamorphosis. Development 124:1757–1769, 1997.
112. White KP, Hurban P, Watanabe T, Hogness DS: Coordination of *Drosophila* metamorphosis by two ecdysone-induced nuclear receptors. Science 276:114–117, 1997.
113. Restifo LL, Guild GM: An ecdysterone-responsive puff site in *Drosophila* contains a cluster of seven differentially regulate genes. J Mol Biol 188:517–528, 1986.
114. Wright LG, Chen T, Thummel CS, Guild GM; Molecular characterization of the 71E late puff in *Drosophila melanogaster* reveals a family of novel genes. J Mol Biol 255:387–400, 1996.
115. Fletcher JC, Thummel CS: The *Drosophila* E74 gene is required for proper stage- and tissue-specific transcription of ecdysone-regulated genes at the onset of metamorphosis. Development 121:1411–1421, 1995.
116. Osterbur DL, Fristrom DK, Natzle JE, Tojo SJ, Fristrom JW: Genes expressed during imaginal disc morphogenesis Imp-L2 a gene expressed during imaginal disc and imaginal histoblast morphogenesis. Dev Biol 129:439–448, 1988.
117. Natzle JE, Robertson JP, Majumdar A, Vesenka GD, Enlow B, Clark KE: Sequence and expression of Imp-L1 an ecdysone-inducible gene expressed during *Drosophila* imaginal disc morphogenesis. Dev Genet 13:331–344, 1992.
118. Natzle JE, Fristrom DK, Fristrom JW: Genes expressed during imaginal disc morphogenesis: IMP-E1, a gene associated with epithelial cell rearrangement. Dev Biol 129:428–438, 1988.
119. Richards G: Sequential gene activation by ecdysone in polytene chromosomes of *Drosophila melanogaster*. IV. The mid-prepupal period. Dev Biol 54:256–263, 1976.
120. Fechtel K, Natzle JE, Brown EE, Fristrom JW: Prepupal differentiation of *Drosophila* imaginal discs: Identification of four genes whose transcripts accumulate in response to a pulse of 20 hydroxyecdysone. Genetics 120:465–474, 1988.
121. Ueda H, Sondoda S, Brown JL, Scott MP, WuC: A sequence-specific DNA-binding protein that activates *fushi tarazu* segmentation gene expression. Genes Dev 4: 624–635, 1990.

122. Lavorgna G, Ueda H, Clos J, Wu C: FTZ-FI, a steroid hormone receptor-like protein implicated in the activation of *fushi tarazu*. Science 252:848–851, 1991.

123. Lavorgna G, Karim FD, Thummel CS, Wu C: A potential role for the embryonic steroid receptor superfamily member in the control of *Drosophila* metamorphosis. Proc Natl Acad Sci USA. 90: 3004–3008, 1993.

124. Wigglesworth VB: The physiology of ecdysis in *Rhodnius prolixus* (Hemiptera) II. Factors controlling moulting and "metamorphosis." Q J Microsc Sci 77:191–222, 1934.

125. Roller H, Dahm KH, Sweeley CC, Trost BM: The structure of the juvenile hormone. Angercandte Chem 6:179–180, 1967.

126. Riddiford LM: In Kercut GA, Gilbert LI (eds): Hormone action at the cellular level. Comprehensive Insect Physiology, Biochemistry and Pharmacology, Vol 8. Oxford: Pergamon, 1985, pp 37–84.

127. Clever U: Genaktivitaten in den Riesenchromosomen von *Chironomus tentans* und ihre Beziehungen zur Entwicklung. IV. Das verhalten der puffs in der larvenhautung. Chromosoma 14:651–675, 1963.

128. Richards G: Sequential gene activation by ecdysone in polytene chromosomes of *Drosophila melanogaster*. VI. Inhibition by juvenile hormones. Dev Biol 66:32–42, 1978.

129. Cherbas L, Koehler MM, Cherbas P: Effects of juvenile hormone on the ecdysone response of *Drosophila* Kc-cells. Dev Genet 10:177–188, 1989.

130. Berger E: The juvenile hormone analogue, methoprene, inhibits ecdysterone induction of small heat shock protein gene expression. Dev Biol 151:410–418, 1992.

131. Fujiwara H, Jindra M, Newitt R, Palli SR, Hiruma K, Riddiford LM: Cloning of an ecdysone receptor homolog from *Manduca sexta* and the developmental profile of mRNA in wings. Insect Biochem Mol Biol 25:845–856, 1995.

132. Jindra M, Huang J-Y, Malone F, Asahina M, Riddiford LM: Identification and mRNA developmental profiles of two Ultraspiracle isoforms in the epidermis and wings of *Manduca sexta*. Insect Mol Biol 6:41–53, 1997.

133. Swevers L, Cherbas L, Cherbas P, Iatrou K: *Bombyx* EcR (BmEcR) and *Bombyx* USP (BmCF1) combine to form a functional ecdysone receptor. Insect Biochem Mol Biol 26:217–221, 1996.

134. Zhao B, Hiruma K, Jindra M, Shinoda T, Segraves WA, Malone F, Riddiford LM: Developmental expression and hormonal regulation of the E75A and E75B homologs in the epidermis of the tobacco hornworm, *Manduca sexta*. Dev Biol, in press.

135. Palli SR, Hiruma K, Riddiford LM: An ecdysteroid-inducible *Manduca* gene similar to the *Drosophila* DHR3 gene, a member of the steroid receptor superfamily. Dev Biol 150:306–318, 1992.

136. Jindra M, Riddiford LM: Expression of ecdysteroid-regulated transcripts in the silk gland of the wax moth, *Gallerial mellonella*. Dev Genes Evolution 206:305–314, 1996.

136a. Riddiford LM: Molecular aspects of juvenile hormone action in insect metamorphosis. In Gilbert LI, Tata JR (eds): Metamorphosis: Post-embryonic reprogramming of gene expression in amphibian and insect cells. San Diego: Academic Press, 1996, pp 223–251.

137. Palli SR, Riddiford LM, Hiruma K: Juvenile hormone and "retinoic acid" receptors in *Manduca* epidermis. Insect Biochem 21:7–15, 1991.

138. Forman BM, Goode E, Chen J, Oro AE, Bradley DJ, Perlmann T, Noonan DJ, Burka LT, McMorris T, Lamph WW, Evans RM, Weinberger C: Identification of a nuclear receptor that is activated by farnesol metabolites. Cell 81:687–693, 1995.

139. Harmon MA, Boehm MF, Heyman RA, Mangelsdorf DJ: Activation of mammalian retinoid X receptors by the insect growth regulator methoprene. Proc Natl Acad Sci USA 92:6157–6160, 1995.

140. Talbot WS, Swyryd EA, Hogness DS: *Drosophila* tissues with different metamorphic responses to ecdysone express different ecdysone receptor isoforms. Cell 73: 1323–1337, 1993.

141. Robinow S, Talbot WS, Hogness DS, Truman JW: Programmed cell death in the *Drosophila* CNS is ecdysone-regulated and coupled with a specific ecdysone receptor isoform. Development 119:1251–1259, 1993.

142. von Kalm L, Crossgrove K, Von Segger D, Guild GM, Beckendorf SK: The *Broad-Complex* directly controls a tissue-specific response to the steroid hormone ecdysone at the onset of metamorphosis. EMBO J 13:3505–3516, 1994.

143. Woodard C, Baeherecke E, Thummel C: A molecular mechanism for the stage specificity of the *Drosophila* prepupal genetic response to ecdysone. Cell 79:607–615, 1994.

144. Danielson M, Hinck L, Ringold GM: Two amino acids within the knuckle of the first zinc finger specify DNA response element activation by the glucocorticoid receptor. Cell 57:1131–1138, 1989.

145. Mader S, Kumar V, de Verneuil H, Chambon P: Three amino acids of the oestrogen receptor are essential to its ability to distinguish an oestrogen from a glucocorticoid-responsive element. Nature 338:271–274, 1989.

146. Umesono K, Evans RM: Determinants of target gene specificity for steroid/thyroid hormone receptors. Cell 57:1139–1146, 1989.

147. Cherbas L, Lee K, Cherbas P: Identification of ecdysone response elements by analysis of the Eip28/29 gene. Genes Dev 5:120–131, 1991.

148. Martinez E, Givel F, Wahli W: A common ancestor DNA motif for invertebrate and vertebrate hormone response elements. EMBO J 10:263–268, 1991.

149. D'Avino PP, Crispi S, Cherbas L, Cherbas P, Furia M: The moulting hormone ecdysone is able to recognize target elements composed of direct repeats. Mol Cell Endocrinol 113:1–9, 1995.

150. Antoniewski C, Mugat B, Delbac F, Lepesant J-A: Direct repeats bind the EcR-USP receptor and mediate ecdysteroid responses in *Drosophila melanogaster*. Mol Cell Biol 16:2977–2986, 1996.

151. Harding HP, Lazar MA: The orphan receptor Rev-ErbAα activates transcription via a novel response element. Mol Cell Biol 13:3113–3121, 1993.

152. Mlodzik M, Hiromi Y, Weber U, Goodman CS, Rubin GM: The *Drosophila* seven-up gene, a member of the steroid receptor gene superfamily, controls photoreceptor cell fates. Cell 60:211–224, 1990.

153. Fisk GJ, Thummel CS: Isolation, regulation, and DNA-binding properties of three *Drosophila* nuclear hormones receptor superfamily members. Proc Natl Acad Sci USA 92:10604–10608, 1995.

154. Zelhof AC, Yao T-P, Chen JD, Evans RM, McKeown M: Seven-up inhibits ultraspiracle-based signaling pathways in vitro and in vivo. Mol Cell Biol 15:6736–6745, 1995.

155. Zelhof AC, Yao T-P, Evans RM, McKeown M: Identification and characterization of a *Drosophila* nuclear receptor with the ability to inhibit the ecdysone response. Proc Natl Acad Sci USA 92:10477–10481, 1995.

156. Ayer S, Walker N, Mosammaparast M, Nelson J, Shilo B, Benyajati C: Activation and repression of *Drosophila* alcohol dehydrogenase distal transcription by two steroid hormone receptor superfamily members binding to a common response element. Nucl Acids Res 21:1619–1627, 1993.

157. Richards G: Switching partners? Curr Biol 2:657–659, 1992.

158. Sutherland JD, Kozlova T, Tzertzinis G, Kafatos FC: *Drosophila* hormone receptor 38: A second partner for *Drosophila* USP suggests an unexpected role for nuclear receptors of the nerve growth factor-induced protein B type. Proc Natl Acad Sci USA 92:7966–7970, 1995.

159. Venkataraman V, O'Mahony PJ, Manzcak M, Jones G: Regulation of juvenile hormone esterase gene transcription by juvenile hormone. Dev Genet 15:391–400, 1994.

160. Zhang J, Saleh DS, Wyatt GR: Juvenile hormone regulation of an insect gene: A specific transcription factor and a DNA response element. Mol Cell Endocrinol 122:15–20, 1996.

4

ESTROGEN CONTROL OF *XENOPUS LAEVIS* EGG YOLK MRNA SYNTHESIS AND DEGRADATION

DAVID J. SHAPIRO, ROBIN E. DODSON, MARIA ACENA, VINCENT A. DIPIPPO, AND HIROSHI KANAMORI

Department of Biochemistry, University of Illinois, Urbana, IL 61801

4.1. INTRODUCTION

The mechanisms controlling the production of egg yolk proteins in *Xenopus laevis* provide an impressive and well-studied example of the respecialization of a previously differentiated cell type under hormonal control. In oviparous vertebrates, the egg yolk precursor protein, vitellogenin, is synthesized in the liver under estrogen control (1–4), secreted into the bloodstream, and taken up in developing oocytes by receptor-mediated endocytosis (5). Within the oocyte, vitellogenin undergoes specific proteolytic cleavage to yield the yolk proteins lipovitellin, phosvitin, and phosvettes (5). The massive induction by estrogen of vitellogenin synthesis in *Xenopus* hepatocytes has been a useful model for hormone-dependent respecialization of hepatocytes, for estrogen action, and for regulation of gene expression at the levels of gene transcription and mRNA stability. This review focuses on two questions about vitellogenin synthesis: What have we learned about the transcriptional and posttranscriptional mechanisms responsible for control of vitellogenin mRNA levels? What are the developmental events by which liver cells gain the competence to respond to estrogen by vitellogenin synthesis?

Hormones and Growth Factors in Development and Neoplasia, Edited by Robert B. Dickson and David S. Salomon
ISBN 0-471-16899-8

4.2. THE *XENOPUS* VITELLOGENIN SYSTEM

Early studies of this system demonstrated that the estrogen induction of vitellogenin synthesis is accompanied by a corresponding increase in the level of vitellogenin mRNA. Controlling the amount of vitellogenin mRNA therefore constitutes the major level at which vitellogenin production is regulated (6,7). In the absence of exogenous estrogen, hepatocytes from male *Xenopus* do not contain detectable vitellogenin mRNA (<1 molecules per cell; [6–9]). Administration of estrogen either *in vivo* or in primary liver fragment cultures results in the syntheses and accumulation of massive amounts of vitellogenin mRNA, which can reach levels of 50,000 molecules per hepatocyte—representing approximately half the cells' mRNA (7,9–11). To achieve this unusually high level of mRNA, hepatocytes regulate vitellogenin mRNA levels at multiple sites. Estrogen increases the level of total nuclear RNA synthesis by $>$20-fold (11). There is also a 17β-estradiol–estrogen receptor-dependent increase of $>$10,000-fold in the specific transcription of the vitellogenin genes (10,11). Vitellogenin gene transcription rises from $>$1 molecule of mRNA per cell per day to approximately 20 molecules per cell per minute (11). In addition to these transcriptional events, there is a striking 17β-estradiol–estrogen receptor-dependent increase in the cytoplasmic stability of vitellogenin mRNA. Vitellogenin mRNA exhibits a half-life of approximately 500 hr (about 3 weeks) when the 17β-estradiol–estrogen receptor complex is present and a half-life of 16 or 30 hr (at 20°C or 25°C, respectively) following removal of estrogen from the medium used to maintain liver fragment cultures (11). Vitellogenin transcripts are complex, with $>$30 introns required to excise correctly in nuclear processing (4,12). Although only about 60% of newly synthesized vitellogenin mRNA reaches the cytoplasm (3,13,14), there is no compelling evidence that nuclear processing and polyadenylation (13,15) control vitellogenin mRNA levels.

Because so much of the cell's synthetic and secretory apparatus is devoted to the synthesis and secretion of vitellogenin, a number of related cellular events are required to allow the deployment of such a large proportion of the cell's resources to this reproductive function. The demands placed on the cell's secretory machinery are such that the synthesis and secretion of albumin are dramatically reduced (16–18). This appears to be brought about by the posttranscriptional destabilization of albumin mRNA (17,18). There are demands for additional lipids, which are required for membrane synthesis and are bound to vitellogenin, a secreted phospholipoglycoprotein. Major lipid biosynthetic enzymes hydroxymethylglutaryl coenzyme A (HMG-CoA) reductase and acetyl coenzyme A (acetyl-CoA) carboxylase are strongly induced in hepatocytes undergoing active vitellogenin synthesis (19,20). The induction of HMG-CoA reductase is achieved by a posttranscriptional mechanism (20). Whether most of these events are primary effects of estrogen or are secondary events in unclear. For example, while estrogen may act directly to regulate HMG-CoA reductase mRNA levels, it is also possible that the depletion of intracellular cholesterol pools due to increased membrane synthesis and secretion of cholesterol bound to vitellogenin results in a feedback loop leading to increased levels of HMG-CoA reductase mRNA. Because it is not presently clear whether most of these responses represent direct or indirect effects of estrogen, they are largely beyond the scope of this review. In contrast, there is compelling evidence that the estrogen induction of vitellogenin gene transcription and the estrogen stabilization of vitellogenin mRNA are direct effects of the 17β-estradiol–estrogen receptor complex (20,21).

4.3. ESTROGEN-REGULATED VITELLOGENIN GENE TRANSCRIPTION

In considering the regulation of vitellogenin gene transcription, several issues must be addressed.

1. While estrogen receptor is widely distributed in somatic cells, vitellogenin mRNA is detectable only in hepatocytes. Although the absence of vitellogenin mRNA in nonhepatic cells may be due in part to a five- to sixfold increase in the degradation rate of vitellogenin mRNA in nonhepatic cells (D. Nielsen, R. Dodson, and D. Shapiro, unpublished observations), a major component of this process must be the inability to transcribe these genes in nonhepatic cells. In contrast to some liver-specific genes that are not transcribed effectively on transient transfection into nonhepatic cells, *Xenopus* vitellogenin promoters show substantial transcriptional activity in a variety of *Xenopus* and mammalian cell lines (22–27). Thus the absence of transcription in nonhepatic cells must be due to a failure to realize the transcriptional potential of genomic vitellogenin genes in these cells.
2. In the absence of exogenous 17β-estradiol, the vitellogenin genes in male hepatocytes are transcriptionally silent ($<$1 molecule of mRNA synthesized per cell per day [11]). In contrast, in transient transfections, the vitellogenin promoter displays substantial transcription in the absence of added estrogen (22–27). A related observation is that cellular vitellogenin gene transcription is completely dependent on the presence of estrogen (10,11,17,23,27). What are the factors responsible for this strong influence of cell context on transcription of the vitellogenin promoters?
3. When estrogen is present, vitellogenin mRNA is synthesized from the four vitellogenin genes at an overall rate of 20 molecules of vitellogenin mRNA per cell per minute (11). This is one of the most rapid transcription rates seen for a vertebrate cellular gene. How do hepatocytes achieve the efficient transcription of the vitellogenin genes?

As an approach to these and related questions, the organization of the four related *Xenopus* vitellogenin genes has been analyzed (28,29). Early studies led to the identification of the sequence (A)GGTCAcagTGACC(T) in the vitellogenin A2 promoter as the consensus palindromic estrogen response element (ERE) (30–32). More recent studies using a variety of methods, including cell-free transcription, transient transfection, and *in vitro* and *in vivo* footprinting, have led to a fairly detailed picture of the structure of the *Xenopus* vitellogenin *B1* gene (23–25,33,34). The critical EREs in the VIT *B1* gene are two imperfect EREs located at −302 to −334. Deletion of these EREs abolishes estrogen-mediated VIT *B1* transcription (22,26).

The vitellogenin *B1* 5′-flanking region is typical of vertebrate promoters in that it contains a complex array of protein binding sites. This region can be divided into three functional subdomains. Immediately upstream of the consensus TATA box is a region of approximately 100 nucleotides that contains a diverse array of functional binding sites. These include two binding sites for (CTF/NF-1) family members, three binding sites for the *Xenopus* (C-EBP) homolog, and three binding sites for liver-enriched (HNF-3) family members (22–25,34). A sequence identified in cell-free transcription as a liver-specific repressor also lies within this region (23). Immediately upstream of this region is an A-T-rich region of intrinsically bent DNA in the *B1* gene (A. Nardulli and D. Shapiro, unpublished observations) and in the closely related *Xenopus* vitellogenin *A2* gene (35), which is devoid of transcription factor binding sites (34). This region has been shown to be a preferred site of nucleosome assembly (36). Upstream of the A-T-rich region at −280 is another CTF/NF-1 site that is adjacent to the imperfect EREs at −302 to −334 (25,31,32). Two additional CTF/NF-1 sites lie in the region from −383 to −405 (34). Thus, the structure of the vitellogenin 5′-flanking region provides a plausible model for the unusually efficient transcription of the vitellogenin genes. Just upstream of the consensus TATA box is a rich series of overlapping binding sites for several general and liver-enriched transcription factors (22–25,34). A static nucleosome assembling on the A-T-rich region of intrinsically bent DNA effectively brings the more distal upstream region containing the EREs and three CTF/NF-1 sites into closer proximity to the activator proteins bound at the sites near the TATA box (36). In addition, the DNA bend induced on binding of the estrogen receptor (ER) to the EREs (37–40) may further potential transcription by bringing the distal binding sites into closer proximity to the sites near the TATA box. This combination of multiple transcription activator sites arrayed in three-dimensional space so that they are close to the TATA box may account for the efficient transcription of the vitellogenin genes. The liver-specific activation of vitellogenin gene transcription is likely to arise from the presence of multiple binding sites for transcription factors enriched in liver in the proximal region near the TATA box (34).

While these factors can account for the efficient liver-specific transcription of the vitellogenin genes, the system that renders the vitellogenin genes transcriptionally silent in the absence of estrogen is less clear. In an important *in vivo* footprinting study, Cardinaux et al (34) found that the VIT *B1* gene exhibited very similar footprints in female nuclei, in which it is transcribed, and in male nuclei, in which it is transcriptionally silent. This is quite different from the reorganization of the chromatin structure and alterations in occupancy of transcription factor binding sites that accompanies estrogen activation of the avian vitellogenin II gene (41) and activation of the mouse mammary tumor virus (MMTV) promoter by glucocorticoids (42). The *in vivo* footprinting reveals a diverse array of proteins that usually act as transcription activators bound to the upstream sites in nuclei from male *Xenopus* hepatocytes (34), in which the gene is transcriptionally silent. It is therefore likely that an active repression mechanism is required to block the activity of these bound transcription factors. One possibility is that a putative negative regulatory element in the upstream region (23) is occupied by an as yet unidentified repressor protein in the absence of ER bound at the EREs. Following ER binding to the EREs, the repressor is displaced and replaced by an activator protein (34). While this model is both obvious and attractive, it remains unsupported by direct experimental data.

The available data point to the occupancy of the imperfect EREs by ER as the key event triggering vitellogenin gene transcription. A number of studies have examined the way in which the properties of the *Xenopus* estrogen receptor (XER) and its regulation may play a role in the regulation of vitellogenin gene transcription (22,43–48). The XER is present at only a few hundred copies per cell in nuclei of male *Xenopus* hepatocytes (43–46). An early effect of estrogen is the induction of additional XER (44–46), perhaps through binding of the liganded receptor to an imperfect ERE in the protein coding region of the XER (49). However, the initial activation of vitellogenin gene transcription can occur in the presence of protein synthesis inhibitors that block the induction of new XER (21). Since the concentration of circulating estrogens in male *Xenopus* is sufficient to load approximately 25% of the XER (50), why doesn't this liganded XER bind to the vitellogenin EREs and activate at least a low level of vitellogenin transcription? It is likely that there is insufficient liganded XER to occupy this site in male hepatocyte nuclei. Binding to the imperfect vitellogenin *B1* EREs, whose affinity for ER and for the ER DNA binding domain is severalfold lower than the affinity of the consensus ERE (25,51,52), is probably facilitated by the presence of an adjacent CTF/NF-1 site. This may explain why the replacement of the VIT *B1* EREs with EREs 20 nucleotides upstream results in a loss of estrogen induction (25). Once the uninduced ER is fully occupied with estrogen, there is sufficient liganded ER to occupy the EREs and initiate at least a low level of vitellogenin transcription (21). It is also possible that the presence of the 17β-estradiol–XER complex induces activation of signal transduction pathways acting through protein kinases, and that such changes play a role in vitellogenin gene activation (53–55). Consistent with this idea is the observation that estrogen administration elicits a rapid increase in intracellular cyclic AMP levels in several types of mammalian cells (54,55). The resulting activation of protein kinases A family members may contribute to phosphorylation of ER and of other proteins important in vitellogenin transcription.

The induction of additional XER by estrogen is necessary for the highly efficient transcription of the vitellogenin genes. Interestingly, transient administration of estrogen sets up a stable autoregulatory loop in which elevated levels of XER are maintained even in the absence of additional exogenous 17β-estradiol (56). This autoregulatory response is distinct from the activation of vitellogenin transcription, which requires the continuous presence of exogenous 17β-estradiol.

4.4. DEVELOPMENTAL REGULATION OF XER AND VITELLOGENIN GENE EXPRESSION

While these data demonstrate that once XER is present it can maintain its level through an autoregulatory mechanism, they do not explain the initial appearance of XER in hepatocytes. The ability of thyroid hormone to induce precocious metamorphosis of *Xenopus* tadpoles makes it an obvious candidate as a developmental regulator of XER levels. While thyroid hormone has been reported to induce XER (57–60), the XER gene does not appear to contain functional thyroid response elements (V.A.

DiPippo and D. Shapiro, unpublished observations). XER mRNA does not appear concomitant with hepatocyte differentiation but occurs at significant levels only late in development near the metamorphic climax (61). It is at this time that the vitellogenin genes gain the ability to be induced by estrogen (61). The failure to induce vitellogenin transcription until late in development is probably not due to the absence of hepatocyte-specific transcription factors. The liver-specific, estrogen-regulated *Xenopus* serum retinol binding protein gene (62,63) is efficiently transcribed early in development, indicating that several transcription factors enriched in liver are present early in development. Nevertheless, the retinol binding protein gene also does not acquire the capacity to be induced by estrogen until the metamorphic climax (61), when significant levels of XER mRNA appear.

4.5. ESTROGEN REGULATION OF VITELLOGENIN mRNA STABILITY

The impressive estrogen induction of vitellogenin gene transcription is insufficient to account for the massive accumulation of vitellogenin mRNA in *Xenopus* hepatocytes. The observation that estrogen administration leads to a nearly linear accumulation of vitellogenin mRNA led to the suggestion that vitellogenin mRNA is unusually stable during estrogen induction (6,64). Using a primary liver fragment culture system, direct measurements of the stability of vitellogenin mRNA were made by pulse labeling the mRNA during estrogen induction and after removal of estrogen from the medium, followed by hybridization of the labeled mRNA to immobilized vitellogenin cDNA clones. Vitellogenin mRNA in liver cells maintained in estrogen is essentially stable and exhibits a half-life of approximately 500 hr. When estrogen is removed from the medium, vitellogenin mRNA exhibits a half-life of approximately 16 hr (at 25°C), which is similar to the half-life of total cell poly(A) mRNA (11). The 30-fold estrogen-mediated stabilization of vitellogenin mRNA is a reversible cytoplasmic effect of estrogen since pulse-labeled cytoplasmic vitellogenin mRNA that is undergoing degradation can be restabilized on addition of estrogen to the culture medium (11). The stabilization of vitellogenin mRNA requires estrogen, ER, and the presence of polysomes associated with the mRNA, but it does not require ongoing protein synthesis (11,65–67). Transient transfections of a liver-derived *Xenopus* cell line led to the demonstration that >90% of the protein coding region of vitellogenin mRNA could be deleted without loss of estrogen-mediated stabilization (67). This work and related studies led us to focus on the 3′-untranslated region of vitellogenin mRNA.

Using RNA gel mobility shift assays, an estrogen-inducible protein was identified in *Xenopus* liver cytosol that binds specifically to a segment of the vitellogenin mRNA 3′-untranslated region. Binding to the vitellogenin mRNA 3′-untranslated region was induced four- to fivefold after estrogen administration. Ultraviolet cross-linking studies indicated that the cross-linked polypeptides have molecular weights of approximately 70,000 and 140,000 (68). Purification of the binding activity indicates that it is the 140-kDa polypeptide that is responsible for binding to the vitellogenin mRNA 3′-region (R. Dodson, H. Kanamori, and D. Shapiro, unpublished observations). While this protein is estrogen inducible, it is also widely distributed, and closely related RNA binding activities have been detected in all vertebrate cells tested (69). Analysis of the protein binding site indicates that this large protein actually binds to and protects an unusually large 70- to 75-nucleotide RNA region upstream of the polyadenylation site (H. Kanamori, R. Dodson, and D. Shapiro, unpublished observations).

These data lead to a testable hypothesis for estrogen-mediated stabilization of vitellogenin mRNA. In this model, prior to estrogen induction of the binding activity, the regulatory region of vitellogenin mRNA is not occupied and cleavage by a sequence-specific cell nuclease can occur. Following estrogen induction of the binding protein, the protein binding site near the 3′ end of vitellogenin mRNA is occupied, and the binding protein physically blocks access to the cleavage sites in this region. While the identity of the enzymes involved in vitellogenin mRNA degradation remains obscure, a ribonuclease (RNase) that appears to be involved in the initial cleavage of estrogen-destabilized albumin mRNA has recently been isolated from *Xenopus* liver (70). Consistent with this model is the observation that this region of vitellogenin mRNA contains sequences homologous to those identified as cleavage sites in the 3′-untranslated region of estrogen-regulation apo very low density lipoprotein (VLDL) II mRNA (71). An alternative model is that the binding of the protein near the 3′ end of vitellogenin mRNA stabilizes interaction of the poly(A) tail with the poly(A) binding protein and thereby blocks poly(A) shortening, which has been shown to be the first step in the degradation of most yeast mRNAs (72).

4.6. FUTURE PROSPECTS

While a good deal is known about the proteins that bind to the vitellogenin promoter, the actual mechanism by which the promoter is silenced in the absence of estrogen and then activated on binding of ER to the EREs remains obscure. Whether this involves architectural changes in the DNA based on receptor-induced DNA bending, displacement of a repressor protein, or some other mechanism remains to be established.

The level of ER in hepatocytes appears to play a key role in their commitment to vitellogenin production. Molecular aspects of the developmental control of XER levels in liver cells remain obscure. The way in which the autoregulation of XER gene expression may contribute to persistent sex-specific differences in XER levels in hepatocytes from male and female *Xenopus* remains to be established.

The estrogen-mediated stabilization of vitellogenin mRNA and destabilization of albumin mRNA have made this a productive system for studying steroid hormone regulation of mRNA stability. While mRNA sequences and proteins that may play a role in this process have been identified, ongoing studies have the formidable task of discriminating between several competing models for the regulation of vitellogenin mRNA stability.

ACKNOWLEDGMENTS

The research from our laboratory described in this review was supported by grants from the National Institute of Health (HD-17882, DK-50080) and the National Science Foundation.

REFERENCES

1. Shapiro DJ, Brock ML, Hayward, MA: Estrogen receptor regulation of vitellogenin gene transcription and chromatin structure. In Roy AK (ed): Gene Regulation by Steroid Hormones II. New York: Springer-Verlag, 1983, pp 61–78.
2. Shapiro DJ: Steroid hormone regulation of vitellogenin gene expression. CRC Crit Rev Biochem 12:187–204, 1982.
3. Shapiro DJ, Barton MC, McKearin DM, Chang T-C, Lew D, Blume J, Nielsen DA, Gould L: Estrogen regulation of gene transcription and mRNA stability. In Clark J (ed): Recent Progress in Hormone Research, Vol. 45 San Diego, CA: Academic Press, 1989, pp 29–64.
4. Wahli W: Evolution and expression of vitellogenin genes. Trends Genet 4:227–232, 1988.
5. Wallace RA: Vitellogeninesis and oocyte growth in nonmammalian vertebrates. In Browder LW (ed): Developmental Biology, Vol 1. New York: Plenum, 1985, pp 127–177.
6. Shapiro DJ, Baker HJ, Stitt DT: *In vitro* translation and estradiol-17β induction of *Xenopus laevis* vitellogenin messenger RNA. J Biol Chem 251:3105–3111, 1976.
7. Baker HJ, Shapiro DJ: Kinetics of estrogen induction of *Xenopus laevis* vitellogenin messenger RNA determined by hybridization to complementary DNA. J Biol Chem 252:8428–8432, 1977.
8. Shapiro DJ, Baker HJ: Purification and characterization of *Xenopus laevis* vitellogenin messenger RNA. J Biol Chem 252:5244–5250, 1977.
9. Baker HJ, Shapiro DJ: Rapid accumulation of vitellogenin messenger RNA during secondary estrogen stimulation of *Xenopus laevis*. J Biol Chem 253:4521–4524, 1978.
10. Brock ML, Shapiro DJ: Estrogen regulates the absolute rate of transcription of the *Xenopus laevis* vitellogenin genes. J Biol Chem 258:5449–5455, 1983.
11. Brock ML, Shapiro DJ: Estrogen stabilizes vitellogenin messenger RNA against cytoplasmic degradation. Cell 34:207–214, 1983.
12. Wahli W, Ryffel GU: *Xenopus* vitellogenin genes. In Browder LW (ed): Oxford Surveys on Eukaryotic Genes, Vol 2. Oxford: Oxford University Press, 1985, pp 96–120.
13. Blume JE, Nielsen DA, Shapiro DJ: Estrogen regulation of vitellogenin mRNA stability. In Granner D, Rosenfeld MG, Chang S (eds): UCLA Symposia, New Series, Transcriptional Control Mechanisms, Vol 52. Alan R Liss, New York: 1987, pp 259–274.
14. Shapiro DJ, Brock ML: Messenger RNA stabilization and gene transcription in the estrogen induction of vitellogenin mRNA. In Litwack G (ed): Biochemical Actions of Hormones, Vol XII. New York: Academic Press, 1985, pp 139–172.
15. Rao MN, Chernokalskaya E, Schoenberg DR: Regulated nuclear polyadenylation of *Xenopus* albumin pre-mRNA. Nucleic Acids Res 24:4078–4083, 1996.

16. May FEB, Ryffel GU, Weber R, Westley BR: Estrogen dramatically decreases albumin mRNA levels and albumin synthesis in *Xenopus laevis* liver. J Biol Chem 257: 13919–13923, 1982.
17. Wolffe AP, Glover JF, Martin SC, Teniswood MPR, Williams JL, Tata JR: Deinduction of transcription of *Xenopus* 74-kD albumin genes and destabilization of mRNA in vivo and in hepatocyte cultures. Eur J Biochem 146:489–496, 1985.
18. Riegel AT, Martin MB, Schoenberg DR: Transcriptional and posttranscriptional inhibition of albumin gene expression by estrogen in *Xenopus* liver. Mol Cell Endocrinol 44: 201–209, 1986.
19. Philipp BW, Shapiro DJ: Estrogen regulation of hepatic 3-hydroxy-3-methylglutaryl coenzyme A reductase and acetyl-CoA carboxylase in *Xenopus laevis*. J Biol Chem 256: 2922–2927, 1981.
20. Chen HC, Shapiro DJ: Nucleotide sequence and estrogen induction of *Xenopus laevis* 3-hydroxy-3-methylglutaryl coenzyme A reductase. J Biol Chem 265:4622–4629, 1990.
21. Hayward MA, Brock ML, Shapiro DJ: Activation of vitellogenin gene transcription is a direct response to estrogen in *Xenopus laevis*. Nucleic Acids Res 10:8273–8274, 1982.
22. Martinez E, Wahli W: Cooperative binding of estrogen receptor to imperfect estrogen-responsive DNA elements correlates with their synergistic hormone-dependent enhancer activity. EMBO J 8:3781–3791, 1989.
23. Corthesy B, Cardinaux JR, Claret F-X, Wahli W: A nuclear factor I-like activity and a liver-specific repressor govern estrogen-regulated in vitro transcription from the *Xenopus* vitellogenin B1 promoter. Mol Cell Biol 9:5548–5562, 1989.
24. Wahli W, Martinez E, Corthesy B, Cardinaux J: Cis- and trans-acting elements of the estrogen-regulated vitellogenin gene B1 of *Xenopus laevis*. J Steroid Biochem 34:17–32, 1989.
25. Chang T-C, Shapiro DJ: An NF1-related vitellogenin activator element mediates transcription from the estrogen regulated *Xenopus laevis* vitellogenin promoter. J Biol Chem 265:8176–8182, 1990.
26. Chang TC, Nardulli AM, Lew D, Shapiro DJ: The role of estrogen response elements in expression of the *Xenopus laevis* vitellogenin B1 gene. Mol Endocrinol 6:346–354, 1992.
27. Marilley D, Mahfoudi A, Wahli W: Gene transfer into *Xenopus* hepatocytes: Transcriptional regulation by members of the nuclear receptor superfamily. Mol Cell Endocrinol 101:227–236, 1994.
28. Wahli W, Dawid IB, Ryffel GU, Weber R: Vitellogeninesis and the vitellogenin gene family. Science 212:298–304, 1981.
29. Wahli W, Dawid IB, Wyler T, Weaver R, Ryffel GU: Comparative analysis of the structural organization of two closely related vitellogenin genes in *Xenopus laevis*. Cell 20:107–117, 1980.
30. Klein-Hitpass L, Schorpp M, Wagner U, Ryffel GU: An estrogen responsive element derived from the 5′-flanking region of the *Xenopus* vitellogenin A2 gene functions in mammalian cells. Cell 46:1053–1061, 1986.
31. Seiler-Tuyns A, Walker P, Martinez E, Merillat AM, Givel F, Wahli W: Identification of estrogen-responsive DNA sequences by transient transfection experiments in a human breast cancer cell line. Nucleic Acids Res 14:8755–8770, 1986.

32. Klein-Hitpass L, Ryffel GU, Heitlinger E, Cato ACB: A 13 bp palindrome is a functional element and interacts specifically with estrogen receptors. Nucleic Acids Res 16:647–663, 1986.
33. Germond JE, ten Heggeler B, Schubiger JL, Walker P, Westley B, Wahli W: Vitellogenin gene B2 in *Xenopus:* Isolation, *in vitro* transcription and relation to other genes. Nucleic Acids Res 11:2979–2997, 1982.
34. Cardinaux J-R, Chapel S, Wahli W: Complex organization of CTF/NF-1, C-EBP and HNF3 binding sites within the promoter of the liver-specific vitellogenin gene. J Biol Chem 269:32947–32956, 1994.
35. Dobelling U, Rob K, Klein-Hitpass L, Morely C, Wagner U, Ryffel GU: A cell-specific activator in the *Xenopus* A2 vitellogenin gene: Promoter elements functioning with rat liver nuclear extracts. EMBO J 7:2495–2501, 1988.
36. Schild C, Claret FX, Wahli W, Wolffe A: A nucleosome-dependent static loop potentiates estrogen-regulated transcription from the *Xenopus* vitellogenin B1 promoter. EMBO J 12:423–433, 1993.
37. Nardulli AM, Shapiro DJ: Binding of the estrogen receptor DNA binding domain to the estrogen response element induces DNA bending. Mol Cell Biol 12:2037–2042, 1992.
38. Sabbah M, Le Ricousse S, Redoiulh G, Baulieu EE: Estrogen receptor induced bending of the *Xenopus* vitellogenin A2 gene hormone response element. Biochem Biophys Res Commun 185:944–955, 1993.
39. Nardulli AM, Greene GL, Shapiro DJ: Human estrogen receptor bound to an estrogen response element bends DNA. Mol Endocrinol 1:331–340, 1993.
40. Nardulli AM, Shapiro DJ: DNA bending by steroid receptors. Receptor 3:247–255, 1993.
41. Bakker O, Philipsen JNJ, Ab G: Estrogen-inducible binding of a nuclear factor to the vitellogenin upstream region. Mol Cell Biol 8:4557–4560, 1988.
42. Cordingly MJ, Riegel AT, Hager GL: Steroid-dependent interaction of transcription factors with the inducible promoter of mouse mammary tumor virus in vivo. Cell 48:261–260, 1987.
43. Westley B, Knowland J: An estrogen receptor from *Xenopus laevis* liver possibly connected with vitellogenin synthesis. Cell 15:367–375, 1978.
44. Westley B, Knowland J: Estrogen causes a rapid, large, and prolonged rise in the level of nuclear estrogen receptor in *Xenopus laevis* liver. Biochem Biophys Res Commun 88: 1167–1172, 1979.
45. Hayward MA, Mitchell T, Shapiro DJ: Induction of estrogen receptor and reversal of the nuclear-cytoplasmic receptor ratio during vitellogenin synthesis and withdrawal. J Biol Chem 255:11308–11312, 1980.
46. Hayward MA, Brock ML, Shapiro DJ: The role of estrogen receptor in *Xenopus laevis* vitellogenin gene expression. Am J Physiol (Cell Physiol) 243:C1–C6, 1982.
47. Weiler IJ, Lew D, Shapiro DJ: The *Xenopus laevis* estrogen receptor: Sequence homology with human and avian receptors and identification of multiple estrogen receptor messenger ribonucleic acids. Mol Endocrinol I:355–362, 1987.
48. Claret F-X, Chapel S, Garcés J, Tsai-Pfugfelder M, Bertholet C, Shapiro DJ, Wittek R, Wahli W: Two functional forms of estrogen receptor translated from a single mRNA species. J Biol Chem 269:14047–14055, 1994.
49. Lee JH, Shapiro DJ: Regulation of estrogen receptor gene expression is mediated by an estrogen response element in the protein coding region. DNA 14:410–430, 1995.

50. Wright J, Knowland J: Partial purification of estradiol receptor from *Xenopus laevis* liver and levels of receptor in relation to estradiol concentration. EMBO J 2:973–977, 1983.
51. Ponglikotmongkol M, White JH, Chambon P: Synergistic activation of transcription by the human estrogen receptor bound to tandem response elements. EMBO J 9:2221–2231, 1990.
52. Green S, Chambon P: The oestrogen receptor: From perception to mechanism. In Parker M (ed): Nuclear Hormone Receptors. Orlando, FL: Academic Press, 1991, pp 15–38.
53. Orti E, Bodwell JE, Munck A: Phosphorylation of steroid hormone receptors. Endocrine Rev 13:105–128, 1992.
54. Le Geoff P, Montano MM, Schodin D, Katzenellenbogen B: Phosphorylation of the human estrogen receptor: Identification of hormone regulated sites and examination of their influence on transcriptional activity. J Biol Chem 269:4458–4466, 1994.
55. Cho H, Katzenellenbogen B: Synergistic activation of estrogen receptor-mediated transcription by estradiol and protein kinase activators. Mol Endocrinol 7:441–442, 1993.
56. Barton MC, Shapiro DJ: Transient administration of estradiol-17β establishes an autoregulatory loop permanently inducing estrogen receptor mRNA. Prod Natl Acad Sci USA 85:7119–7123, 1988.
57. Rabelo EML, Baker BS, Tata JR: Interplay between thyroid hormone and estrogen in modulating expression of their receptor and vitellogenin genes during *Xenopus* development. Mech Dev 45:49–57, 1994.
58. Rabelo EM, Tata JR: Thyroid hormone potentiates estrogen activation of vitellogenin genes and autoinduction of estrogen receptor in adult *Xenopus* hepatocytes. Mol Cell Endocrinol 96:37–44, 1993.
59. Ulisse S, Tata JR: Thyroid hormone and glucocorticoids independently regulate the expression of estrogen receptor in male *Xenopus* liver cells. Mol Cell Endocrinol 105:45–53, 1994.
60. Tata JR: Autoregulation and crossregulation of nuclear receptor genes. Trends Exp Med 5:283–290, 1994.
61. Gould L: Development and estrogen regulated expression of *Xenopus laevis* retinol binding protein gene expression. PhD thesis, University of Illinois, 1989.
62. McKearin DM, Barton MC, Keller MC, Shapiro DJ: Estrogen induces transcription of the *Xenopus laevis* serum retinol binding protein gene. J Biol Chem 262:4939–4942, 1987.
63. McKearin DM, Keller MC, Shapiro DJ: Persistent estrogen induction of *Xenopus laevis* serum retinol binding protein gene transcription. J Biol Chem 263:3261–3266, 1988.
64. Ryffel GV, Wahli W, Weber R: Quantitation of vitellogenin messenger RNA in the liver of male *Xenopus* toads during primary and secondary stimulation by estrogen. Cell 11:213–221, 1977.
65. Blume JE, Shapiro DJ: Ribosome loading, but not protein synthesis, is required for estrogen stabilization of vitellogenin mRNA. Nucl Acids Res 17:9003–9014, 1989.
66. Nielsen DA, Shapiro DJ: Estradiol and estrogen receptor dependent stabilization of a mini-vitellogenin mRNA lacking 5100 nucleotides of coding sequence. Mol Cell Biol 10:171–176, 1990.
67. Nielsen DA, Shapiro DJ: Insights into hormonal control of mRNA stability. Mol Endocrinol 4:953–957, 1990.
68. Dodson R, Shapiro DJ: An estrogen-inducible protein binds specifically to the 3′-untranslated region of vitellogenin mRNA. Mol Cell Biol 14:3130–3138, 1994.

69. Dodson R, Accna M, Shapiro DJ: Tissue distribution, hormone regulation and evidence for a human homologue of the estrogen-inducible vitellogenin mRNA binding protein. J Steroid Biochem Mol Biol 52:505–515, 1995.
70. Dompenciel R, Garnepudi VR, Schoenberg DR: Purification and characterization of an estrogen-regulated *Xenopus* liver polysomal nuclease involved in the estrogen destabilization of albumin mRNA. J Biol Chem 270:6108–6118, 1995.
71. Binder R, Hwang R, Ratnasabapathy R, Williams DL: Degradation of apolipoprotein II mRNA occurs via endonucleolytic cleavage at 5′-AAU-3′/5′UAA-3′ elements in single stranded domains of 3′-noncoding region. J Biol Chem 264:16910–16918, 1989.
72. Beelman CA, Parker R: Degradation of mRNA in eukaryotes. Cell 81:179–183, 1995.

PART II

GROWTH FACTORS AND HORMONES IN MAMMALIAN DEVELOPMENT

5

GROWTH FACTORS IN MAMMALIAN PRE- AND POSTIMPLANTATION EMBRYOS

DANIEL A. RAPPOLEE
Department of Obstetrics and Gynecology and Department of Cell and Molecular Biology, Northwestern University Medical School, Chicago, IL 60611

5.1. INTRODUCTION

The role of growth factors in murine preimplantation and pregastrulation development will be discussed in this review. It is an exciting time in this field from the perspectives of reproductive and basic sciences. Loss-of-function studies have shown that three growth factors are essential for embryonic and uterine events during the preparation for and process of implantation. Other circumstantial evidence shows that growth factors are important during the entire span of development before gastrulation.

From a clinical standpoint, this period of development represents the greatest challenge to *in vitro* fertilization programs. In addition, in humans and some farm animals, the greatest loss of life occurs in the peri-implantation period (1).

The period between ovulation and gastrulation, which occurs about 6.5 days after fertilization (E6.5) in the mouse, can be divided into three segments: 1) from fertilization to compaction (the first differentiation event), 2) postcompaction through implantation and the formation of both yolk sac endoderm layers, and 3) from postimplantation to gastrulation. These segments will be described below. Current evidence shows that growth factors are essential in the middle of this period from cavitation through the early postimplantation events (E2.5–5.5; Fig. 5.1). This is the

Hormones and Growth Factors in Development and Neoplasia, Edited by Robert B. Dickson and David S. Salomon
ISBN 0-471-16899-8

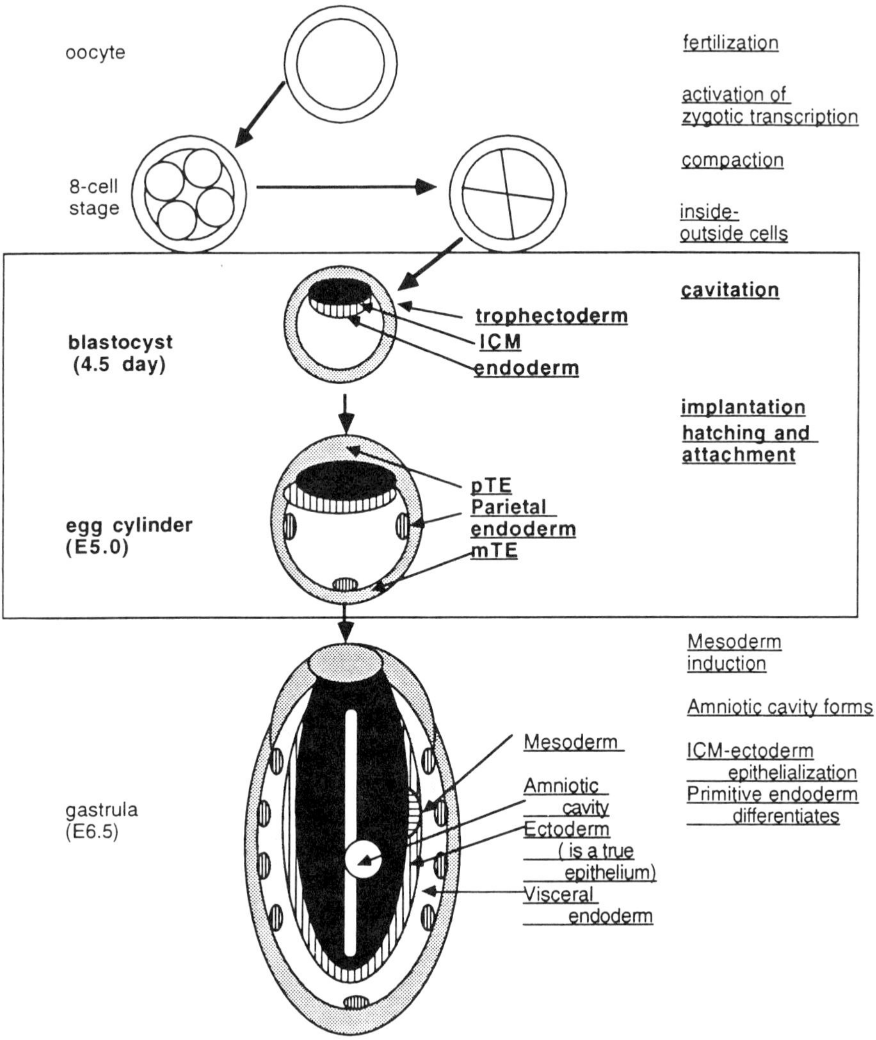

Fig. 5.1. Development of mouse embryos from fertilization through the start of gastrulation. Top to bottom: the progression of time. The column on the left shows the stage of the embryo. The second column shows the morphology of the embryo. The third column describes newly arisen cell types in the embryo. The column on the right describes key events in embryogenesis. The boxed area shows the middle stage of pregastrulation development when growth factors are essential to life. pTE and mTE are polar and mural trophectoderm, respectively.

period when fibroblast growth factor (FGF-4), leukemia inhibitory factor (LIF), and the epidermal growth factor (EGF) receptor are essential for life (Table 5.1). In this period, the embryo undergoes cleavage divisions until implantation occurs and then increases in size after implantation. Extraembryonic endoderm for the yolk sac and extraembryonic ectoderm for the placenta are first made during this period.

Growth factors have less apparent influence during the first segment of time: early cleavage divisions through compaction. First, the embryo differentiates at the eight-cell stage in forming the first epithelium, an event termed *compaction* (at E2.0). This initiates the period when growth factors influence growth and differentiation. The last period of time, from E5.5 to E6.5, is influenced by growth factors, but there is no clear evidence for *esssential* growth factor function that starts in the second and third segments. Experiments in *Xenopus* and the chick, and knockouts in mouse, suggest that endoderm is essential for the nutrition and patterning of the embryo in the third period. In the mouse, superimposed on the nutritive and protective functions of the endoderm adjacent to the inner embryonic cells, there is a likely inductive function required for successful patterning at gastrulation. It is probable that these endodermal functions are mediated by growth factors.

After fertilization in the oviduct, cleavage divisions begin. After compaction, in the next three or four divisions, cells are produced that have no surface facing the outside environment (inner cell mass [ICM] cells). These cells are stem cells for the embryo and the extraembryonic yolk sac endoderm. The outer cells will form extraembryonic ectoderm, a cell lineage in the placenta. By the 16-cell stage, the outer epithelial cells express Na^+/K^+-ATPase on the basolateral surface and the embryo begins cavitating. This event can be influenced by growth factors. The embryo implants after six or seven cell divisions. At this time, it hatches from an acellular zona pellucida, attaches to the epithelial cells of the uterus, causes programmed cell death in the epithelium, and then attaches to the basement membrane. By this time, essential extraembryonic endoderm cells have also formed adjacent to the ICM. Embryonic cells, extraembryonic endoderm, and extraembryonic ectoderm must grow and differentiate in a coordinated fashion so that productive interactions with maternal tissue can occur. Parietal endoderm cells migrate from primitive endoderm (its predecessor) to contact adjacent trophectoderm (Fig. 5.1). Together, parietal endoderm and trophectoderm maintain and remodel a protective, filtering basement membrane (Reichardt's membrane). It is likely that ICM-derived FGF-4 regulates the growth of ICM, trophectoderm, and parietal endoderm. EGF family members may regulate growth of the ICM and trophectoderm. Parathyroid hormone-related peptide (PtHRP) mediates a trophectodermal modulation of parietal endoderm function that may later switch to an autocrine regulation within endoderm.

Summary Growth factors may be involved from the moment the oocyte is ovulated in the nutritional maintenance of the embryo. Factors in the oviduct and uterus and secreted luminal fluids may be important in maintaining growth. Embryos exhibit density-dependent growth in culture. Presumably, the conditioning of oviductal and uterine secretions mediates a similar control of embryonic vitality during preimplantation development *in vivo*. But there are no data that suggest direct control of early cleavage divisions, activation of the zygotic genome, or compaction by growth factors. There is strong evidence that growth factors modulate cavitation; early maintenance of ICM, endoderm, and trophectoderm growth; maternal preparation of the uterus; and some embryonic-to-maternal events such as endocrine recognition of pregnancy and paracrine preparation of the uterus. Maternal factors are important in general maintenance of embryonic growth. Preimplantation events such as hatching, activation, attachment, and implantation may also be partially mediated by growth factors. In the next few years, the role of growth factors in endoderm-to-embryonic ectoderm signaling will become apparent.

5.2. DIRECTIONALITY OF SIGNALING

Growth factor signaling occurs in every direction: embryo to maternal, maternal to embryo, embryo to embryo, and maternal to maternal. It can also be paracrine, endocrine, or autocrine.

It is clear from studies with loss-of-function mutants that one essential pathway is maternal-to-maternal, where leukemia inhibitory factor and LIF receptor are both required in the uterus, but not the embryo, for implantation of the embryo (2–4); Table 5.2. A second essential pathway is embryo-to-embryo, where FGF-4 null embryos die *in utero* shortly after implantation despite ample amounts of maternal FGF (5–7).

There is evidence for paracrine embryo-to-maternal signaling. Four genes are expressed in the uterus near the site of implantation in a blastocyst-dependent way: heparin binding-epidermal growth factor (HB-EGF), FGF-2, amphiregulin, and Cox-2 (8–10). In ruminants, tau interferons function as the trophoblast-produced maternal "recognition of pregnancy" protein (11,12).

Maternal-to-embryo endocrine (e.g., ovarian steroids) and paracrine (e.g., uterine HB-EGF) may mediate maternal control of blastocyst implantation in normal pregnancy or in delayed implantation that is normal or experimentally inducible in many mammalian species (13). As mentioned, the density dependence of growth of preimplantation embryos *in vitro* reflects the role of maternal growth factors *in vivo*.

As yet, embryo-to-maternal and maternal-to-embryo pathways have no loss-of-function evidence to indicate the *necessity* of the pathways. *In vitro* fertilization and preimplantation at the blastocyst stage show that maternal factors are not required by preimplantation embryos to produce a normal pregnancy.

5.3. NULL MUTANTS

In reviewing and thinking about the role of growth factors in early development, three types of evidence in the literature are evaluated. First, the strongest case for the importance of growth factor-mediated events is made when loss-of-function experiments show essential need. FGF-4, EGF receptor, and LIF are the factors that have been reported to be essential for the life of the embryo before gastrulation. Second, gain-of-function experiments also indicate the possible importance for life before gastrulation. When a loss-of-function experiment shows that a growth factor is not essential, a positive gain-of-function result leaves the possibility that the factor is involved in a process that has compensatory or redundant pathways. Such factors may still have clinical and developmental importance. Important factors in this category are colony stimulating factor (CSF)-1, insulin, PtHRP, and HB-EGF. Third, myriad studies on the expression of ligands and receptors indicate the possibility of specific signal-to-source pathways. Transcription of mRNA does not always lead to synthesis of functional protein. Therefore, studies of expression are strongest when transcriptional, translational, and posttranslational processing steps are shown so that synthesis of functional protein products can be inferred.

The types of lethal phenotypes beginning or ending before gastrulation indicate the types of necessary events that the embryo must perform. Included in these is the formation of the first epithelium (*E. cadherin:* ref. 14), formation of endoderm at implantation (*Evx-1:* ref. 15; *HNF-4:* ref. 16, *Hβ-58:* ref. 17), nutrition acquirement at the peri-implantation stage (reviewed refs. 18 and 19), suppression of some growth control mechanisms (*Mdm-2/p53:* ref. 20), and coordination of growth and differentiation at implantation (*FGF-4, EGFR, β1-integrin:* refs. 21–23). The expression of these essential genes may be modulated by growth factors.

5.4. LESSONS FROM LOWER VERTEBRATES

Several molecular and cellular mechanisms that regulate pregastrulation development have evolved. *Xenopus* and chick embryos use growth factors made by endoderm to pattern the ectoderm prior to gastrulation (e.g., FGFs, wnts, Vg-1/activin). In the three-signal hypothesis, the endoderm produces signals (including FGF) that pattern ventral mesoderm and the Neiuwkoop center in *Xenopus* (24). In the chick, a nutrition-acquiring endoderm also patterns the embryonic ectoderm in preparation for gastrulation.

Mammalian development, as represented by a mouse model (the best-studied system), has several unique aspects when compare to *Xenopus* and chick. One obvious difference is the placenta, an organ generated by the embryo to obtain nutrients from maternal tissue after implantation of the blastocyst. Although placentation in mammals is variable, the mouse is representative of chorioallantoic placentation and is similar to placentation in the human (24). Two types of interacting membranes function during implantation and placentation. The yolk sac endoderm acquires nutrients early after implantation, and the trophoblast forms part of the placenta, which becomes the major and final nutrient-acquiring tissue by E12. FGF signaling is used by *Xenopus* only when preparation for gastrulation starts; however, in the mouse, FGF signaling is used before preparation for gastrulation in coordinating the growth of extraembryonic trophectoderm and endoderm (as well as embryonic cells).

5.5. GROWTH FACTOR FAMILIES

5.5.1. Insulin-like Growth Factors

Insulin-like growth factors (IGFs) are a family of four related polypeptides, but IGF-I, IGF-II, and insulin have insulin-like activity, whereas relaxin does not. The three polypeptides that functionally resemble insulin are 51–75 amino acids long. IGF-II is thought to be the *embryonic isoform* due to its early embryonic expression and early restriction of expression (25). In the mouse, IGF-II transcripts are first expressed after activation of zygotic transcription at the two-cell stage, and loss of expression begins before parturition (26) (Table 5-1). Insulin transcripts are not expressed until the pancreas forms in the mouse. IGF-I transcripts are not expressed until after implantation, although there is one report of expression of insulin by early visceral endoderm and of maternal expression of IGF-I in the oocyte (26–28). It is likely that only the endogenous IGF-II acts as a fetal regulator of embryogenesis in pregastrulation mouse embryos.

The transient ablation of IGF-II leads to slower growth without obvious morphologic deficits in preimplantation mouse embryos (26). Ablation of the IGF-II or IGF-I receptor by gene targeting leads to a reduction in the size of both embryo and placenta at parturition (25). Taken together, the evidence suggests that early embryonic growth, but not differentiation, is regulated by IGF-II.

From the time of compaction on, mouse embryos transcribe and translate mRNA for two type I tyrosine kinase-containing receptors, insulin and IGF-I receptors, and for the type II IGF-II/mannose-6-phosphate receptor (26). Binding studies indicate that blastocysts bind insulin, IGF-I, and IGF-II through specific receptors and that trophectodermal outgrowths from the cultured blastocyst bind insulin (25). Functional studies indicate that the mouse ICM binds and responds specifically to insulin and IGF-I (29–31) and may respond to endogenous IGF-II. Insulin is transported across the trophectoderm by transcytosis (32,33). The loss of maintenance of ICM stem cells in parthenogenotes is rescued by overexpression of IGF-1 receptor (34). These studies are limited by the small cell number (<100) in the ICM or trophectoderm in the mouse blastocyst, but other studies in sheep suggest that trophectoderm also responds to insulin (25). It has been suggested that soon after implantation, endogenous IGF-II may regulate trophectodermal growth (25).

The three IGF receptors are transcribed within two cell divisions of the onset of zygotic transcription at the two-cell stage in the mouse (Table 5-2). They are ubiquitously expressed in the ICM and trophectoderm. The IGF-I receptor and the insulin receptor are related transmembrane tyrosine kinases, and the IGF-II/mannose-6-phosphate receptor lacks intrinsic kinase activity but has a ligand-dependent Gi_2-activating moiety on its short cytosolic surface. The functional significance of this Gi_2 binding is not known, but interestingly, Gi_2 is the major form of G protein in the preimplantation mouse embryo (25).

Although IGF-II binds all three receptors with high affinity ($K_D = 10^{-9}$ M), it is likely to transduce a signal through two or perhaps only one of these. Preimplantation mouse embryos grown in culture from the two-cell stage to the blastocyst stage have more cells in the blastocyst if exogenous IGF-II or IGF-I is added. However, a mutant form of IGF-II (^{27}Leu IGF-II), which binds only the IGF-II receptor, has no effect on embryonic growth (26). This suggests that embryonic IGF-II regulates growth by type I receptors. It is likely that endogenous IGF-II enhances early embryonic growth by activating the IGF-I receptor, to which it binds with high affinity, or the insulin receptor, to which it binds less avidly (26). Biochemical data suggest

TABLE 5.1. Growth Factor (GF) Expression and Function in Mouse Embryos Before Gastrulation

GF Ligand	Preimplanation	Postimplantation	Reference
FGF-4	mRNA: maternal, zygotic polypeptide: ICM	mRNA primitive streak; knockout has an early postimplantation lethal phenotype	5, 6, 39
IGF-II	mRNA: zygotic polypeptide: every cell IGF-II protein is growth inducing through class I receptor	Expressed in the allantois	25, 26, 30
PhRP	Found first in TE. Then in endoderm	Receptor in parietal endoderm, ligand in primitive endoderm	73, 74
HGF	HGF receptor in blastocyst	Ligand in allantois, growth in chorion	78
EGF	Negative in blastocyst: effects on cavitation and TE outgrowth increase implantation rate; ligand and R are correlated with TE extension in the pig	Receptor in trophectoderm and giant cells, giant cells are responsive, Knockout in one background has a post-implantation phenotype	46, 53, 81–84
Crypto	Blastocyst library		45
TGF-α	mRNA: maternal zygotic; polypeptide every cell		38, 45, 46
TGF-β1	mRNA: zygotic; polypeptide: every cell; effects on cultured embryos		38, 51, 56, 68
TGF-β2	mRNA: blastocyst; polypeptide: trophectoderm, then ICM		67, Rappolee, unpub.
TGF-β3	mRNA: blastocyst		Rappolee, unpub, 85
PDGF A	mRNA: maternal and zygotic; polypeptide: every cell in blastocyst		25, 38, 43
IL-6	mRNA in blastocyst; polypeptide: activity		57, Rappolee, unpub.
IL-7	mRNA in blastocyst		86
IL-1β	mRNA in blastocyst		86
SCF	mRNA in blastocyst		87
JE	mRNA in blastocyst		Rappolee, unpub.
TNF-α	mRNA in blastocyst		Rappolee, unpub.

(continued)

TABLE 5.1. ***(Continued)***

GF Ligand	Preimplanation	Postimplantation	Reference
BA activin	mRNA: maternal and zygotic; polypeptide: ICM, then trophectoderm	Not expressed in embryo, expressed in decidua	69
BB activin	mRNA: maternal and zygotic; polypeptide: ICM, then TE	Not in embryo, in decidua	
α inhibin	mRNA: maternal only		69
Placental lactogen-1	Postmitotic TE cells, blastocyst	Continues to be expressed in trophoblast giant cells through E8.5	76, 77, 89
IL-1α	Negative in blastocyst library	First trimester trophoblasts respond by invading matrix *in vitro*	75, 86
IL-2	Negative in blastocyst library		86
IL-3	Negative in blastocyst		57, 86
IL-4	Negative in blastocyst		86
IL-5	Negative in blastocyst		86
G-CSF	Negative in blastocyst		38
GM-CSF	Negative in blastocyst library		86
FGF-2	Negative in blastocyst		38
FGF-5		mRNA in gastrulation, ectoderm, streak	89
Nodal	Negative in blastocyst	mRNA in gastrulation, ectoderm, streak	25
FGF-3	Negative throughout preimp. dev.	mRNA in gastrulation, streak, mesoderm	Rappolee et al., unpub, 90
EGF	Negative in blastocyst; affects cavitation, outgrowth of TE		38, 45, 81, 92
NGF	Negative in blastocyst		38
CSF-1	Negative exp. throughout preimp. dev.; affects number of TE in preimp. culture		63, 88, 91
Insulin	Negative throughout preimp. dev.; in culture, increases blastocyst cell number (ICM) and metabolism		26, 30
Tau interferon	In ruminants, TE cells synthesize antiluteolytic "recognition of pregnancy" protein		11

TABLE 5.1. (***Continued***)

GF Ligand	Preimplanation	Postimplantation	Reference
IGF binding proteins	Expressed in preimp. human, mouse		35, 93
Amphiregulin	Negative in blastocyst library; blastocyst-dependent expression in uterus		9, 45
IGF-1	Present in oocyte	mRNA postimplantation	26, 28

that, in certain cell culture models, IGF-II activates the IGF-I receptor (and perhaps the insulin receptor), but the IGF-II receptor acts negatively in delivering IGF-II to the lysosome, where it is destroyed. A genetic mutant, which imprints only the mouse IGF-II receptor and results in an IGF-II receptor-negative mouse, is larger at birth (25).

The expression and function of IGF binding proteins (BPs), which act negatively or positively in defined growth assays in cultures, may be of critical importance in modulating the effects of paracrine and maternal IGFs. A subset of IGF-BPs are expressed in preimplantation mouse embryos (35). Insulin is not bound by IGF binding proteins and may have a direct effect on cells of the pregastrulation mouse embryo (25). Since insulin does not cross the placenta, maternal insulin may have a direct effect on the embryo only before placentation occurs.

Insulin response substrate-1 (IRS-1) is an 185-kDa tyrosine phosphorylation target of both the IGF-I receptor and the insulin receptor. It is expressed at high levels in all cells of the implanting embryo (36).

Summary IGF-II mediates early growth of the preimplantation embryo through type I receptors. IGF-II mediates later trophectodermal and placental growth. Maternal IGF-I and insulin may also influence embryonic growth.

5.5.2. Fibroblast Growth Factors

Nine FGF ligands are expressed in the mouse (FGF 1–9), and three of these have been found in *Xenopus* (37). These are heparin binding growth factors 145–210 amino acids long.

FGF-4, but not FGF-2 or FGF-3, is expressed in preimplantation mouse development (6,38,39). Interestingly, FGF-4 mRNA and polypeptide are localized to the ICM of the mouse blastocyst (6). Expression of FGF-4 in the ICM is probably dependent on the transcription factors oct-3 and Sox-2, which activate the FGF-4 promoter (40–42). This localization is distinct in that transforming growth factor (TGF)-β1, platelet-derived growth factor (PDGF) A, TGF-α, and IGF-II proteins are expressed throughout the mouse blastocyst (38,42,43). Mouse embryos cultured from the two-cell stage through the blastocyst stage have no mitogenic response to

TABLE 5.2. Growth Factor (GF) Receptor Expression and Function in the Mouse Embryo Before Gastrulation

GF Receptor: Function	Preimplanation	Postimplantation	Reference
FGFR-1	MRNA: blastocyst	mRNA is in TE, ICM; outgrowth increase in cell number after adding FGF-4 knockout lethal has early gestation	7, 25, 94
FGFR-2	mRNA:blastocyst	mRNA: primitive ecto at 5.5 day p.c.	7, 95
FGFR-3	mRNA:blastocyst		6, 7
FGFR-4	mRNA:blastocyst		6, 7
Syndecan	mRNA:blastocyst, polypeptide		7, 25
IGF-IR	mRNA: 8-cell stage through blastocyst; polypeptide binding and response		25, 26
IGF-IIR	mRNA: 2- to 8-cell stage through blastocyst; polypeptide; in every cell		25, 26, 30
Insulin R	mRNA: 8-cell stage through blastocyst; protein; binding and response		25, 26
PDGF A R	mRNA: maternal and zygotic		25, 43
PDGF B R	mRNA: maternal and zygotic		25, 43
EGF R	mRNA: 2-cell stage through blastocyst; protein: trophectoderm response, rescue survival ICM, basolateral TE mouse, In human preimp. correlated with delayed implantation	Knockout postimplantation produces lethal in one background; ICM/ectodermal defects	13, 21, 46, 47, 49, 50–52
Activin R	Rescue of two-cell block; inferred from response		71, 72
CSF-1 R	mRNA: 2-cell stage through blastocyst		25
SCF/c-kit	mRNA: 2-cell stage through blastocyst		25
TGF-β1	Binding in blastocyst		60
TGF-β2	Binding in blastocyst		60
TGF-β3 betaglycan	Binding in blastocyst		60
GH	Receptor inferred from response in culture		92
LIF receptor	Human preimplantation embryo		59
IL-1	Regulates human TE invasion		75
IL-6	Negative		25

exogenous FGF-4. However, the addition of FGF-4 to ICMs, isolated by immunosurgery, results in an increase in the number of outgrowing parietal endoderm cells and in the amount of 92-kDa gelatinase that these cells secrete (6). Blastocysts cultured on fibronectin grow out more trophectoderm in the presence of FGF-4 (7).

In the mouse, FGF-4 null mice die *in vitro* or *in vivo* just after implantation (5). A dominant negative FGF receptor (dnFGFR) causes ICM and trophectoderm cells to cease growing at the fifth cell division, after compaction and cavitation but before hatching and implantation (7). These data suggest that FGF signaling (perhaps due to ICM-derived FGF-4) coordinates growth of embryonic ICM stem cells and stem cells for extraembryonic endoderm and trophectoderm. FGF does not have a mitogenic effect on intact embryos *in vitro,* and maternal FGF does not rescue FGF-4 null embryos *in vivo.* These data suggest that the embryo does not "hear" outside FGF. Therefore, the FGF signal is used by the embryo to coordinate its own development without maternal input.

Four high-affinity receptors ($KD = 10^{-11}\ M$) are expressed in the mouse (FGFR1-4). Transcripts for all of these are found in the blastocyst (7). From implantation through gastrulation, FGF receptors are widely expressed in all embryonic cells (6,25). Heparin sulfate proteoglycans expressed in the blastocyst, represented by the syndecans and perlecans, are also required for transduction of the mitogenic signal by the high-affinity FGF receptors (25,44).

FGFs may have other vital functions after the time of the lethal phenotype of the FGF-4 ablation of dnFGFR. A dnFGFR lacking the intracellular tyrosine kinase domain and forming nonfunctional ligand-dependent dimers with endogenous FGF receptors has been expressed in pregastrulation *Xenopus* embryos. These embryos have a deficit in posterior/ventral mesoderm, whereas overexpression of FGF results in an overabundance of ventral/posterior mesoderm (25). In the chick, where the endoderm can program the organization and/or determination of mesoderm at stage XIII pregastrulation embryos, FGF can mimic the effects of the endoderm (24,25). In the mouse, the patterning effects of FGF may be hard to detect due to the essential role of FGF earlier in development.

Summary These data suggest that ICM-derived FGF-4 may directly affect the growth of the ICM and extraembryonic ectoderm. ICM-derived FGF-4 may also indirectly affect the parietal endoderm by its action on ICM or primitive endoderm. Since the embryo does not "hear" maternal FGF, FGF mediates an embryonic checkpoint for implantation.

5.5.3. Epidermal Growth Factors

There are five related EGF ligands and four related EGF tyrosine kinase receptors. The ligands show variability in size, and the precursors are transmembrane molecules from 53–1216 amino acids long. There is great cross-reactivity between the four receptors and five ligands. TGF-α mRNA is expressed as a maternal mRNA and is reexpressed as zygotic mRNA at the eight-cell through blastocyst stage of the mouse, but EGF mRNA is not expressed (25,38). Crypto is expressed in the mouse blastocyst, but amphiregulin and EGF are not expressed in preimplantation mouse blastocyst (45). In human preimplantation embryos, EGF and TGF-α are co-expressed (46). In the mouse, TGF-α polypeptide is expressed in every cell of the blastocyst, although mainly in the ICM and trophectoderm (38,47). Binding studies and immunocytochemical techniques indicate that the four-cell-stage mouse embryo expresses EGF receptor, and EGF stimulates the synthesis of EGF receptor in early mouse embryos (48,49). The receptor is active after compaction since EGF rescues embryos growing in dilute culture and increases the rate of cavitation in compacted morulae (50,51). Antisense oligonucleotides or injected antisense mRNA specific for the EGF receptor decrease the rate of cavitation without affecting the growth rate of these morulae (25). It is likely that endogenous embryonic TGF-α and crypto, and HB-EGF in the uterine epithelium, are involved with the induction and maintenance of the epithelial nature of trophectoderm. A shift in the preponderance of the expression of EGF receptor from the apical to the basolateral surface of the trophectoderm indicates that uterine epithelial homologs and endogenous TGF-α/crypto may have timed phases of input on the functional state of the trophectoderm (47,50,52). However, EGF can also regulate the production of gelatinase B by trophoblast giant cells in blastocyst outgrowths (8,53). This suggests that these cells may be responding to maternal EGF.

A knockout of one of the four related EGF receptors has a peri-implantation lethal phenotype in one of three genetic backgrounds (21). After implantation, the first observable defect leading to lethality is in the ICM. Trophectoderm cells persist *in utero* after embryonic ICM cells die. The embryo contains TGF-α and crypto, and the uterine epithelium expresses HB-EGF in a blastocyst-dependent manner (8,38,45). However, a TGF-α knockout has no peri-implantation phenotype, so embryonic TGF-α is not essential. The distribution of cross-reacting ligands and receptors is so general that, from these data, it is not clear where or what the first defect in signal input is.

Summary EGF signaling can be essential for embryonic survival during implantation. The ICM requires EGF signaling, but maternal HB-EGF stimulation of trophectoderm, or stimulation by embryonic crypto or TGF-α, also may be required. EGF-related ligands also modulate cavitation and metabolic function in the preimplantation embryo.

5.5.4. Platelet-Derived Growth Factor

There are four related PDGF ligands: PDGF A, PDGF B, vascular endothelial growth factor/vascular permeability factor, and an uncharacterized factor that cross-reacts with PDGF (25). The PDGF A and B chains form homo- and heterodimers. The A isoform also has alternate splices producing slightly different proteins, though functional differences between the chains are not known (54).

There are also two receptors, the alpha and beta forms, which are tyrosine kinases activated allosterically to dimerize by dimeric ligand (25). These receptors have different affinities for the PDGF hetero- and homodimers (25).

The PDGF A chain mRNA is a maternal mRNA that is transcribed zygotically from the two-cell stage through the blastocyst stage. The PDGF A polypeptide is found in every cell of the blastocyst (38). The PDGF B chain is found in the human blastocyst (6) but appears later during mouse embryogenesis (43). The PDGF alpha receptor is expressed as a maternal mRNA as well as in the mouse blastocyst, but the PDGF beta receptor is expressed at lower levels (26,43). In other mammals, PDGF has different effects: for example, PDGF BB (B chain homodimer) relieves the developmental block at the eight-cell-stage embryo of cows but has no effects on the early rabbit embryo (25,55). In postimplantation development, mouse embryos respond to PDGF AA (A chain homodimer) with a rapid induction of fos mRNA (25), whereas the PDGF B chain homodimer induces a weaker response.

Summary These data suggest that PDGF has autocrine and paracrine roles in mammalian pregastrulation development. Knockouts in the PDGF A ligand and receptors (patched mutation) have no early phenotype during mouse development (25,56).

5.5.5. Leukemia-Inhibiting Factor, Interleukin-6, Colony Stimulating Factor-1, and Stem Cell Factor

These factors are discussed together for two reasons: 1) interleukin (IL)-6 and stem cell factor (SCF) can act to maintain embryonic stem cells in a mitotic undifferentiated state; 2) LIF and CSF-1 are likely to act in a maternal-to-embryo direction.

LIF and IL-6 transcripts are expressed in the preimplantation mouse embryo (57,58). The receptors for these factors share a common signal transduction molecule: gp130. The mouse blastocyst lacks an IL-6 receptor (25), but both gp130 and a receptor for LIF exist in the blastocyst. Mice targeted for LIF do not support implantation via a maternal defect, although null LIF blastocysts can implant and produce pups in normal mothers (2). Also, a null mutation for the receptor leads to maternal infertility (19). These data suggest that LIF and its receptor mediate a maternal checkpoint for the start of pregnancy. This contrasts with FGF-4, which mediates an embryonic checkpoint for the start of pregnancy that operates independently of maternal FGF. Interestingly, overexpression of LIF results in overproliferation of ectoderm followed by degeneration but a lack of mesoderm induction (58). LIF and its receptor are expressed in human endometrium, and LIF receptor is expressed in the embryo during implantation (59). Other gain-of-function studies show that LIF can affect preimplantation growth and implantation processes in the embryo itself (53). This may occur through an indirect effect on extraembryonic ectoderm.

CSF-1 transcripts are not found at any stage of the preimplantation mouse embryo, although the CSF-1 receptor is present (25). Both LIF and CSF-1 are present at low levels in the mouse uterus during the 4.5 days between ovulation and implantation, but there is a surge of LIF, CSF-1, and TGF-α (as well as TGF-β-1,2 and IGF-I) transcription and translation, with the estrogen spike that occurs concurrently with implantation (25,60–62). Addition of CSF-1 to preimplantation culture causes an increase in embryonic progression and number of trophectoderm cells (63). CSF-1 null embryos implant normally. These data suggest that CSF-1 can act in a maternal-to-embryo direction that is redundant with other factors.

Only uterine epithelial expression of LIF seems essential for implantation, as TGF-α null mice and the op/op CSF-1 null mothers both accommodate implanting blastocysts (2,25). A fourth growth factor, SCF, is also found in the postovulation oviduct and uterine epithelium and the transcript for the receptor, c-kit, is found in the preimplantation mouse embryo (25), but CSF receptor/c-kit null mice do implant.

Summary LIF was the first growth factor identified to be essential in the pregastrulation development of mice. LIF is required for maternal fertility, but gain-of-function studies show that LIF also plays a role in embryonic development. CSF-1 is expressed on uterine epithelium and presumably interacts with trophectodermal CSF-1 receptor/c-fms to promote growth; however, CSF-1 is not essential for any pregastrulation events in laboratory mice.

The SCF receptor is expressed in the blastocyst and SCF is expressed in the uterus, but the three effects of the SCF ligand or receptor null mutants are on lineages functioning after gastrulation. IL-6 is synthesized in the preimplantation embryo, but there is no transcript for its high-affinity receptor, so it may be having an effect on maternal tissue. If this effect is true, then IL-6 resembles another tau-interferon that mediates the recognition of pregnancy in other mammalian species (12). Although IL-6 mediates stem cell proliferation, no such function has been demonstrated in the early embryo.

5.5.6. Transforming Growth Factor-β/Activins/VG-1-Related Growth Factor

Transforming growth factor (TGF)-β and activins are related molecules that form homo- and heterodimers of cysteine disulfide-bonded polypeptides. The nascent polypeptide is proteolytically cleaved, and the conserved C terminus of about 110–114 amino acids dimerizes into a latent form that can be purified in presence of a latent TGF-β binding protein. The activation of TGF-β may be a means for localizing the effects of the ligand (25). There are three mammalian TGF-β (TGF-β1, -β2, and -β3) and three related activins and inhibin (activin βA, activin βB, and inhibin α). All TGF-β ligands, activins, and inhibins have been knocked out without a pregastrulation phenotype (19,64–66; and references cited therein), suggesting redundancy or lack of critical function. Preimplantation mouse embryos synthesize TGF-β1 mRNA as a zygotic transcript, and the polypeptide is found in most cells of the blastocyst (38). TGF-β2 and TGF-β3 are also found in the preimplantation mouse embryo. TGF-β2 is initially found at the four-cell stage, and the polypeptide is localized to the ICM and then to the trophectoderm. Later, it is localized within the visceral endoderm (67). TGF-β3 mRNA has been reported in the blastocyst (25). TGF-β1 aids in the rescue of sparsely cultured preimplantation mouse embryos and is mitogenic for these embryos (51,68).

Activin is related to TGF-β (25). Both activin βA and activin βB, as well as inhibin α chain mRNA, are expressed in the unfertilized mouse oocyte. After loss of these mRNA during the degradation of maternal mRNA at the two-cell stage, zygotic βA and βB, but not inhibin α chain mRNA, are found again at the blastocyst stage. The polypeptide is found in the ICM of the blastocyst but only in extraembryonic ectoderm 1 day later (69,70). Activins are expressed only in the decidua in the early postimplantation stages (71). During culture of preimplantation mouse embryos, activin can alleviate the two-cell block. This suggests that the receptor is functioning early and that the endogenous ligand may be having autocrine effects (72).

There are three TGF-β receptors. The type I receptor facilitates the enzymatic activity of the type II receptor. The third receptor, called *betaglycan,* does not mediate the TGF-β signal but may facilitate the binding of TGF-β to the first two receptors. TGF-β receptor, like the activin receptor, is a serine/threonine kinase (25). All three receptors are expressed in the preimplantation mouse embryos at the blastocyst stage (60).

A dominant negative activin receptor, expressed in pregastrulation *Xenopus* embryos, results in loss of induction of all mesoderm (25). It is likely that activin is the inductive signal of Spemann's organizer, but Vg-1 is likely to mediate the pregastrulation induction of Nieuwkoop's center in *Xenopus* (24). During the development of the chick, FGF and activin appear to be important in the induction and patterning of the mesoderm at gastrulation, when FGF and activin are expressed in the inductive endoderm of the stage XIII embryo (25).

Summary It seems clear from gain-of-function and expression studies that TGF-β and activin can have a role in preimplantation mouse development. Since activins are not expressed in the postimplantation embryo before gastrulation, their role in mouse gastrulation may be different than in the *Xenopus*. Loss-of-function studies show that all family members have redundant functions.

5.5.7. Parathyroid Hormone-Related Peptide

Parathyroid hormone-related peptides (PtHRP) is first made in trophectoderm and then in the primitive endoderm at the time of implantation. The receptor of PtHRP is strongly expressed in the parietal endoderm (73). When added to the ICM outgrowth cultures, PtHRP facilitates the outgrowth of visceral and especially parietal endoderm and makes parietal endoderm more varied in the types of extracellular matrix molecules it can adhere to and migrate on. PtHRP does this by stimulating receptors in the endoderm that, in turn, stimulate the expression of a greater variety of integrins in endoderm (73,74). PtHRP is thought to mediate some of the secreted stimulus of endoderm that conditioned media from tropectoderm are known to do. However, this is not a required pathway, as both the PtHRP and PtHRP receptor knockouts have no phenotype in the peri-implantation embryo (19).

5.5.8. Other Growth Factors on the Horizon

Several growth factors and hormones may have effects on peri-implantation mouse and human embryos. IL-1 stimulates trophoblast outgrowth in cultured first trimester human placenta (75). Placenta lactogen (PL)-1 is a major secreted product of postmitotic trophoblast giant cells in blastocysts through late gastrulation and in blastocyst outgrowths (76,77 and references cited therein). Hepatocyte growth factor/scatter factor can influence the growth of extraembryonic ectoderm at implantation, but lethality due to placental failure occurs only at E12 and later (78–81).

5.6. SUMMARY

Growth factors are essential for embryonic life before and after implantation. Specific factors have been shown to mediate required maternal (LIF) and embryonic (FGF-4) checkpoints for implantation and growth. In one genetic background, the peri-implantation embryo requires EGF signaling, although the source of and the direct requirement of EGF have not been shown. Other growth factors can mediate cavitation (EGF), trophectoderm growth (FGF, CSF-1), endoderm growth and migration (FGF-4, P+HRP), and ICM growth (FGF and EGF signal input). Some growth factors are likely to mediate induction of the ectodermal pattern by the endoderm.

ACKNOWLEDGMENT

I thank Crystal Beecher for editorial assistance. This research is supported by grants from the National Institutes of Health (NICHD HD30739-02), the Feinberg Cardiovascular Research Institute at Northwestern University Medical School, the American Heart Association (96014080), and the Markey Charitable Trust.

REFERENCES

1. Cross JC, Werb Z, Fisher SJ: Implantation and the placenta: Key pieces to the development puzzle. Science 266:1508–1518, 1994.
2. Stewart CL, Kaspar P, Brunet LJ, Bhatt H, Gadi I, Kontgen F, Abbondanzo SJ: Blastocyte implantation depends on maternal expression of leukaemia inhibitory factor. Nature 359:76–79, 1992.
3. Stewart CL: Leukaemia inhibitory factor and the regulation of pre-implantation development of the mammalian embryo. Mol Reprod Dev 39:233–238, 1994.
4. Yoshida K, Taga T, Saito M, Suematsu S, Kumanogoh A, Tanaka T, Fujiwara H, Hirata M, Yamagami T, Nakahata T, Hirabayashi T, Yoneda Y, Tanaka K, Wang WZ, Mori C, Shiota K, Yoshida N, Kishimoto T: Targeted disruption of gp130, a common signal transducer for the interleukin-6 family of cytokines, leads to myocardial and hematological disorders. Proc Natl Acad Sci USA 93:407–411, 1996.
5. Feldman B, Poueymirou W, Papaioannou VE, DeChiara TM, Goldfarb M: Requirement of FGF-4 for postimplantation mouse development. Science 267:246–249, 1995.
6. Rappolee DA, Basilico CA, Patel Y, Werb Z: Expression and function of FGF-4 in peri-implantation development of mouse embryos. Development 120:2259–2269, 1994.
7. Chai N, Patel Y, Jacobson K, McMahon A, Rappolee DA: Fibroblast growth factor is essential for growth of placental and embryonic stem cells at the fifth cell division in the mouse. Submitted to Developmental Biology.

8. Das SK, Wang XN, Paria BC, Damm D, Abraham JA, Klagsbrun M, Andrews GK, Dey SK: Heparin-binding EGF-like growth factor gene is induced in the mouse uterus temporally by the blastocyst solely at the site of its apposition: A possible ligand for interaction with blastocyst EGF-receptor in implantation. Development 120:1071–1083, 1994.
9. Das SK, Chakraborty I, Paria BC, Wang XN, Plowman G, Dey SK: Amphiregulin is an implantation-specific and progesterone-regulated gene in the mouse uterus. Mol Endocrinol 9(6):691–705, 1995.
10. Carlone DL, Rider V: Embryonic modulation of basic fibroblast growth factor in the rat uterus. Biol Reprod 49:653–665, 1993.
11. Imakawa K, Anthony RV, Kazemi M, Marotti KR, Polites HG, Roberts RM: Interferon-like sequence of ovine trophoblast protein secreted by embryonic trophectoderm. Nature 330:377–379, 1987.
12. Roberts RM, Leaman DW, Cross JC: Role of interferons in maternal recognition of pregnancy in ruminants. Proc Soc Exp Biol Med 200:7–18, 1992.
13. Paria BC, Das SK, Andrews GK, Dey SK: Expression of the epidermal growth factor receptor gene is regulated in mouse blastocysts during delayed implantation. PNAS 90:55–59, 1993.
14. Larue L, Ohsugi J, Hirchenhain J, Kemler R: E-cadherin mutant embryos fail to form a trophectoderm epithelium. Proc Natl Acad Sci USA 91:8263–8267, 1994.
15. Spyropoulos DD, Capecchi M: Targeted disruption of the *even-skipped* gene, *evx-1*, causes early postimplantation lethality of the mouse conceptus. Genes Dev. 8:1949–1961, 1994.
16. Chen WS, Manova K, Weinstein DC, Duncan SA, Plump AS, Prezioso VR, Bachvarova RF, Darnell JE Jr: Disruption of the HNF-4 gene, expressed in visceral endoderm, leads to cell death in embryonic ectoderm and impaired gastrulation of mouse embryos. Genes Dev 8:2466–2477, 1994.
17. Lee JJ, Radice G, Perkins CP, Costantini F: Identification and characterization of a novel, evolutionarily conserved gene disrupted by the murine Hβ-58 embryonic lethal transgene insertion. Development 115:277–288, 1992.
18. Rappolee DA: What knockouts tell us about early mouse development. Submitted to Mol Reprod Dev, 1992.
19. Copp AJ: Death before birth: Clues from gene knockouts and mutations. Trends Genet 11:87–93, 1995.
20. Jones SN, Roe AE, Donehower LA, Bradley A: Rescue of embryonic lethality in Mdm-2-deficient mice by absence of p53. Nature 378:206–208, 1996.
21. Threadgill DW, Dlugosz AA, Hansen LA, Tennenbaum T, Lichti U, LaMantia C, Mourton T, Herrup K, Harris RC, Barnard JA, Yupsa SH, Coffey RJ, Magnuson T: Targeted disruption of mouse EGF receptor: Effect of genetic background on mutant phenotype. Science 269:230–234, 1995.
22. Sibilia M, Wagner EF: Strain-dependent epithelial defects in mice lacking the EGF receptor. Science 269:234–238, 1995.
23. Stephens LE, Sutherland AE, Klimanskaya IV, Andrieux A, Meneses J, Pedersen RA, Damsky CH: Deletion of β1 integrins in mice results in inner cell mass failure and peri-implantation lethality. Genes Dev 9:1883–1895, 1995.
24. Slack JMW: From Egg to Embryo: Regional Specification in Early Development, ed. 2. Cambridge: Cambridge University Press, 1991.
25. Rappolee DA, Werb Z: The role of growth factors in early mammalian development. Adv Dev Biol 4:41–70, 1994.

26. Rappolee DA, Schultz GA, Pedersen RA, Sturm KA, Werb Z: IGF-II acting through the IGF-I receptor forms an endogenous growth circuit regulated by imprinting in early mouse embryos. Genes Dev 6:639–652, 1992.
27. Telford NA, Hogan A, Franz CR, Schultz GA: Expression of genes for insulin and insulin-like growth factors in the early post-implantation period. Mol Reprod Dev 27: 81–92, 1992.
28. Doherty AS, Temeles GL, Schultz RM: Temporal pattern of IGF-I expression during mouse preimplantation embryogenesis. Mol Reprod Dev 37:21–26, 1994.
29. Harvey MB, Kaye PL: Insulin increase the cell number of the ICM and stimulates morphological development of the mouse blastocyst in vitro. Development 110:963–967, 1990.
30. Harvey MB, Kaye PL: Mediation of the actions of insulin and insulin-like growth factor-1 on preimplantation mouse embryos in vitro. Mol Reprod Dev 33:270–275, 1992.
31. Dunglison GF, Kaye PL: Insulin regulates protein metabolism in mouse blastocysts. Mol Reprod Dev 36(1):42–48, 1993.
32. Heyner S, Rao LV, Jarrett L, Smith RM: Preimplantation mouse embryos internalize maternal insulin via receptor-mediated endocytosis: Pattern of uptake and functional correlations. Dev Biol 145:48–58, 1989.
33. Smith RM, Garside WT, Aghayan M, Shi CZ, Shah N, Jarett L, Heyner S: Mouse preimplantation embryos exhibit receptor-mediated binding and transcytosis of maternal insulin-like growth factor I. Biol Reprod 49:1–12, 1993.
34. Newman-Smith ED, Werb Z: Stem cell defects in parthenogenetic peri-implantation embryos. Development 121:2069–2077, 1995.
35. Hahnel A, Schultz GA: Insulin-like growth factor binding proteins are transcribed by preimplantation mouse embryos. Endocrinology 134:1956–1959, 1994.
36. Puscheck EE, Dreschler J, Patel Y, Rappolee DA: Expression of insulin receptor substrate-1 in the mouse blastocyst. Submitted to Mol Reprod Dev, 1997.
37. Basilico C, Moscatelli D: The FGF family of growth factors and oncogenes. Adv Cancer Res 59:115–165, 1992.
38. Rappolee DA, Brenner CA, Schultz C, Mark D, Werb Z: Developmental expression of PDGF, TGF-alpha, and TGF-beta genes in preimplantation mouse embryos. Science 241:1823–1825, 1988.
39. Niswander L, Martin GR: Fgf-4 expression during gastrulation, myogenesis, limb and tooth development in the mouse. Development 114:755–768, 1992.
40. Yuan H, Corbi N, Basilico C, Dailey L: Developmental-specific activity of the FGF-4 enhancer requires the synergistic action of Sox2 and Oct-3. Genes Dev 9:2635–2645, 1995.
41. Scholer HR, Balling R, Hatzopoulos AK, Suzuki N, Gruss P: Octamer binding proteins confer transcriptional activity in early mouse embryogenesis. EMBO J 8:2551–2557, 1989.
42. Scholer HR, Hatzopoulos AK, Balling R, Suzuki N, Gruss P: A family of octamer-specific proteins present during mouse embryogenesis: Evidence for germline-specific expression of an Oct factor. EMBO J 8:2543–2550, 1989.
43. Palmieri SL, Payne J, Stiles CD, Biggers JD, Mercola M: Expression of PDGF-A and PDGF α receptor during pre- and post-implantation development: Evidence for a developmental shift from an autocrine to a paracrine mode of action. Mech Dev 39:181–191, 1992.

44. Carson DD, Tang JP, Julian J: Heparan sulfate proteoglycan (perlecan) expression of mouse embryos during acquisition of attachment competence. Dev Biol 155(1):97–106, 1993.
45. Johnson SE, Rothstein JL, Knowles BB: Expression of epidermal growth factor family gene members in early mouse development. Dev Dyn 201:216–226, 1994.
46. Chia CM, Winston RM, Handyside AH: EGF, TGF-alpha and EGFR expression in human preimplantation embryos. Development 121(2):299–307, 1995.
47. Dardik A, Smith RM, Schultz RM: Colocalization of transforming growth factor-alpha and a functional epidermal growth factor receptor (EGFR) to the inner cell mass and preferential localization of the EGFR on the basolateral surface of the trophectoderm in the mouse blastocyst. Dev Biol 154:396–409, 1992.
48. Adamson ED: Activities of growth factors in preimplantation embryos. J Cell Biochem 53:280–287, 1993.
49. Wiley LM, Wu JX, Haran I, Adamson ED: Epidermal growth factor receptor mRNA and protein increases after the four-cell preimplantation in murine development. Dev Biol 149:247–260, 1992.
50. Dardik A, Schultz RM: Blastocoel expansion in the preimplantation mouse embryo: Stimulatory effect of TGF-α and EGF. Development 113:919–930, 1991.
51. Paria BC, Dey SK: Preimplantation embryo development in vitro: Cooperative interaction among embryos and role of growth factors. PNAS 87:4756–4760, 1990.
52. Dardik A, Doherty AS, Schultz RM: Protein secretion by the mouse blastocyst: Stimulatory effect on secretion into the blastocoel by transforming growth factor-alpha. Mol Reprod Dev 34:396–401, 1993.
53. Harvey MB, Leco KJ, Arcellana-Panlilio MY, Zhang X, Edwards DR, Schultz GA: Proteinase expression in early mouse embryos is regulated by leukaemia inhibitory factor and epidermal growth factor. Development 121:1005–1014, 1995.
54. Mercola M, Wang CY, Kelly J, Brownlee C, Judine-Gruyler C, Stilcs CD, Bowen-Pope D. Selective expression of PDGF A and its receptor during early mouse embryogenesis. Dev Biol 138:114–122, 1990.
55. Larson RC, Ignotz GG, Currie WB: Platelet-derived growth factor (PDGF) stimulates development of bovine embryos during the fourth cell cycle. Development 115:821–826, 1992.
56. Bostrom H, Willetts K, Pekny M, Leveen P, Lindahl P, Hedstrand H, Pekna M, Hellstrom M, Gebre-Medhin S, Schalling M, Nilsson M, Kurland S, Tornell J, Heath JK, Betsholtz C: PGFD A signaling is a critical event in lung alveolar myofibroblast development and alveogenesis. Cell 85:863–874, 1996.
57. Murray R, Lee V, Chiu CP: The genes for leukemia inhibitory factor and interleukin-6 are expressed in mouse blastocysts prior to the onset of hemopoiesis. Mol Cell Biol 10:4953–4956, 1990.
58. Conquet F, Peyrieras N, Tivet L, Brulet P: Inhibited gastrulation in mouse embryo overexpressing the leukemia inhibitory factor. PNAS 89:8195–9, 1992.
59. Charnock-Jones DS, Sharkey AM, Fenwick P, Smith SK: Leukaemia inhibitory factor mRNA concentration peaks in human endometrium at the time of implantation and the blastocyst contains mRNA for the receptor at this time. J Reprod Fertil 101:421–426, 1994.
60. Paria BC, Jones K, Flanders K, Dey SK: Localization and binding of transforming growth factor-beta isoforms in mouse preimplantation embryos and in delayed and activated blastocysts. Dev Biol 151:91–104, 1992.

61. Das SK, Flanders KC, Andrews GK, Dey SK: Expression of transforming growth factor-beta isoforms (beta 2 and beta 3) in the mouse uterus: Analysis of the peri-implantation period and the effects of steroids. Endocrinology 130:3459–3466, 1992.
62. Tamada H, Das SK, Andrews GK, Dey SK: Cell type specific expression of transforming growth factor-alpha in the mouse uterus during the peri-implantation period. Biol Reprod 45:365–372, 1991.
63. Bhatnagar P, Papaioannou VE, Biggers JD: CSF-1 and mouse preimplantation development in vitro. Development 121:1333–1339, 1995.
64. Shull MM, Doetschman T: Transforming growth factor-beta in reproduction and development. Mol Reprod Dev 39:239–246, 1994.
65. Matzuk MM, Kumar TR, Vassalli A, Bickenbach JR, Roop DR, Jaenisch R, Bradley A: Functional analysis of activins during mammalian development. Nature 374:354–356, 1995.
66. Matzuk MM, Kumar TR, Bradley A: Different phenotypes for mice deficient in either activins or activin receptor type II. Nature 374:356–360, 1995.
67. Slager HG, Lawson KA, Van Den Eijnden-Van Raaij AJ, De Laat SW, Mummery CL: Differential localization of TGF-β2 in mouse preimplantation and early postimplantation development. Dev Biol 145:205–218, 1991.
68. Lim J, Bongso A, Ratnam S: Mitogenic and cytogenetic evaluation of transforming growth factor-beta on murine preimplantation embryonic development in vitro. Mol Reprod Dev 36:482–487,1993.
69. Albano RM, Groome N, Smith JC: Activins are expressed in preimplantation mouse embryos and in ES and EC cells and are regulated on their differentiation. Development 117(2):711–723, 1993.
70. Paulusma CC, Van der Kruijssen CM, Van den Eijnden-Van Raaij AJ: Localization of activin subunits in early murine development determined by subunit-specific antibodies. J Immunol Meth 169:143–152, 1994.
71. Manova K, Paynton RV, Bachvarova RF: Expression of activins and TGF-beta 1 and beta 2 mRNAs in early postimplantation mouse embryos and uterine decidua. Mech Dev 36:141–151, 1992.
72. Lu R, Shiota K, Toyoda Y, Takahashi M: Activin A (EDF) releases the two cell block of mouse embryos in culture. Jpn J Anim Reprod 36:127–131, 1990.
73. Van de Stolpe A, Karperien M, Lowik CWGM, Juppner J, Segre GV, Abou-Samra A, de Laat SW, Defize LHK: Parathyroid hormone-related peptide as an endogenous inducer of parietal endoderm differentiation. J Cell Biol 120:235–243, 1993.
74. Behrendtsen O, Alexander CM, Werb Z: Cooperative interactions between extracellular matrix, integrins and parathyroid hormone-related peptide regulate parietal endoderm differentiation in mouse embryos. Development 121:4137–4148, 1995.
75. Librach CL, Feigenbaum SL, Bass KE, Cui TY, Verastas N, Sadovsky Y, Quigley JP, French DL, Fisher SJ: Interleukin-1 beta regulates human cytotrophoblast metalloproteinase activity and invasion in vitro. J Biol Chem 269:17125–17131, 1994.
76. Carney EW, Prideaux V, Lye SJ, Rossant J: Progressive expression of trophoblast-specific genes during formation of mouse trophoblast giant cells in vitro. Mol Reprod Dev 34:357–368, 1993.
77. Nieder GL, Nagy F: Initiation of placental lactogen-1 production by blastocysts. J Exp Zool 260:247–257, 1989.

78. Uehara Y, Minowa O, Mori C, Shiota K, Kuno J, Noda T, Kitamura N: Placental defect and embryonic lethality in mice lacking hepatocyte growth factor/scatter factor. Nature 373:702–705, 1995.

79. Bladt F, Riethmacher D, Isenmann S, Aguzzi A, Birchmeier C: Essential role for the c-met receptor in the migration of myogenic precursor cells into the limb bud. Nature 376:768–771, 1995.

80. Rappolee DA, Iyer A, Patel Y: A role for hepatocyte growth factor/scatter factor during implantation and early placental development of the mouse. Submitted to Mol Reprod Dev.

81. Bass KE, Morrish D, Roth I, Bhardwaj D, Taylor R, Zhou Y, Fisher SJ: Human cytotrophoblast invasion is up-regulated by epidermal growth factor: Evidence that paracrine factors modify this process. Dev Biol 164:550–561, 1994.

82. Morita Y, Tsutsumi O, Taketani Y: In vitro treatment of embryos with epidermal growth factor improves viability and increases the implantation rate of blastocysts transferred to recipient mice. Am J Obstet Gynecol 171:406–409, 1994.

83. Vaughan TJ, James PS, Pascall JC, Brown KD: Expression of genes for transforming growth factor-alpha, epidermal growth factor and EGF receptor during early pig development. Development 116:663–669, 1992.

84. Babalola GO, Schultz RM: Modulation of gene expression in the preimplantation mouse embryo by TGF-alpha and TGF-beta. Mol Reprod Dev 41:133–139, 1995.

85. Campbell WJ, Miller KA, Anderson TM, Shull JD, Rizzino A: Expression of fibroblast growth factor receptors by embryonal carcinoma cells and early mouse embryos. In Vitro Cell Dev Biol 28A:61–62, 1992.

86. Rothstein J, Johnson D, Deloia JA, Skowronski J, Solter D, Knowles B: Gene expression during preimplantation mouse development. Genes Dev 6:1190–1201, 1992.

87. Arceci RJ, Pampfer S, Pollard JW: Expression of CSF-1/c-fms and SF/c-kit mRNA during preimplantation mouse development. Dev Biol 151:1–8, 1992.

88. Schuller AG, van Neck JW, Lindenbergh-Kortleve DJ, Groffen C, de Jong I, Zwarthoff EC, Drop SL: Gene expression of the IGF binding proteins during post-implantation embryogenesis of the mouse; comparison with the expression of IGF-I and -II and their receptors in rodent and human. Adv Exp Med Biol 343:267–277, 1993.

89. Hebert JM, Basilico C, Goldfarb M, Haub O, Martin G: Isolation of cDNAs encoding four mouse FGF family members and characterization of their expression patterns during embryogenesis. Dev Biol 138:454–463, 1990.

90. Wilkinson DG, Peters G, Dickson C, McMahon AP: Expression of the FGF-related proto-oncogene int-2 during gastrulation and neurulation in the mouse. EMBO J 7:691–695, 1988.

91. Pampfer S, Tabibsadeh S, Chuan FS, Pollard J: Expression of colony stimulating factor (CSF)-1 messenger RNA in human endometrial glands during the menstrual cycle: Molecular cloning of a novel transcript that predicts a cell surface form of CSF-1. Mol Endocrinol 5:1931–1938, 1992.

92. Drakakis P, Loutradis D, Milingos S, Michalas S, Kallianidis K, Bletsa R, Aravantinos D, Kiessling AA: A preliminary study of the effect of growth hormone on mouse preimplantation embryo development in vitro. Gynecol Obstet Invest 40:222–226, 1995.

93. Schuller AG, van Neck JW, Lindenbergh-Kortleve DJ, Groffen C, de Jong I, Zwarthoff EC, Drop SL: Gene expression of the IGF binding proteins during post-implantation em-

bryogenesis of the mouse; comparison with the expression of IGF-I and -II and their receptors in rodent and human. Adv Exp Med Biol 343:267–277, 1993.

94. Yamaguchi TP, Harpal K, Henkemeyer M, Rossant J: fgfr-1 is required for embryonic growth and mesodermal patterning during mouse gastrulation. Genes and Dev 8: 3032–3044, 1994.

95. Orr-Urtreger A, Givol D, Yayon A, Yarden Y, Lonai P: Developmental expression of 2 murine FGF receptors flg and bek. Development 113:1419–1434, 1991.

6

TRANSCRIPTIONAL REGULATION IN AN *IN VITRO* MODEL SYSTEM FOR MAMMALIAN EMBRYOGENESIS

ANGIE RIZZINO

Eppley Institute for Research in Cancer and Allied Diseases, University of Nebraska Medical Center, Omaha, NE 68198-6805

6.1. INTRODUCTION

Due to a concerted effort during the past 15 years, our understanding of the molecular mechanisms that control transcription of eukaryotic genes has increased considerably. However, far less is known about the mechanisms by which differentiation regulates batteries of genes that specify functional differences between various cell types. Armed with a general understanding of gene transcription, we now understand how differentiation regulates specific genes, in particular genes that are differentially regulated during mammalian embryogenesis. Understanding how differentiation alters the transcription of specific genes will enable us to learn from nature one of the most important lessons about the regulation of any gene, namely, how a gene is turned on and turned off. In addition to demonstrating how gene expression is regulated temporally and spatially during development, the lessons learned will help in the design of clinical therapies that regulate gene expression with minimal side effects. In this connection, two of the largest challenges in gene replacement therapy are to determine how to deliver a gene of interest and how to ensure that the gene is expressed at the appropriate level in the correct tissue. Unless we understand fully how genes are regulated, we may learn how to deliver genes, but we may not be able to control their expression adequately.

Hormones and Growth Factors in Development and Neoplasia, Edited by Robert B. Dickson and David S. Salomon
ISBN 0-471-16899-8

Until the late 1980s, it was difficult to monitor the expression of individual genes during embryogenesis due to the small size of the mammalian embryo. With the development of the polymerase chain reaction (PCR) and the coupling of PCR with reverse transcription (RT), it became possible to monitor the expression of virtually any cloned mouse gene at the mRNA level. RT-PCR was used in mouse embryos initially to monitor the expression of mRNA for several growth factors and several growth factor receptors. These studies demonstrated that a number of growth factor and growth factor receptor genes, including fibroblast growth factor-4 (FGF-4), transforming growth factor-α (TGF-α), platelet-derived growth factor-A (PDGF-A), TGF-β1, and receptors for insulin, IGF-I and IGF-II, are expressed before the morula stage (1,2). More recently, *in situ* hybridization has been used to monitor the expression of specific genes at the blastocyst stage, in particular by the cells of the inner cell mass (ICM), which contain the totipotent cells of the early embryo (3).

Given the small size of the mammalian embryo, many investigators interested in transcriptional regulation during early development turned to the use of embryonal carcinoma (EC) cells. These cells are used widely because they resemble closely, both morphologically and biochemically, the cells of the ICM (reviewed in ref. 4). Moreover, EC cells appear to express the same growth factors and growth factor receptors expressed by mouse embryos at the blastocyst stage, including FGF-4, PDGF-A, TGF-β1, TGF-α, and receptors for insulin, IGF-I and IGF-II (reviewed in ref. 5). In addition, several growth factor and growth factor receptor genes have been shown to be regulated at the transcriptional level when EC cells undergo differentiation, including FGF-4 (6–8), TGF-β2 (9), and the type II TGF-β-receptor (10).

This review focuses on the transcriptional regulation of the FGF-4 gene in EC cells, because FGF-4 appears to be the first FGF expressed during mammalian development (2) and because FGF-4 has been shown by gene targeting to be essential for development beyond the stage of implantation (11). Furthermore, the FGF-4 gene contains several interesting structural features, including a powerful distal enhancer located downstream of the transcriptional start site in the untranslated region of its last exon. Equally important, the expression and regulation of the FGF-4 gene in EC cells parallels closely the expression of the FGF-4 gene by the ICM and its early cellular derivatives.

6.2. EXPRESSION OF FIBROBLAST GROWTH FACTORS BY EARLY MOUSE EMBRYOS AND EC CELLS

Thus far, 14 members of the FGF family have been identified. Several are closely related and appear to be functionally equivalent in many biologic systems (reviewed in ref. 12), whereas others exhibit relatively low sequence similarity, and their receptor specificity and biologic effects have not been identified (13,14). The expression patterns of several FGFs during mammalian embryogenesis have been investigated in depth, but the expression patterns of the newer members of the FGF family during early development have not been examined. FGF-4 has been detected in mouse embryos as early as the four-cell stage by RT-PCR (2), whereas FGF-2 (15), FGF-5 (16), and FGF-3 (17) are expressed at slightly later developmental stages. At the late blastocyst stage, *in situ* hybridization revealed that *FGF-4* is expressed by the ICM but not by the trophectoderm (3). In addition, visceral extraembryonic endoderm and parietal extraembryonic endoderm, which are derived from the ICM, express *FGF-5* and *FGF-3*, respectively, but not *FGF-4* (16,17).

Several years after the initial report of FGF expression by EC cells (18), retinoic acid was shown to induce the differentiation of EC cells into parietal extraembryonic endoderm-like cells and to repress the expression of FGF activity (19). With the cloning of the FGF-4 gene and its cDNA, it soon became apparent that EC cells express FGF-4, whereas FGF-4 mRNA and protein expression is repressed when EC cells are induced to differentiate into cells that exhibit the properties of parietal extraembryonic endoderm-like cells (6–8). Other studies using FGF-4 neutralizing antibodies have shown that EC cells appear to express primarily FGF-4 (20). If other FGFs are expressed, they are produced at very low levels.

6.3. TRANSCRIPTIONAL REGULATION OF THE FGF-4 GENE

The structural portion of the FGF-4 gene is relatively small by comparison to many eukaryotic genes. The FGF-4 gene consists of three exons and two introns in a region of approximately 4 kb. In addition to a short 5′ untranslated region (UTR) in exon 1, exon 3 contains a typically sized 3′ UTR of approximately 1.3 kb. Thus far, approximately 11 kb of the mouse FGF-4 gene and flanking regions have been sequenced, including 4 kb upstream of the transcription start site and 2.8 kb downstream of the last exon (21–24). Hence, unless long-range regulatory elements are involved directly in the expression of the FGF-4 gene, the entire gene has been sequenced.

Run-on experiments were employed initially to demonstrate that *FGF-4* expression is repressed, at least in part if not entirely, at the transcriptional level when EC cells undergo differentiation to parietal extraembryonic endoderm-like cells (7). Strong support for this conclusion was provided by the finding that *FGF-4* promoter/receptor gene constructs are expressed actively in EC cells but not in their differentiated cells (23,25). Early work with *FGF-4* promoter/receptor gene constructs also identified another important feature of the FGF-4 gene. Its expression at the transcriptional level is dependent on cis-regulatory elements in two regions of the gene, namely, several cis-regulatory elements in the 5′ flanking region of the gene and the powerful distal enhancer located approximately 3 kb downstream of the transcription start site in the untranslated region of the last exon of the gene (Table 6.1).

The presence of important cis-regulatory elements in the promoter region of the FGF-4 gene plus the presence of the essential distal enhancer element 3 kb downstream of the transcription start site raise an important question: How do the two regulatory regions function together and contribute to the transcriptional expression of the FGF-4 gene? Several years ago, a looping model was proposed to explain the expression of the FGF-4 gene in EC cells (26). This model proposes that transcrip-

TABLE 6.1. Expression of FGF-4 Promoter/Reporter Gene Constructs in EC Cells

FGF-4 Promoter/Reporter Construct[1]	Activity[2]
2112	1
2112 + E	30- to 100-fold higher than 2112
2112 + O	7-fold higher than 2112
267 + E	5-fold higher than 2112 + E
61 + E	5-fold lower than 267 + E

[1]Design and construction of the *FGF-4* promoter/reporter gene constructs have been described (23). Construct 2112 contains the promoter region of the mouse *FGF-4* gene inserted upstream of a promoterless CAT reporter cassette. The *FGF-4* insert in construct 2112 contains the region of the *FGF-4* gene that extends from -2112 bp to +63 relative to the transcription start site. Construct 2112 + E was derived by inserting the *FGF-4* enhancer region (316 bp) downstream of the CAT reporter cassette in construct 2112. Construct 2112 + O was derived by inserting a 48-bp sequence, which contains the octamer motif and 20 bp on each side of the motif, downstream of the CAT reporter cassette in construct 2112. Construct 267 + E contains an insert from the *FGF-4* gene located just upstream of the CAT reporter construct that extends from −267 to +63 relative to the transcription start site. Construct 61 + E contains an insert upstream of the CAT reporter construct that extends from −61 to +63 relative to the transcription start site. Constructs 267 + E and 61 + E both contain the 316-bp *FGF-4* enhancer fragment inserted downstream of the CAT reporter cassette.
[2]Modified from the studies reported in ref. 23.

tion factors that bind to cis-regulatory elements in the promoter region of the gene interact with transcription factors that bind to cis-regulatory elements located in the downstream enhancer (Fig. 6.1). This model is updated and described in further detail in this review. This interesting feature of the FGF-4 gene makes it an excellent candidate for understanding how distal regulatory regions influence gene transcription in general.

6.4. CIS-REGULATORY ELEMENTS LOCATED IN THE PROMOTER REGION OF THE FGF-4 GENE

A series of *FGF-4* promoter/reporter gene constructs was employed to map the approximate positions of cis-regulatory elements in the 5′ flanking region of the

Fig. 6.1. Looping models explaining the transcription of the *FGF-4* gene in EC cells and the repression of this gene after differentiation of EC cells. (a) Transcription of the *FGF-4* gene in EC cells is proposed to involve cooperative interactions between transcription factors that bind to cis-regulatory elements in the promoter region of the gene and cis-regulatory elements in the distal enhancer region. The transcription start site of the gene is shown by the arrow. The exons of the gene are indicated by the boxes. The 5′ UTR is shown by the nonshaded region of exon 1, and the 3′ UTR is indicated by the nonshaded region of exon 3. The *FGF-4* enhancer is located within the 3′ UTR and is not drawn to scale. (b) Repression of the *FGF-4* gene after the differentiation of EC cells is proposed to result, at least in part, from the disappearance of OCT-3 and SOX-2 and the concomitant destabilization of the gene's topography, which is needed for the promotion of transcription.

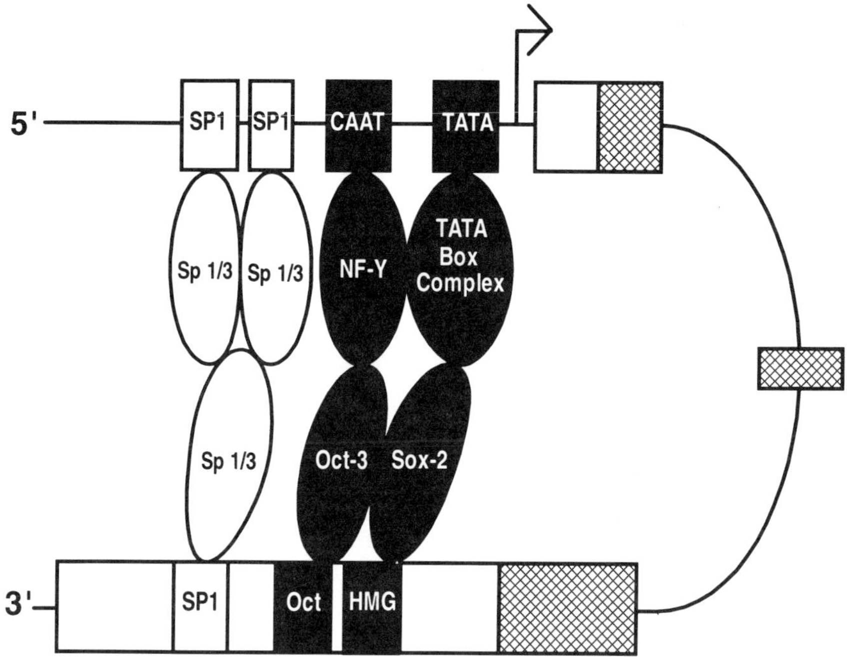
5'
SP1
SP1
CAAT
TATA
Sp 1/3
Sp 1/3
NF-Y
TATA Box Complex
Sp 1/3
Oct-3
Sox-2
3'
SP1
Oct
HMG

(a)

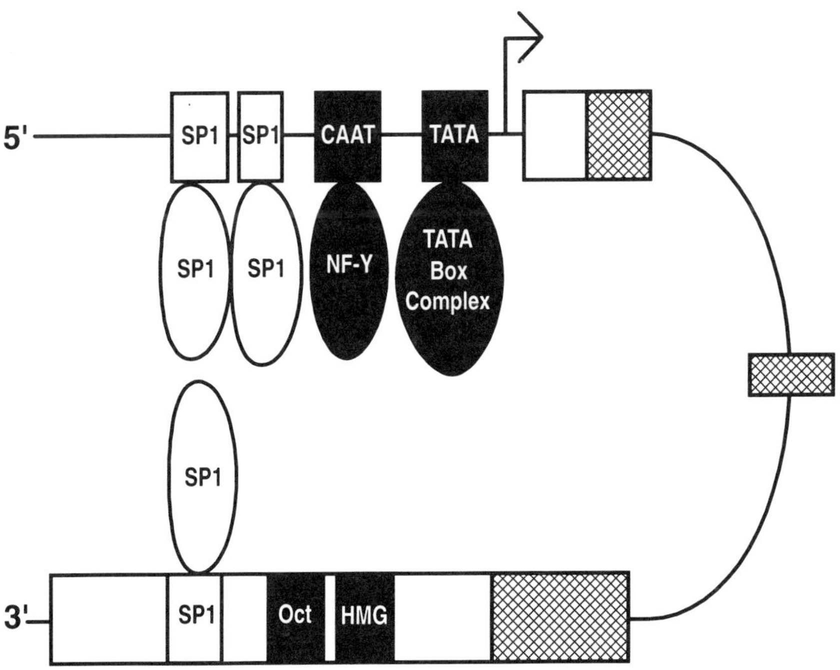
5'
SP1
SP1
CAAT
TATA
SP1
SP1
NF-Y
TATA Box Complex
SP1
3'
SP1
Oct
HMG

(b)

mouse FGF-4 gene (23,27). In addition to a TATA box located 24 bp upstream of the transcription state site, the 5′ flanking region of the mouse FGF-4 gene contains a negative regulatory region located between −1394 and −427 (relative to the transcription start site) and a positive regulatory region located between −267 and −61 (23; Table 6.1). Thus far, the precise location of the putative negative regulatory element(s) has not been determined, but a comparable region has been identified in the 5′ flanking region of the human FGF-4 gene. Embedded in the positive regulatory region of the FGF-4 gene are two Sp1 motifs and a CCAAT box motif. The CCAAT box motif, located approximately 140 bp upstream of the transcription start site, has been implicated strongly in the regulation of the FGF-4 gene. *FGF-4* promoter/reporter gene constructs with a scrambled CCAAT box motif are expressed at levels significantly lower than those of their unmodified counterparts (28,29). Similarly, one or both Sp1 motifs are likely to influence the transcription of the FGF-4 gene since the sequence of both Sp1 motifs and the 20 bp that separate them are conserved perfectly in the mouse and human FGF-4 genes. However, the role of the Sp1 motifs in the expression of the FGF-4 gene has not been tested directly by site-directed mutagenesis.

6.5. IDENTIFICATION OF TRANSCRIPTION FACTORS THAT BIND TO CIS-REGULATORY ELEMENTS LOCATED IN THE PROMOTER REGION OF THE FGF-4 GENE

Understanding fully how a gene is regulated requires identifying the transcription factors that bind to each cis-regulatory element. For this purpose, gel mobility shift analysis and supershift analysis with antibodies raised against specific transcription factors are used to identify nuclear factors that bind to the putative cis-regulatory elements *in vitro*. As one might expect, this analysis has shown that the transcription factor Sp1 binds to both Sp1 motifs in the 5′ flanking region of the FGF-4 gene. Interestingly, this analysis demonstrated that the related transcription factor Sp3 also binds to both Sp1 motifs. Currently, it is unclear whether only Sp1 or Sp3 plays a role in the regulation of the FGF-4 gene.

Gel mobility supershift analysis has also shown that the trimeric transcription factor complex NF-Y binds to the CCAAT box motif *in vitro*. NF-Y is believed to bind DNA with high affinity only as a complex of three subunits: NF-YA, NF-YB, and NF-YC (also known as CBF-B, CBF-A, and CBF-C, respectively) (30). To test the function of NF-Y in the *in vivo* expression of the FGF-4 gene, a dominant-negative form of NF-YA, NF-Y-A29, has been employed. NF-Y-A29 contains a 3-bp deletion in its DNA binding domain. Although NF-Y-A29 is believed to form complexes with NF-YB and NF-YC, complexes containing NF-Y-A29 do not bind DNA. Consequently, ectopic expression of NF-Y-A29 in EC cells should lower the *in vivo* levels of NF-YB and NF-YC that are available for binding to DNA. Using this approach, NF-Y has been implicated strongly in the expression of the FGF-4 gene (31). In fact, ectopic expression of NF-Y-A29 in EC cells lowers expression of *FGF-4* promoter/reporter gene constructs to levels comparable to those observed with FGF-4 promoter/reporter gene constructs containing a scrambled CCAAT box motif.

An important question that remains to be addressed is whether changes in transcription factors that bind to the Sp1 motifs and/or the CCAAT box motif play critical roles in the repression of the FGF-4 gene when EC cells undergo differentiation. Thus far, there is little indication that differentiation affects significantly the levels of Sp1, Sp3, or NF-Y that are able to bind to their respective DNA binding motifs. In this regard, gel shift analysis indicates that nuclear extracts prepared from EC cells and their RA-induced differentiated counterparts contain significant amounts of all three transcription factors. However, it is unclear whether differentiation of EC cells alters the ability of these factors to function *in vivo*. For example, phosphorylation of some transcription factors is known to affect their binding to DNA or their ability to activate transcription (transactivate) when bound to DNA (reviewed in ref. 32). Similarly, glycosylation of Sp1 has been reported to modulate the action of Sp1 (33). In addition, changes in the redox state of the transcription factors could alter their function (reviewed in ref. 34). Alternatively, changes in chromatin structure at or in the vicinity of critical cis-regulatory elements during and after differentiation could alter the binding of essential transcription factors. These possibilities are discussed further in Section 6.9.

6.6. MULTIPLE CIS-REGULATORY ELEMENTS ACCOUNT FOR THE FUNCTION OF THE DISTAL *FGF-4* ENHANCER

Evidence accumulated during the past several years argues strongly that the enhancer region located in the last exon of the FGF-4 gene is critical for both the expression of the FGF-4 in EC cells and the repression of the gene when these cells undergo differentiation. Moreover, it is now clear that the activity of the enhancer is dependent on multiple cis-regulatory elements. Early work focused on an octmer motif located in the enhancer region of both the mouse FGF-4 gene and the human FGF-4 gene. In the case of the human *FGF-4* enhancer, deletion of the octamer motif reduced the effect of the enhancer by approximately half (35). This implicated the octamer motif in the expression of the FGF-4 gene, but it also suggested that other cis-regulatory elements might contribute to the function of the *FGF-4* enhancer.

Clear support for the role of additional cis-regulatory elements was derived initially from efforts to determine the minimum size of the enhancer region. In one study, the mouse *FGF-4* enhancer fragment (approximately 300 bp in length) used in most *FGF-4* promoter/reporter gene constructs was replaced by a smaller sequence containing the octamer motif and 20 bp of flanking sequence on each side of the octamer motif (48 bp in all). Although the 48-bp enhancer sequence elevated the expression of an *FGF-4* construct containing the promoter region of the gene, the increase observed was much lower than that typically attained with the larger mouse *FGF-4* enhancer fragment (23; Table 6.1). Further support for the role of additional cis-regulatory elements was obtained from saturation mutagenesis of the mouse *FGF-4* enhancer region (36). These studies not only confirmed the importance of the octamer motif but also identified two additional putative cis-regulatory elements: a nonconsensus Sp1 motif located 46 bp downstream of the octamer motif and a high-mobility group (HMG) motif located just 3 bp upstream of the octamer motif.

6.7. IDENTIFICATION OF TRANSCRIPTION FACTORS THAT BIND TO CIS-REGULATORY ELEMENTS LOCATED IN THE *FGF-4* ENHANCER

Gel mobility shift analysis was used to determine the identity of the transcription factors that bind to each of the putative cis-regulatory elements located in the mouse *FGF-4* enhancer. As in the case of the Sp1 motifs located in the promoter region of the FGF-4 gene, the nonconsensus Sp1 motif in the *FGF-4* enhancer binds the transcription factors Sp1 and Sp3 (37). In addition, gel mobility shift analysis demonstrated that differentiation does not overtly influence the binding of either factor to the Sp1 motif in the enhancer.

In contrast, differentiation has a major effect on the binding of transcription factors to the octamer motif. Octamer motifs are known to be bound by a family of homeodomain-containing transcription factors, including OCT-1, OCT-2, OCT-3, and OCT-6 (38). Gel mobility shift analysis demonstrated that OCT-1, OCT-3, and OCT-6 are present in the nuclear extracts prepared from EC cells and that each can bind to the octamer motif in the *FGF-4* enhancer (37). Interestingly, the mouse *FGF-4* enhancer region also contains a second near-consensus octamer motif (located 46 bp upstream of the octamer motif discussed above), but this upstream sequence does not bind octamer binding transcription factors with high affinity (37), and mutagenesis of this sequence does not appear to influence the activity of the *FGF-4* enhancer (36). This is interesting because the two octamer-like sequences in the mouse *FGF-4* enhancer differ by a single base pair. The sequence of the functional downstream octamer motif is ATGCTAAT, whereas the sequence of the upstream near-consensus octamer motif is ATGCCAAT. Thus, the presence of a cytosine in place of a thymine at the fifth position strongly decreases the affinity of octamer binding transcription factors.

Early studies established that differentiation of EC cells alters the expression of octamer binding transcription factors. Although OCT-1 continues to be expressed after EC cells differentiate, the expression of OCT-3 and OCT-6 is greatly diminished (38). Hence, it was not surprising to detect by gel mobility shift analysis only OCT-1 binding when nuclear extracts prepared from EC-derived differentiated cells were incubated with a DNA probe containing the *FGF-4* enhancer octamer motif (35,37). Nonetheless, this finding led to two important predictions. First, it suggested that the disappearance of OCT-3 and/or OCT-6 that accompanies differentiation contributes to the repression of the FGF-4 gene when EC cells undergo differentiation; thus, it suggested that OCT-3 and/or OCT-6 is likely to play a key role in the transcription of the gene. In the case of OCT-3, this possibility was tested directly by ectopic expression of OCT-3 in EC-derived differentiated cells, where the FGF-4 gene is not expressed (23). In this study, the differentiated cells were transfected with an *FGF-4* promoter/reporter gene construct along with an *Oct-3* expression plasmid. Ectopic expression of OCT-3 elevated the expression of the reporter gene construct 7- to 10-fold both in EC-derived differentiated cells and in the parietal endoderm-like cell line PYS-2, which also does not express the FGF-4 gene (23). Subsequent studies by other investigators confirmed the function of OCT-3 and demonstrated that OCT-1 is unlikely to substitute for OCT-3 (39). Based on these studies, it appears that OCT-3 plays an important role in the expression of the FGF-4 gene in EC cells and in the repression of the FGF-4 gene when EC cells undergo differentiation. However, it should be borne in mind that OCT-6 may also contribute to the transcription of the FGF-4 gene and thus warrants consideration.

Recent efforts have focused on the identification of the transcription factor(s) that binds to the HMG motif. HMG motifs are known to bind a wide range of related proteins, including transcription factors originally identified as low molecular weight, acid-soluble, nonhistone chromosomal proteins. Gel mobility shift analysis of nuclear extracts prepared from EC cells demonstrated specific binding of a nuclear factor to the FGF-4 enhancer HMG motif. Equally important, it was apparent that differentiation decreases the binding of the HMG binding factor. Attempts to characterize this HMG transcription factor suggested that the factor is relatively heat stable and insensitive to oxidation (37,40). Heating EC cell nuclear extracts to 90°C or treating the extracts with the oxidizing agent diamide had little effect on the ability of the factor to bind to the HMG motif. Interestingly, these properties are not shared by most transcription factors, including Sp1, Sp3, OCT-1, and OCT-3 (40).

Recently, the transcription factor SOX-2 was identified as a likely candidate for binding to the *FGF-4* enhancer HMG motif (39). SOX-2 is a member of the *Sry*-related HMG box-containing gene family (*Sox*), which contains a well-conserved 80-bp DNA binding domain known as an *HMG box*. Thus far, numerous *Sox* genes have been identified in vertebrates, including approximately 20 in mammals. The human homolog of *Sox-2* was originally cloned several years ago (41). More recently, *Sox-2* cDNA was cloned by RT-PCR from mRNA expressed by EC cells (39,42). Several lines of evidence suggest that SOX-2 can bind to the HMG motif of the *FGF-4* enhancer. For example, a recombinant fragment of SOX-2 containing its DNA binding domain binds *in vitro* to the *FGF-4* enhancer HMG motif (42). Moreover, ectopic expression of SOX-2 and OCT-3 in HeLa cells, which do not normally express FGF-4, SOX-2, or OCT-3, stimulates the activity of a heterologous promoter linked to multiple tandem repeats of the HMG and octamer motifs (39). Equally important, differentiation of EC cells decreases the level of *Sox-2* mRNA. Thus, it appears that SOX-2 plays an important role, along with OCT-3, in the expression of the FGF-4 gene in EC cells and the repression of the FGF-4 gene when EC cells undergo differentiation. However, additional studies discussed below are needed before this attractive conclusion can be accepted fully.

6.8. MODEL FOR THE TRANSCRIPTION AND REPRESSION OF THE *FGF-4* GENE

Taken together, the evidence suggests that at least six cis-regulatory elements contribute to the transcriptional regulation of the FGF-4 gene: an HMG motif, an octamer motif and an Sp1 motif in the *FGF-4* enhancer, and a CCAAT box motif and two Sp1 motifs in the promoter region of the gene. In addition, transcription factors that are likely to bind to each of these sites *in vivo* have been identified. With this information in hand, an updated looping model for the transcriptional regulation of the FGF-4 gene is presented wherein specific transcription factors that bind to the cis-regulatory elements in the promoter region of the gene interact with transcription factors that bind to cis-regulatory elements located in the enhancer region (Fig. 6.1a).

This looping model proposes that Sp1 or Sp3 forms multimeric complexes linking one or both of the Sp1 motifs in the promoter region with the Sp1 motif in the enhancer region. This part of the model is supported by several findings. Sp1 is known to form multimeric structures with itself. (This remains to be determined for Sp3). In addition, multimeric complexes of Sp1 bound to widely separated motifs can promote DNA looping *in vitro* (43–45), and widely separated Sp1 motifs have been shown to act together to increase synergistically the activity of *TK-CAT* promoter/reporter gene constructs (46). This model further proposes that transcription factors that bind to the HMG and octamer motifs play key roles in the transcription of the FGF-4 gene by helping to stabilize the loop structure of the gene and by providing proper alignment of transactivation domains, which is needed to increase the frequency at which the gene is transcribed. Precedent for the envisioned interactions is drawn from the findings that octamer binding proteins can interact functionally with CCAAT box binding proteins (47), and HMG binding proteins can interact with octamer binding proteins (48) as well as with proteins that make up the basal transcription machinery (49). Finally, this model proposes that the transcriptional repression of the FGF-4 gene when EC cells undergo differentiation may result from the disappearance of the transcription factors that bind to the HMG and the octamer motifs. Specifically, the lack of OCT-3 and SOX-2 binding could alter the overall topography of the gene by destabilizing the loop structure (Fig. 6.1b). However, other possibilities should also be considered; these are discussed in the final section of the review.

If the disappearance of OCT-3 and SOX-2 is crucial to repression of the FGF-4 gene, how are the levels of these transcription factors regulated by differentiation? Thus far, little is known about this important issue. However, differentiation leads to a reduction in the steady-state levels of *Oct-3* and *Sox-2* mRNAs (39,50). Critical to this discussion is the turnover of the transcription factors themselves. Currently, the mechanisms responsible for degradation of transcription factors remain to be determined. However, one study may shed light on this matter. Recently, it has been shown that the differentiation of EC cells to parietal extraembryonic endoderm-like cells leads to the appearance of nuclear protease activity (51). This protease(s), which exhibits the properties of a cysteine protease, is not detected in the cytoplasm, and it can degrade a wide range of transcription factors, including ATF-1,

Sp1, NF-Y, and the octamer binding proteins OCT-1 and OCT-3. However, it does not appear to readily degrade several other proteins, including serum albumin, myosin, phosphorylase B, and carbonic anhydrase. Although these findings are highly suggestive, it is unclear whether this protease(s) plays a role in the turnover of transcription factors during the early stages of EC cell differentiation. In this connection, the steady-state levels of *FGF-4* mRNA decrease before the nuclear-associated protease activity is first detected, which occurs approximately 2 days after EC cells begin to differentiate. However, it is possible that the indirect protease assay used in this study was not sufficiently sensitive to detect the protease when it first appeared.

6.9. FUTURE PERSPECTIVES

Although many of the cis-regulatory elements that control the FGF-4 gene have been identified, several important questions remain to be addressed. Do the promoter Sp1 motifs influence the transcription of the FGF-4 gene, and are both motifs utilized? What is the precise location of the negative regulatory element present in the 5′ flanking region of the FGF-4 gene, and which transcription factor(s) binds to this site? Perhaps the most challenging questions relate to the interactions between the different transcription factors, in particular the interactions between transcription factors that bind to the cis-regulatory elements in the promoter and enhancer regions. Currently, relatively little is known about the transcription factor domains involved in promoting protein–protein interactions and transactivation. One can begin to map the location of the essential domains by using the yeast two-hybrid system and "pull-down" experiments with recombinant transcription factors with specific regions deleted. Equally important questions relate to the functions of OCT-3 and SOX-2. Are these factors essential for transcription of the FGF-4 gene in EC cells, and is the repression of the FGF-4 gene due solely to the disappearance of OCT-3 and SOX-2? One way to address the latter question is to express both transcription factors stably in EC-derived differentiated cells.

Although downregulation of OCT-3 and SOX-2 is likely to play an important role in the repression of the FGF-4 gene, other mechanisms may also play important roles. For example, repression of the FGF-4 may also involve posttranslational modifications of proteins, such as Sp1 and NF-Y, that bind to other cis-regulatory elements in the gene. Alternatively, differentiation may alter the placement of nucleosomes or promote other changes in chromatin structure, which prevent the binding of transcription factors to critical cis-regulatory elements. In this regard, the theoretical algorithm developed by Calladine and Drew predicts the presence of approximately 20 favorable nucleosome-forming regions scattered throughout the FGF-4 gene (24). Interestingly, placement of a nucleosome with its dyad axis at the most favorable position in the 5′ flanking region of the FGF-4 gene would leave the important CCAAT box motif just outside of the nucleosome but would include within the nucleosome the two conserved promoter Sp1 motifs. In addition, the Sp1 motif nearest the transcriptional start site would be rotationally positioned to expose its major groove on the surface of the nucleosome. On the other hand, the HMG and octamer motifs in the *FGF-4* enhancer are located in a region with a low potential for nucleosome formation.

Lastly, a full understanding of the transcriptional regulation of the FGF-4 gene will require understanding how this gene is regulated temporally and spatially. To answer this question will require determining whether the same cis-regulatory elements regulate the FGF-4 gene in all tissues. Based on work with other genes, including the TGF-β2 gene (9,52), different but overlapping combinations of cis-regulatory elements are likely to control the FGF-4 gene in different tissues. The most precise way to address this complex question is to use gene targeting to modify specific cis-regulatory elements in the FGF-4 gene and to examine their function *in vivo* (53). Given the importance of FGF-4 during early development (11) and the interesting structural features of the FGF-4 gene, further study of its transcriptional regulation is warranted. These studies will not only help understand how the FGF-4 gene is regulated, but will also help us to understand how other genes are regulated by distal enhancers and the process of differentiation.

ACKNOWLEDGMENTS

Dave Kelly, Kim Lamb, and Lance Johnson are thanked for reading this manuscript, and Heather Rizzino is thanked for editorial assistance. Work in the author's laboratory was supported by grants from the National Institute of Child Health and Human Development (HD 19837) and the National Cancer Institute (CA 74771) and by a core grant from the National Cancer Institute (Laboratory Cancer Research Center Support Grant, CA 36727). Finally, present and past members of the author's laboratory (Kristen Brigman, Qing-Sheng Gao, Lance Johnson, Dave Kelly, Michelle Kingsley-Kallesan, Kim Lamb, Kim Lickteig, Yan-Gao Ma, Ed Rosfjord, Beáta Scholtz, Jay Tiesman, and Phillip Wilder) are acknowledged for their excellent work, which contributed substantially to our understanding of transcriptional regulation during differentiation.

REFERENCES

1. Rappolee DA, Brenner CA, Schultz R, Mark D, Werb Z: Developmental expression of PDGF, TGF-α, and TGF-β genes in preimplantation mouse embryos. Science 241: 1823–1825, 1988.
2. Rappolee DA, Sturm K, Schultz G, Pedersen R, Werb Z: Expression of growth factor ligands and receptors in preimplantation mouse embryos. In Heyner S, Wiley L (eds): Early Embryo Development and Paracrine Relationships. New York: Wiley-Liss, 1990, pp 11–25.
3. Niswander L, Martin G: FGF-4 expression during gastrulation, myogenesis, limb and tooth development in the mouse. Development 114:755–768, 1992.
4. Martin GR: Teratocarcinomas and mammalian embryogenesis. Science 209:768–776, 1980.

5. Rizzino A: Understanding the roles of growth factors in carcinogenesis: Modulation of autocrine growth control by differentiation. Int J Dev Biol 37:61–65, 1993.
6. Yoshida T, Muramatsu H, Muramatsu T, Sakamoto H, Katoh O, Sugimura T, Terada M: Differential expression of two homologous and clustered oncogenes, Hstl and Int-2, during differentiation of F9 cells. Biochem Biophys Res Comm 157:618–625, 1988.
7. Velcich A, Delli-Bovi P, Mansukhani A, Ziff EB, Basilico C: Expression of the K-fgf protooncogene is repressed during differentiation of F9 cells. Oncogene Res 5:31–37, 1989.
8. Tiesman J, Rizzino A: Expression and developmental regulation of the k-FGF oncogene in human and murine embryonal carcinoma cells. In Vitro Cell Dev Biol 25:1193–1198, 1989.
9. Kelly D, O'Reilly MA, Rizzino A: Differential regulation of the TGF-β2 gene promoter in embryonal carcinoma cells and their differentiated cells. Dev Biol 153:172–175, 1992.
10. Kelly D, Rizzino A: Unpublished results.
11. Feldman B, Poueymirou W, Papaiannou VE, DeChiara TM, Goldfarb M: Requirement of FGF-4 for postimplantation mouse development. Science 267:246–249, 1995.
12. Miller K, Rizzino A: Developmental regulation and signal transduction pathways of fibroblast growth factors and their receptors. In Nilsen-Hamilton M (ed): Growth Factors and Signal Transduction in Development. New York: Wiley-Liss, 1994, pp 19–49.
13. Yamasaki M, Miyake A, Tagashira S, Itoh N: Structure and expression of the rat mRNA encoding a novel member of the fibroblast growth factor family. J Biol Chem 271: 15918–15921, 1996.
14. Smallwood PM, Munoz-Sanjuan I, Tong P, Macke JP, Hendry SHC, Gilbert DJ, Copeland NG, Jenkins NA, Nathans J: Fibroblast growth factor (FGF) homologous factors: New members of the FGF family implicated in nervous system development. Proc Natl Acad Sci USA 93:9850–9857, 1996.
15. Hebert JM, Basilico C, Goldfarb M, Haub O, Martin GR: Isolation of cDNAs encoding four mouse FGF family members and characterization of their expression patterns during embryogenesis. Dev Biol 138:454–463, 1990.
16. Haub O, Goldfarb M: Expression of the fibroblast growth factor-5 gene in the mouse embryo. Development 112:397–406, 1991.
17. Jakobovits A, Shackleford GM, Varmus HE, Martin GR: Two proto-oncogenes implicated in mammary carcinogenesis, int-1 and int-2 are independently regulated during mouse development. Proc Natl Acad Sci USA 83:7806–7810, 1986.
18. Heath JK, Isake CM: PC13 embryonal carcinoma-derived growth factor. EMBO J 3: 2957–2962, 1984.
19. Rizzino A, Kuszynski C, Ruff E, Tiesman J: Production and utilization of growth factors related to fibroblast growth factor by embryonal carcinoma cells and their differentiated cells. Dev Biol 129:61–71, 1988.
20. Miller K, Rizzino A: Characterization of fibroblast growth factor activity secreted by embryonal carcinoma cells. In Vitro Cell Dev Biol 32:531–533, 1996.
21. Peters G, Brookes S, Smith R, Placzek M, Dickson C: The mouse homolog of the hst/k-FGF gene is adjacent to int-2 and is activated by proviral insertion in some virally induced mammary tumors. Proc Natl Acad Sci USA 86:5678–5682, 1989.
22. Tiesman J, Rizzino A: Nucleotide sequence of the 5′-flanking region of the mouse k-FGF oncogene exhibits an alternating purine:pyrimidine motif with the potential to form Z-DNA. Gene 96:311–312, 1990.
23. Ma Y-G, Rosfjord E, Huebert C, Wilder P, Tiesman J, Kelly D, Rizzino A: Transcriptional regulation of the murine k-FGF gene in embryonic cell lines. Dev Biol 154:45–54, 1992.

24. Wilder P, Mountjoy C, MacLeod MC, Rizzino A: DNA sequence and nucleosome placement on the muringe fibroblast growth factor-4 gene. DNA Sequence 7:117–121, 1997.
25. Curatola AM, Basilico C: Expression of the K-fgf proto-oncogene is controlled by 3′ regulatory elements which are specific for embryonal carcinoma cells. Mol Cell Biol 10:2475–2484, 1990.
26. Rizzino A, Rosfjord E: Transcriptional regulation of the murine k-FGF gene. Mol Reprod Dev 39:106–111, 1994.
27. Koda T, Hasan S, Sasaki A, Arimura Y, Kakinuma M: Regulatory sequences required for *hst-1* promoter in embryonal carcinoma cells. FEBS Lett 342:71–75, 1994.
28. Hasan S, Koda T, Kakinuma M. An upstream NF-Y-binding site is required for transcriptional activation from the *hst* promoter in F9 embryonal carcinoma cells. J Biol Chem 269:25042–25048, 1994.
29. Bryans M, Lucas J, Knobloch T, Wilkie N, Lang J: Regulation of FGF-4 of enhancer activity by transcription factor NF-Y. Biochem Biophys Res Commun 211:519–527, 1995.
30. Maity SN, Sinah S, Ruteshouser EC, deCrombruhge B: Three different polypeptides are necessary for DNA binding of the mammalian heteromeric CCAAT binding factor. J Biol Chem 267:16574–16580, 1992.
31. Lamb K, Johnson L, and Rizzino A: NF-Y binds to the CCAAT box motif of the FGF-4 Gene and promotes FGF-4 expression *in vivo*. Molecular Reproduction and Development, in press.
32. Meek DW, Street AJ: Nuclear protein phosphorylation and growth control. Biochem J 287:1–15, 1992.
33. Jackson SP, Tjian R: *O*-glycosylation of eukaryotic transcription factors: Implications for mechanisms of transcriptional regulation. Cell 55:125–133, 1988.
34. Pahl HL, Baeuerle PA: Oxygen and the control of gene expression. BioEssay 16:497–502, 1994.
35. Schoorlemmer J, Kruijer W: Octamer-dependent regulation of the kFGF gene in embryonal carcinoma and embryonic stem cells. Mech Dev 36:75–86, 1991.
36. Dailey L, Yuan H, Basilico C: Interaction between a novel F9-specific factor and octamer-binding proteins is required for cell-type-restricted activity of the fibroblast growth factor 4 enhancer. Mol Cell Biol 14:7758–7769, 1994.
37. Lamb K, Rosfjord E, Brigman K, Rizzino A: Binding of transcription factors to widely separated cis-regulatory elements of the murine FGF-4 gene. Mol Reprod Dev 44:460–471, 1996.
38. Schöler HR, Hatzopoulos AK, Balling R, Suzuki N, Gruss P: A family of octamer-specific proteins present during mouse embryogenesis: Evidence for germline-specific expression of an Oct factor. EMBO J 8:2543–2550, 1989.
39. Yuan H, Corbi N, Basilico C, Dailey L: Developmental-specific activity of the FGF-4 enhancer requires the synergistic action of Sox2 and Oct-3. Genes Dev 9:2635–2645, 1995.
40. Lickteig K, Lamb K, Brigman K, Rizzino A: Effects of oxidation and reduction on the binding of transcription factors to cis-regulatory elements in the FGF-4 gene. Mol Reprod Dev 44:146–152, 1996.
41. Stevanovic M, Lovell-Badge R, Collignon J, Goodfellow PN: The cDNA sequence and chromosomal location of the human *SOX-2* gene. Mammalian Genome 5:640–642, 1994.
42. Gao QS, Rizzino A: Unpublished results

43. Mastrangelo IA, Courey AJ, Wall JS, Jackson SP, Hough PVC: DNA looping and Sp1 multimer links: A mechanism for transcriptional synergism and enhancement. Proc Natl Acad Sci USA 88:5670–5674, 1991.
44. Pascal E, Tjian R: Different activation domains of Sp1 govern formation of multimers and mediate transcriptional synergism. Genes Dev 5:1646–1656, 1991.
45. Su W, Jackson S, Tjian R, Echols H: DNA looping between sites for transcription activation: Self-association of DNA-bound Sp1. Genes Dev 5:820–826, 1991.
46. Courey AJ, Holtzman DA, Jackson SP, Tjian R: Synergistic activation by the glutamine-rich domains of human transcription factor Sp1. Cell 59:827–836, 1989.
47. Hinkley C, Perry M: Histone H2B gene transcription during *Xenopus* early development requires functional cooperation between properties bound to the CCAAT and octamer motifs. Mol Cell Biol 12:4400–4411, 1992.
48. Zwilling S, König H, Wirth T: High mobility group protein 2 functionally interacts with the POU domains of octamer transcription factors. EMBO J 14:1198–1208, 1995.
49. Shykind BM, Kim J, Sharp PA: Activation of the TFIID-TFIIA complex with HMG-2. Genes Dev 9:1354–1365, 1995.
50. Lamb K, Rizzino A: Unpublished results.
51. Sholtz B, Lamb K, Rosfjord E, Kingsley M, Rizzino A: Appearance of a nuclear protease activity after embryonal carcinoma cells undergo differentiation. Dev Biol 173:420–427, 1996.
52. O'Reilly MA, Geiser AG, Bruggeman LA, Luu AX, Roberts AB, Sporn MB: Identification of a CRE/ATF-like element in the human transforming growth factor-β2 gene that is essential for promoter activity. J Biol Chem 267:19938–19943, 1992.
53. Wilder PJ, Rizzino A: Mouse genetics in the 21st century: Using gene targeting to create a cornucopia of mouse mutants possessing precise genetic modifications. Cytotechnology 11:79–99, 1993.

7

CELLULAR INTERACTIONS MEDIATED BY TYROSINE KINASE RECEPTORS DURING DEVELOPMENT: DRIVING FORCES FOR GROWTH, MOTILITY, AND DIFFERENTIATION

CARMEN BIRCHMEIER AND WALTER BIRCHMEIER

Departments of Medical Genetics (C.B.) and Cell Biology (W.B.), Max-Delbrück-Center for Molecular Medicine, Berlin, Germany

7.1. INTRODUCTION

Tyrosine kinase receptors and their ligands can form both autocrine and paracrine signaling systems. Recent evidence demonstrates that a paracrine signaling mode is typical of several receptor-ligand pairs that affect epithelia. For example, scatter factor (SF), also known as hepatocyte growth factor (HGF), was initially described as a fibroblast-derived factor that scatters epithelial cells (1,2). The specific receptor of SF/HGF was found to be c-met (3,4), which is expressed by epithelia, cells of the myogenic lineage, and by neuronal cells. *In situ* hybridization analysis suggested a paracrine signaling mode for SF/HGF and c-met in the developing mouse: SF/HGF is expressed by various types of mesenchymal cells located in the vicinity of epithelial or muscle cells that express c-met (5). Neuregulin, also known as *Neu differentiation factor (NDF), heregulin, glial growth factor (GGF),* or *acetylcholine receptor-inducing activity (ARIA),* is a member of the EGF family of growth factors

Hormones and Growth Factors in Development and Neoplasia, Edited by Robert B. Dickson and David S. Salomon
ISBN 0-471-16899-8

(6). Neuregulin binds with high affinity to the erbB-3 and erbB-4 receptor tyrosine kinases and requires erbB2 as co-receptor for signaling (7). Neuregulin is produced predominantly by mesenchymal and neuronal cells, whereas the receptors are expressed in epithelia and in cells of the nervous system (8–10). Fibroblast growth factors (FGFs) were initially identified as potent growth factors for fibroblasts and other cell types, as oncogenes activated by mouse mammary tumor virus (MMTV) or isolated in transformation assays using NIH 3T3 cells. Four receptors for FGFs exist in mammals that have overlapping specificities for the many FGFs (11). FGFs and their receptors frequently form paracrine signaling systems in embryogenesis that are known to regulate independent developmental events. The c-ret receptor was identified because of its oncogenic potential when mutated (12). The ligand of c-ret was recently found to correspond to glial-derived growth factor (GDNF) (13–15). The c-ret receptor is expressed in epithelia of the kidney and in cells of the peripheral nervous system, whereas GDNF is produced by kidney mesenchyme and at sites of development of the sympathetic and enteric nervous systems (16–18). Other epithelial receptor tyrosine kinases exist to which no ligand has yet been assigned, such as c-ros (19). Indirect evidence suggests that this receptor also receives paracrine signals (see 20 for a review).

7.2. THE ROLE OF SF/HGF AND NEUREGULIN IN THE DEVELOPMENT OF BREAST EPITHELIA

SF/HGF induces various cellular responses on epithelial and other cells in culture: 1) it promotes the growth of primary hepatocytes and other epithelial cells; 2) it can induce increased motility of epithelial and endothelial cells; 3) it has an unusual activity, the ability to induce branching morphogenesis of MDCK (kidney) and other epithelial cells in three-dimensional collagen matrices; and 4) it promotes outgrowth of axons from developing motoneurons (1,2,21–27). Neuregulin also induces morphologic changes in breast carcinoma cells that are accompanied by terminal differentiation, i.e., accumulation of fat droplets and β-casein secretion (28). Other carcinoma cells respond by growing (29). In addition, skeletal muscle and cells of neuronal origin (i.e., glial cells, oligodendrocytes, and retinal neurons) have been reported to respond to neuregulin in cell culture (8,30–37).

We recently studied the role of SF/HGF and neuregulin in development of the mammary epithelium, using organ culture of mouse mammary glands (38). *In vivo,* the epithelia of the mammary glands first undergo a period of growth and branching during puberty; later, during pregnancy, alveolar differentiation and milk production occur. Both processes are under central hormonal control. Branching and alveolar morphogenesis can be recapitulated in organ culture. Addition of exogenous SF/HGF to mammary glands in organ culture strongly promotes epithelial growth and branching, whereas exogenous neuregulin induces alveolar maturation followed by terminal differentiation characterized by milk production (Fig. 7.1a–d). Corresponding antisense oligonucleotides prevent branching morphogenesis or alveolar differentiation; additional exogenous factor overcomes the inhibitory effect of the antisense oligonucleotides. Interestingly, both factors are specifically expressed in the mesenchyme of the mammary gland during development: SF/HGF during puberty and neuregulin during pregnancy (Fig. 7.1e). Neuregulin is not expressed at all during puberty and involution (when the lactating glands are remodeled and return to a prepregnancy state) but is specifically turned on late in pregnancy. It is then expressed at high levels. In contrast, the receptors of both of these factors (i.e., the c-met receptor and the members of the c-erbB family) do not appear to be developmentally regulated in the mammary gland. This suggests that central hormonal control acts by regulating the production of local mesenchymal factors during two phases of mammary gland development: branching and alveolar maturation. The mesenchymal factors then promote growth and differentiation of the epithelium in the mammary gland (38).

The two phases of epithelial morphogenesis in the mammary gland, branching and alveolar morphogenesis, were recently recapitulated in cell culture. On addition of SF/HGF, EpH 4 mammary gland epithelial cells form branched tubules in a matrigel matrix. Subsequent exposure to neuregulin induces alveolar-like bulbs along the tubules (C. Niemann, V. Brinkmann, and W.B. Brinkmann, submitted). Branching morphogenesis was also induced by the recently identified substrate of c-Met, Gab1 (39), whereas alveolar development was promoted by NGF in cells expressing a trk-erbB2 hybrid (40, Niemann et al., submitted). Apparently, mammary gland epithelial cells respond to signals given by different factors and mediated by different tyrosine kinase receptors in a specific manner. Thus, signaling specificity of these tyrosine kinase receptors exists in the mammary gland.

7.3. SF/HGF AND THE C-MET RECEPTOR CONTROL GROWTH AND MOTILITY OF CELLS *IN VIVO*

Targeted mutagenesis in embryonal stem (ES) cells by homologous recombination was employed to ablate the SF/HGF and the c-*met* genes (41–43). The mutant ES cells were then used to generate mice that carry the mutation. Both mutations show the same phenotype when present in a homozygous state: They lead to embryonal death between days 13 and 16 of development. The identical phenotypes are genetic evidence of an absence of redundancy in the SF/HGF and c-met signaling systems: SF/HGF appears to be the only functional ligand for c-met, and c-met the only receptor for SF/HGF.

The targeted mutations of c-*met* and SF/HGF affect development of the embryonal liver, the placenta, and distinct skeletal muscles (43). The liver is reduced in size, and histologic damage of variable severity is observed in the parenchyme (Fig. 7.2). That the hepatocellular compartment, and not the embryonal hematopoietic system (which is also located in the liver), is primarily affected by the mutation was verified by the analysis of chimeric animals: c-*met*-/- cells cannot contribute to the adult liver but participate normally in the generation of blood and various hematopoietic cell types (43). The placenta, particularly the labyrinth layer, is also affected by a lack of SF/HGF or c-met (41–43). In addition, a complete absence of

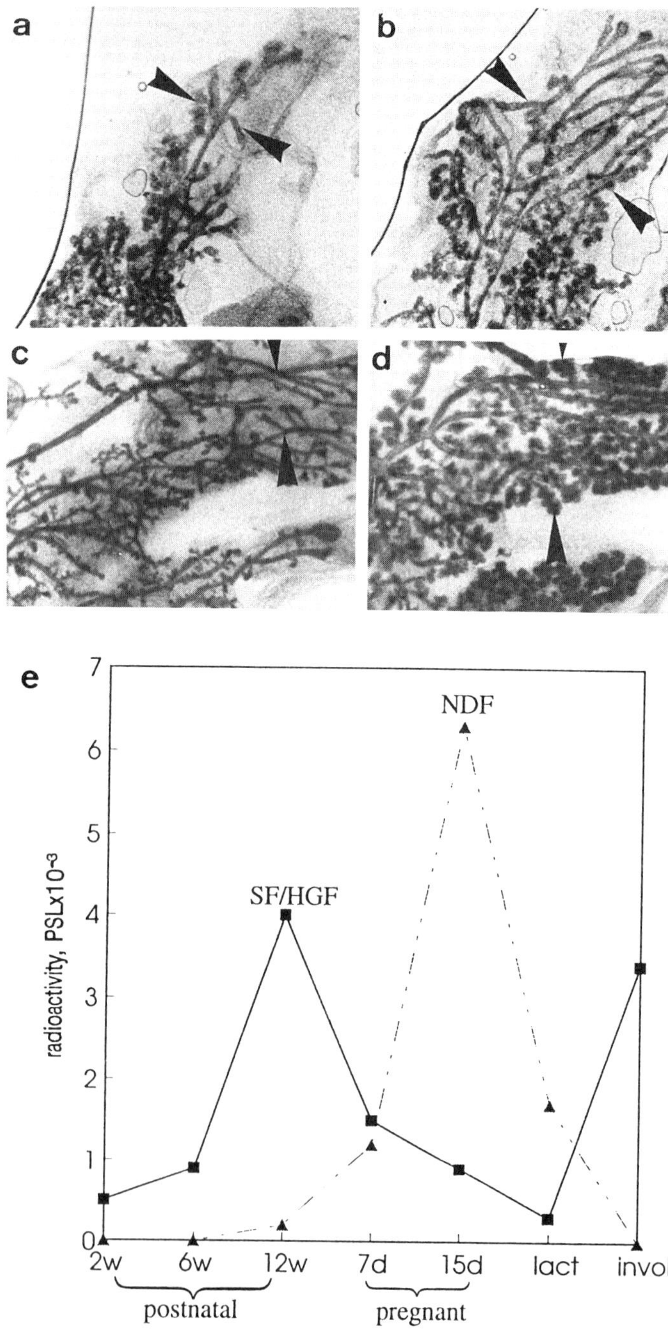
a
b
c
d
e
NDF
SF/HGF
radioactivity, PSLx10⁻³
0
1
2
3
4
5
6
7
2w
6w
12w
7d
15d
lact
invol
postnatal
pregnant

Fig. 7.1. SF/HGF and neuregulin control the morphogenesis of mammary glands. Pairs of mammary glands of 4-week-old female mice were subjected to organ culture in the presence (b) and absence (a) of SF/HGF and in the presence (d) and absence (c) of neuregulin. SF/HGF promotes branching of the glands, as indicated by the arrowheads in (b). Neuregulin induced the production of alveoli, which are assembled into lobules, as indicated by the arrowheads in (d). For details, see (38). The expression of SF/HGF and neuregulin (NDF) during the development of mouse mammary gland was examined by RNase protection (e). Quantitative analysis shows that SF/HGF is strongly expressed during the virginal phase of development (maximum at 12 weeks), weakly expressed during pregnancy and lactation, and strongly expressed again during involution (when the lactating glands are remodeled and return to the adult stage). In contrast, NDF is selectively expressed during late pregnancy (38).

particular muscles was observed in the mutant embryos (43): Skeletal muscles of the limb, the diaphragm, and the internal lingual muscles are lacking, whereas all other skeletal muscles can form. The primary cause of this defect is an absence of migration of muscle precursor cells from the dermomyotomes of the somites to the anlagen of the limb, diaphragm, or tongue. As a consequence, muscles at these sites cannot form (43). The muscle precursor cells located in the ventrolateral dermomyotomes express c-*met* (Fig. 7.3). The signal that induces a mesenchymal-epithelial conversion of the somite cells and migration to the limb bud is known to be given by limb mesenchyme. Indeed, the medial mesenchyme of the limb anlage expresses SF/HGF (Fig. 7.3). Thus, SF/HGF acts *in vivo* as a paracrine and as a locally acting factor during development of various organs. All phenotypes in the mutant mice reflect disturbances in the interactions between different cell types. Different cellular functions are affected: growth and morphogenesis of epithelial cells in the liver and placenta, and migration of myogenic precursor cells that originate from the somites.

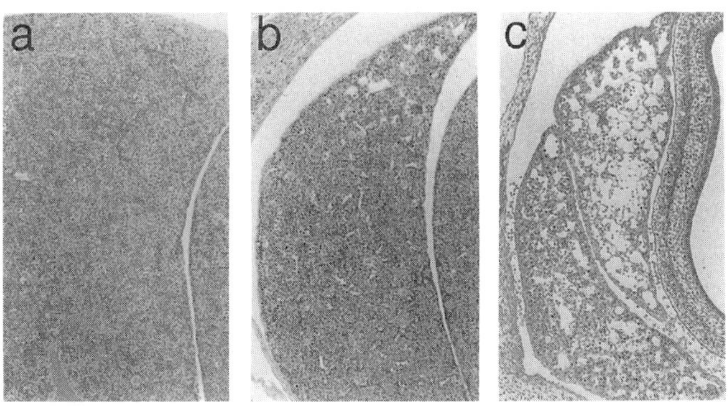

Fig. 7.2. SF/HGF and c-*met* are essential for liver development. The livers of control (SF/HGF +/−) embryos (a) and homozygous mutant (SF/HGF −/−) embryos (b,c) on day 14.5 (E14.5) of development. Note the decreased size of the livers of mutant animals and the damage to liver parenchyma, which is variable in severity. At higher magnification, apoptotic nuclei can be observed in areas of liver damage. Livers from c-*met* −/− embryos have an indistinguishable appearance. (From Bladt et al., (1995) Nature 376, 768–771.).

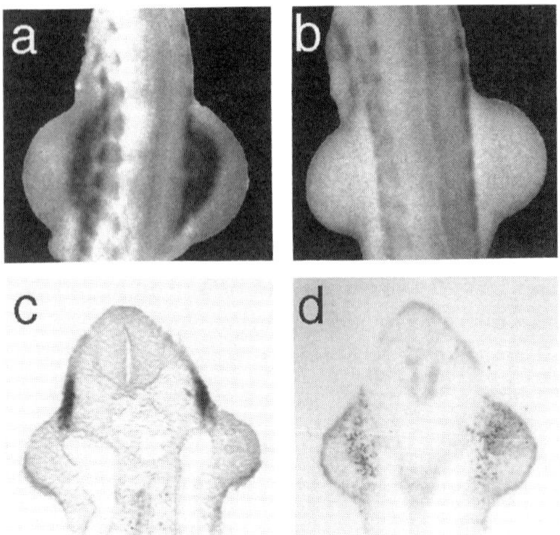

Fig. 7.3. SF/HGF and c-*met* are essential for migration of muscle precursor cells. *In situ* hybridization of c-*met* −/− (a) and control c-*met* +/− (b) embryos on E9.5 with a c-*met*-specific probe that hybridizes to myogenic precursor cells in the somite (a,b) and to migrating muscle precursor cells in the developing limb bud (a). Such migrating cells are absent in the homozygous mutant embryos (b), which express a mutant c-*met* transcript lacking functionally important sequences. Sections of wild-type embryos hybridized with c-*met* (c) and SF/HGF (d) probes, demonstrating that c-*met* is expressed in the myogenic precursor cells in the dermomyotome, whereas SF/HGF is produced in the proximal mesenchyme of the limb bud. The SF/HGF signal emanating from the limb mesenchyme thus induces the migration of the myogenic cells from the somite into the limb. (From Bladt et al. (1995), Nature 376, 768–771.).

7.4. NEUREGULIN AND THE C-ERBB RECEPTOR TAKE OVER MULTIPLE FUNCTIONS IN DEVELOPMENT

The functions of neuregulin and its receptors, c-erbB2 and c-erbB4, has also been analyzed genetically in mice (44–46). Multiple independent defects in development of the heart and peripheral nervous system were observed in such mutant animals. The homozygous mutant embryos die early in development, i.e., on day 10 of gestation. Embryonal death is most likely caused by a defect in development of the myocardium. During development, the endothelial lining of the heart (the endocardium) produces neuregulin, which is an essential signal for the appropriate development of myocardium. In the absence of this signal, trabeculation of the heart ventricle does not occur (44). In the heart, the neuregulin signal is received by the erbB-4 and erbB-2 receptors: Mutations in the genes encoding erbB2, erbB-4 or neuregulin all show identical defects in development of the heart muscle (44–46). Since erbB-4 but not erbB-2 can bind neuregulin directly, erbB-2 functions *in vivo* as an essential coreceptor for erbB-4. Moreover, severe defects in the development of the peripheral nervous system are detected in animals that lack a functional neuregulin gene (44). These include defects in the development of cranial ganglia and Schwann cells. Interestingly, erbB-4 is not involved in neuregulin signaling in the development of the peripheral nervous system since *erbB*-4 mutant mice do not display the same defects as neuregulin mutant mice. Instead, the erbB-3 receptor, again together with erbB-2 as coreceptor, mediates the essential signals of this growth factor in the peripheral nervous system.

From the neuregulin gene, various isoforms are produced that are expressed in distinct and dynamic patterns during development and in the adult (47–49). This indicates that the neuregulin isoforms take over additional functions in embryogenesis and in the adult. Experiments in cell culture demonstrate that neuregulin affects Schwann cell precursors, as well as immature and mature Schwann cells, and influences cell fate decisions, growth, survival, and maturation of cells in this lineage (8,50). Direct axonal contact has long been known to induce DNA synthesis in Schwann cell precursors. Antibodies against erbB-2 inhibit this axon-driven DNA proliferation, indicating that the neuronal signal is given by an axon-bound isoform of neuregulin (51). Moreover, experimental evidence indicates that neuregulin provides an important signal for maturation of the neuromuscular junction (30,34) and for alveolar differentiation of the mammary gland epithelium (28,38; see also above). The strong expression of neuregulin observed in the central nervous system of the perinatal animal implies additional functions in the development and/or maintenance of the brain or spinal cord.

The lethality of embryos that carry mutations in the neuregulin or SF/HGF signaling systems has prevented us from investigating other additional roles, particularly the expected function in epithelial development of the mammary gland. Isoform-specific mutations (6,49), as well as tissue-specific or inducible ablation (52,53) of the genes encoding factors or receptors, will allow us to address these questions in the future.

7.5. FGFs AND THEIR RECEPTORS FUNCTION IN DEVELOPMENT OF EPITHELIA

The FGF family of growth factors and receptors take over many important functions, including development of the limbs and skeleton (54). Here we will discuss in detail only the functional roles in the development of epithelia. A transdominant FGF2 receptor was used to analyze the functional role of this receptor, which binds keratinocyte growth factor (KGF) but also other members of the FGF family of growth factors. This receptor mutant lacks the kinase domain, forms inactive heterodimers, and thus prevents cross-phosphorylation and signaling through the intact receptor. When expressed under a lung-specific promoter (surfactant promoter) in the mouse, this receptor interferes with branching of the lung epithelium and formation of alveoli. Instead, two undifferentiated tubes are formed that extend from the bifurcation of the tracheae (55). The transdominant receptor also interferes with development of the skin when expressed in keratinocytes *in vivo* and prevents normal wound healing (56,57). When the transdominant receptor is expressed under the control of the keratin 6 promoter, which directs the transdominant receptor to the suprabasal layer of the skin, epidermal thickening and disrupted organization of epidermal keratinocytes are observed. When this receptor is expressed under the control of the keratin 14 promoter, which directs the transdominant receptor to undifferentiated keratinocytes in the basal level of the skin, epidermal atrophy and a severe delay in wound healing is observed.

The function of FGFs and their receptors in branching morphogenesis of the lung has been well conserved in evolution (58). The *Drosophila* gene *branchless* is essential for determining the tracheal branching pattern and encodes a homolog of mammalian FGFs. It is expressed dynamically in clusters of cells surrounding the developing tracheal system at each position where a new branch will form. Localized misexpression of *branchless* can direct branch formation and outgrowth to new positions (58). The work in *Drosophila* thus represents a clear example of the important role of a growth factor in determining branching morphogenesis and the branching patterns *in vivo*.

7.6. THE FUNCTION OF THE C-RET RECEPTOR AND ITS LIGAND *IN VIVO*

The c-ret receptor was identified originally as a gene able to induce malignant transformation when mutated (59). Mutations in c-*ret* have been identified in thyroid cancers and in inherited multiple endocrine neoplasia (MEN) 2A and 2B (60,61); the germline mutations found in the *ret* gene of MEN2A and MEN2B patients were demonstrated to upregulate the tyrosine kinase activity of the receptor, rendering it independent of the presence of the putative ligand (62). The c-ret receptor is expressed only in a few distinct cells: epithelia of the meso- and metanephros (i.e., Wolffian duct and ureter), migrating neural crest cells and neurons of the peripheral nervous system (63). A targeted mutation in the c-*ret* gene of the mouse has demonstrated essential roles of c-ret in development of the kidney and the enteric nervous system (64). In the kidney, c-ret is produced initially in the entire ureter and consequently in the very tips of the ureter bud, where branching morphogenesis occurs. Animals that carry two mutant c-*ret* genes have renal agenesis or severe dysgenesis, presumably due to a defect in growth and morphogenesis of the ureter. In addition, enteric ganglia that derive from the neural crest do not form, and as a consequence, intestinal movements are absent and the intestine does not function. Heterozygous animals are healthy. Interestingly, in humans, an inherited disorder of enteric ganglia, Hirschsprung's disease, is caused by mutations in the c-*ret* gene (65,66). Here one copy of an intact allele is apparently not sufficient for c-ret function, and consequently Hirschsprung's disease is dominantly inherited.

The ligand for c-ret was recently identified (13–15): It is glial-derived neurotrophic factor (GDNF), a factor of the transforming growth factor (TGF)β family found because of its potent neurotrophic functions (67). Mutation of the GDNF gene in mice causes the same developmental defects as the mutation in c-*ret* (68,69). However, the GDNF receptor is complex and requires, in addition to the c-ret protein, a coreceptor protein, GDNFR-α. This coreceptor is a glycosylphosphatidylinositol-linked cell surface receptor that binds GDNF specifically and mediates activation of the c-ret protein-tyrosine kinase (70,71).

7.7. CONCLUSIONS AND FUTURE PERSPECTIVES

We have discussed the recent identification of the physiologic functions of a collection of tyrosine kinase receptors *in vivo*. These receptors were originally identified as oncogenes and have been implicated in the development of human epithelial tumors (carcinomas). Recent evidence demonstrates that these receptors have different roles in the growth or survival of cells, in the morphogenesis of epithelia, and in the motility of muscle and neural crest precursor cells. Particularly revealing is the genetic finding of the importance of tyrosine kinase receptors, which mediate cellular interactions, such as those between mesenchyme and epithelia. The list of receptors discussed here is by no means exhaustive, and many more such signaling systems are currently being studied. In all of the cases discussed here, the targeted mutation have led to embryonal or perinatal death and thus have yet not revealed the physiologic roles of the signaling systems in the adult. In the future, conditional genetic alterations will allow the analysis of additional functions in other development stages.

REFERENCES

1. Stoker M, Gherardi E, Perryman M, Gray J: Scatter factor is a fibroblast-derived modulator of epithelial cell mobility. Nature 327:239–242, 1987.
2. Weidner KM, Arakaki N, Hartmann G, Vandekerckhove J, Weingart S, Rieder H, Fonatsch C, Tsubouchi H, Hishida T, Daikuhara Y, Birchmeier W: Evidence for the identity of human scatter factor and human hepatocyte growth factor. Proc Natl Acad Sci USA 88:7001–7005, 1991.
3. Bottaro DP, Rubin JS, Faletto DL, Chan AM, Kmiecik TE, Vande WG, Aaronson SA: Identification of the hepatocyte growth factor receptor as the c-met proto-oncogene product. Science 251:802–804, 1991.
4. Naldini L, Weidner KM, Vigna E, Gaudino G, Bardelli A, Ponzetto C, Narsimhan RP, Hartmann G, Zarnegar R, Michalopoulos GK, Birchmeier W, Comoglio PM: Scatter factor and hepatocyte growth factor are indistinguishable ligands for the MET receptor. EMBO J 10:2867–2878, 1991.
5. Sonnenberg E, Weidner KM, Birchmeier C: Scatter factor/hepatocyte growth factor and its receptor, the c-met tyrosine kinase, can mediate a signal exchange between mesenchyme and epithelia during mouse development. J Cell Biol 123:223–235, 1993.
6. Peles E, Yarden Y: Neu and its ligands: From an oncogene to neural factors. Bioessays 15:815–824, 1993.
7. Carraway K 3, Cantley LC: A neu acquaintance for erbB3 and erbB4: A role for receptor heterodimerization in growth signaling. Cell 78:5–8, 1994.
8. Marchionni MA, Goodearl AD, Chen MS, Bermingham MO, Kirk C, Hendricks M, Danehy F, Misumi D, Sudhalter J, Kobayashi K, Wroblewski D, Lynch C, Baldessare M, Hiles I, Davis JB, Hsuan JJ, Totty NF, Otsu M, McBurney RN, Waterfield MD,

Stroobant P, Gwynne D: Glial growth factors are alternatively spliced erbB2 ligands expressed in the nervous system. Nature 362:312–318, 1993.

9. Orr UA, Trakhtenbrot L, Ben LR, Wen D, Rechavi G, Lonai P, Yarden Y: Neural expression and chromosomal mapping of Neu differentiation factor to 8p12–p21. Proc Natl Acad Sci USA 90:1867–1871, 1993.
10. Meyer D, Birchmeier C: Distinct isoforms of neuregulin are expressed in mesenchymal and neuronal cells during mouse development. Proc Natl Acad Sci USA 91:1064–1068, 1994.
11. Johnson DE, Williams LT: Structural and functional diversity in the FGF receptor multigene family. Adv Cancer Res 60:1–41, 1993.
12. Takahashi M, Ritz J, Cooper GM: Activation of a novel human transforming gene, *ret*, by DNA rearrangement. Cell 42:581–588, 1985.
13. Treanor JJ, Goodman L, De SF, Stone DM, Poulsen KT, Beck CD, Gray C, Armanini MP, Pollock RA, Hefti F, Phillips HS, Goddard A, Moore MW, Buj BA, Davies AM, Asai N, Takahashi M, Vandlen R, Henderson CE, Rosenthal A: Characterization of a multicomponent receptor for GDNF. Nature 382:80–83, 1996.
14. Vega QC, Worby CA, Lechner MS, Dixon JE, Dressler GR: Glial cell line-derived neurotrophic factor activates the receptor tyrosine kinase RET and promotes kidney morphogenesis. Proc Natl Acad Sci USA 93:10657–10661, 1996.
15. Durbec P, Marcos GC, Kilkenny C, Grigoriou M, Wartiowaara K, Suvanto P, Smith D, Ponder B, Costantini F, Saarma M, Sariola H, Pachnis V: GDNF signalling through the Ret receptor tyrosine kinase. Nature 381:789–793, 1996.
16. Pachnis V, Mankoo B, Costantini F: Expression of the c-*ret* proto-oncogene during mouse embryogenesis. Development 119:1005–1017, 1993.
17. Choi LD, Bohn MC: Ontogeny and distribution of glial cell line-derived neurotrophic factor (GDNF) mRNA in rat. Brain Res Dev Brain Res 85:80–88, 1995.
18. Trupp M, Ryden M, Jornvall H, Funakoshi H, Timmusk T, Arenas E, Ibanez CF: Peripheral expression and biological activities of GDNF, a new neurotrophic factor for avian and mammalian peripheral neurons. J Cell Biol 130:137–148, 1995.
19. Sonnenberg E, Godecke A, Walter B, Bladt F, Birchmeier C: Transient and locally restricted expression of the *ros1* protooncogene during mouse development. EMBO J 10:3693–3702, 1991.
20. Birchmeier C, Birchmeier W: Molecular aspects of mesenchymal–epithelial interactions. Annu Rev Cell Biol 9:511–540, 1993.
21. Nakamura T, Nishizawa T, Hagiya M, Seki T, Shimonishi M, Sugimura A, Tashiro K, Shimizu S: Molecular cloning and expression of human hepatocyte growth factor. Nature 342:440–443, 1989.
22. Miyazawa K, Tsubouchi H, Naka D, Takahashi K, Okigaki M, Arakaki N, Nakayama K, Hirono S, Sakiyama O, Takahashi K, Gohda E, Daikuhara Y, Kitamura N: Molecular cloning and sequence analysis of cDNA for human hepatocyte growth factor. Biochem Biophys Res Commun 163:967–973, 1989.
23. Grant DS, Kleinman HK, Goldberg ID, Bhargava MM, Nickoloff BJ, Kinsella JL, Polverini P, Rosen EM: Scatter factor induces blood vessel formation in vivo. Proc Natl Acad Sci USA 90:1937–1941, 1993.
24. Bussolino F, Di RM, Ziche M, Bocchietto E, Olivero M, Naldini L, Gaudino G, Tamagnone L, Coffer A, Comoglio PM: Hepatocyte growth factor is a potent angiogenic factor which stimulates endothelial cell motility and growth. J Cell Biol 119:629–641, 1992.

25. Montesano R, Matsumoto K, Nakamura T, Orci L: Identification of a fibroblast-derived epithelial morphogen as hepatocyte growth factor. Cell 67:901–908, 1991.

26. Brinkmann V, Foroutan H, Sachs M, Weidner KM, Birchmeier W: Hepatocyte growth factor/scatter factor induces a variety of tissue-specific morphogenic programs in epithelial cells. J Cell Biol 131:1573–1586, 1995.

27. Ebens A, Brose K, Leonardo ED, Hanson MG, Bladt F, Birchmeier C, Barres B, Tessier-Lavigne M: Hepatocyte growth factor/scatter factor is an axonal chemoattractant and a neurotrophic factor for spinal motor neurons. Neuron 17:1157–1172, 1996.

28. Peles E, Bacus SS, Koski RA, Lu HS, Wen D, Ogden SG, Levy RB, Yarden Y: Isolation of the neu/HER-2 stimulatory ligand: A 44 kd glycoprotein that induces differentiation of mammary tumor cells. Cell 69:205–216, 1992.

29. Holmes WE, Sliwkowski MX, Akita RW, Henzel WJ, Lee J, Park JW, Yansura D, Abadi N, Raab H, Lewis GD, Shepard HM, Kuang W-J, Wood WI, Goeddel DV, Vandlen RL: Identification of heregulin, a specific activator of p185erbB2. Science 256:1205–1210, 1992.

30. Falls DL, Rosen KM, Corfas G, Lane WS, Fischbach GD: ARIA, a protein that stimulates acetylcholine receptor synthesis, is a member of the neu ligand family. Cell 72:801–815, 1993.

31. Shah NM, Marchionni MA, Isaacs I, Stroobant P, Anderson DJ: Glial growth factor restricts mammalian neural crest stem cells to a glial fate. Cell 77:349–360, 1994.

32. Dong Z, Brennan A, Liu N, Yarden Y, Lefkowitz G, Mirsky R, Jessen KR: Neu differentiation factor is a neuron-glia signal and regulates survival, proliferation, and maturation of rat Schwann cell precursors. Neuron 15:585–596, 1995.

33. Morrissey TK, Levi AD, Nuijens A, Sliwkowski MX, Bunge RP: Axon-induced mitogenesis of human Schwann cells involves heregulin and p185erbB2. Proc Natl Acad Sci USA 92:1431–1435, 1995.

34. Jo SA, Zhu X, Marchionni MA, Burden SJ: Neuregulins are concentrated at nerve-muscle synapses and activate ACh-receptor gene expression. Nature 373:158–161, 1995.

35. Trachtenberg JT, Thompson WJ: Schwann cell apoptosis at developing neuromuscular junctions is regulated by glial growth factor. Nature 379:174–177, 1996.

36. Canoll PD, Musacchio JM, Hardy R, Reynolds R, Marchionni MA, Salzer JL: GGF/neuregulin is a neuronal signal that promotes the proliferation and survival and inhibits the differentiation of oligodendrocyte progenitors. Neuron 17:229–243, 1996.

37. Bermingham MO, McCabe KL, Reh TA: Effects of GGF/neuregulins on neuronal survival and neurite outgrowth correlate with erbB2/neu expression in developing rat retina. Development 122:1427–1438, 1996.

38. Yang Y, Spitzer E, Meyer D, Sachs M, Niemann C, Hartmann G, Weidner KM, Birchmeier C, Birchmeier W: Sequential requirement of hepatocyte growth factor and neuregulin in the morphogenesis and differentiation of the mammary gland. J Cell Biol 131:215–226, 1995.

39. Weidner KM, Dicesare S, Sachs M, Brinkmann V, Behrens J, Birchmeier WA: Interaction between Gab1 and the c-met receptor tyrosine kinase is responsible for epithelial morphogenesis. Nature 384:173–176, 1996.

40. Sachs M, Weidner KM, Brinkmann V, Walther I, Obermeier A, Ullrich A, Birchmeier W: Motogenic and morphogenic activity of epithelial receptor tyrosine kinases. J Cell Biol 133:1095–1107, 1996.

41. Schmidt C, Bladt F, Goedecke S, Brinkmann V, Zschiesche W, Sharpe M, Gherardi E, Birchmeier C: Scatter factor/hepatocyte growth factor is essential for liver development. Nature 373:699–702, 1995.
42. Uehara Y, Minowa O, Mori C, Shiota K, Kuno J, Noda T, Kitamura N: Placental defect and embryonic lethality in mice lacking hepatocyte growth factor/scatter factor. Nature 373:702–705, 1995.
43. Bladt F, Riethmacher D, Isenmann S, Aguzzi A, Birchmeier C: Essential role for the c-met receptor in the migration of myogenic precursor cells into the limb bud. Nature 376:768–771, 1995.
44. Meyer D, Birchmeier C: Multiple essential functions of neuregulin in development. Nature 378:386–390, 1995.
45. Gassmann M, Casagranda F, Orioli D, Simon H, Lai C, Klein R, Lemke G: Aberrant neural and cardiac development in mice lacking the ErbB4 neuregulin receptor. Nature 378:390–394, 1995.
46. Lee KF, Simon H, Chen H, Bates B, Hung MC, Hauser C: Requirement for neuregulin receptor erbB2 in neural and cardiac development. Nature 378:394–398, 1995.
47. Corfas G, Rosen KM, Aratake H, Krauss R, Fischbach GD: Differential expression of ARIA isoforms in the rat brain. Neuron 14:103–115, 1995.
48. Chen MS, Bermingham MO, Danehy FJ, Nolan C, Scherer SS, Lucas J, Gwynne D, Marchionni MA: Expression of multiple neuregulin transcripts in postnatal rat brains. J Comp Neurol 349:389–400, 1994.
49. Meyer D, Yamaai T, Garratt A, Rietmacher-Sonnenberg E, Kramer R, Theill LE, Birchmeier C: Isoform-specific expression and function of neuregulin. Development, in press.
50. Dong Z, Brennan A, Liu N, Yarden Y, Lefkowitz G, Mirsky R, Jessen KR: Neu differentiation factor is a neuron-glia signal and regulates survival, proliferation, and maturation of rat Schwann cell precursors. Neuron 15:585–596, 1995.
51. Morrissey TK, Levi AD, Nuijens A, Sliwkowski MX, Bunge RP: Axon-induced mitogenesis of human Schwann cells involves heregulin and p185erbB2. Proc Natl Acad Sci USA 92:1431–1435, 1995.
52. Kuhn R, Schwenk F, Aguet M, Rajewsky K: Inducible gene targeting in mice. Science 269:1427–1429, 1995.
53. Gu H, Marth JD, Orban PC, Mossmann H, Rajewsky K: Deletion of a DNA polymerase beta gene segment in T cells using cell type-specific gene targeting. Science 265:103–106, 1994.
54. Yamaguchi TP, Rossant J: Fibroblast growth factors in mammalian development. Curr Opin Genet Dev 5:485–491, 1995.
55. Peters K, Werner S, Liao X, Wert S, Whitsett J, Williams L: Targeted expression of a dominant negative FGF receptor blocks branching morphogenesis and epithelial differentiation of the mouse lung. EMBO J 13:3296–3301, 1994.
56. Werner S, Smola H, Liao X, Longaker MT, Krieg T, Hofschneider PH, Williams LT: The function of KGF in morphogenesis of epithelium and reepithelialization of wounds. Science 266:819–822, 1994.
57. Werner S, Weinberg W, Liao X, Peters KG, Blessing M, Yuspa SH, Weiner RL, Williams LT: Targeted expression of a dominant-negative FGF receptor mutant in the epidermis of transgenic mice reveals a role of FGF in keratinocyte organization and differentiation. EMBO J 12:2635–2643, 1993.

58. Sutherland D, Samakovlis C, Krasnow MA: Branchless encodes a *Drosophila* FGF homolog that controls tracheal cell migration and the pattern of branching. Cell 87:1091–1101, 1996.

59. Takahashi M, Ritz J, Cooper GM: Activation of a novel human transforming gene, ret, by DNA rearrangement. Cell 42:581–588, 1985.

60. Grieco M, Santoro M, Berlingieri MT, Melillo RM, Donghi R, Bongarzone I, Pierotti MA, Della PG, Fusco A, Vecchio G: PTC is a novel rearranged form of the *ret* proto-oncogene and is frequently detected in vivo in human thyroid papillary carcinomas. Cell 60:557–563, 1990.

61. Mulligan LM, Kwok JB, Healey CS, Elsdon MJ, Eng C, Gardner E, Love DR, Mole SE, Moore JK, Papi L, Ponder MA, Telenius H, Tunnacliffe A, Ponder BAJ: Germ-line mutations of the *RET* proto-oncogene in multiple endocrine neoplasia type 2A. Nature 363:458–460, 1993.

62. Santoro M, Carlomagno F, Romano A, Bottaro DP, Dathan NA, Grieco M, Fusco A, Vecchio G, Matoskova B, Kraus MH, Di Fore PP: Activation of *RET* as a dominant transforming gene by germline mutations of MEN2A and MEN2B. Science 267:381–383, 1995.

63. Pachnis V, Mankoo B, Costantini F: Expression of the c-*ret* proto-oncogene during mouse embryogenesis. Development 119:1005–1017, 1993.

64. Schuchardt A, D'Agati V, Larsson BL, Costantini F, Pachnis V: Defects in the kidney and enteric nervous system of mice lacking the tyrosine kinase receptor Ret. Nature 367:380–383, 1994.

65. Romeo G, Ronchetto P, Luo Y, Barone V, Seri M, Ceccherini I, Pasini B, Bocciardi R, Lerone M, Kaariainen H, Martucciello G: Point mutations affecting the tyrosine kinase domain of the *RET* proto-oncogene in Hirschsprung's disease. Nature 367:377–378, 1994.

66. Edery P, Lyonnet S, Mulligan LM, Pelet A, Dow E, Abel L, Holder S, Nihoul FC, Ponder BA, Munnich A: Mutations of the *RET* proto-oncogene in Hirschsprung's disease. Nature 367:378–380, 1994.

67. Henderson CE, Phillips HS, Pollock RA, Davies AM, Lemeulle C, Armanini M, Simmons L, Moffet B, Vandlen RA, Simpson TSL, Moffet B, Vandlen RA, Koliatsos VE, Rosenthal A: GDNF: A potent survival factor for motoneurons present in peripheral nerve and muscle. Science 266:1062–1064, 1994.

68. Moore MW, Klein RD, Farinas I, Sauer H, Armanini M, Phillips H, Reichardt LF, Ryan AM, Carver MK, Rosenthal A: Renal and neuronal abnormalities in mice lacking GDNF. Nature 382:76–79, 1996.

69. Pichel JG, Shen L, Sheng HZ, Granholm AC, Drago J, Grinberg A, Lee EJ, Huang SP, Saarma M, Hoffer BJ, Sariola H, Westphal H: Defects in enteric innervation and kidney development in mice lacking GDNF. Nature 382:73–76, 1996.

70. Jing S, Wen D, Yu Y, Holst PL, Luo Y, Fang M, Tamir R, Antonio L, Hu Z, Cupples R, Louis JC, Hu S, Altrock BW, Fox GM: GDNF-induced activation of the ret protein tyrosine kinase is mediated by GDNFR-alpha, a novel receptor for GDNF. Cell 85:1113–1124, 1996.

71. Treanor JJ, Goodman L, De SF, Stone DM, Poulsen KT, Beck CD, Gray C, Armanini MP, Pollock RA, Hefti F, Phillips HS, Goddard A, Moore MW, Buj BA, Davies AM, Asai N, Takahashi M, Vandlen R, Henderson CE, Rosenthal A: Characterization of a multicomponent receptor for GDNF. Nature 382:80–83, 1996.

8

DEVELOPMENTAL AND PHYSIOLOGIC ROLES OF ErbB RECEPTORS AND THEIR LIGANDS IN MAMMALS

NOA BEN-BARUCH, IRIS ALROY, AND YOSEF YARDEN

Department of Molecular Cell Biology, The Weizmann Institute of Science, Rehovot, 76100 Israel

8.1. INTRODUCTION

Intercellular communication plays a pivotal role in the establishment of cell lineage diversity in mammalian and other organisms. Two classes of molecules are considered crucial mediators of this communication. These are insoluble components, in the form of the extracellular matrix and adhesion molecules at the cell surface, and soluble hormone-like messengers, often called *growth factors*. Signaling by all polypeptide growth factors to the cytoplasm and to the nucleus is mediated by membranal receptors, whose extracellular portions bind the ligand messengers; their intracellular parts associate with appropriate cytoplasmic effector molecules, often enzymes or adapters of multimolecular complexes.

Many growth factors bind to cell surface receptors whose intracellular domain carries an enzymatic activity, a tyrosine-specific kinase (1). The family of receptor tyrosine kinases (RTKs) includes more than a dozen subgroups, each comprising 2–15 homologous members that bind related growth factors and maintain intrasubgroup physical interactions (2). The extracellular ligand binding domains are the most diverse portions of RTKs and may be considered the variable portions, by

Hormones and Growth Factors in Development and Neoplasia, Edited by Robert B. Dickson and David S. Salomon
ISBN 0-471-16899-8

analogy to immunoglobulins. In contrast, the cytoplasmic portions share a relatively constant domain, the tyrosine kinase catalytic core, but this is flanked at both ends by variable amino acid sequences that enable interaction with receptor-specific effector and regulatory proteins in the cytoplasm. In addition to their structural homology, all RTKs have a common mechanism of signal transduction. Crucial to understanding this mechanism is the realization that RTKs are in fact allosteric molecules whose allosteric regulators are the ligands themselves. As in other allosteric enzyme systems, the monomeric receptor molecule is inactive. However, a ligand-driven minimal oligomerization, namely, the formation of receptor dimers, is sufficient for full receptor activation (3). Within receptor dimers a rapid phosphorylation reaction occurs, generating phosphotyrosine moieties on the cytoplasmic receptors' portions. Presumably, phosphorylation is mediated by an exclusive intermolecular mechanism (transphosphorylation), which explains the absolute necessity for receptor dimerization. In the next step, the phosphotyrosine residues act as docking sites for a protein module, called the *Src homology 2 (SH2) domain,* that recognizes a phosphotyrosine in the context of a few neighboring amino acids (4). Many regulatory cytoplasmic proteins, including enzymes, adaptor proteins, and transcription factors, share the SH2 domain. Therefore their association with the activated RTKs enables simultaneous stimulation of diverse intracellular processes, which explains the pleiotropic biochemical response of cells to growth factors.

Type 1 RTKs are among the most extensively studied signaling systems, probably because the prototypic receptor of this family, the epidermal growth factor (EGF) receptor, was the first receptor to be isolated and molecularly cloned (5). Another reason this group of receptors attracted attention is their involvement in human cancer, particularly tumors of nerve and epithelial cells (carcinomas), an aspect that is covered in other chapters of this volume. Figure 8.1a presents a schematic view of the known ErbB proteins and their respective ligands. In addition to the EGF receptor (also called ErbB-1), subtype 1 RTKs include two receptors for neuregulins, ErbB-3 (6,7) and ErbB-4 (8), and a fourth member, ErbB-2 (also called HER2 and Neu), whose ligand, if it exists, is still unknown (9).

8.2. ErbB PROTEINS AND LIGANDS

The four ErbB receptors share significant structural homology. All possess a large, glycosylated extracellular ligand binding domain with two cysteine-rich stretches, a single hydrophobic transmembrane domain, and a long cytoplasmic tail containing the kinase domain and C-terminally located autophosphorylation sites. The ligand binding cleft, at least in the case of ErbB-1, is confined to the portion linking the two cysteine-rich domains (10). Remarkably, this region is relatively conserved within the family, although the four proteins bind distinct ligands. Receptor specificity is not confined to the ligands but extends to effector proteins and substrates. Each receptor couples to a unique set of SH-2-containing signaling proteins whose identities are determined by the receptor's autophosphorylation sites. The ensuing

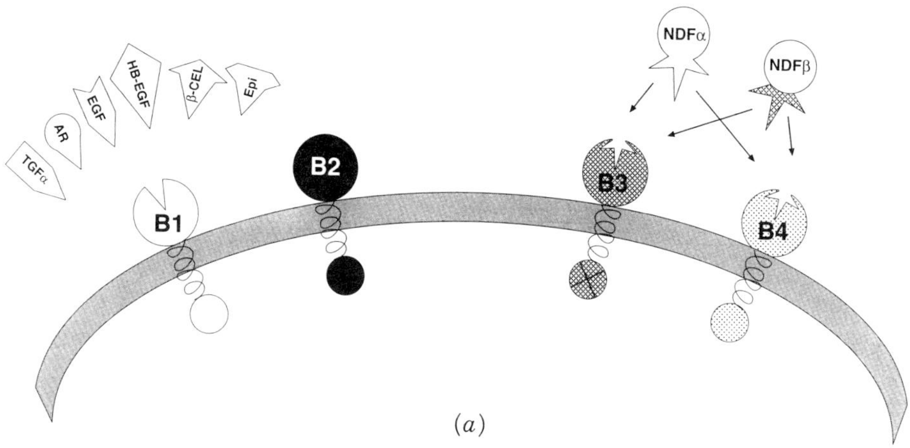

(*a*)

EGF	Human	1	NSDSEC-PLSHDGYCLHDGVCMYIEALD----------KYACNCVVGYIGERCQYRDLK
TGFα	Human	1	SHFNDC-PDSHTQFCFH-GTCRFLVQED----------KPACVCHSGYVGARCEHADLL
Amphiregulin	Human	41	KKKNPC-NAEFQNFCIH-GECKYIEHLE----------AVTCKCQQEYFGERCGEK
HB-EGF	Human	30	KKRDPC-LRKYKDFCIH-GECKYVKELR----------APSCICHPGYHGERCHGLSLP
Betacellulin	Mouse	132	THFSRC-PKQYKHYCIH-GRCRFVVDEQ----------TPSCICEKGYFGARCERVDLF
Epiregulin	Mouse	56	VQITKC-SSDMDGYCLH-GQCIYLVDMR----------EKFCRCEVGYTGLRCEHFFLT
NDFα	Human	177	SHLVKC-AEKEKTFCVNGGECFMVKDLSNPS-------RYLCKCQPGFTGARCTENVPM
NDFβ	Human	177	SHLVKC-AEKEKTFCVNGGECFMVKDLSNPS-------RYLCKCPNEFTGDRCQNYVMA
VGF	Vaccinia	19	PAIRLC-GPEGDGYCLH-GDCIHARDID----------GMYCRCSHGYTGIRCQHVVLV
SFGF	Shope Fibroma	26	KHVKVC-NHDYENYCLNNGTCFTIALDNVSI-------TPFCVCRINYEGSRCQFINLV
MGF	Myxoma	30	KRIKLC-NDDYKNYCLNNGTCFTVALNNVSL-------NPFCACHINYVGSRCQFINLI
Vein	*Drosophila*	560	ASGIPC----NFDYCFHNGTCRMIPDIN----------EVYCRCPTEYFGNRCENKWPD
Gurken	*Drosophila*	177	IQMLPCSEAYNTSFCINGGHCFQHPMVNNTV-------FHSCLCVNDYDGFRCAYKSWN
Argos	*Drosophila*	373	RYLFAC-SPLTRLRCQRKQPCKLFTVRKRQEFLDEVNINSLCQCPK---GHRCPSHHTQ
Spitz	*Drosophila*	74	FPTYKCPETFDAWYCLNDAHCFAVKIADLP--------VYSCECAIGFMGQRCEYKEID
Lin-3	*C. elegans*	149	LKEAKC-----KDYCHHNATCHVEVIFREDRVSAV---VPSCHCPQGWEGTRCDRHYVQ

(*b*)

Fig. 8.1. ErbB proteins and their ligands. (a) Schematic representation of the four ErbB proteins and their respective ligands. Bilobular structures that cross the plasma membrane represent the extracellular and cytoplasmic portions of each ErbB protein. Note that the kinase function of ErbB-3 (crossed domain) is inactive. The following ligands of ErbB-1 are shown: EGF, TGFα, amphiregulin (AR), heparin-binding EGF-like growth factor (HB-EGF), betacellulin (β-CEL), and epiregulin (Epi). The cross-interaction of betacellulin with ErbB-4 is not shown. All of the isoforms of NDF/neuregulin are represented by the two classes, α and β. (b) Amino acid sequence alignment of mammalian and invertebrate ErbB ligands. The sequences of all of the factors represented in (a) are shown in the upper part. Also aligned are the sequences of three viral proteins: the vaccinia growth factor (VGF), the factor of Shope fibroma virus (SFGF), and the Myxoma virus factor (MGF). The four known *Drosophila* ligands and the *C. elegans* ligand encoded by the *lin-3* gene are shown in the lower part. The number of the first amino acid of each of the EGF-like motifs is indicated. The six cysteine residues are marked, as are other residues that are conserved in the family. Note the relatively long B-loop and short C-loop of Argos, and the absence of one of the conserved glycines in both Argos and Spitz.

selectivity provides a basis for signaling specificity (11). Remarkably, the set of signaling proteins that associate with each receptor are largely overlapping, with only a few known targets that are not shared by at least another ErbB protein. One example is the c-Cbl proto-oncogenic protein (12). Unlike other ErbB proteins, the catalytic activity of ErbB-3 is severely impaired (13). Another feature that is unique to each ErbB protein is its rate of endocytosis after ligand binding (a process termed *downregulation*) (14,15). Lastly, the patterns of expression of type 1 RTKs are distinct: ErbB-2 is the most widely expressed protein, found on derivatives of all three embryonic germ layers, whereas ErbB-4 expression is relatively limited to nerve and other cell types. In summary, each of the four RTKs displays a unique set of structural and functional features that are consistent with nonredundant physiologic and developmental roles.

The allosteric regulators of all ErbB proteins are relatively small single-chain polypeptides that share a module called the *EGF motif* (Fig. 8.1b). This 45- to 55-amino-acid-long motif includes six cysteine residues that are typically spaced and form three covalently held loops. The cysteine residues, along with some glycines and a critical arginine, are crucial for the formation of a double β-sheet structure that appears to be shared by all ligands (16). The prototype ligand of ErbB-1 is EGF, a factor that was originally isolated from the submaxillary glands of mice or from human urine. Another ErbB-1 ligand is transforming growth factor α (TGFα), which was isolated from the medium of cultured transformed cells (17). Other ligands of ErbB-1 include amphiregulin (18), betacellulin (19), epiregulin (20), and the heparin-binding EGF (21). In addition, three viral proteins bind to ErbB-1 and activate it to some extent. These are the vaccinia virus growth factor (VGF), the Shopes sarcoma virus growth factor (SGF), and the Myxoma virus growth factor (MGF).

Betacellulin was originally identified as an ErbB-1 ligand, but it has been recently shown to cross-react with ErbB-4 (22). Otherwise, the specificity barriers of ErbB proteins are quite well-defined, at least at physiologic ligand concentrations. Both ErbB-4 and ErbB-3 share their ligand specificity, although their binding pockets are immunologically distinct (23). All known ligands of these two receptors are encoded by a single gene and arise by alternative RNA splicing. Currently, there are 15 known isoforms, collectively denoted *neuregulins (NRGs),* that fall into two groups, α and β, differing in the most C-terminal loop of the EGF-like motif (Fig. 8.1b) (24). NRGs, like all other mammalian ErbB ligands, are synthesized as large transmembrane precursors, whose extracellular N-terminal domains contain a juxtamembrane EGF-like module and various other motifs. In the NRG family of isoforms, these other motifs include an almost ubiquitous immunoglobulin (Ig)-like loop that is flanked at its amino-terminal side by various sequences, including a kringle-like domain in the glial growth factor (GGF) (25). In another isoform, the acetylcholine receptor-inducing activity (ARIA), a short, unrelated sequence, is found in the N terminus (26). Yet, the Ig-like domain is replaced by a cysteine-rich domain in another neuronal isoform called *sensory and motor neuron-derived factor (SMDF)* (27). All mesenchymal isoforms of NRG, including heregulin (28) and the Neu differentiation factor (NDF) (29), share a unique glycosylation domain that is not found in the neuronal isoforms. Otherwise, the various isoforms differ in the length of their cytoplasmic tails and in the putative proteolytic processing site connecting the EGF-like motif with the transmembrane domain.

8.3. THE ErbB SIGNALING NETWORK

Originally, isoforms of NRGs were isolated in the course of a search for ligands of ErbB-2 by using a kinase assay. However, the isolated ligands, termed *NDF* or *heregulin,* were unable to stimulate ErbB-2 phosphorylation in certain cells, which implied that the direct receptor was not ErbB-2 (30). Later studies identified the direct receptors as ErbB-3 and ErbB-4 (31–33) and suggested that ErbB-2 was activated by NRG only in trans. Likewise, activation of ErbB-2 by ligands of ErbB-1 is due to heterodimer formation between the two proteins (34,35). In fact, all 10 possible homo- and heterodimers of ErbB proteins exist, including a homodimer of ErbB-2 whose formation is driven by an oncogenic mutation in the transmembrane domain (36). Moreover, the intrafamily interactions display a definite hierarchy with a preference for heterodimers, especially those containing ErbB-2 (37). The best example of an interreceptor network is provided by ErbB-3. This kinase-defective receptor forms both homo- and heterodimers with all other members of its family. However, whereas homodimers are completely devoid of biologic activity, heterodimers are active; the most potent is the ErbB-3/ErbB-2 heterodimer (15). Possibly this represents a mechansim of stringent control over ErbB-3, a protein characterized by multiple docking sites for SH2 proteins since both a ligand and a heterodimer partner are necessary for receptor activation. Analysis of individual homo- and heterodimers of ErbB proteins revealed a graded pattern of mitogenic signals, implying that the complex network enables diversification of signaling through combinatorial interactions (15,38).

Although ErbB-2 has no known direct ligand of its own, it emerges as a common auxiliary subunit of the ErbB network. Inactivation of ErbB-2 in cultured cells, by inhibiting its expression at the cell surface, significantly reduces signaling by EGF and NRG (39,40). Consistently, ErbB-2-containing heterodimers exhibit higher mitogenic activity than other receptor combinations. The mechanism of signal amplification by ErbB-2 involves a significant reduction in the rate of dissociation of both types of ligands from heterodimeric complexes containing ErbB-2 (40). Possibly ErbB-2 acts as a signal amplifier because it is a potent activator of the MAP-kinase pathway (41) and also because it prolongs signaling by all ErbB ligands. This realization led us to propose that ErbB-2 may not function as a bone fide receptor, but rather as a signaling subunit of other receptors (40). Independent of the exact role of ErbB-2, the existence of extensive interreceptor interactions within the ErbB family, and perhaps in other subtypes of RTKs, extends the classic pathway of vertical signaling by homodimeric receptors to a heterodimer-driven lateral type of signaling. Together the two mechanisms provide enormous potential for signal diversification in developmental processes.

Another layer of signal diversification is encoded at the level of the ligands. Potentially, the existence of multiple ErbB ligands that share receptor specificity could lead to redundant functions. However, this is probably not the case because each ligand molecule displays unique temporal and spatial patterns of expression and contains, in addition to the EGF module, diverse signals that determine its processing and retention by the extracellular matrix. Moreover, ligands may differ in their ability to form receptor heterodimers, as implied by the formation of ErbB-3/ErbB-1 heterodimers by NRGβ but not by α isoforms of this ligand (42). As a system, the ErbB signaling network comprises three layers of signal diversification: two dozen ligands, 10 different dimers of receptors, and many distinct sets of SH2 proteins that are activated by the various receptor dimers.

8.4. DEVELOPMENTAL ROLES OF ErbB IN INVERTEBRATE MODEL SYSTEMS

The ErbB family of receptors is evolutionary conserved in lower organisms such as *Caenorhabditis elegans* and the fruit fly *Drosophila*. The availability of many mutants of these relatively simple organisms provides significant insights that are relevant to the role of ErbB in mammals. Although the lessons learned in model systems are reviewed in other chapters in this volume, it is worthwhile to concentrate on selected aspects that are relevant to our later discussion.

8.4.1. Receptors

In the simpler organism, only one ErbB member is known. This is the product of the *C. elegans let-23* gene (43). The protein shares all structural motifs in the extracellular domain, as well as in the cytoplasmic region, with mammalian ErbB proteins. Amino acid homology is greater in the cytoplasmic tyrosine kinase sequences than elsewhere in the receptor. Likewise, only a single ErbB-like homolog is known in *Drosophila*. This is the product of the *DER* gene, which encodes two alternatively spliced transcripts differing in ectodomain-coding sequences (44). The existence of four ErbB proteins in mammals but only one in lower organisms suggests that gene duplication and diversification occurred relatively late in evolution (less than 800 million years ago), probably to accommodate the increased cellular heterogeneity and the more complex physiology of mammalian organisms.

8.4.2. Ligands

A similar conclusion applies to the respective ligand molecules of invertebrate ErbB proteins. Whereas only one putative ligand was identified in *C. elegans,* four EGF-like ligands are known in *Drosophila*. The *lin-3* gene of *C. elegans* encodes an EGF-like protein with a transmembrane domain and a short cytoplasmic tail (45). Interestingly, as with mammalian neuregulins, two Lin-3 transcripts exist. They differ in the stretch of amino acids that links the EGF-like motif with the transmembrane domain. Two of the *Drosophila* ligands, Grk and Spi, like Lin-3, resemble

TGFα in that their precursors are transmembrane proteins that carry no recognizable motif other than the EGF-like motif (46,47). A third ligand, Vein, is similar to NRGs because it contains an Ig-like motif and a relatively long N-terminal sequence with potential glycosylation sites (48). However, unlike neuregulins and other ErbB ligands, it carries no transmembrane domain. Nevertheless, all ligands are characterized by conservation of the canonical cysteine, glycine, and arginine residues of the receptor-binding type of EGF motifs (Fig. 8.1b), suggesting that they indeed function as ErbB ligands. Evolutionary conservation of the transmembrane topology of most ErbB ligands, as well as the membrane proximal localization of the EGF motif and conservation of non-EGF-like motifs, imply that these characteristics also play a biologic role in mammals, but its exact nature is still unknown. Another open question in both mammals and *Drosophila* relates to the multiplicity of ligands. Evidence from *C. elegans* suggests that limited availability of the ligand is essential for its localized action. Similarly, unlike the wide expression of several RTKs in the entire circumference of *Drosophila* embryos, their activation is specific to certain regions. This is probably mediated by spatially restricted ligands (49). The mechanisms underlying limitation of ligand availability are not well characterized, but genetic evidence implies that the *Drosophila* Rhomboid membrane protein is somehow involved in processing of the transmembrane Spi precursor. The latter may be active only in its soluble form (50). Presumably, the diverse non-EGF motifs of the multiple ligands determine their rates of processing by an extracellular protease, and also affect signal termination by ligand clearance through binding to heparan sulfate proteoglycans and other matrix components. Termination of ligand action is an issue that has not been studied in depth in mammals, but evidence from *Drosophila* implies that a specific mechanism was developed to limit ligand availability. A fourth EGF-like ligand, the product of the *argos* gene, appears to act as a receptor antagonist whose synthesis is upregulated on receptor activation (51). No known mammalian ligand is functionally analogous to Argos. This ligand is synthesized as a soluble secreted protein, and its antagonist action may be due to an unusually long B-loop of the EGF motif.

8.4.3. Intracellular Signaling

Strikingly, and in contrast to the diversity of extracellular interactions of invertebrate ErbB receptors, their downstream signaling pathways are similar and shared with other receptor tyrosine kinases. In *C. elegans,* Let-23 activation by its ligand results in recruitment of the Ras-Raf pathway (Let-60 and Lin-45, respectively) through stimulation of the Sem-5 adapter protein, the homolog of mammalian Grb-2 (52). Genetic and biochemical lines of evidence indicate that several *Drosophila* receptors, including DER, Torso and Sevenless, couple to the analogous pathway and implicate Drk, a Sem-5 homolog, and Sos, a guanine-nucleotide releasing factor, in the linear cascade leading to stimulation of the MAP-kinase pathway (53). Considering the large variety of SH-2 proteins that physically associate with mammalian ErbB proteins, as well as with other receptors (2), the uniformity of signaling by invertebrate RTKs is surprising.

8.4.4. Developmental Roles

Perhaps the best-characterized developmental role of an ErbB protein and its ligand is provided by vulval induction in *C. elegans*. The vulva of the nematode is used for egg laying and copulation and is generated by six precursor cells. A gonadal anchor cell secretes a ligand, Lin-3, that binds to the Let-23 homolog of ErbB, thereby directing the closest three precursor cells to undergo a series of divisions and morphogenetic changes culminating in vulva generation. The other three cells, which do not receive the ligand, differentiate as a nonspecialized hypodermis (54). Interestingly, in the vulva, Let-23 stimulates three rounds of cell division that are coupled to cell differentiation. However, the same receptor induces differentiation, without cell growth, of bipotent neuroectoblasts in *C. elegans* (P11 and P12 cells), indicating that cell fate determination is variably linked to cell division. As will be discussed later, this feature is also reflected in mammals, but the long and extensive use of *in vitro* experimental systems and established cell lines often avoided recognition of ErbB ligands, and other so-called growth factors, as differentiation inducers. A full description of the developmental roles of the *Drosophila* ErbB homolog and its ligands is beyond the scope of this review, and the specific aspects of insect embryogenesis may not be relevant to mammals. Nevertheless, one important lesson learned by isolation of about 50 mutant alleles of the *Drosophila* receptor is that the DER signaling pathway is necessary for the development of multiple and seemingly unrelated tissues and organs, and its requirement coincides with more than one developmental phase (55,56). A role in cell fate determination is exemplified in oogenesis. The anterodorsal-localized Gurken ligand of the oocyte appears to bind to dorsal follicle cells that express DER, thereby establishing dorsoventral polarity of the egg. Similarly, during embryogenesis and larva development, the receptor relays establishment of ventral ectodermal fates, differentiation of midline glial cells, Malphigian tubule development, head development, vein and bristle formation, and differentiation of photoreceptor cells. However, DER and its ligands are involved not only in differentiation but also in cell proliferation. For example, during imaginal disc formation, DER is essential for proliferation of disc cells, so that mutant adult animals are generally smaller than wild-type flies. In conclusion, the same ligand-receptor system is repeatedly used during embryonic development and later in adult life to mediate unrelated cellular interactions, the majority of which do not involve mitosis.

8.5. DEVELOPMENTAL AND PHYSIOLOGIC ROLES OF ErbB RECEPTORS AND LIGANDS IN MAMMALS

8.5.1. ErbB-1 and Its Ligands

As early as 1962, Stanley Cohen injected EGF purified from the mouse submaxillary gland into newborn mice and rats. After a week of daily injections, he observed acceleration of incisor eruption and eyelid opening (57). Gene targeting experiments that eliminated expression of ErbB-1, and therefore blocked the action of

EGF as well as that of other ErbB-1 ligands, have shown that the major ensuing effect is on epithelial cells, including those involved in tooth growth and eyelid formation (58–60). Remarkably, absence of ErbB-1 expression is lethal, but the exact phase at which the embryos or newborn animals die is dependent on the genetic background. With specific strains of mice, animals survived for up to 3 weeks, enabling more extensive phenotype analysis. In addition to delayed and defective epithelial development, the mutant mice were relatively small, indicating an effect on cell proliferation, which is supported by observations made with aggregation chimeras. Obviously, the phenotype of erbB-$1^{-/-}$ mice cannot resolve the specific function of each of the multiple ligands of ErbB-1. However, targeting of the TGFα gene and the consequent realization that the murine waved-1 *(wa-1)* mutation reduces TGFα expression (61,62) predict that milder phenotypes are reflected on targeting of specific ligands. Unlike erbB-1 knockedout mice, mice homozygous for a disrupted TGFα gene are healthy and fertile, but display eye abnormalities and waviness of the whiskers and fur due to derangement of hair follicles. These characteristics are shared by erbB-$1^{-/-}$ mice, and also by *wa-2* mutant mice whose ErbB-1 protein carries a mutation in the kinase domain resulting in reduced catalytic activity (63,64). It is important to note that the phenotypic effect of TGFα gene knockout reflected only part of the abnormalities of erbB-$1^{-/-}$ mice. Thus, the limited penetrance of TGFα mutations, and confinement of the phenotype to the skin and eye, suggest that each ErbB-1 ligand has a distinct functional role and tissue specificity, by analogy to the different roles played by each of the DER ligands in insect development. The following discussion summarizes the known physiologic roles of the ErbB-1 signaling module by integrating *in vivo* and *in vitro* observations.

Placenta and Early Embryo Development Transcripts of TGFα and ErbB-1 are detectable in preimplantation mouse embryos, EGF is present in oviductal fluid, and the egg contains a maternally derived mRNA of TGFα (65). In addition, heparin-binding (HB)-EGF is expressed in the mouse uterine luminal epithelium just prior to implantation and only at the site of blastocyst apposition (66). Furthermore, preimplantation embryos respond *in vitro* to ErbB-1 ligands, and the blastocysts of this stage adhere to cells manipulated to express the transmembrane form of HB-EGF (67), suggesting that ErbB-1 plays a role in blastocyst attachment to the uterus. Consistent with this possibility, in one strain of erbB-$1^{-/-}$ mutant mice, blastocysts exhibited peri-implantation failure due to degeneration of the inner cell mass (59). Strain specificity of the effect, however, suggests that other mechanisms may be operative, depending on the genetic background. Alternatively, a maternally derived ErbB-1 may explain why, in certain strains, development proceeds in the absence of ErbB-1. Another role of ErbB-1 in early development is to regulate the first embryonic epithelium, the trophectoderm. After implantation in the mouse, this cell layer forms the spongiotrophoblast layer, a highly invasive tissue that anchors the placenta to the uterine wall. Size reduction and disorganization of the spongiotrophoblast layer were observed in ErbB-$1^{-/-}$ embryos, along with smaller placentas and reduced embryonic-maternal adhesiveness (58).

Hair and Skin Numerous *in vitro* studies established the role of TGFα as a major autocrine regulator of the epidermis. This ligand is more potent than EGF in promoting keratinocyte proliferation and migration (68), and its expression is remarkably elevated in two hyperproliferative skin diseases, psoriasis and (EHK) lesions (69,70). Transgenic mice expressing TGFα under the control of a keratin promoter confirmed a regulatory role in skin; epidermal thickening was observed in regions where the skin is normally thick and hair follicle density is low (71,72). The outer epidermal layers of the transgenic mice were flaky, and hair growth was stunted. Interestingly, the mutant mice were smaller than their normal littermates and exhibited other abnormalities, including skin papillomas. Evidence from gene targeting experiments also supports the role of TGFα in skin. Although the dermis of TGF$\alpha^{-/-}$ mice exhibited no histologic abnormalities, these animals displayed altered hair shafts and follicles in both skin and whiskers (61,62). Similarly, very thin skin with fewer and disorganized hair follicles characterized ErbB-1-deficient mice. Apparently, the absence of ErbB-1 in these mice retarded both proliferation and maturation of epidermal cells. Taken together, a wealth of evidence supports a major role for ErbB-1 in skin and hair development and attributes a specific role to one of its ligands, namely, TGFα.

Lung Development EGF, TGFα, and ErbB-1 are expressed in lung epithelium of mice, rats, and humans, and treatment of fetal lung organ explants with exogenous EGF stimulated branching morphogenesis and cell proliferation (73). Pharmacologic treatment of rabbits, as well as other animals, with EGF resulted in accelerated cellular differentiation, DNA synthesis, and surfactant production in the lungs. Blocking EGF activity *in vivo* by injecting an antibody to this ligand into pregnant mice decreased the number of inclusion bodies in type II epithelial cells (74). These and other early lines of evidence established a role for the ErbB-1 module in controlling the lung epithelium, a possibility that is supported by gene targeting experiments. The lungs of erbB-1$^{-/-}$ newborn fetuses appeared premature, condensed with collapsed alveoli, and showed undifferentiated epithelium. The mutant animals had breathing difficulties, probably due to a thicker and more cellular alveolar epithelium that had less surfactants (60).

Gastrointestinal Tract Both EGF and TGFα inhibit gastric acid secretion and promote healing of gastric ulcers, but localized expression of TGFα precedes that of EGF, suggesting that it is the more relevant ligand of gastric ErbB-1. The role of the ErbB-1 module in gastric epithelial function and renewal cannot be addressed in depth with the knockout animals due to their early death. Nevertheless, despite an overall normal morphology, the mutant mice exhibited disorganized crypt columns and villi, and DNA synthesis was reduced in the basal layer of the esophageal epithelium (59,60). Administration of TGFα and EGF to suckling rats resulted in a rapid increase in stomach and intestinal weights, along with signficant stimulation of epithelial cell proliferation (75). Overexpression of TGFα in transgenic mice resulted in features that are consistent with a proliferative role in several types of epithelium, including that of the stomach, small intestine, and cecum (76). Hyperplasia in TGFα transgenics was not accompanied by a major alteration in organ architecture, but in the pancreas it altered acinar cell differentiation. The gastric mucosa of another trangenic line developed remarkably large nodular thickening; stomach secretions contained no detectable gastric acid, primarily due to loss of parietal cells (77). Cellular analysis of the transgenic stomach identified the surface mucous cells as the expanded population, whereas the base-located glandular population shrank, indicating an effect of TGFα on progenitor cell growth (78). Thus, the effect of TGFα on the stomach is cell specific, reinforcing its role in epithelial cell renewal in adult life.

Other Systems Obviously, due to early mortality, many aspects of the ErbB-1 module cannot be revealed by using a gene knockout strategy. For example, ErbB-1 and two of its ligands appear to be involved in mammary gland development during pregnancy (79). Plasma EGF levels are elevated during pregnancy and decrease before the onset of lactation; the factor is mitogenic to ductal and alveolar epithelium of virgin mice. Consistent with epithelial mitogenicity, transgenic mice overexpressing TGFα displayed a variety of mammary tissue abnormalities, including hyperplasia, adenoma, and adenocarcinoma (80,81). Other types of epithelium that are probably controlled by the ErbB-1 module are those of the kidney and pancreas, as well as certain hepatic cells. As predicted by earlier observations, proliferation of the mandibular epithelium is regulated by EGF, and corresponding antisense oligonucleotides blocked initiation of odontogenesis in embryonic mandibular explants (82). Remarkably, dental hypoplasia characterized Rieger syndrome, an autosomal dominant disorder that is linked to the EGF locus on human chromosome 4 (83). Other systemic anomalies that are associated with the syndrome, and therefore are likely effects of EGF, are involution of the periumbilical skin and mild craniofacial dysmorphism. Not only epithelia are affected by ErbB-1 signaling, but also nerve cell survival, especially in the anterior part of the cerebral cortex and the olfactory bulb of erbB-1$^{-/-}$ animals (59). EGF acts *in vitro* as a mitogen for neural progenitor cells (84), and TGFα drives differentiation of the multipotent progenitor cells of the retina into a glial fate (85), implying that neural development, in addition to epithelial differentiation, is regulated to some extent by ErbB-1.

8.5.2. Neuregulins and Their Receptors

Signaling by the NRG receptors is apparently more complex than signaling by ErbB-1 since it involves two receptors, as well as a coreceptor, and at least a dozen isoforms of the ligand. Already at the time of their discovery, pleiotropicity of NRGs and their major cellular targets was implied because the factors were initially isolated on the basis of their action on mammary cells (28,29,86), Schwann cells (25), and the neuromuscular synapse (26). Currently available are the phenotypes of NRG$^{-/-}$, erbB-4$^{-/-}$, and erbB-2$^{-/-}$ mice, but the phenotype of erbB-3$^{-/-}$ animals is still unknown (87–89). On the basis of these genetic lines of evidence and many *in vitro* studies, it seems that the NRG signaling module affects three major targets: epithelial organs, for which most of the data relate to the mammary gland; effectors of the nervous system, such as neuromuscular synapses and Schwann cells; and muscle cells of heart trabeculae. Common to the observed phenotypes is the conclusion that NRGs act primarily in a paracrine mode; their range of effect is limited to the vicinity of the sites of ligand synthesis and deposition.

Heart Development Perhaps the most surprising phenotypic aspect of $NRG^{-/-}$ embryos (87,88), which is shared with mouse embryos lacking either erbB-4 (89) or erbB-2 (90), is the absence of the finger-like extensions of the ventricular myocardium called *trabeculae*. Although other anatomic features are normal, the developing hearts of mutant embryos isolated at E10.5 are characterized by an irregular heartbeat, an enlarged common ventricle, and reduced blood flow, causing lethality at E10.5–E11.5. NRG is expressed in the endocardium, an endothelial ventricular lining, whereas ErbB-4 is present in the underlying muscular portion of the ventricle and atrium (myocardium). It is therefore likely that the endocardium-derived NRG activates trabeculae formation by the ErbB-4-expressing myocardium, thereby initiating ventricular differentiation. The endocardial cushion of embryos defective in either NRG or erbB-4 displayed less prominent morphologic defects, but its size was reduced. This portion of the heart develops into valves, and its mesenchymal cells express ErbB-3 but not NRG or ErbB-4. Taken together, NRG interaction with ErbB-4 appears essential for trabeculation, but its binding to ErbB-3 is involved in cushion differentiation. Apparently, both actions are mediated by NRG isoforms that contain an Ig-like domain, but not by the SMDF type of isoforms, because heart defects were observed in mutant mice whose Ig-containing isoforms were specifically targeted (88).

Neuromuscular Synapse Close apposition of axons and muscle cells is known to generate trophic effects on the latter, even in the absence of physical contact. At least part of the myotrophic action of neurons may be attributed to NRG isoforms. In myoblast cultures the ligand elevates creatine kinase, a marker for myogenic differentiation, with practically no concomitant cell proliferation (91). The effect of nerve-derived NRG on muscle is best characterized in the formation of the neuromuscular junction. During synapse formation, axonal molecules inductively instruct muscle cells to synthesize and concentrate several molecules that are essential for synapse integrity and function. NRG is synthesized by motor neurons (92,93) transported along axons and deposited in the synaptic basal lamina (94,95). Its interaction with muscle cell receptors induces at least two molecules, the acetylcholine receptor (AChR) and the voltage-gated sodium channel (96). Consistent with a causative effect in synapse formation, NRG expression coincides with contact formation between muscle cells and neurons in the developing chick embryo. Both the epsilon and alpha subunits of AChR are transcriptionally induced by NRG (95,97,98). Agrin, another nerve-derived synaptic molecule, complements enhanced synthesis by concentrating existing AChR molecules in the synaptic site and thereby securing synaptic transmission. Importantly, not only AChRs are concentrated in the synaptic site but also three ErbB proteins, namely, ErbB-2,

ErbB-3, and ErbB-4 (99). Although ErbB-4 may not be essential for the myogenic response (98), its localized expression increases during postnatal development, suggesting that all three NRG receptors are involved in synapse signaling (99).

Schwann and Glia Cells NRGs appear to act as major regulators of the myelin-forming glia cells of both the central and peripheral nervous systems. Embryonic precursors of Schwann cells (SCs) are derived from the neural crest, migrate along newly formed axons, and undergo differentiation to myelin-forming arrested cells. This process involves extensive programmed cell death that establishes the one-to-one ratio characteristic of mature nerves. In $NRG^{-/-}$ embryos, ErbB-3-expressing precursor SCs of the trunk are reduced in number (87). Because in wild-type animals these cells eventually ensheath NRG-producing motor and sensory neurons, it is likely that the nerve-derived NRGs play a role in SC proliferation and maturation *in vivo*. Indeed, several *in vitro* studies have now established this possibility. Precursor SCs isolated from E14 rat sciatic nerve escaped apoptosis in the presence of NRG but not other growth factors (100). Likewise, postnatal sciatic nerve SCs undergo apoptosis *in vitro* on serum withdrawal, but a beta isoform of NRG prevents their death (101). Axotomy increases apoptosis of SCs *in vitro,* but as in developmentally induced apoptosis, NRG markedly restricts it (102). Interestingly, SC survival involves no increase in DNA synthesis, reinforcing a significant differentiation effect of the ligand. Nevertheless, a concentration-dependent response to NRGs was demonstrated on SCs isolated from newborn rats: At low concentrations the factor was chemotactic, but at saturating concentrations it induced a proliferative response (103). The relevance of these observations to living animals has been demonstrated by neonatal denervation of the hind limb muscle in the rat. Repeated injections of NRG into living treated rats prevented apoptotic loss of SCs from the neuromuscular junction (104). Conceivably, apoptotic death of SCs during development and after denervation may be regulated by competition for a limiting amount of the available NRG, so that SCs that do not receive this nerve-derived factor do not survive. Wherever investigated, the effect of NRG on SCs appears to be mediated by ErbB-3, presumably in association with ErbB-2 or ErbB-1 but not ErbB-4. Importantly, NRG action on cells of the nervous system is not restricted to SCs. The factor is extensively expressed in the central nervous system (92,93,105,106), and when applied to primary embryonic brain cultures, its most remarkable effect is acceleration of astrocyte maturation, with no detectable effect on neurons (105). Another type of central nervous system glia cell whose formation is accelerated *in vitro* by NRG is the myelin-forming oligodendrocyte cell (107). Evidence that NRG instructs a glial fate, rather than selectively promoting survival and proliferation of already committed cells, was provided by using multipotent neural crest stem cells (108); the factor promoted glial differentiation while suppressing differentiation along the neuronal pathway. In summary, by affecting differentiation, survival, proliferation, and migration, NRGs act as potent instructors of the glial lineage, but their exact roles in the central nervous system remain largely unknown.

Cranial Nerves Disruption of all or only the Ig-containing isoforms of NRG severely impaired development of cranial ganglia (87,88). Despite specificity to cranial nerves, distinct ganglia were altered to different extents: Trigeminal nerves and mandibular projections were lost, and petrose and nodose ganglia were reduced in size and their projections disorganized. By contrast, targeting of one of the NRG receptors, ErbB-4, did not impair cranial sensory ganglia but did alter their innervation from and to the hindbrain (89). Remarkably, axons that connect ganglia V (trigeminal) and VII/VIII (facial and acoustic) with the hindbrain through rhombomers 2 and 4, respectively, were partially mistargeted to rhombomers 4 and 2, respectively. Because ErbB-4 is expressed in the intervening rhombomer 3, it was proposed that this receptor mediates a repulsive signal for innervating axons. The differences between the phenotypes of $NRG^{-/-}$ and $erbB\text{-}4^{-/-}$ embryos suggest that ErbB-4 binds an additional ligand whose activity differs from that of NRG. On the other hand, the cranial phenotype of $NRG^{-/-}$ mice suggests that the Ig-containing isoforms of the ligand regulate development or migration of ErbB-3-expressing neural crest cells from the hindbrain, a conclusion supported by the observed impairment of SCs. Whether or not NRGs directly affect neurons is unclear at present. ErbB-4 is expressed on the surface of neurons, and NRGs promote survival and neurite outgrowth of cultured neonatal retinal cells (109). Likewise, overexpression of ErbB-2 in the extensively used model system for neuronal differentiation, namely, the PC12 pheochromocytoma cells, allowed NRGs to induce neurite outgrowth (110). Other studies, however, implicate an indirect effect of NRG on neurons. Incubation of mature SCs with NRG induced secretion of activities that promote sympathetic neuron survival and outgrowth (103). One of the nerotrophic factors involved may be nerotrophin 3 (NT-3) because a reciprocal loop in which NRG is synthesized by sympathetic neuroblasts and elevates NT-3 secretion by adjacent nonneuronal cells has been demonstrated in coculture experiments (111).

Mammary Gland Due to the *in utero* death of mice defective in NRG or its receptors, the role of this signaling module in mammary and other types of epithelial cell differentiation cannot be approached by a simple gene targeting strategy. It is clear, however, that the factor is synthesized by mesenchymal stromal cells that underlie many types of epithelia, and its receptors, in most cases ErbB-3 and ErbB-2, are expressed on the opposed epithelium. This configuration and the apparent absence of NRG receptors on mesenchymal cells enables short-range paracrine intercellular interactions that probably play a role in mesenchyme-epithel inductive processes. For example, experiments using whole glands from hormonally primed mice demonstrated a marked effect of NRG on lobulo-alveolar budding and milk production (112). Furthermore, NRG that is synthesized by connective tissue surrounding ductal and alveolar structures is upregulated at pregnancy, in parallel with the major phase of lobulo-alveolar development (112). Targeted overexpression of NRG in the mammary gland of transgenic mice, however, inhibited terminal differentiation of the terminal end bud structures and led to hyperplasia or adenocarcinoma

(113). Experiments performed with various mammary cell lines also resulted in phenotypic differentiation and growth arrest or cell proliferation, implying that the effect of NRG on the mammary gland is subtle. Several breast cancer cell lines undergo growth arrest in response to NRG. Concomitantly, the arrested cells assume a differentiated phenotype that includes synthesis of milk proteins and lipids, as well as elevation of the intercellular adhesion molecule 1 (ICAM-1) (114). HC11 normal mouse mammary cells respond to NRG by accelerated proliferation, but their sensitivity to a differentiation-inducing mixture of hormones is also increased (115). Most other normal or transformed breast cell lines undergo proliferation in response to NRG stimulation (116), but growth arrest is often observed when high ligand concentrations are used or the cells overexpress ErbB-2. Taken together, these results implicate NRG in mammary gland development, but due to variation in the origins and nature of cultured mammary cell lines, the exact mode of NRG action is still unclear. Other types of epithelial cells also respond to NRG in a subtle manner that probably reflects a role in mesenchyme-epithel inductive morphogenesis. For example, in a rabbit ear model of excisional wound repair, NRG-α induced reepithelialization and epidermal thickening that were associated with enhanced expression of keratinocyte differentiation markers (117). Cultured keratinocytes, however, do not undergo differentiation by NRGs; whereas the beta isoform is strongly mitogenic for them, the alpha isoform acts as a survival factor (118).

8.6. THE ErbB-2 CORECEPTOR

The biochemical function of ErbB-2 is not yet clear; while it may act as a receptor for a still unknown EGF-like molecule, recent analyses of signal transduction imply that ErbB-2 may function as a shared subunit of the operational receptors for all ErbB ligands (40). Remarkably, the phenotype of erbB-2$^{-/-}$ mice shares characteristics with both NRG$^{-/-}$ mice (absence of trabeculae and defects in cranial neural crest-derived sensory nerves) and erbB-4$^{-/-}$ embryos (absence of trabeculae) (90). Should ErbB-2 bind its own direct ligand, whose functions may not overlap with those of the known EGF-like molecules, unique features of erbB-2$^{-/-}$ embryos would be expected. Indeed, erbB-2$^{-/-}$ embryos display abnormal development of trigeminal and facial motor nerves, a defect that has not been reported in NRG- or erbB-4-deficient embryos. However, the observed motor nerve defect may well be secondary to abnormal development of sensory ganglia. Consistent with a role of ErbB-2 in regulation of neural crest-derived cells, a monoclonal antibody to ErbB-2 partially blocks the ErbB-3-mediated proliferative effect of NRG on human SCs (119). Another line of evidence that supports an auxilliary role of ErbB-2 in growth factor signaling is derived from observations made with human cancers. Overexpression of the putative coreceptor is frequently observed in several types of epithelial cell tumors, such as breast and ovarian cancer, and this correlates with an overall poor prognosis (120). Experiments performed with cultured cells showed that overexpression of ErbB-2 alone can cause cell transformation, but coexpression of either ErbB-1 (121) or ErbB-3 (122) synergizes the oncogenic action of ErbB-2.

Thus the developmental role of ErbB-2 may be to augment signaling by NRGs, as well as to prolong the activity of ErbB-1 ligands.

8.7. PERSPECTIVES

Currently, we are witnessing a transitional phase in the field of growth factor research: from *in vitro* to *in vivo* experimentation. Whereas most growth factors, including ErbB ligands, were isolated from and assayed on established cultured cell lines, this *in vitro* experimental system is severely limited when biologic activities are analyzed. Due to a strong selection pressure for continuous growth in culture, the excessive use of cultured cells may have obscured the major action of polypeptide growth factors, namely, determination of cell fate. Instructive signals for lineage selection inevitably involve both cell proliferation (aimed at enlarging the pool of committed cells) and cell differentiation, but cultured cells, almost by definition, reflect only the former. This important lesson perfectly applies to the ErbB family of receptors and their ligands. Because of an opportunistic use of the proliferative action of this family in cancer, as well as in other hyperproliferative diseases, and due to limitation of *in vivo* experimental strategies, mostly the mitogenic aspects of ErbB ligands were recognized. However, recent application of gene targeting and other genetic strategies is providing a more balanced view of ErbB as a machinery that controls development of a variety of organs and cell types. Specifically, development of several different types of epithelia is controlled by the underlying mesenchyme through the ErbB-1 module of the network, whereas the other module is used primarily by nerve cells to regulate SCs and muscle cells. Interestingly, unlike other families of polypeptide morphogens (e.g., the BMP family) that regulate early developmental phases, ErbB ligands appear to act at relatively late stages (i.e., middle and late gestation), when the primary derivatives of the three embryonic germ layers are already developed and organogenesis is at its peak.

From a biochemical point of view, the ErbB family may emerge as a prototypic machinery for diversity generation. Both diversification and control of signal transduction are generated through a network of combinatorial protein–protein interactions operating at all levels of signal transmission (i.e., ligand, receptor, and effector). The network consists of a relatively small number of protein components whose expression is not ubiquitous but strictly controlled. However, by maintaining hierarchical and highly specific interactions, the network provides enormous potential for diversity and tuning of biologic signals. Whether this very large potential is fully exploited *in vivo* is currently an open question. However, because the network is only partially developed in lower organisms, its elaboration to the present state may have taken part in establishing the extremely large cell lineage diversity of mammals. Since the ErbB family is but one of the large group of RTK subtypes, similar networks of ligands and receptors may exist (e.g., insulin and fibroblast growth factor receptor subfamilics), thereby increasing the diversity potential by at least one order of magnitude.

Due to the critical developmental roles of the ErbB network, and especially its involvement in heart development, existing genetic technologies may not allow further developmental examination. However, tissue- and phase-specific gene targeting technologies that are currently being developed may allow more detailed understanding of the many developmental roles of the network, especially the specific role played by each of its many ligands.

REFERENCES

1. Yarden Y, Ullrich A: Growth factor receptor tyrosine kinases. Annu Rev Biochem 57:443–478, 1988.
2. van der Geer P, Hunter T, Lindberg RA: Receptor protein-tyrosine kinases and their signal transduction pathways. Annu Rev Cell Biol 10:251–337, 1994.
3. Yarden Y, Schlessinger J: Epidermal growth factor induces rapid, reversible aggregation of purified epidermal growth factor receptors. Biochemistry 26:1443–1445, 1987.
4. Koch AC, Anderson D, Moran MF, Ellis C, Pawson T: SH-2 and SH-3 domains: Elements that control interactions of cytoplasmic signaling proteins. Science 252: 668–674, 1991.
5. Ullrich A, Coussens L, Hayflick JS, Dull TJ, Gray A, Tam AW, Lee J, Yarden Y, Libermann TA, Schlessinger J, Downward J, Mayes ELV, Whittle N, Waterfield MD, Seeburg PH: Human epidermal growth factor receptor cDNA sequence and aberrant expression of the amplified gene in A431 epidermoid carcinoma cells. Nature 309:418–425, 1984.
6. Plowman GD, Whitney GS, Neubauer MG, Green JM, McDonald VI, Todaro GJ, Shoyab M: Molecular cloning and expression of an additional epidermal growth factor receptor-related gene. Proc Natl Acad Sci USA 87:4905–4909, 1990.
7. Kraus MH, Issing M, Popescu NC, Aaronson SA: Isolation and characterization of *Erb*B-3, a third member of the *Erb*B/epidermal growth factor receptor family: Evidence for overexpression in a subset of human mammary tumors. Proc Natl Acad Sci USA 86:9193–9197, 1989.
8. Plowman GD, Culouscou J-M, Whitney GS, Green JM, Carlton GW, Foy L, Neubauer MG, Shoyab M: Ligand-specific activation of HER-4/p180*erb*B4, a fourth member of the epidermal growth factor receptor family. Proc Natl Acad Sci USA 90:1746–1750, 1993.
9. Dougall WC, Qian X, Peterson NC, Miller MJ, Samanta A, Greene MI: The neu-oncogene: Signal transduction pathways, transformation mechanisms and evolving therapies. Oncogene 9:2109–2123, 1994.
10. Lax I, Bellot F, Howk R, Ullrich A, Givol D, Schlessinger J: Functional analysis of the ligand binding site of EGF-receptor utilizing chimeric/human receptor molecules. EMBO J 8:421–427, 1989.
11. Carraway KL, Cantley LC: A neu acquaintance for ErbB3 and ErbB4: A role for receptor heterodimerization in growth signaling. Cell 78:5–8, 1994.
12. Levkowitz G, Klapper LN, Tzahar E, Freywald A, Sela M, Yarden Y: Coupling of the c-Cbl protooncogene product to ErbB-1/EGF-receptor but not to other ErbB proteins. Oncogene 12:1117–1125, 1996.
13. Guy PM, Platko JV, Cantley LC, Cerione RA, Carraway KL: Insect cell-expressed

p180ErbB3 possesses an impaired tyrosine kinase activity. Proc Natl Acad Sci USA 91:8132–8136, 1994.

14. Baulida J, Kraus MH, Alimandi M, Di Fiore PP, Carpenter G: All ErbB receptors other than the epidermal growth factor receptor are endocytosis impaired. J Biol Chem 271:5251–5257, 1996.
15. Pinkas-Kramarski R, Soussan L, Waterman H, Levkowitz G, Alroy I, Klapper L, Lavi S, Seger R, Ratzkin B, Sela M, Yarden Y: Diversification of Neu differentiation factor and epidermal growth factor signaling by combinatorial receptor interactions. EMBO J 15:2452–2467, 1996.
16. Jacobsen NE, Abadi N, Sliwkowski MX, Reilly D, Skelton NJ, Fairbrother WJ: High-resolution solution structure of the EGF-like domain of heregulin-alpha. Biochemistry 35:3402–3417, 1996.
17. Marquardt H, Hunkapiller MW, Hood LE, Todaro GJ: Rat transforming growth factor type 1: Structure and relation to epidermal growth factor. Science 223:1079–1082, 1984.
18. Shoyab M, McDonald VL, Bradley JG, Todaro GJ: Amphiregulin: A bifunctional growth modulating glycoprotein produced by the phorbol12-myristate13-acetate-treated human breast adeno-carcinoma cell line MCF-7. Proc Natl Acad Sci USA 85:6528–6532, 1988.
19. Sasada R, Ono Y, Taniyama Y, Shing Y, Folkman J, Igrashi K: Cloning and expression of cDNA encoding human betacellulin, a new member of the EGF family. Biochem Biophys Res Commun 190:1173–1179, 1993.
20. Toyoda H, Komursaki T, Uchida D, Takayama Y, Isobe T, Okuyama T, Hanada K: Epiregulin, a novel epidermal growth factor with mitogenic activity for rat primary hepatocytes. J Biol Chem 270:7495–7500, 1995.
21. Higashiyama S, Abraham JA, Miller J, Fiddes JC, Klagsbrun M: A heparin-binding growth factor secreted by macrophage-like cells that is related to EGF. Science 251:936–939, 1991.
22. Riese DJ, Bermingham Y, van Raaij TM, Buckley S, Plowman GD, Stern DF: Betacellulin activates the epidermal growth factor receptor and erbB-4, and induces cellular response patterns distinct from those stimulated by epidermal growth factor or neuregulin-beta. Oncogene 12:345–353, 1996.
23. Chen X, Levkowitz G, Tzahar E, Karunagaran D, Lavi S, Ben-Baruch N, Leitner O, Ratzkin BJ, Bacus SS, Yarden Y: An immunological approach reveals biological differences between the two NDF/heregulin receptors, ErbB-3 and ErbB-4. J Biol Chem 271:7620–7629, 1996.
24. Peles E, Yarden Y: Neu and its ligands: From an oncogene to neural factors. BioEssays 15:815–824, 1993.
25. Marchionni MA, Goodearl ADJ, Chen MS, Bermingham-McDonogh O, Kirk C, Hendricks M, Denehy F, Misumi D, Sudhalter J, Kobayashi K, Wroblewski D, Lynch C, Baldassare M, Hiles I, Davis JB, Hsuan JJ, Totty NF, Otsu M, McBury RN, Waterfield MD, Stroobant P, Gwynne D: Glial growth factors are alternatively spliced *erb*B-2 ligands expressed in the nervous system. Nature 362:312–318, 1993.
26. Falls DL, Rosen KM, Corfas G, Lane WS, Fischbach GD: ARIA, a protein that stimulates acetylcholine receptor synthesis, is a member of the neu ligand family. Cell 72:801–815, 1993.

27. Ho WH, Armanini MP, Nuijens A, Phillips HS, Osheroff PL: Sensory and motor neuron-derived factor. A novel heregulin variant highly expressed in sensory and motor neurons. J Biol Chem 270:14523–14532, 1995.
28. Holmes WE, Sliwkowski MX, Akita RW, Henzel WJ, Lee J, Park JW, Yansura D, Abadi N, Raab H, Lewis GD, Shepard M, Wood WI, Goeddel DV, Vandlen RL: Identification of heregulin, a specific activator of p185*erb*B2. Science 256:1205–1210, 1992.
29. Wen D, Peles E, Cupples R, Suggs SV, Bacus SS, Luo Y, Trail G, Hu S, Silbiger SM, Ben-Levy R, Luo Y, Yarden Y: Neu differentiation factor: A transmembrane glycoprotein containing an EGF domain and an immunoglobulin homology unit. Cell 69:559–572, 1992.
30. Peles E, Ben-Levy R, Tzahar E, Liu N, Wen D, Yarden Y: Cell-type specific interaction of Neu differentiation factor (NDF/heregulin) with Neu/HER-2. EMBO J 12:961–971, 1993.
31. Plowman GD, Culouscou JM, Whitney GS, Green JM, Carlton GW, Foy L, Neubauer MG, Shoyab M: Ligand-specific activation of HER4/p180erbB4, a fourth member of the epidermal growth factor receptor family. Proc Natl Acad Sci USA 90:1746–1750, 1993.
32. Carraway KL, Sliwkowski MX, Akita R, Platko JV, Guy PM, Nuijens A, Diamonti AJ, Vandlen RL, Cantley LC, Cerione RA: The *erb*B3 gene product is a receptor for heregulin. J Biol Chem 269:14303–14306, 1994.
33. Tzahar E, Levkowitz G, Karunagaran D, Yi L, Peles E, Lavi S, Chang D, Liu N, Yayon A, Wen D, Yarden Y: ErbB-3 and ErbB-4 function as the respective low and high affinity receptors of all Neu differentiation factor/heregulin isoforms. J Biol Chem 269: 25226–25233, 1994.
34. Goldman R, Ben-Levy R, Peles E, Yarden Y: Heterodimerization of the *erb*B-1 and *erb*B-2 receptors in human breast carcinoma cells: A mechanism for receptor transregulation. Biochemistry 29:11024–11028, 1990.
35. Wada T, Qian X, Greene MI: Intermolecular association of the p185^{neu} protein and EGF receptor modulates EGF receptor function. Cell 61:1339–1347, 1990.
36. Weiner DB, Liu J, Cohen JA, Williams WV, Greene MI: A point mutation in the *neu* oncogene mimics ligand induction of receptor aggregation. Nature 339:230–231, 1989.
37. Tzahar E, Waterman H, Chen X, Levkowitz G, Karunagaran D, Lavi S, Ratzkin BJ, Yarden Y: A hierarchical network of inter-receptor interactions determines signal transduction by NDF/neuregulin and EGF. Mol Cell Biol 16:5276–5287, 1996.
38. Riese DJ, van Raaij TM, Plowman GD, Andrews GC, Stern DF: The cellular response to neuregulins is governed by complex interactions of the ErbB receptor family. Mol Cell Biol 15:5770–5776, 1995.
39. Graus-Porta D, Beerli RR, Hynes NE: Single-chain antibody-mediated intracellular retention of ErbB-2 impairs Neu differentiation factor and epidermal growth factor signaling. Mol Cell Biol 15:1182–1191, 1995.
40. Karunagaran D, Tzahar E, Beerli RR, Chen X, Graus-Porta D, Ratzkin BJ, Seger R, Hynes NE, Yarden Y: ErbB-2 is a common auxiliary subunit of NDF and EGF receptors: Implications for breast cancer. EMBO J 15:254–264, 1996.
41. Ben-Levy R, Paterson HF, Marshall CJ, Yarden Y: A single autophosphorylation site confers oncogenicity to the Neu/ErbB-2 receptor and enables coupling to the MAP-kinase pathway. EMBO J 13:3302–3311, 1994.

42. Pinkas-Kramarski R, Shelly M, Glathe S, Ratzkin BJ, Yarden Y: Neu differentiation factor/neuregulin isoforms activate distinct receptor combinations. J Biol Chem 271: 19029–19032, 1996.

43. Aroian RV, Koga M, Mendel JE, Ohshima Y, Sternberg PW: The let-23 gene necessary for *Caenorhabditis elegans* vulval induction encodes a tyrosine kinase of the EGF receptor subfamily. Nature 348:693–699, 1990.

44. Livneh E, Glazer L, Segal D, Schlessinger J, Shilo BZ: The *Drosophila* EGF receptor gene homolog: Conservation of both hormone binding and kinase domains. Cell 40: 599–607, 1985.

45. Hill RJ, Sternberg PW: The gene lin-3 encodes an inductive signal for vulval development in *C. elegans*. Nature 358:470–476, 1992.

46. Neuman-Silberberg FS, Schupbach T: The *Drosophila* gene *gurken* produces a dorsally-localized RNA and encodes a TGFα-like protein. Cell 75:165–174, 1993.

47. Rutledge BJ, Zhang K, Bier E, Jan YN, Perrimon N: The *Drosophila* spitz gene encodes a putative EGF-like growth factor involved in dorsal-ventral axis formation and neurogenesis. Genes Dev 6:1503–1517, 1992.

48. Schnepp B, Grumbling G, Donaldson T, Simcox A: Vein is a novel component in the *Drosophila* epidermal growth factor pathway with similarity to the neureguils. Genes Dev 15:2302–2313, 1996.

49. Stein DS, Stevens LM: Establishment of dorsal-ventral and terminal pattern in the Drosophila embryo. Curr Opin Genet Dev 1:247–254, 1991.

50. Schweitzer R, Shaharabany M, Seger R, Shilo B-Z: Secreted Spitz triggers the DER signaling pathway and is a limiting component in embryonic ventral ectoderm determination. Genes Dev 10:1518–1529, 1995.

51. Schweitzer R, Howes R, Smith R, Shilo B-Z, Freeman M: Inhibition of *Drosophila* EGF receptor activation by the secreted Argos protein. Nature 376:699–702, 1995.

52. Han M, Golden A, Han Y, Sternberg PW: *C. elegans lin-45 raf* gene participate in *let-60 ras*-stimulated vulval differentiation. Nature 363:133–140, 1993.

53. Brunner D, Oellers N, Szabad J, Biggs W3, Zipursky SL, Hafen E: A gain-of-function mutation in *Drosophila* MAP kinase activates multiple receptor tyrosine kinase signaling pathways. Cell 76:875–888, 1994.

54. Sternberg PW, Horvitz R: Signal transduction during *C. elegans* vulval induction. Trends Genet 7:366–371, 1991.

55. Shilo B-Z, Raz E: Developmental control by the *Drosophila* EGF receptor homolog DER. Trends Genet 7:388–392, 1991.

56. Clifford R, Schuppbach T: Molecular analysis of the *Drosophila* EGF receptor homolog reveals that several genetically defined classes of alleles cluster in subdomains of the receptor protein. Genetics 137:531–550, 1994.

57. Cohen S: Isolation of a mouse submaxillary gland protein accelerating incisor eruption and eyelid opening in new-born animal. J Biol Chem 237:1555–1562, 1962.

58. Sibilia M, Wagner EF: Strain-dependent epithelial defects in mice lacking the EGF receptor. Science 269:234–238, 1995.

59. Threadgill DW, Dlugosz AA, Hansen LA, Tennenbaum T, Lichti U, Yee D, LaMantia C, Mourton T, Herrup K, Harris RC: Targeted disruption of mouse EGF receptor: Effect of genetic background on mutant phenotype. Science 269:230–234, 1995.

60. Miettinen PJ, Berger JE, Meneses J, Phung Y, Pedersen RA, Werb Z, Derynck R: Ep-

ithelial immaturity and multiorgan failure in mice lacking epidermal growth factor receptor. Nature 376:337–341, 1995.

61. Mann GB, Fowler KJ, Gabriel A, Nice EC, Williams RL, Dunn AR: Mice with null mutations of the TGFα gene have abnormal skin architecture, wavy hair, and curly whiskers and often develop corneal inflammation. Cell 73:249–261, 1993.
62. Luetteke NC, Qiu TH, Peiffer RL, Oliver P, Smithies O, Lee DC: TGFα deficiency results in hair follicles and eye abnormalities in targeted and Waved-1 mice. Cell 73:263–278, 1993.
63. Luetteke NC, Phillips HK, Qiu TH, Copeland NG, Earp HS, E, Jenkins NA, Lee DC: The mouse *waved-2* phenotype results from a point mutation in the EGF receptor tyrosine kinase. Genes Dev 8:399–413, 1994.
64. Fowler KJ, Walker F, Alexander W, Hibbs ML, Nice EC, Bohmer RM, Mann GB, Thumwood C, Maglitto R, Danks JA, Chetty R, Burgess AW, Dunn AR: A mutation in the epidermal growth factor receptor in waved-2-mice has a profound effect on receptor biochemistry that results in impaired lactation. Proc Natl Acad Sci USA 92:1465–1469, 1995.
65. Wiley LM, Wu JX, Harari I, Adamson ED: Epidermal growth factor receptor mRNA and protein increase after the 4-cell preimplantation stage in murine development. Dev Biol 149:247–260, 1992.
66. Das SK, Wang X-N, Paria BC, Damm D, Abraham JA, Klagsbrun M, Andrews GK, Dey SK: Heparin-binding EGF-like growth factor gene is induced in the mouse uterus temporally by the blastocyst solely at the site of its apposition: A possible ligand for interaction with blastocyst EGF-receptor in implantation. Development 120:1071–1083, 1994.
67. Raab G, Kover K, Paria BC, Dey SK, Ezzel RM, Klagsbrun M: Mouse implantation blastocysts adhere to cells expressing the transmembrane form of heparin-binding EGF-like growth factor. Development 122:637–645, 1996.
68. Barrandon Y, Green H: Cell migration is essential for sustained growth of keratinocyte colonies: The roles of transforming growth factor-α and epidermal growth factor. Cell 50:1131–1137, 1987.
69. Finzi E, Harkins R, Horn T: TGF-α is widely expressed in differentiated as well as hyperproliferative skin epithelium. J Invest Dermatol 96:328–393, 1991.
70. Elder JT, Fisher JG, Lindquist PB, Bennet GL, Pittelw MR, Coffey RJJ, Ellingsworth L, Derynck R, Voorhees JJ: Overexpression of transforming growth factor α in psoriatic epidermis. Science 243:811–814, 1989.
71. Dominey AM, Wang X-J, King LEJ, Nanney LB, Gange TA, Sellheyer K, Bundman DS, Longley MA, Rothnagel JA, Greenhalgh DA, Roop DR: Targeted overexpression of transforming growth factor α in the epidermis of transgenic mice elicits hyperplasia, hyperkeratosis, and spontaneous, squamous papillomas. Cell Growth Different 4: 1071–1082, 1993.
72. Vassar R, Fuchs E: Transgenic mice provide new insights into the role of TFG-α during epidermal development and differentiation. Genes Dev 5:714–727, 1991.
73. Warburton D, Seth R, Shum L, Horcher PG, Hall FL, Werb Z, Slavkin HC: Epigenetic role of epidermal growth factor expression and signaling in embryonic mouse lung morphogenesis. Dev Biol 149:123–133, 1992.
74. Yasui S, Nagai A, Oohira A, Iwashita M, Konno K: Effects of anti-mouse EGF antiserum on prenatal lung development in fetal mice. Pediatr Pulmon 15:251–256, 1993.

75. Hormi K, Lehy T: Transforming growth factor-α in vivo stimulates epithelial cell proliferation in digestive tissues of suckling rats. Gut 39:532–538, 1996.

76. Sandgren EP, Luetteke NC, Palmiter RD, Brinster RL, Lee DC: Overexpression of TGFα in transgenic mice: Induction of epithelial hyperplasia, pancreatic metaplasia and carcinoma of the breast. Cell 61:1121–1135, 1990.

77. Tagaki H, Jhappen C, Sharp R, Merlino G: Hypertrophic gastropathy resembling Menterier's disease in transgenic mice overexpressing transforming growth factor α in the stomach. J Clin Invest 90:1161–1167, 1992.

78. Sharp R, Babyatsky MW, Takagi H, Tagerud S, Wang TC, Bockman TE, Brand SJ, Merlino G: Transforming growth factor α disrupts the normal program of cellular differentiation in the gastric mucosa of transgenic mice. Development 121:149–161, 1995.

79. Oka T, Yoshimura M, Lavandero S, Wada K, Obha Y: Control of growth and differentiation of the mammary gland by growth factors. J Dairy Sci 74:2788–2800, 1991.

80. Jhappan C, Stahle C, Harkins RN, Fausto N, Smith GH, Merlino GT: TGFα overexpression in transgenic mice induces liver neoplasia and abnormal development of the mammary gland and pancreas. Cell 61:1137–1146, 1991.

81. Matsui M, Halter SA, Holt JT, Hogan BLM, Coffey RJ: Development of mammary hyperplasia and neoplasia in MMTV-TGFα transgenic mice. Cell 61:1147–1155, 1991.

82. Kronmiller JE, Upholt WB, Kollar EJ: EGF antisense oligonucleotides block murine odontogenesis *in vitro*. Dev Biol 147:485–488, 1991.

83. Murray JC, Bennett SR, Kwitek AE, Small KW, Schnizel A, Alward WLM, Weber JL, Bell GI, Buetow KH: Linkage of Rieger syndrome to the region of epidermal growth factor gene on chromosome 4. Nature Genet 2:46–49, 1992.

84. Kitchens DL, Snyder EY, Gottlieb DI: FGF and EGF are mitogens for immortalized neural progenitors. J Neurobiol 25:797–807, 1994.

85. Lillian L: Changes in retinal cell fate induced by overexpression of EGF receptor. Nature 377:158–162, 1995.

86. Peles E, Bacus SS, Koski RA, Lu HS, Wen D, Ogden SG, Ben-Levy R, Yarden Y: Isolation of the *neu*/HER-2 stimulatory ligand: A 44 kd glycoprotein that induces differentiation of mammary tumor cells. Cell 69:205–216, 1992.

87. Meyer D, Birchmeier C: Multiple essential functions of neuregulin in development. Nature 378:386–390, 1995.

88. Kramer R, Bucay N, Kane DJ, Martin LE, Tarpley JE, Theill LE: Neuregulins with an Ig-like domain are essential for mouse myocardial and neuronal development. Proc Natl Acad Sci USA 93:4833–4838, 1996.

89. Gassmann M, Casagranda F, Orioli D, Simon H, Lai C, Klein R, Lemke G: Aberrant neural and cardiac development in mice lacking the ErbB4 neuregulin receptor. Nature 378:390–394, 1995.

90. Lee KF, Simon H, Chen H, Bates B, Hung MC, Hauser C: Requirement for neuregulin receptor erbB2 in neural and cardiac development. Nature 378:394–398, 1995.

91. Florini JR, Samuel DS, Ewton DZ, Kirk C, Sklar RM: Stimulation of myogenic differentiation by a neuregulin, glial growth factor 2. Are neuregulins the long-sought muscle trophic factors secreted by nerves? J Biol Chem 271:12699–12702, 1996.

92. Orr-Urtreger A, Trakhtenbrot L, Ben-Levi R, Wen D, Rechavi G, Lonai P, Yarden Y: Neural expression and chromosomal mapping of Neu differentiation factor to 8p12-p21. Proc Natl Acad Sci USA 90:1867 1871, 1993.

93. Corfas G, Rosen KM, Aratake H, Krauss R, Fischbach GD: Differential expression of ARIA isoforms in the rat brain. Neuron 14:103–115, 1995.
94. Goodearl AD, Yee AG, Sandrock AW Jr, Corfas G, Fischbach GD: ARIA is concentrated in the synaptic basal lamina of the developing chick neuromuscular junction. J Cell Biol 130:1423–1434, 1995.
95. Jo SA, Zhu X, Marchionni MA, Burden SJ: Neuregulins are concentrated in nerve-muscle synapses and activate Ach-receptor gene expression. Nature 373:158–161, 1995.
96. Corfas G, Fischbach GD: The number of Na^+ channels in cultured chick muscle is increased by ARIA, an acetylcholine receptor-inducing activity. J Neurosci 13: 2118–2125, 1993.
97. Chu GC, Moscoso LM, Sliwkowski MX, Merlie JP: Regulation of the acetylcholine receptor epsilon subunit gene by recombinant ARIA: An in vitro model for transsynaptic gene regulation. Neuron 14:329–339, 1995.
98. Altiok N, Bessereau JL, Changeux JP: ErbB3 and ErbB2/neu mediate the effect of heregulin on acetylcholine receptor gene expression in muscle: Differential expression at the endplate. EMBO J 14:4258–4266, 1995.
99. Zhu X, Lai C, Thomas S, Burden SJ: Neuregulin receptors, erbB3 and erbB4, are localized at neuromuscular synapses. EMBO J 14:5842–5848, 1995.
100. Dong Z, Brennan A, Liu N, Yarden Y, Lefkowitz G, Mirsky R, Jessen KR: Neu differentiation factor is a neuron-glia signal and regulates survival, proliferation, and maturation of rat Schwann cell precursors. Neuron 15:585–596, 1995.
101. Syroid DE, Maycox PR, Burrola GP, Liu N, We D, Lee K-F, Lemke G, Kilpatrick TJ: Cell death in the Schwann cell lineage and its regulation by neuregulin. Proc Natl Acad Sci USA 93:9229–9234, 1996.
102. Grinspan JB, Marchionni MA, Reeves M, Coulaloglou M, Scherer SS: Axonal interactions regulate Schwann cell apoptosis in developing peripheral nerve: Neuregulin receptors and the role of neureguins. J Neurosci 16:6107–6118, 1996.
103. Mahanthappa NK, Anton ES, Matthew WD: Glial growth factor 2, a soluble neuregulin, directly increases Schwann cell motility and indirectly promotes neurite outgrowth. J Neurosci 16:4673–4683, 1996.
104. Trachtenberg JT, Thompson WJ: Schwann cell apoptosis at developing neuromuscular junctions is regulated by glial growth factor. Nature 379:174–177, 1996.
105. Pinkas-Kramarski R, Eilam R, Spiegler O, Lavi S, Liu N, Chang D, Wen D, Schwartz M, Yarden Y: Brain neurons and glial cells express Neu differentiation factor/heregulin: A survival factor for astrocytes. Proc Natl Acad Sci USA 91:9387–9391, 1994.
106. Chen MS, Bermingham-McDonogh O, Danehy FT Jr, Nolan C, Scherer SS, Lucas J, Gwynne D, Marchionni MA: Expression of multiple neuregulin transcripts in postnatal rat brains. J Comp Neurol 349:389–400, 1994.
107. Vartanian T, Corfas G, Li Y, Fischbach GD, Stefansson K: A role for the acetylcholine receptor-inducing protein ARIA in oligodendrocyte development. Proc Natl Acad Sci USA 91:11626–11630, 1994.
108. Shah NM, Marchionni MA, Isaacs I, Stroobant P, Anderson DJ: Glial growth factor restricts mammalian neural crest stem cells to a glial fate. Cell 77:349–360, 1994.
109. Bermingham-McDonogh O, McCabe KL, Reh TA: Effects of GGF/neuregulins on neuronal survival and neurite outgrowth correlate with erbB2/neu expression in developing rat retina. Development 122:1427–1438, 1996.

110. Gamett DC, Greene T, Wagreich AR, Kim HH, Koland JG, Cerione RA: Heregulin-stimulated signaling in rat pheochromocytoma cells. Evidence for ErbB3 interactions with Neu/ErbB2 and p85. J Biol Chem 270:19022–19027, 1995.

111. Verdi JM, Groves AK, Farinas I, Jones K, Marchionni MA, Reichardt LF, Anderson DJ: A reciprocal cell–cell interaction mediated by NT-3 and neuregulins controls the early survival and development of sympathetic neuroblasts. Neuron 16:515–527, 1996.

112. Yang Y, Spitzer E, Meyer D, Sachs M, Niemann C, Hartmann G, Weidner KM, Birchmeier C, Birchmeier W: Sequantial requirement of hepatocyte growth factor and neuregulin in the morphogenesis and differentiation of the mammary gland. J Cell Biol 131:215–226, 1995.

113. Krane IM, Leder P: NDF/heregulin induces persistence of terminal end buds and adenocarcinomas in the mammary glands of transgenic mice. Oncogene 12:1781–1788, 1996.

114. Bacus SS, Gudkov AV, Zelnick CR, Chin D, Stern R, Stancovski I, Peles E, Ben-Baruch N, Farbstein H, Lupu R, Wen D, Sela M, Yarden Y: Neu differentiation factor (heregulin) induces expression of intercellular adhesion molecule 1: Immplications for mammary tumors. Cancer Res 53:5251–5261, 1993.

115. Marte BM, Jeschke M, Graus-Porta D, Taverna D, Hofer P, Groner B, Yarden Y, Hynes NE: Neu differentiation factor/heregulin modulates growth and differentiation of HC11 mammary epithelial cells. Mol Endocrinol 9:14–23, 1995.

116. Lewis GD, Lofgren JA, McMurtrey AE, Nuijens A, Fendly BM, Bauer KD, Sliwkowski MX: Growth regulation of human breast and ovarian tumor cells by heregulin: Evidence for the requirement of ErbB2 as a critical component in mediating heregulin responsiveness. Cancer Res 56:1457–1465, 1996.

117. Danilenko DM, Ring BD, Lu JZ, Tarpley JE, Chang D, Liu N, Wen D, Pierce GF: Neu differentiation factor upregulates epidermal migration and integrin expression in excisional wounds. J Clin Invest 95:842–851, 1995.

118. Marikovsky M, Lavi S, Pinkas-Kramarski R, Karunagaran D, Liu N, Wen D, Yarden Y: ErbB-3 mediates differential mitogenic effects of NDF/heregulin isoforms on mouse keratinocytes. Oncogene 10:1403–1411, 1995.

119. Morrissey TK, Levi AD, Nuijens A, Sliwkowski MX, Bunge RP: Axon-induced mitogenesis of human Schwann cells involves heregulin and p185erbB2. Proc Natl Acad Sci USA 92:1431–1435, 1995.

120. Slamon DJ, Godolphin W, Jones LA, Holt JA, Wong SG, Keith DE, Levin WJ, Stuart SG, Udove J, Ullrich A, Press MF: Studies of the HER-2/*neu* proto-oncogene in human breast and ovarian cancer. Science 244:707–712, 1989.

121. Kokai Y, Myers JN, Wada T, Brown VI, LeVea CM, Davis JG, Dobashi K, Greene MI: Synergistic interaction of p185c-*neu* and the EGF receptor leads to transformation of rodent fibroblasts. Cell 58:287–292, 1989.

122. Wallasch C, Weiss FU, Niederfellner G, Jallal B, Issing W, Ullrich A: Heregulin-dependent regulation of HER2/neu oncogenic signaling by heterodimerization with HER3. EMBO J 14:4267–4275, 1995.

9

THE ESTROGEN RECEPTOR IN MAMMALIAN DEVELOPMENT

SYLVIA W. CURTIS AND KENNETH S. KORACH

Receptor Biology Section, Laboratory of Reproductive and Developmental Toxicology, National Institute of Environmental Health Sciences, National Institutes of Health, Research Triangle Park, NC 27709

9.1. INTRODUCTION

Prenatal sexual development of mammals is the culmination of interactions and feedback among numerous hormones and peptides. The development of the reproductive tract, in particular, is interesting: Two ductal systems, the Mullerian and Wolffian, develop in both sexes. However, development of the testis causes an androgen-dependent factor, Mullerian inhibiting substance (MIS), to be produced, leading to regression of the Mullerian duct structures. Testosterone supports the development of the Wolffian duct into the epididymis, seminal vesicles, and vas deferens. In the absence of androgen from the testis in the female or following removal of the testis or ovaries, the Mullerian ducts develop into the female reproductive structures and the Wolffian ducts regress (1,2). In addition, estrogen has roles in the development of other organs, including the mammary gland, where it directs epithelial duct proliferation and branching, and in the brain, where it programs physiologic responses through imprinting.

We have been interested in the role of estrogens in the development and function of the rodent reproductive tract. Within responsive tissues, estrogen transmits its signal via a specific receptor protein, the estrogen receptor (ER). The ER has specific high-affinity binding sites for estrogens, as well as specific DNA enhancer sequences (EREs) that regulate estrogen-responsive genes (3,4). The ER also binds to transcriptional regulators and has two domains that are important for regulation of transcriptional activity (AF1 and AF2) (5–8).

Hormones and Growth Factors in Development and Neoplasia, Edited by Robert B. Dickson and David S. Salomon
ISBN 0-471-16899-8

9.2. REPRODUCTIVE TRACT DEVELOPMENT IN THE ERKO

The evidence thus far indicates that androgen has the dominant role in reproductive development and that its absence results in a "default" development of the female phenotype (1). It is unclear whether this default development requires the ER or is independent of receptor. The role of estrogen in reproductive tract development is not apparent and may be masked by androgen action. Estrogen is important, however, in the growth and maintenance of reproductive tract organs in the sexually mature female. To assess the role of estrogen in the rodent reproductive tract, we evaluated the effects of estrogen signaling disruption on rodent reproductive tissue development and function using the estrogen receptor knockout (ERKO) mouse.

ERKO males have grossly normal reproductive tracts, indicating that estrogen and ER are not necessary for normal prenatal development (9). Their circulating testosterone levels are moderately elevated (10), which probably accounts for the normal reproductive tract development. Adult ERKO males are infertile, however, with a low sperm count and nonfertile sperm, which may indicate a role for estrogen in sperm maturation. The low sperm count was found to be due to a progressive degeneration of the seminiferous tubules, which is quite prominent by the time of sexual maturity (10). This may indicate a role for estrogen in postnatal maturation or maintenance in the testis.

ERKO females have grossly normal reproductive tracts at birth as well, again indicating the lack of a role for estrogen in prenatal female reproductive tract development. Adult ERKO females are also infertile (9). Ovariectomized ERKO females have uteri that are one quarter the size of wild-type littermates and show no growth in response to estrogen (11). The nonresponsiveness of the uterus to estrogen is shown by the lack of induction of lactoferrin and progesterone receptor mRNAs (12). The ERKOs do not ovulate and the ovaries become hemorrhagic, cystic, and enlarged after puberty (9,12, unpublished data). Thus, removal of estrogen action by genetic disruption of the *ER* gene has allowed us to show that estrogen does not have a major role in prenatal development of reproductive structures in the mouse. In contrast, estrogen plays an essential role at the time of puberty in growth, maintenance, and development of reproductive capability in both males and females.

9.3. EGF ACTION IN THE MOUSE UTERUS

We also have an interest in the role of epidermal growth factor (EGF) in the growth of the mouse uterus. The effects of estrogen in the rodent reproductive tract include the initiation of hypertrophy and mitosis (13,14). As these tissues have growth factor receptors, a role for cytokines in the process has also been hypothesized. Estrogen has been shown to increase the level of both EGF and its receptor (EGF-R) (15–22). Treatment with exogenous EGF has also been shown to mimic estrogen-induced effects in the uterus, resulting in mitosis, induction of lactoferrin, and modified cell morphology (23). Treatment with an EGF-specific antibody resulted in a 60–70% reduction in the estrogen-induced epithelial proliferation in the uterus and vagina (23), illustrating the probable role of EGF in estrogen action, as well as cross-talk between nuclear ER and membrane-bound EGF-R cellular signaling pathways. Further evidence of EGF-estrogen cross-talk was provided by showing that treatment with the pure estrogen antagonist ICI164,384, which reduces the level of uterine ER, greatly diminished the uterine response following EGF treatment (24). This indicates the necessity for the presence of ER for the EGF to mimic estrogen. Further studies in Ishikawa cells, a human endometrial carcinoma cell line devoid of ER, indicated that an estrogen-responsive *CAT* reporter gene could be activated by EGF only after cotransfection with an ER-expression plasmid. Furthermore, an ER deletion mutant that lacked a hormone binding domain could still be activated by EGF, while a deletion mutant lacking the N-terminal transcription activation domain could not (25,26). These studies support a model in which EGF can play a role in ER-mediated events even in the absence of estrogen.

9.4. USE OF ERKO TO STUDY EGF/ER CROSS-TALK

To prove the mediation by ER itself, EGF effects were studied in the uteri of ERKO mice. Since estrogen upregulates the EGF-R in the mouse uterus and removal of the ER during development might alter the level of EGF-R, it was necessary to determine the level of EGF-R in the ERKO mouse before proceeding with the studies of EGF action. An ^{125}I EGF binding assay showed that EGF-R was present in membranes from the uteri and liver of ERKO animals, although at a level 35% and 15% lower, respectively, than the wild-type levels (27). The EGF-R also undergoes autophosphorylation in both tissues in response to ligand (27), indicating that EGF-R can be activated by EGF in the ERKO mouse. To determine if the level of EGF-R in the ERKO mouse was sufficient to elicit a gene response, c-*fos* mRNA induction was measured following a 30-min EGF treatment. The wild-type uterus had an 18-fold induction and the ERKO had a 13-fold increase in c-*fos* mRNA (27). This response in both genotypes indicates that although the EGF-R level is lower in the ERKO, the EGF-R signaling pathway is intact and able to mediate gene activation in ERKO mice.

The ERKO uterus lacks a growth response following estradiol treatment, as measured by a uterotropic bioassay and ^{3}H-thymidine incorporation (9,12). In addition, after an 18-hr EGF treatment, wild-type uteri showed a 4.3-fold increase in ^{3}H-thymidine incorporation over placebo-treated animals, whereas ERKO mice showed no increase in DNA synthesis compared to the placebo controls, indicating that ER is essential for EGF to induce a growth response in the uterus.

The *PR* gene is also estrogen regulated in the rodent uterus (28,29); however, it is not induced by estrogen in the ERKO (12). EGF did not induce *PR* mRNA in the ERKO, although the PR message increases after EGF treatment in the wild-type animals. In addition, the anti-estrogen ICI 182,780 is able to block the EGF induction of *PR* in the wild type and has no effect on the ERKO (27).

9.5. MECHANISM OF EGF-ER CROSS-TALK

Studies from our group in combination with those of others (23–25,30) have led us to propose a model of cross-talk or interaction between the EGF and estrogen signaling pathways such that EGF-R activation can result in activation of ER target genes in the absence of estrogen (Fig. 9.1). In this model, the EGF-R and ER pathways can initiate

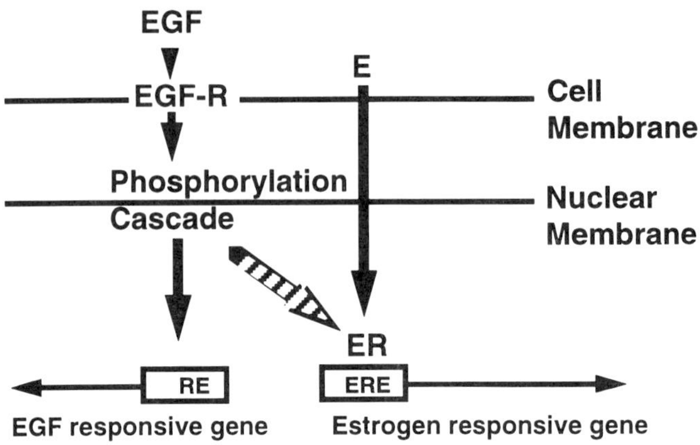

Fig. 9.1. Cross-talk between EGF-R/ER signaling pathways. The figure depicts the separate EGF- and estrogen-inducible pathways. EGF binds to its membrane-bound receptor, which initiates a cascade of phosphorylation modulating the activities of cellular kinases and nuclear transcription factors (depicted as the phosphorylation cascade in the figure). The activation of transcription factors results in regulation of responsive genes. In contrast, estrogen (E) diffuses to the nucleus, where it binds and activates the nuclear ER, thereby modulating transcription of estrogen-responsive genes. Cross-talk between the pathways (depicted by the dashed arrow) occurs when EGF initiates the phosphorylation cascade, which activates the ER, presumably via ER phosphorylation, regardless of the presence of estrogen. (From [27]).

distinct and separable signals that modulate responses of their respective target genes. However, it is also possible to couple the two pathways when, as depicted by the dashed arrow in the figure, the EGF-R signal activates an ER-mediated response of estrogen targets. The ERKO mouse is the definitive model to demonstrate the role of ER in this mechanism since removal of the ER should disable the cross-talk. Indeed, both DNA synthesis and PR mRNA, which are estrogen responsive, could be induced by EGF in wild-type mice but not in ERKO mice. Induction of c-*fos,* which required only the EGF-R pathway, occurred in both genotypes. Since EGF-R activates several pathways and since the precise pathway involved in EGF induction via the ER has not been determined, it is only possible to conclude definitely that the c-*fos* induction mechanism is intact. However, it appears that ER-mediated responses, including EGF-R-initiated cross-talk with ER, but not processes dependent only on the EGF-R signaling pathway, are disrupted in the ERKO.

The mechanism of this EGF-R-mediated ER activation is unknown, but *in vitro* studies using a truncated ER illustrated the importance of ER amino terminal sequences for activation by EGF (25). The ER mediates transcriptional activities by utilizing two transcription activation functions (AFs) (5–8). AF-1 is a ligand-independent activation function in the N terminus, and AF-2 is a ligand-dependent activation function in the ligand binding domain. An ER deletion mutant lacking the hormone binding domain was activated by EGF, whereas a deletion mutant lacking the AF-1 was not (25). Phosphorylation in the AF-1 region of the ER is a likely mechanism of EGF-mediated activity because EGF-R activation results in phosphorylation and consequent modulation of the activities of several cellular kinases (31–33), as well as subsequent phosphorylation and activation of nuclear transcription factors (31–33). Serines 122, 156, 158, and 298 of the mouse ER are known to be phosphorylated in COS-1 cells (34). Serines 104, 106, and 167 of the human ER have been identified as phosphorylation sites as well (35–37). Mutation of serine 118 to alanine, which is located in AF-1 of the human ER, eliminated phosphorylation of this serine and resulted in reduction of transcriptional activity by 40–75% in HeLa and COS-1 cells (35), indicating that phosphorylation of serine 118 may mediate transcriptional activity of ER and therefore could be a target for EGF-R-activated kinase. Furthermore, recent *in vitro* data show that serine 118 of the human ER can be phosphorylated by the MAP kinase pathway (38), resulting in increased estrogen-dependent activity in transfected COS-1 cells. Estrogen-independent activation of the human ER by EGF via the MAP kinase pathway and phosphorylation of serine 118 have been shown in HeLa and SK-Br-3 cells (39). However, other serines in the AF-1 region or other kinases may be required. Alternatively, phosphorylation of substrates other than ER, such as steroid receptor coactivators or nuclear transcription factors, may be required for ligand-independent activity.

9.6. PHYSIOLOGIC ROLE FOR CROSS-TALK

The importance of estrogen-independent ER action to normal physiology is unclear. The levels of EGF and EGF-R are both increased by estradiol (15), which indicates that EGF-R may play a role at a time when estrogen is already present. Perhaps during proestrus, when estrogen is low, EGF is able to mediate and maintain a constitutive level of estrogen-responsive gene expression. In addition, EGF may play a greater role during fetal development or postnatally before estrogen levels increase. The recent development of the aromatase knockout mouse, which retains functional ER but lacks the ability to synthesize estrogen, may shed light on the role of ligand-independent activity of ER. Interestingly, the aromatase knockout males are fertile (E. Simpson, personal communication), suggesting that ligand-independent activity may compensate for the lack of ER in a way that allows the males to develop into fertile adults. Similarly, the aromatase knockout females do not develop cystic ovaries (E. Simpson, personal communication), again suggesting that ligand-independent activity of the ER can help direct development of adult females in the absence of estrogen.

Alternatively, EGF-R-mediated ER activation may synergize with the estrogen hormone signal by increasing the activity of ER in terms of both the level of gene response and broadening the spectrum of activated genes. Studies *in vitro* have suggested that AF-1 and AF-2 activities are promoter specific (40) and that an increase in AF-1 activity as a result of EGF-R activation may allow for an enhanced response by genes requiring AF-1 activity.

9.7. CONCLUSIONS

The ERKO mouse has provided the first definitive link between growth factor pathways and ER *in vivo*. However, other membrane receptor pathways have also been shown to activate nuclear steroid receptors. Insulin and IGF-1 both increase growth and induce an estrogen-responsive CAT reporter in human neuroblastoma cells that contain ER (41). Membrane-permeable agents that increase cellular phosphorylation, such as 8-bromo-cAMP and okadaic acid, also enhance progesterone-dependent gene transcription (42). Dopamine, a neurotransmitter, has been shown to induce ligand-independent activation of several nuclear receptors, including the chicken progesterone receptor (43,44) and the human ER (45). Together with the activation of mouse ER by EGF, these results confirm the existence of conserved cross-talk mechanisms between the signaling pathways of membrane receptors and nuclear receptors.

REFERENCES

1. Jost A: Recent Prog Horm Res 8:379–418, 1953.
2. Wilson JD, George FW, Renfree MB: The endocrine role in mammalian sexual differentiation. In Bardin CW (ed): Recent Progress in Hormone Research. San Diego, CA: Academic Press, 1995, pp 349–364.
3. Parker MG, Arbuckle N, Dauvois S, Danielian P, White R: Structure and function of the estrogen receptor. Ann NY Acad Sci 684:119–126, 1993.
4. Gronemeyer H: Transcription activation by estrogen and progesterone receptors. Annu Rev Genet 25:89–123, 1991.
5. Green S, Chambon P: The oestrogen receptor: From perception to mechanism. In Parker MG (ed): Nuclear Hormone Receptors. San Diego, CA: Academic Press, 1991, pp 15–38.
6. Tora L, White J, Brou C, Tasset D, Webster N, Scheer E, Chambon P: The human estrogen receptor has two independent nonacidic transcriptional activation functions. Cell 59:477–487, 1989.
7. Lees JA, Fawell SE, Parker MG: Identification of constitutive and steroid-dependent transactivation domains in the mouse oestrogen receptor. J Steroid Biochem 34:33–39, 1989.
8. Lees JA, Fawell SE, Parker MG: Identification of two transactivation domains in the mouse oestrogen receptor. Nucleic Acids Res 17:5477–5488, 1989.
9. Lubahn DB, Moyer JS, Golding TS, Couse JF, Korach KS, Smithies O: Alteration of reproductive function but not prenatal sexual development after insertional disruption of the mouse estrogen receptor gene. Proc Natl Acad Sci USA 90:11162–11166, 1993.
10. Eddy EM, Washburn TF, Bunch DO, Goulding EH, Gladen BC, Lubahn DB, Korach KS: Targeted disruption of the estrogen receptor gene in male mice causes alteration of spermatogenesis and infertility. Endocrinology 137:4796–4805, 1996.
11. Korach KS: Insights from the study of animals lacking functional estrogen receptor. Science 266:1524–1527, 1994.
12. Korach KS, Couse JF, Curtis SW, Washburn TF, Lindzey J, Kimbro KS, Eddy EM, Migiaccio S, Snedeker SM, Lubahn DB, Schomberg DW, Smith EP: Estrogen receptor gene disruption: Molecular characterization and experimental and clinical phenotypes. In Conn P (ed): Recent Progress in Hormone Research. Bethesda MD: Endocrine Society, 1996, pp 159–188.
13. Couse JF, Curtis SW, Washburn TF, Lindzey J, Golding TS, Lubahn DB, Smithies O, Korach KS: Analysis of transcription and estrogen insensitivity in the female mouse after targeted disruption of the estrogen receptor gene. Mol Endocrinol 9:1441–1454, 1995.
14. Quarmby VE, Fox Davies C, Korach KS: Estrogen action in the mouse uterus: The influence of the neuroendocrine-adrenal axis. Endocrinology 114:108–115, 1984.

15. Quarmby VE, Korach KS: The influence of 17 beta-estradiol on patterns of cell division in the uterus. Endocrinology 114:694–702, 1984.
16. Das SK, Tsukamura H, Paria BC, Andrews GK, Dey SK: Differential expression of epidermal growth factor receptor (EGF-R) gene and regulation of EGF-R bioactivity by progesterone and estrogen in the adult mouse uterus. Endocrinology 134:971–981, 1994.
17. Stancel GM, Chiapetta C, Gardner RM, Kirkland JL, Lin TH, Lingham RB, Loose Mitchell DS, Mukku VR, Orengo CA: Regulation of the uterine epidermal growth factor receptor by estrogen. Prog Clin Biol Res 322:213–226, 1990.
18. Stancel GM, Gardner RM, Kirkland JL, Lin TH, Lingham RB, Loose Mitchell DS, Mukku VR, Orengo CA, Verner G: Interactions between estrogen and EGF in uterine growth and function. Adv Exp Med Biol 230:99–118, 1987.
19. Mukku VR, Stancel GM: Regulation of epidermal growth factor receptor by estrogen. J Biol Chem 260:9820–9824, 1985.
20. Lingham RB, Stancel GM, Loose Mitchell DS: Estrogen regulation of epidermal growth factor receptor messenger ribonucleic acid. Mol Endocrinol 2:230–235, 1988.
21. Gardner RM, Verner G, Kirkland JL, Stancel GM: Regulation of uterine epidermal growth factor (EGF) receptors by estrogen in the mature rat and during the estrous cycle. J Steroid Biochem 32:339–343, 1989.
22. DiAugustine RP, Petrusz P, Bell GI, Brown CF, Korach KS, McLachlan JA, Teng CT: Influence of estrogens on mouse uterine epidermal growth factor precursor protein and messenger ribonucleic acid. Endocrinology 122:2355–2363, 1988.
23. Huet Hudson YM, Chakraborty C, De SK, Suzuki Y, Andrews GK, Dey SK: Estrogen regulates the synthesis of epidermal growth factor in mouse uterine epithelial cells. Mol Endocrinol 4:510–523, 1990.
24. Nelson KG, Takahashi T, Bossert NL, Walmer DK, McLachlan JA: Epidermal growth factor replaces estrogen in the stimulation of female genital-tract growth and differentiation. Proc Natl Acad Sci USA 88:21–25, 1991.
25. Ignar Trowbridge DM, Nelson KG, Bidwell MC, Curtis SW, Washburn TF, McLachlan JA, Korach KS: Coupling of dual signaling pathways: Epidermal growth factor action involves the estrogen receptor. Proc Natl Acad Sci USA 89:4658–4662, 1992.
26. Ignar Trowbridge DM, Teng CT, Ross KA, Parker MG, Korach KS, McLachlan JA: Peptide growth factors elicit estrogen receptor-dependent transcriptional activation of an estrogen-responsive element. Mol Endocrinol 7:992–998, 1993.
27. Ignar-Trowbridge DM, Pimentel M, Parker MG, McLachlan JA, Korach KS: Peptide growth factor-cross-talk with the estrogen receptor requires the A/B domain and occurs independently of protein kinase C or estradiol. Endocrinology.
28. Curtis SW, Washburn T, Sewall C, DiAugustine R, Lindzey J, Couse JF, Korach KS: Physiological coupling of growth factor and steroid receptor signaling pathways: Estrogen receptor knockout mice lack estrogen-like response to epidermal growth factor. Proc Natl Acad Sci USA 93:12626–12630, 1996.
29. Aronica SM, Katzenellenbogen BS: Progesterone receptor regulation in uterine cells: Stimulation by estrogen, cyclic adenosine 3′,5′-monophosphate, and insulin-like growth factor I and suppression by antiestrogens and protein kinase inhibitors. Endocrinology 128:2045–2052, 1991.
30. Cho H, Aronica SM, Katzenellenbogen BS: Regulation of progesterone receptor gene expression in MCF-7 breast cancer cells: A comparison of the effects of cyclic adenosine 3′,5′-monophosphate, estradiol, insulin-like growth factor-I, and serum factors. Endocrinology 134:658–664, 1994.

31. Sumida C, Pasqualini JR: Antiestrogens antagonize the stimulatory effect of epidermal growth factor on the induction of progesterone receptor in fetal uterine cells in culture. Endocrinology 124:591–597, 1989.
32. Heldin C-H: Dimerization of cell surface receptors in signal transduction. Cell 80:213–233, 1995.
33. Hill CH, Treisman R: Transcriptional regulation by extracellular signals: Mechanisms and specificity. Cell 80:199–211, 1995.
34. Marshall CJ: Specificity of receptor tyrosine kinase signalling: Transient versus sustained extracellular signal-regulated kinase activation. Cell 80:179–185, 1995.
35. Lahooti H, White R, Danielian PS, Parker MG: Characterization of ligand-dependent phosphorylation of the estrogen receptor. Mol Endocrinol 8:182–188, 1994.
36. Ali S, Metzger D, Bornert JM, Chambon P: Modulation of transcriptional activation by ligand-dependent phosphorylation of the human oestrogen receptor A/B region. EMBO J 12:1153–1160, 1993.
37. Arnold SF, Obourn JD, Jaffe H, Notides AC: Serine 167 is the major estradiol-induced phosphorylation site on the human estrogen receptor. Mol Endocrinol 8:1208–1214, 1994.
38. Le Goff P, Montano MM, Schodin DJ, Katzenellenbogen BS: Phosphorylation of the human estrogen receptor. Identification of hormone-regulated sites and examination of their influence on transcriptional activity. J Biol Chem 269:4458–4466, 1994.
39. Kato S, Endoh H, Masuhiro Y, Kitamoto T, Uchiyama S, Sasaki H, Masushige S, Gotoh Y, Nishida E, Kawashima H, Metzger D, Chambon P: Activation of the estrogen receptor through phosphorylation by mitogen-activated protein kinase. Science 270: 1491–1494, 1995.
40. Bunone G, Briand P-A, Miksicek RJ, Picard D: Activation of the unliganded estrogen receptor by EGF involves the MAP kinase pathway and direct phosphorylation. EMBO J 15:2174–2183, 1996.
41. Metzger D, Losson R, Bornert JM, Lemoine Y, Chambon P: Promoter specificity of the two transcriptional activation functions of the human oestrogen receptor in yeast. Nucleic Acids Res 20:2813–2817, 1992.
42. Ma ZQ, Santagati S, Patrone C, Pollio G, Vegeto E, Maggi A: Insulin-like growth factors activate estrogen receptor to control the growth and differentiation of the human neuroblastoma cell line SK-ER3. Mol Endocrinol 8:910–918, 1994.
43. Denner LA, Weigel NL, Maxwell BL, Schrader WT, O'Malley BWW: Regulation of progesterone receptor-mediated transcription by phosphorylation. Science 250: 1740–1743, 1990.
44. Power RF, Lydon JP, Conneely OM, O'Malley BWW: Dopamine activation of an orphan of the steroid receptor superfamily. Science 252:1546–1548, 1991.
45. Power RF, Mani SK, Codina J, Conneely OM, O'Malley BW: Dopaminergic and ligand-independent activation of steroid hormone receptors. Science 254:1636–1639, 1991.
46. Smith CL, Conneely OM, O'Malley BW: Modulation of the ligand-independent activation of the human estrogen receptor by hormone and antihormone. Proc Natl Acad Sci USA 90:6120–6124, 1993.

10

PROGESTERONE AND DEVELOPMENT

J.K. LEIGHTON

Center for Veterinary Medicine, Food and Drug Administration, Rockville, MD 20855

L.L. WEI

Department of Physiology and Biophysics, Lombardi Cancer Center, Georgetown University, Washington, DC 20007

10.1. INTRODUCTION

Progesterone is an essential hormone required for normal growth and development of female reproductive organs such as the uterus and mammary gland (1,2). Most notable is its role in preparing the uterus for blastocyst implantation and the differentiation of glandular elements of the mammary gland required for milk production. Although progesterone has known roles in physiologic processes of the uterus, mammary gland, and other reproductive organs, the precise temporal pattern and cellular basis for its actions remain ill defined. Moreover, almost always, estrogen production precedes or coincides with progesterone synthesis, and thus the actions of the two hormones are often intertwined and difficult to distinguish. Of the two hormones, estrogen, in conjunction with other growth-promoting factors, has been considered the major determinant in stimulating cell proliferation in the uterus and mammary gland in postnatal life (1). Whether or not estrogen and progesterone play a role in the fetal development of these organs remains to be clearly determined.

The purpose of this chapter is to provide an overview of our present understanding of the action of progesterone or progestins on several female target organs at

Hormones and Growth Factors in Development and Neoplasia, Edited by Robert B. Dickson and David S. Salomon
ISBN 0-471-16899-8

various developmental stages: embryonic development, puberty, and, in the case of the mammary gland, in preparation for milk production. Great advances have been made in delineating the molecular basis of progesterone action through its receptor, its effects on organogenesis, and the regulation of the secretory phase of the uterus during the reproductive cycle. In spite of these advances, the actual temporal events and cellular physiology for progestin's actions, particularly in the mammary gland, remain very fragmented. In this chapter, we will present data from some commonly used as well as newly emerging model systems as they relate to the study of progestin's actions, as well as their limitations, and define possible contributing factors that may confound the interpretation of results. Since the majority of data focus on the uterus and breast, most of the chapter will be based on studies from these two organ systems.

10.1.1. Molecular Basis for Progesterone Action

Progestins, like other steroids, mediate their effects via intracellular receptors (3–5). Progesterone receptors (PRs), as well as estrogen receptors, belong to a superfamily of ligand-activated transcriptional factors that contain a highly conserved DNA binding domain (DBD) consisting of two zinc fingers and a carboxy-terminal hormone binding domain (HBD) (4,5). In addition, these functional domains include regions that are involved in nuclear localization, dimerization, and transcriptional activation. Although many of these properties have identifiable specific locations or sequences, not all of these domains have been definitely mapped. In general, the amino terminus contains a transcriptional activation function (TAF-1). A central DBD, which also has a receptor dimerization function, confers specificity for DNA target sequences. The ligand-binding ability resides in the C terminus, along with a second transcriptional activation function (TAF-2). In humans, chicks, and rodents, PRs consist of multiple progestin-binding proteins; in the rabbit uterus, one PR protein appears to be sufficient for full biologic activity (6–9). In humans, there exists a 116,000-dalton B-receptor, an N-terminal truncated 94,000-dalton A-receptor, and a third, smaller 60,000-dalton C-receptor (10,11).

The functional significance of multiple PR isoforms that have similar binding affinities for progestins and antiprogestins is unknown. Historically, the two larger isoforms (A- and B-receptors) were first identified. The physiologic importance of these two nearly identical human PR isoforms was bewildering since both were coordinately upregulated by estrogens, downregulated by progestins, and exist in near equimolar ratios (12,13). However, in the chick oviduct, it was noted that the B:A receptor ratio was not always constant but that A-receptors disappeared in winter and were the major form in estrogen-withdrawn chicks (14,15). Other than these two observations, the hormonal regulation of the two larger receptor forms was considered similar, and neither protein was thought to have any biologic activity distinct from the other form. As to the origin of A- and B-receptors, it was controversial whether the two receptor isoforms arose as alternate products from a single gene or whether the A-receptor was a proteolytic cleavage product of the larger B-receptor protein. With the cloning of the PR gene and the use of oligonucleotide probes to map progesterone receptor mRNAs, we and others demonstrated that the smaller N-terminally truncated A-receptor is a PR isoform distinct from the larger B-receptor (16,17). The two proteins are derived from multiple mRNAs transcribed from one gene, and the A-receptor is not a proteolytic fragment of the larger B form, as was once postulated (9,16,17).

In 1988, Tora et al. provided evidence that A- and B-receptors can have different transcriptional activity (18). The distinct functional properties of A- and B-receptors were promoter and cell-type dependent. B-receptors were more potent transcriptional activators than A-receptors on the mouse mammary tumor virus (MMTV) promoter in HeLa cells but were equally efficient on this same promoter in chick fibroblasts. These studies suggested that progestin-induced responses were not simply modulated by the receptor protein alone, but that other tissue-specific protein factors and specific DNA sequences also determine the eventual biologic outcome of progesterone stimulation.

Our understanding of progesterone receptor function is further complicated by the formation of dimer complexes. Progesterone receptors, like other members of the steroid hormone receptor family, form dimers on hormone binding. This step is obligatory for tight DNA binding. In the case of A- and B-receptors, three dimer combinations are possible (A:A, A:B, B:B) (19). It has been postulated that each of these dimer combinations can lead to different biologic outcomes even from a single target gene (18–21).

The discovery of a third progesterone receptor isoform further complicates this paradigm (11). A structural comparison of the 116,000-dalton human B-receptor and the 60,000-dalton N-terminally truncated human PR protein, the C-receptor, reveals that the C-receptor is missing the first 594 amino acids of the larger B-protein (16). Consequently, the C-receptor lacks the first zinc finger of the DBD but contains the second finger and has a complete hormone binding region. In addition, this truncated form of PR contains sequences for nuclear localization and receptor dimerization. In transfection studies utilizing a reporter gene construct, we found that the C-receptor has novel transcriptional potentiating properties. In the presence of progestins, C-receptors can enhance the transcriptional activity of either A- or B-receptors (22). C-receptors are found in both normal and cancerous cells, indicating that this truncated PR isoform is not peculiar to malignant cells but is an intergral component of PR-mediated pathways (L.L.W., unpublished data). So far, preliminary data also indicate that C-receptors do not enhance the transcriptional potential of glucocorticoid receptors but may modulate estrogen receptor activity in a cell-dependent manner. Thus, multiple PR isoforms provide a mechanism by which a defined set of transcription factors may lead to different levels of regulation on a given pool of hormone-dependent genes.

10.1.2. Progestin Action and The Uterus

Hormonal Regulation of Uterine Development during Embryogenesis Although the genetic sex of an embryo is determined at fertilization, the actual early genital structures are very similar in the two sexes (23,24). During this time, called the *indifferent stage,* both sexes have similar sexual organs: gonads with distinct cortical and medullary areas, two sets of genital ducts, and comparable external genitalia. Between the 5th and 10th weeks of gestation, the internal and external genitalia develop. The uterus forms during this period of early embryogenesis by fusion of the Mullerian ducts. The apparent gender of the fetus does not become fully evident until about the 12th week of life.

Determination of whether male or female reproductive organs develop is dependent, in part, on the expression of testis determining factor, a product of the SRY gene located on chromosome Y, in the early gonad (25,26). The SRY protein binds to sequences upstream from the gene coding for antimullerian hormone, a primary constituent responsible for regression of the mullerian ducts in male embryos (26). If the SRY gene product is not expressed, the early gonad develops into an ovary. In addition, the normal development of male reproductive organs requires the presence of both testosterone and dihydrotestosterone, thus implicating the presence and role of functional androgen receptors in this process (27). Maternal estrogen production also is low during early organogenesis (28). In further support of the idea that androgens are critical in normal male development are the findings that (1) masculization of some genital structures fails because of a deficiency of 5-alpha reductase (an enzyme that converts testosterone to dihydrotesterone) (27) and (2) abnormal development of male external genitalia occurs in immature alligators exposed to the antiandrogen (DDE) and in males with nonfunctional androgen receptors (29,30). Thus, in genotypic males, if androgens and antimullerian hormone are not present at the appropriate time in organogenesis, the structures derived from the urogenical sinus have female characteristics. Together, these observations suggest that estrogens and perhaps progestins do not have a prominent role, at least, in the normal development of the uterus during embryogenesis.

Although female reproductive tissues may not require estrogens for their natural development, this does not exclude the possibility that fetal tissues may still be estrogen responsive. Recently, Hou and Gorski demonstrated that mRNA for the estrogen receptor, the primary mediator of estrogen action, is present in blastocysts as early as the four-cell stage (31). Progesterone receptor mRNA was also detected in the blastocyst stage (31). Clinical evidence also suggests that fetal estrogen receptors exist and are functional. Diethylstilbestrol (DES) is a nonsteroidal estrogen first synthesized in 1938 and used clinically from the early 1940s to the early 1970s (32). At that time, the practicing dogma was that certain adverse outcomes of pregnancy could be prevented if sufficient progesterone was available. DES was thought to stimulate progesterone synthesis. DES also binds to estrogen receptors with high affinity, and is both a weak carcinogen and a potent teratogen in laboratory animals and humans (33). In children exposed *in utero* to DES, the presence of the compound is associated with development of clear cell adenocarcinoma of the vagina and cervix, as well as other abnormalities of the uterus, vagina, and cervix (33). These studies have been corroborated by investigations of the prenatal effects of DES in the fetal mouse model (32,34–36). Since DES mediates its actions primarily through the estrogen receptor, these studies indicate that estrogen receptor-containing fetal cells may be responsive to estrogens. This is further complicated by the recent discovery of another estrogen receptor form, ER β, which is distinct from the previously described classical 65,000-dalton ER, now designated ER α (37). At this time, the evidence that progesterone affects fetal development is less certain (see below), and is primarily suggested by the presence of PR in fetal cells and the close association of estrogen and progesterone activities.

Anatomy and Physiology of the Mature Uterus The uterus is the muscular organ where implantation of the embryo occurs (23). It consists of three parts: the fundus, the part above the entrance of the uterine tubes; the body; and the cervix, the narrow vaginal end. The uterine wall consists of the endometrium (the site of implantation), and is composed of a basal layer and a more superficial functional (luminal) layer; the myometrium (the muscular layer of the uterus); and a peritoneal layer that covers the uterine body (also referred to as the *serosa*). The functional layer of the endometrium consists of columnar epithelial cells and connective tissue stroma. This layer expands and sloughs off during the menstrual cycle in response to changes in endocrine stimuli. These changes are regulated by the complex interaction of hormones from several different endocrine organs, such as the ovary, pituitary, hypothalamus, and the placenta if implantation occurs. Autocrine and paracrine factors also play a role. Of particular and direct importance in this process are the ovarian hormones estrogen and progesterone (see below). The functional layer of the endometrium is renewed by estrogen-stimulated proliferation of epithelial cells from the basal layer, which, along with the growing stromal cells, contributes to the cyclical thickening of the endometrium.

Progesterone-Induced Changes in the Uterus during the Menstrual Cycle During the proliferative (follicular) phase, endometrial epithelial cells are tall and columnar, with numerous microvilli (24). This phase is primarily regulated by estrogen secreted from the ovary. After ovulation, under the influence of progesterone synthesized by the corpus luteum, the endometrial glands convert to a secretory (luteal) phase, so named because of the presence of eosinophilic, protein-rich secretory products. During this phase and in the presence of progesterone, the epithelial cells become longer, branched, and convoluted structures. In contrast to the endometrial epithelial cells, the stroma remains unchanged histologically during the early secretory phase. At the midsecretory stage, the stroma undergoes a change resulting in a progressive increase in edema. During the late secretory stage, the epithelial glandular cells are filled with glycogen-containing vacuoles and become engorged. Progesterone output from the corpus luteum decreases during the late secretory phase due to a decrease in stimulation from the pituitary hormones. During this period the spiral arteries, which supply blood to the endometrium, alternate between engorged and constricted. The lack of constant blood flow may explain the eventual degeneration of the endometrium and epithelial breakdown. The entire luminal layer and mucus are lost during menstruation. The withdrawal of progesterone stimulation is considered of primary importance in the degeneration of the endometrium. However, the mechanism by which decreased serum progesterone levels translate to a downstream event at the cellular level is unknown at this time.

Biochemically, estrogen, estriol, and other short-acting estrogens are known to stimulate a number of early effects in the human uterus including hyperemia, calcium influx, histamine release, and increased RNA and protein synthesis (27). However, not all of these early effects result in uterine growth. In fact, it appears that these early uterotropic responses may be supportive but not necessarily obligatory for true uterine growth. On the other hand, progesterone clearly inhibits estrogen-induced endometrial proliferation and converts endometrial cells to a secretory state. Based on identical findings in other uterine model systems (i.e., rat and mouse uteri), it has been concluded that progestins are natural antagonists of estrogen-stimulated endometrial cell proliferation (1). This finding was extended to mammary gland physiology and is currently being reevaluated (see below). Nevertheless, the protective effect of progestins on estrogen's growth-promoting action in the uterus remains firmly established and provides the clinical basis for the use of progestins in the treatment of endometrial cancer and the clinical evaluation of progestins with estrogens for hormone replacement therapy in postmenopausal women (1,38).

In order to better understand the direct actions of progesterone on the uterus, investigators have focused on correlating PR expression in specific uterine cell types and associated progestin-induced changes. Several investigational groups, using immunohistochemical techniques, have found that progesterone receptor levels are not stagnant but fluctuate throughout the reproductive cycle (39–41). The most dramatic changes in PR expression occur in the glandular epithelial cells of the endometrium. The highest expression is present during the late proliferative and early secretory changes (39,40). This correlates well with inhibition of estrogen-induced mitosis and differentiation of cells to a secretory phenotype. In the late secretory phase, PR levels drop both in the relative amount per cell and in the total number of glandular epithelial cells expressing PR. These changes parallel estrogen upregulation of PR and the autologous downregulation of PR by its ligand (12,42). In contrast, stromal cells and smooth muscle cells contain PR, but the PR levels remain relatively stable throughout the cycle (41). This observation of PR in smooth muscle cells is consistent with the finding that progesterone inhibits myometrial contractions during implantation and gestation (39). Despite the differences of PR regulation in the three cell types, PR expression is highest in all cells during the proliferative phase; this is probably attributable to high local estradiol concentrations (41).

As to the role and distribution of the three PR isoforms in the different uterine cell types throughout the menstrual cycle, very little is known. However, recently it was shown in normal human endometrial samples that the levels of A- and B-receptor protein during the menstrual cycle were not always equimolar (43). In fact, the A:B ratio could be as high as 50:1, and on some days of the cycle B-receptors were undetectable. More detailed studies are required for us to understand the role of each PR isoform in controlling the eventual biological outcome.

Progesterone Carrier Proteins Serum progesterone is bound primarily to transcortin (corticosteroid binding globulin, CBG) and to albumin, with a relative ratio of 17% and 80% to total serum progesterone, respectively. Approximately 2.5% of the total serum progesterone is found in an unbound or free state (44). Generally, albumin has high capacity and low affinity for steroid hormones, whereas the specific carrier proteins (e.g., CBG) have lower capacity but higher affinity. The relative binding affinity of progesterone for CBG is 36% compared to 100% for cortisol. Synthetic progestins also bind to CBG, but with different affinities relative to that for the natural hormone progesterone (45).

CBG is primarily synthesized and secreted by the liver, and the circulatory concentration increases throughout pregnancy (44,46,47). A posttranslationally modified form of CBG may be expressed during pregnancy (48). In rabbits and perhaps other species, CBG found in the maternal circulation can be transferred to the fetus and to the amniotic fluid (49). The role this plays in modifying maternal or fetal free levels of either cortisol or progesterone remains uncertain. Fetal hepatocytes can also synthesize CBG (47,50–52). In hepatocytes, the level of CBG is regulated by the interplay of triiodothyronine (T3), retinoic acid, cortisol, and estradiol (52). CBG also is expressed in a complex time and spatial manner during development (46,47). In spite of our increasing knowledge of CBG, its physiologic role in regulating progesterone concentrations in target tissues remains unclear; this is an important area of investigation.

10.1.3. The Mammary Gland

Anatomy of the Human Mammary Gland Mammary glands are essential for infant nutrition in all mammalian species (53). The glands are present at about the fifth week of gestation in humans and remain relatively undeveloped until puberty (54). Embryonically, the ectodermal primitive milk streak is evident in the 5-week-old human fetus (24,54). By the seventh to eighth week, the rudimentary mammary gland thickens and invades the chest wall. At 16 weeks, epithelial buds form and branch; eventually, these branches develop into the adult secretory alveoli. In the third trimester, canalization of the epithelial tissues occurs in response to circulating placental hormones; this continues until the end of the 32nd week of gestation. By the time of parturition, there is a fourfold increase in total glandular mass and the nipple areolas develop pigmentation.

The immature breast consists of two layers of epithelial cells: a basal cuboidal cell layer and a superficial flat epithelial layer. At puberty, estrogen induces cell proliferation and results in the differentiation of three alveolar cell types: luminal A cells, chief B cells, and myoepithelial cells (54). The B cells are the predominant cell type. However for full ductal-lobular-alveolar development, progesterone is necessary (1).

At puberty, the glands begin to respond to ovarian hormones (see below). Each mature gland consists of tubuloalveolar glands embedded in connective tissue stroma surrounded by adipose tissue. Collecting ducts converge and drain into the nipple. Approximately 20 large ducts empty into each nipple. The largest ducts are called *galactophores* or *lactiferous ducts;* these become progressively smaller and originate from individual lobules containing clusters of cells that secrete milk.

Mammary Gland Changes during the Female Reproductive Cycle Cyclic changes in estrogen and progesterone levels during the menstrual cycle profoundly influence the cellular physiology of the mammary gland (54). During the follicular phase, increasing levels of circulating estrogen stimulate both epithelial proliferation and sprouting. Biochemically, this is demonstrated in an increase in mitoses, RNA synthesis, and in the size and number of various organelles such as ribosomes and mitochondria. With progestins present during the luteal phase, the mammary ducts now become dilated and the terminal end buds (alveolar buds) differentiate to a secretory state. Together with other hormones, estrogen and progesterone stimulate the accumulation of lipid droplets. For a more detailed description of the structural microanatomy or morphology of the breast, refer to Russo and Russo (55).

***Endocrine Regulation of the Mammary Gland* in Utero** Based on the inhibitory actions of progesterone on epithelial cell growth in the uterus, it was assumed that progesterone would likewise inhibit mammary epithelial cell proliferation. This original presumption is now being questioned. In 1988, Haslam reported two very interesting findings: (1) estrogen, not progesterone, acts as a potent mitogen in stimulating DNA synthesis in the mammary gland of 5-week-old immature mice; (2) in striking contrast, progesterone, not estradiol, is responsible for the marked increase in DNA synthesis in the mammary gland of 10-week-old adult mice (56). These two observations indicated that the response of the mammary gland to progesterone is highly dependent on the developmental stage of the gland. Moreover, progesterone promotion of DNA synthesis was found to be restricted to the epithelial component, and did not include the stroma of the mammary gland. Later, Haslam and colleagues demonstrated that the PR in epithelial cells is estrogen inducible, whereas the PR in mammary stroma is estrogen independent (57). Progesterone stimulated DNA synthesis only in cells containing estrogen-inducible PR. Interestingly, in the immature mammary gland (at 5 weeks of postnatal development) that contains only estrogen-independent PR, there is no response to progesterone. These studies imply that other factors, perhaps tissue-specific factors, also regulate estrogen and progesterone action (58–60). Molecular studies on PR function support this conclusion (58–60). These studies highlight the likely possibility that the response to progestins may vary among cell types within a single organ and between different tissues.

In comparison to the numerous studies analyzing PR expression in the various cell types within the uterus, surprisingly few experiments have focused on the cellular distribution of PR in the normal breast (1,61 and references therein). In part this can be attributed to the relatively low level of PR in normal breast compared to that found in breast cancer cells and tumors. PRs are expressed in normal breast and, like the myometrial and stromal components of the uterus, appear to remain relatively unchanged over the course of the reproductive cycle (61,62). Thus, it appears that progesterone receptors in normal breast tissue may not have the same autoregulation by progestins demonstrated in breast cancer cells. Further careful studies are required to verify these findings, as well as to determine which PR isoforms are present in each of the constituent cell types in the normal mammary gland.

10.1.4. Model Systems

Understanding the role of progesterone in development has been hampered until recently by the lack of appropriate model systems that can directly address the complex regulatory mechanisms in an animal. Although much information has been gained by the use of *in vitro* cell culture systems and organ culture systems, it is difficult at times to infer clearly from these models the precise events responsible for the biologic responses in the whole animal. For instance, effects of progestins and other steroid hormones on some cell types (e.g., cells of the neuronal or immune systems) and the temporal nature of gene transcription have not been amenable in these previous experimental models. Moreover, these systems do not lend themselves to studies on endocrine interactions or the effect of serum binding proteins or metabolism of specific ligands. In animal experiments, hormones generally have been administered at supraphysiologic concentrations to determine the toxicologic properties of the hormones. These experiments were designed to address the growing concerns raised by women receiving progestins and/or estrogens for hormone replacement therapy or in contraceptives. The biologic endpoints used in these experiments focused on estrogen and progestin effects on tumor incidence and growth rather than on the developmental effects of progestins. Epidemiologic studies have also been conducted principally with this endpoint in mind (63).

Recently, new transgenic mice that lack the progesterone receptor have been developed (64). These knockout mice were generated bearing a null mutation within the coding region of the PR gene. A neomycin resistance gene was inserted in the first exon of the PR gene downstream of the translation initiation codons for B- and A-receptor isoforms. This site was specifically selected so that transcription of both isoforms would fail. Based on the absence of PR-specific transcripts in progestin target tissues from these PR null mice, it appears very likely that the third, smaller PR isoform, the C-receptor, is also not expressed in these animals. This is further supported by the absence of specific progestin binding activity.

Externally, both sexes of adult PR null mice were found not to be drastically different from their wild-type counterparts. Gross morphologic examination revealed that the homozygous knockout mice and wild-type mice had normal-appearing external genitalia; no other gross abnormalities were observed in either sex. Because of progesterone's central role in blastocyst implantation, it was not surprising to find that the female homozygous knockout mice were infertile. On the other hand, the lack of PR did not alter the fertility of adult male homozygous knockout mice.

Closer examination of the ovaries from female homozygous knockout mice revealed the presence of "preovulatory" follicles and a complete absence of corpora lutea in estrogen- and progesterone-challenged animals. In normal and wild-type animals, estrogen/progesterone treatment results in ovulation, which leads to the appearance of several corpora lutea and the lack of preovulatory follicles. These studies indicate that PRs are necessary for successful ovulation and are required for the formation of corpora lutea.

Initial gross analysis of the uterus also did not reveal any dramatic peculiarities in either the myometrial or endometrial layers. However, in animals exposed to estradiol (1 μg) and progesterone (1 mg), PR-deficient mice displayed highly abnormal uteri compared to those found in control animals. Uteri were enlarged and fluid-filled. Examination of the uterine fluid demonstrated an infiltration of polymorphonuclear (PMN) leukocytes (an inflammatory response). Cross-sectional analyses of the uterine wall demonstrated increased growth of mucosal and glandular epithelia and the presence of extracellular edema. The epithelium appeared hyperplastic and was disorganized compared to that of control animals. Endometrial glands were also distended. These histologic observations closely resemble previous reports of uteri from control mice treated only with estradiol (65,66).

Mammary glands from 6-week-old PR null animals versus control mice displayed no profound differences in glandular development. However, on hormone treatment (1 μg estradiol and 1 mg progesterone), glands from the wild-type animals exhibited the classic complex of ductal side branching and extension into the mammary fat pad. The knockout mice had dramatically fewer side branches and ductal extensions. Additionally, there was a lack of lobuloalveolar development at the end of each duct.

When the lordosis behavior between PR-deficient mice and normal wild-type females was compared, progesterone was found to be required for normal sexual responsivity. Thus, PR is not only important for the anticipated process of blastocyst implantation but is also necessary for successful ovulation, appropriate mammary gland development, and the coordination of normal sexual behavior.

In sum, PR knockout mice have provided a unique system to address the role of progestins in well-recognized progestin-regulated reproductive processes and also in other physiologic processes, including embryonic development. The studies discussed herein strongly support the idea that embryonic-derived PR and ER are not essential for embryonic survival and that PR is not a key prerequisite for proper development of the female reproductive tract *in utero*. In contrast, PR is essential for ovulation and also appears to be involved in modulating immune responses (i.e., altered numbers of PMN leukocytes). In addition, the stunted ductal development in the knockout mouse mammary gland, even in the presence of estradiol, supports strongly a mitogenic role for progesterone in the developing mammary gland rather than the inhibitory effect observed in the uterus. In further support of the growth-promoting properties of progesterone on mammary gland epithelium, antiprogestins inhibit the proliferative effects of progesterone on some experimental mammary tumor systems, thereby suggesting that progesterone via PR has proliferative actions (67). These findings, progesterone's proliferative effect on normal mammary gland development and mammary tumor progression, may have long-term implications for the use of progestins in hormone replacement therapy and in contraception. Further studies are necessary to precisely define progestin's role in these known processes, as well as in other unrecognized processes.

ACKNOWLEDGMENT

This work was supported by Public Health Service grants HD29793 and CA40011C (L.L.W.), Grosvenor International (Atlantic) Limited (L.L.W.), and the Food and Drug Administration, Office of Women's Health (J.K.L.).

REFERENCES

1. Clarke CL, Sutherland RL: Progestin regulation of cellular proliferation. Endocrine Rev 11:266–301, 1990.
2. Baulieu EE: Contragestation and other clinical applications of RU486, antiprogesterone at the receptor. Science 245:1351–1357, 1989.
3. Evans RM: The steroid and thyroid hormone receptor superfamily. Science 240: 889–895, 1988.
4. O'Malley B: The steroid receptor superfamily: More excitement predicted for the future. Mol Endocrinol 4:363–369, 1990.
5. Wei LL: New models and insights in steroid hormone action. In Pavlik EJ (ed): Estrogens, Progestins, and Their Antagonists. Boston: Birkhauser, 1997, pp 102–121.
6. Schrader WT, O'Malley BW: Progesterone-binding components of chick oviduct IV. Characterization of purified subunits. J Biol Chem 247:51–59, 1972.
7. Horwitz KB, Alexander PS: In situ photolinked nuclear progesterone receptors of human breast cancer cells: Subunit molecular weights after transformation and translocation. Endocrinology 113:2195–2201, 1983.
8. Schneider W, Ramachandran C, Satyaswaroop PG, Shyamala G: Murine progesterone receptor exists predominantly as the 83-kilodalton "A" form. J Steroid Biochem Mol Biol 38:285–291, 1991.
9. Loosfelt H, Logeat F, Vu Hai MT, Milgrom E: The rabbit progesterone receptor: Evidence for a single steroid-binding and characterization of receptor mRNA. J Biol Chem 259:14196–14202, 1984.
10. Wei LL, Horwitz KB: The structure of progesterone receptors. Steroids 46:677–695, 1985.
11. Wei LL, Miner R: Evidence for the existence of a third progesterone receptor protein in human breast cancer cell line T47D. Cancer Res 54:340–343, 1994.
12. Wei LL, Krett NL, Francis MD, Gordon DF, Wood WM, O'Malley BW, Horwitz KB: Multiple human progesterone receptor ribonucleic acids and their autoregulation by progestin agonists and antagonists in breast cancer cells. Mol Endocrinol 2:62–72, 1988.
13. Horwitz KB, Wei LL, Francis MD: Structural analyses of progesterone receptors. J Steroid Biochem 24:109–117, 1986.
14. Spelsberg TC, Halberg F: Circannual rhythms in steroid receptor concentration and nuclear binding in the chick oviduct. Endocrinology 107:1234–1244, 1980.
15. Boyd-Leinen PA, Fournier D, Spelsberg TC: Nonfunctioning progesterone receptors in the developed oviducts from estrogen-withdrawn immature chicks and in aged nonlaying hens. Endocrinology 111:30–36, 1982.

16. Wei LL, Gonzalez-Aller C, Wood WM, Miller LA, Horwitz KB: 5′-Heterogeneity in human progesterone receptor transcripts predicts a new amino-terminal truncated "C"-receptor and unique A-receptor messages. Mol Endocrinol 4:1833–1840, 1990.
17. Kastner P, Krust A, Turcotte B, Stropp U, Tora L, Gronemeyer H, Chambon P: Two distinct estrogen-regulated promoters generate transcripts encoding the two functionally different human progesterone receptor forms A and B. EMBO J 9:1603–1614, 1990.
18. Tora L, Gronemeyer H, Turcotte B, Gaub MP, Chambon P: The N-terminal region of the chicken progesterone receptor specifies target gene activation. Nature 333:185–188, 1988.
19. Edwards DP, DeMarzo AM, Onate SA, Beck CA, Estes PA, Nordeen SK: Mechanism controlling steroid receptor binding to specific DNA sequences. Steroids 56:271–278, 1991.
20. Vegeto E, Shahbaz MM, Wen DX, Goldman ME, O'Malley BW, McDonnell DP: Human progesterone receptor A form is cell- and promoter-specific repressor of human progesterone receptor B function. Mol Endocrinol 7:1244–1255, 1993.
21. Tung L, Mohamed MK, Hoeffler JP, Takimoto GS, Horwitz KB: Antagonist-occupied human progesterone B-receptors activate transcription without binding to progesterone response elements and are dominantly inhibited by A-receptors. Mol Endocrinol 7:1256–1265, 1993.
22. Wei LL, Hawkins P, Baker C, Norris B, Sheridan PL, Quinn PG: An amino-terminal truncated progesterone receptor isoform, PR_C, enhances progestin-induced transcriptional activity. Mol Endocrinol 10:1379–1387, 1996.
23. Anderson JR, Genadry R: Anatomy and embryology. In Berek JS, Adashi EY, Hillard PA (eds): Novak's Gynecology. Baltimore: Williams & Wilkins, 1996, pp 71–121.
24. Johnson KE: Histology: Microscopic Anatomy and Embryology. New York: Wiley, 1982, pp 306–317.
25. Nordqvist K: Sex differentiation—gonadogenesis and novel genes. Int J Dev Biol 39:727–736, 1995.
26. Harley VR, Goodfellow PN: The biochemical role of SRY in sex determination. Mol Reprod Dev 39:184–193, 1994.
27. Clark JH, Schrader WT, O'Malley BW: Mechanism of action of steroid hormones. In Wilson JD, Foster DW (eds): Williams' Textbook of Endocrinology. Philadelphia: WB Saunders, 1995, pp 35–90.
28. Carr BR: Fertilization implantation, and endocrinology of pregnancy. In Griffin JE, Ojeda SR (eds): Textbook of Endocrine Physiology. New York: Oxford University Press, 1992, pp 189–209.
29. Guillette LJ Jr, Pickford DB, Crain DA, Rooney AA, Percival HF: Reduction in penis size and plasma testosterone concentrations in juvenile alligators living in a contaminated environment. Gen Compar Endocrinol 101:32–42, 1996.
30. Brown TR, Lubahn DB, Wilson EM, French FS, Migeon CJ, Corden JL: Functional characterization of naturally occurring mutant androgen receptors from subjects with complete androgen insensitivity. Mol Endocrinol 4:1759–1772, 1990.
31. Hou Q, Gorski J: Estrogen receptor and progesterone receptor genes are expressed differentially in mouse embryos during development. Proc Natl Acad Sci USA 90: 9460–9464, 1993.
32. Greco TL, Cuello TM, Gorski J: Estrogen receptors, estradiol, and diethylstilbestrol in early development: The mouse as a model for the study of estrogen receptors and estro-

gen sensitivity in embryonic development of male and female reproductive tracts. Endocrine Rev 14:59–71, 1993.

33. Mittendorf R: Teratogen update: Carcinogenesis and teratogenesis associated with exposure to diethylstilbestrol (DES) in utero. Teratology 51:435–445, 1995.
34. Haney AF, Newbold RR, Fetter VF, McLachlan JA: Paraovarian cysts associated with prenatal diethylstilbestrol exposure. Am J Pathol 124:405–411, 1986.
35. McLachlan JA, Newbold RR, Bullock BC: Long-term effects of the female mouse genital tract associated with prenatal exposure to diethylstilbestrol. Cancer Res 40: 3988–3999, 1980.
36. Takasugi N, Bern HA, DeOme KB: Persistent vaginal cornification in mice. Science 138:438–439, 1962.
37. Kuiper GGJM, Enmark E, Pelto-Huikko M, Nilsson S, Gustafsson J-A: Cloning of a novel estrogen receptor expressed in rat prostate and ovary. Proc Natl Acad Sci USA 93:5925–5930, 1996.
38. Key TJA, Pike MC: The dose–effect relationship between "unopposed" oestrogens and endometrial mitotic rate: Its central role in explaining and predicting endometrial cancer risk. Br J Cancer 57:205–212, 1988.
39. Lessey BA, Killiam AP, Metzger DA, Haney AF, Greene GL, McCarty KS: Immunohistochemistry analysis of uterine estrogen and progesterone receptors throughout the menstrual cycle. J Clin Endocrinol Metab 67:334–340, 1988.
40. Press MF, Udove JA, Greene GL: Progesterone receptor distribution in the human endometrium. Analysis using monoclonal antibodies to the human progesterone receptor. Am J Pathol 131:112–124, 1988.
41. Snijders MPML, deGoeij AFPM, Debets-TeBaert MJC, Rousch MJM, Koudstaal J, Bosman FT: Immunocytochemical analysis of oestrogen receptors and progesterone receptors in the human uterus throughout the menstrual cycle and after menopause. J Reprod Fertil 94:363–371, 1992.
42. Nardulli AM, Greene GL, O'Malley BW, Katzenellenbogen BS: Regulation of progesterone receptor messenger ribonucleic acid and protein levels in MCF-7 cells by estradiol: Analysis of estrogen's effect on progesterone receptor synthesis and degradation. Endocrinology 122:935–944, 1988.
43. Wiehle RD, Mangal R, Poindexter AN III, Weigel N: Human progesterone receptor A and B expression during the menstrual cycle. Abstract #P1-446. Presented at the 77th Annual Endocrine Society Meeting, Washington, DC, June 14–17, 1995.
44. Orth DN, Kovacs WJ, DeBold CR: The adrenal cortex. In Wilson JD, Foster DW (eds): Williams' Textbook of Endocrinology. Philadelphia: WB Saunders, 1995, pp 489–619.
45. Juchem M, Pollow K: Binding of oral contraceptive progestogens to serum proteins and cytoplasmic receptor. Am J Obstet Gynecol 163:2171–2183, 1990.
46. Underhill DA, Hammond GL: Cis-regulatory elements within the proximal promoter of the rat gene encoding corticosteroid-binding globulin. Gene 162:205–211, 1995.
47. Burdusco ET, Yang K, Hammond GL, Challis JR: Corticosteroid-binding globulin (CBG) production by hepatic and extra-hepatic sites in the ovine fetus; effects of CBG on glucocorticoid negative feedback on pituitary cells in vitro. J Endocrinol 146:121–130, 1995.
48. Avvakumov GV, Strel'chonok OA: Properties and serum levels of pregnancy-associated variant of human transcortin. Biochim Biophys Acta 925:11–16, 1987.

49. Seraline GE, Underhill CM, Smith CL, Nguyen VT, Hammond GL: Biological half-life and transfer of maternal corticosteroid-binding globulin to amniotic fluid in the rabbit. Endocrinology 125:1321–1325, 1989.
50. Challis JR, Berdusco ET, Jeffray TM, Yang K, Hammond GL: Corticosteroid-binding globulin (CBG) in fetal development. J Steroid Biochem Mol Biol 53:523–527, 1995.
51. Scrocchi LA, Orava M, Smith CL, Han VK, Hammond GL: Spatial and temporal distribution of corticosteroid-binding globulin and its messenger ribonucleic acid in embryonic and fetal mice. Endocrinology 132:903–909, 1993.
52. Elfahime EL, Felix JM, Koch B: Antagonistic effects of retinoic acid and triiodothyronine in the expression of corticoid-binding globulin (CBG) by cultured fetal hepatocytes. J Steroid Biochem Mol Biol 48:467–474, 1994.
53. Vorherr H: The Breast: Morphology, Physiology and Lactation. New York: Academic Press, 1974.
54. Osborne MP: Breast development and anatomy. In Harris JR, Lippman ME, Morrow M, Hellman S (eds): Diseases of the Breast. Philadelphia: Lippincott-Raven, 1996, pp 1–14.
55. Russo J, Russo IH: Development of the human mammary gland. In Neville MC, Daniel CW (eds): The Mammary Gland. New York: Plenum, 1987, pp 67–93.
56. Haslam SZ: Progesterone effects on deoxynucleic acid synthesis in normal mouse mammary gland. Endocrinology 122:464–470, 1988.
57. Wang S, Counterman LJ, Haslam SZ: Progesterone action in normal mouse mammary gland. Endocrinology 127:2183–2189, 1990.
58. Onate SA, Tsai SY, Tsai M-J, O'Malley BW: Sequence and characterization of a coactivator for the steroid hormone receptor superfamily. Science 270:1354–1357, 1995.
59. Halachmi S, Marden E, Martin G, MacKay H, Abbondanza C, Brown M: Estrogen receptor-associated proteins: Possible mediators of hormone-induced transcription. Science 264:1455–1458, 1994.
60. Cavailles V, Dauvois S, L'Horset F, Lopez G, Hoare S, Kushner PJ, Parker MG: Nuclear factor RIP 140 modulates transcriptional activation by the estrogen receptor. EMBO J 14:3741–3751, 1995.
61. Clarke CL, Sutherland RL: Progestin regulation of cellular proliferation: Update 1993. Endocrine Rev 1:132–135, 1993.
62. Elashry D, Zava DT, Speers WC, Edwards DP: Immunocytochemical localization of progesterone receptor in breast cancer with anti-human receptor monoclonal antibodies. Cancer Res 48:6462–6474, 1988.
63. Klein R, Berlin L: Benefits and risks of hormone replacement therapy. In Pavlik EJ (ed): Estrogens, Progestins, and Their Antagonists, Vol 1. Boston: Birkhauser, 1997, pp 3–50.
64. Lydon JP, Mayo FJ, Funk CR, Mani SK, Hughes AR, Montgomery CA, Shyamala G, Conneely OM, O'Malley BW: Mice lacking progesterone receptor exhibit pleiotropic reproductive abnormalities. Genes Dev 9:2266–2278, 1995.
65. Martin L, Finn CA, Trinder GL: Hypertrophy and hyperplasia in the mouse uterus after estrogen treatment. An autoradiogram study. J Endocrinol 56:133–144, 1973.
66. Quarmby VE, Korach KS: The influence of 17β-estradiol on patterns of cell division in the uterus. Endocrinology 114:694–702, 1984.
67. Michna H, Schneider MR, Nishino Y, El Etreby MF: Antitumor activity of the antiprogestins ZK 98.299 and RU 38.486 in hormone dependent rat and mouse mammary tumors: Mechanistic studies. Breast Cancer Res Treat 14:275–288, 1989.

PART III

POSTNATAL DEVELOPMENT PROCESSES IN THE ADULT: REPRODUCUTIVE TRACTS AND MAMMARY GLANDS

11

PROLACTIN IN DEVELOPMENT OF THE MAMMARY GLAND AND REPRODUCTIVE TRACT

BARBARA K. VONDERHAAR

Molecular and Cellular Endocrinology Section, Laboratory of Tumor Immunology and Biology, National Cancer Institute, National Institutes of Health, Bethesda, MD 20892

11.1. INTRODUCTION

Prolactin (PRL) is a peptide hormone primarily produced by the anterior pituitary. It plays pivotal roles in immune responses, osmoregulation, steroidogenesis, and reproduction, as well as normal growth, development, and differentiation of the mammary gland (1). PRL belongs to a superfamily that includes human growth hormone (GH) and human placental lactogen. In humans, the pituitary PRL gene is made up of five exons and four introns and is located on chromosome 17. Human PRL was first characterized as a 22- to 25-kDa pituitary hormone whose biologic activity is altered by posttranslational modifications such as phosphorylation and glycosylation.

Extrapituitary production of PRL has long been suspected. Following surgical removal of the pituitary in humans, circulating levels of PRL remain at 30–80% of the presurgical levels for as long as 10 months in breast cancer patients who received a total hypophysectomy (2); the circulating levels of all other pituitary hormones became undetectable. Patients given bromocriptine plus somatostatin to achieve a "chemical" hypophysectomy persistently maintained low levels of circulating PRL (3). While these data could result from other PRL-like molecules circulating in the blood, it now appears that PRL itself is produced by peripheral tissues. Placenta is the richest extrapituitary source of PRL. The hormone is also produced

Hormones and Growth Factors in Development and Neoplasia, Edited by Robert B. Dickson and David S. Salomon
ISBN 0-471-16899-8

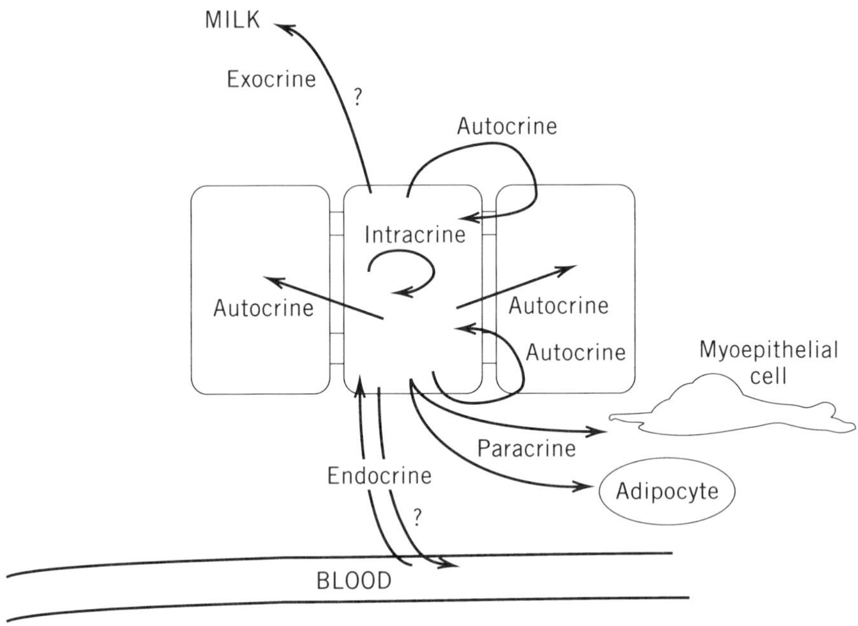

Fig. 11.1 Schematic of PRL action in the mammary gland via several mechanisms.

by the decidua, the uterine endometrium, the brain, and the spinal cord, as demonstrated by the presence of the message (reviewed in [4]). In addition, recent evidence indicates that PRL is synthesized by a variety of cell types of the immune system (reviewed in [4]) and by the mammary gland (5), suggesting that in addition to acting as an endocrine factor, PRL is an autocrine/paracrine hormone in several systems (Fig. 11.1).

11.2. PROLACTIN IN THE MAMMARY GLAND

11.2.1. Development of the Normal Gland

PRL is both a mitogen and a differentiation factor in the mammary gland. Functional differentiation of the mammary gland, as measured by milk protein expression both *in vivo* and *in vitro,* requires PRL. In humans, excess PRL in adults results in inappropriate lactation (reviewed in [1]).

The mitogenic action of PRL in the mammary gland is less well recognized than its differentiative action. *In vivo,* a variety of hormones act on the mammary gland throughout development. Although the ovarian steroids, estrogen and progesterone, are primarily involved in mammary ductal growth and branching, evidence strongly suggests that for lobuloalveolar development and extensive growth of the alveolar cells, PRL is the primary player (1). In the human, PRL is essential for proliferation and maintenance of ducts; the absence of PRL during puberty results in underdevelopment of the gland. Progesterone acts synergistically with PRL to induce mitogenesis during mammary gland development, possibly by increasing the PRL receptor levels in these glands (6). Under controlled conditions of whole organ culture, further identification of the role of the individual hormones in mammary development is possible. As shown in Figure 11.2, lobuloalveolar development occurs when a mammary gland from an estrogen- and progesterone-primed animal is cultured in a serum-free medium in the presence of insulin (I), PRL, aldosterone (A), and hydrocortisone (H) for 6 days. For full lobuloalveolar development of the gland *in vitro*, the presence of PRL in the medium is absolutely required. The action of PRL is enhanced by the addition of thyroid hormones to the medium (1). Thyroid hormones act in part by increasing the binding of PRL to the mammary epithelial cells (7). This results in increased cell division and eventually in enhanced milk protein production (8). At high concentrations, ovine GH can partially replace PRL to produce lobuloalveolar development, as well as casein gene expression (9). Addition of epidermal growth factor (EGF) to the cultures with either PRL or growth hormone results in maximal development of the gland. Apoptosis is triggered and involution of the alveolar structure is observed if PRL is removed after full lobuloalveolar development and functional differentiation are achieved (10). These regressed glands can be induced for a second cycle of development and differentiation in the presence of IAH + EGF and PRL (11).

11.2.2. PRL Receptors in Mammary Gland

The presence of specific receptors for PRL has been demonstrated in both normal and malignant mammary glands. Specific PRL receptors have been demonstrated in over 70% of malignant breast biopsy samples (12–15). PRL receptors belong to the cytokine hematopoietic family of receptors (16). The members of this superfamily are single membrane-spanning receptors organized into three domains comprising an extracellular ligand binding domain, a hydrophobic transmembrane domain, and an intracellular domain containing a proline-rich motif. There are three different forms of the PRL receptor that differ in their cytoplasmic sequence: all three forms of the receptor have been found in mammary tissues (16,17). The long (90-kDa) and short (40-kDa) forms of the receptor are thought to derive from the same gene (18), either by the use of different initiation or polyadenylation sites or by alternative splicing of the common primary transcript (16,19). Thus they differ only in the length of the cytoplasmic domain (18). The intermediate (68-kDa) form of the receptor is a deletion mutant of the long form, lacking 198 amino acids in its cytoplasmic region. This form of the receptor is more sensitive to PRL both in its affinity and in its activity compared to the other forms of the PRL receptor. It is found predominantly in Nb2 rat lymphoma cells.

The long and intermediate forms of the PRL receptor are able to induce differentiation, as measured by induction of milk protein gene expression (20). Both forms have also been shown to induce mitosis (21). We have shown that the short form of the mouse PRL receptor is also able to induce mitosis (22). To define the functions of the various forms of the PRL receptor more clearly, we transfected either the

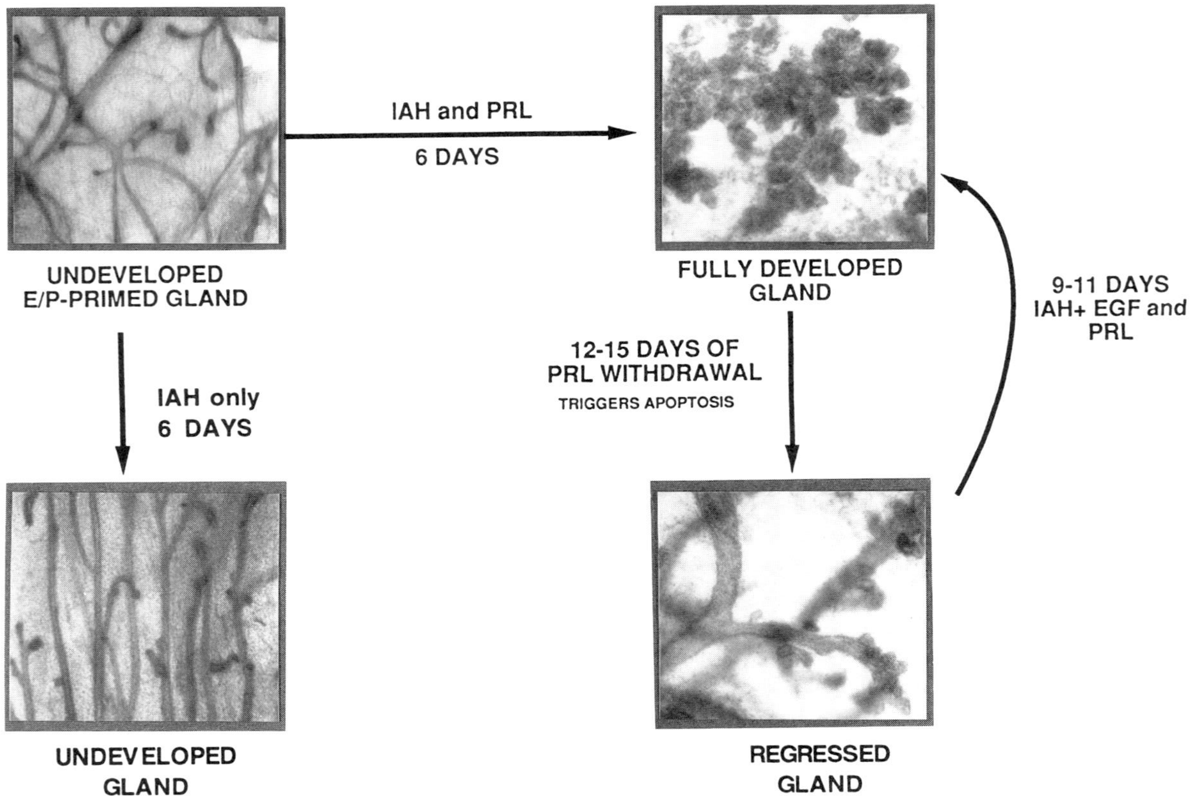

Fig. 11.2. Development of the mouse mammary gland in whole organ culture. (From [78]).

long form of the human or the short form of the mouse PRL receptor genes into NIH-3T3 cells devoid of any PRL receptors. The resulting stable transfectants LFH (long form) and SFM (short form) express PRL receptors, as demonstrated by immunocytochemistry. These cell lines, both LFH and SFM, bind ovine PRL and respond to its mitogenic signal in a concentration-dependent manner. The long form of the PRL receptor is able to induce the expression of β-casein-CAT activity (22), which is a differentiative function of PRL. However, in agreement with others, we found that the short form of the receptor is unable to induce β-casein-CAT activity. Thus the three forms of the PRL receptors, although they bind PRL, appear to function in a different manner, inducing the cells to either differentiate or proliferate.

11.2.3. Synthesis of PRL by Mammary Cells

While synthesis and secretion of PRL by mammary tissue have been suspected for some time, direct evidence has only recently been provided. By *in situ* hybridization, Steinmetz et al. (23) showed that PRL gene transcripts are present in secretory mammary epithelial cells in pregnant rats. PRL message has also been demonstrated by Northern blot analysis and reverse transcription-polymerase chain reaction (RT-PCR) in mammary glands from lactating rats (24), goats, and sheep (25). More recently, the presence of PRL message in some primary human breast carcinomas and in human breast cancer cells in culture (5,17,26) was demonstrated by RT-PCR using primers spanning all four introns of genomic pituitary PRL DNA. We found that 83% of all breast cancer cell lines tested contain message for PRL (26); 75% of primary breast cancer biopsy specimens contain message for PRL. When human breast cancer cells are grown as solid tumors in nude mice, the resulting tumors contain PRL gene transcripts.

In our laboratory, conditioned media prepared from T47Dco human breast cancer cells were shown to contain biologically active PRL. The level of biologic activity in the conditioned medium is equivalent to 0.7 μg/ml (14.5 pg PRL per cell) of pituitary PRL, as measured by the Nb2 assay, and is 30% of the amount normally produced by the rat pituitary cell line, GH3 (27,28). By a specific radioimmunoassay for human pituitary PRL, 0.35 μg/ml of PRL protein is found to be present in conditioned media. The activity in the conditioned media, like that of the human pituitary PRL, is abolished when pretreated with some anti-PRL monoclonal antibodies (5). The nature of the posttranslational alteration of PRL synthesized by the mammary gland is unknown. Our observation that monoclonal antibodies directed against pituitary PRL vary in their ability to recognize mammary PRL (5) suggests that there are distinct differences in posttranslational modifications of the hormone, depending on the tissue source. Whether other hormones and growth factors, such as EGF, progesterone, and estrogens, which stimulate proliferation of normal and/or malignant mammary epithelial cells, also affect the synthesis and secretion of autocrine PRL is an important question that needs to be answered.

That mammary-derived PRL acts in an autocrine manner to stimulate cell proliferation is suggested by our studies utilizing either a panel of anti-PRL monoclonal antibodies or antisense RNA directed against the gene encoding for pituitary PRL.

In both cases, proliferation of human breast cancer cell lines is inhibited by 20–90% (5,26). Subsequently, Mershon et al. (29) showed that cells derived from N-nitroso methylurea (NMU)-induced rat mammary tumors placed into culture respond to the mitogenic signal of PRL, and specific antibodies to rat PRL inhibit their growth.

The physiologic significance of PRL as an autocrine factor is also suggested by the presence of mRNA in normal breast tissue. LeProvost and colleagues (25) showed, by Northern blot analysis and RT-PCR, that mammary glands from late pregnant and lactating sheep and goats express the PRL gene. The sequence of the cDNA generated from these tissues differs from pituitary transcripts by only three mutations, two of which are silent. The PRL gene from the sheep mammary gland appears to be transcribed from the same promoter as that in the pituitary (25). The highest level of PRL message is present in mammary glands from virgin rabbits; the level diminishes throughout pregnancy and rises slightly during early lactation. The PCR signal is very weak by midlactation and absent in the postweaning mammary gland (30). In our laboratory, preliminary data suggest that expression of PRL in the mouse mammary gland may follow a different developmental pattern, and the gene sequence may be less homologous to that found in the pituitary (26).

11.3. PROLACTIN IN THE FEMALE REPRODUCTIVE TRACT

Human decidual tissue, the specialized endometrium of the luteal phase of the menstrual cycle and pregnancy, also synthesizes and secretes PRL (31). That the decidua is the major source of PRL within the uterus has been clearly demonstrated by *in situ* hybridization histochemistry (32).

Endometrial stromal decidualization is essential for implantation of the developing blastocyst. Decidual PRL, unlike pituitary PRL, is rapidly released after synthesis and is not stored in dense secretory granules (33). The synthesis and release of PRL appear to be regulated locally by other factors such as insulin (34), insulin-like growth factor I (35), and relaxin (36,37), which are produced by placenta, decidua, and fetal membranes. Progesterone, which induces decidualization of the endometrium, also plays a key role since PRL was first detected during the late luteal phase of the menstrual cycle and in endometrial tissue of women receiving progesterone, alone or in combination with estrogen (38,39). Both PRL and the progesterone receptor have been colocalized by immunohistochemistry to the same stromal cells of human decidua (40). In the presence of estradiol and medroxyprogesterone acetate (E/MPA), PRL is detected beginning on day 9 of culture and gradually increases through day 18. In cells treated with prostaglandin E2 (PGE2), PRL is detected on day 3. On days 9–12 of culture, the PRL levels are 13- to 14-fold higher than the levels in E/MPA-treated cultures. In addition, the PRL mRNA content of the PGE2-treated cells on day 12 of culture is 4.6-fold greater than that in the E/MPA-treated cultures (41). Human endometrial stromal cells induced to decidualize in culture differentially regulate the synthesis and release of PRL, depending on the extracellular matrix proteins present (42). At the peak of PRL release, decidual fibroblasts plated at an initial cell density of 0.25×10^6 cells per well release 30 ng PRL per well into the conditioned medium over a 24-hr period (43). Like pituitary PRL, decidual PRL is synthesized and released as both a glycosylated and an unglycosylated protein (44). In the human, the amino acid sequences of decidual and pituitary PRL are identical (45,46). However, there is a significant difference in the 5′ untranslated region of the mRNAs (47) due to an extra exon in the decidual PRL cDNA (48). The regulatory element involved in decidual PRL gene expression is distinct from pit-1 utilized in the pituitary (47). The decrease in decidual PRL content and in the PRL concentration of amniotic fluid at term pregnancy is due to a decrease in the transcription of the decidual PRL mRNA (32).

A paracrine role for this PRL was originally suggested since maternal and fetal pituitary PRL make only a minor contribution to the total amount of PRL in the amniotic fluid (49). The presence of local PRL is paralleled by the presence of specific lactogenic hormone receptors in fetal membranes, decidua, and placenta (50). Subcellular localization studies show that the bulk of the PRL receptors in the human placental chorion are in the microsomal fraction of membranes. Both the long and short forms of the PRL receptor are present in the placental chorion and decidual tissue, as determined by Western blot analysis (50–52). Covalent cross-linking studies followed by sodium dodecyl sulfate polyacrylamide gel electrophoresis and gel filtration chromatography suggest that a short form of the receptor is dominant in the placental chorion (52). Preliminary data from our laboratory, using RT-PCR, suggest the presence of an intermediate form in placenta as well (E. Ginsburg and B.K. Vonderhaar, unpublished). Also present in cytosol and medium from amnion, chorion, and decidua and in amniotic fluid are a 63-kDa, and a 30-kDa PRL receptor-related protein (50). The relationship of the 30-kDa protein to the PRL-binding protein described in human milk (53) has not been established.

The physiologic role of decidual PRL is still unclear, but studies in rhesus monkeys suggest that PRL has an osmoregulatory function within the maternal-fetal-placental compartment (54). Amniotic fluid PRL may play a role in regulating amniotic fluid volume and ion content, as well as growth and surfactant synthesis in the fetus, inhibiting uterine contractility, suppressing the immune response, and maintaining T-cell immunocompetence (55,56). In addition, local PRL inhibits membrane-associated phosphatidylinositol kinase activity in human endometrial fibroblasts (57) and regulates transcription and translation of specific metalloproteinases in late pregnancy that are important during parturition (58).

11.4. PROLACTIN IN THE MALE REPRODUCTIVE TRACT

In the male, PRL is known to influence reproductive functions, but the significance and mechanisms of PRL action in male accessory reproductive tissues are poorly understood. PRL binding sites on developing rat testicular tissue do not significantly change during sexual maturation (59). However, PRL regulates testicular function by increasing the binding of luteinizing hormone to Leydig cells (60), thus increasing testosterone production.

In normal prostate, PRL is essential for development, growth, and differentiation (59,61,62), both by enhancing the effects of testosterone and by acting independently of it. The immature prostate is particularly sensitive to the trophic action of PRL (62). In the mature prostate, PRL acts in synergy with testosterone to stimulate 5α-reductase activity (63). It also directly regulates citrate production. Testosterone increases the rate of citrate production, while PRL inhibits citrate oxidation (61,64). In the lateral lobe of the rat prostate, PRL selectively stimulates ornithine decarboxylase activity (65). PRL is also believed to be involved in regulating the growth of neoplastic prostate (66–68). Several prostate cancer cell lines respond to the mitogenic action of PRL in a dose-dependent manner (69).

Organ cultures of human prostate (70), as well as rat dorsolateral (71,72) and ventral prostate (73), maintain hormone responsiveness and specific tissue function. In culture, the ventral prostate responds to low levels of PRL with cell growth and preservation of the columnar shape of the glandular epithelium. High concentrations of PRL are growth inhibitory (74). The dorsal and lateral lobes of the rat prostate are considered most homologous with human prostate (75) and in the rat are the sites that give rise to spontaneous and experimental tumors with varied hormone responsiveness. In organ culture of rat dorsal and lateral prostate, PRL has a direct effect on morphology, on DNA labeling with ^{3}H-thymidine, and on the expression of tissue-specific genes encoding secretory proteins (71). The cultured human tissue responds to PRL with induction of a typical stratified morphology with round microlumina formed irregularly within the epithelium. This also is associated with an increase in DNA synthesis and cell proliferation (76). Autoradiographic studies show that the labeling of DNA occurs in the prostatic epithelium, which also contains the PRL receptors.

Both the long and short forms of the PRL receptor are expressed in the dorsal and lateral lobes of the rat prostate (72) and in the human prostate (76). Covalent cross-linking studies using membranes from both cultured and uncultured dorsal and lateral rat prostate tissue confirm that both forms of the receptor bind the hormone (72). In the rat, the level of prostatic PRL receptors is relatively low in the immature gland and rises as the animal matures (59). Multiple transcripts for both forms of the receptor are present in both rat and human tissue. However, all of the transcripts are not necessarily translated into proteins. In the mature rat dorsolateral prostate, testosterone, estradiol, and PRL itself regulate the level of the message for the long form (72). The different mRNA splicing variants for the long form are variably regulated by all three hormones in the dorsal prostate; only testosterone does so in the lateral prostate. Hormonal regulation of the mRNA for the short form of the receptor is modest, and posttranscriptional regulation of the receptors cannot be ruled out (72).

PRL receptors in rat dorsolateral (72) and human (76) prostate tissue were localized by immunohistochemistry using a polyclonal anti-rat PRL receptor antibody, R120. This antibody recognizes the extracellular domain of both the long and short forms of the PRL receptors (72). The immunohistochemical stain is localized mainly to the apical surface of the secretory epithelium; the nuclei do not stain. In the rat prostate, basal cells do not stain with the anti-PRL receptor antibody; no staining was noted in prostatic stroma. Pituitary PRL, circulating in the blood, would be expected to bind on the basolateral cell membranes of prostatic epithelial cells. PRL receptors located on the apical surface of the prostatic epithelial cells are not accessible to circulating PRL because of tight junctions between the cells. Apical receptors, however, would be accessible to a secreted ligand; this is compatible with an autocrine/paracrine mechanism of PRL in this tissue as well.

In studies paralleling those done in the mammary gland described above, synthesis and secretion of PRL by the prostate have been demonstrated (76). Human prostatic epithelium was stained for PRL with a polyclonal anti-human PRL antibody. The stained cells are scattered irregularly within the prostatic epithelium, with the most intense staining localized to the secretory epithelial cells. Using the same primers developed for the mammary gland studies, RT-PCR confirmed the presence of mRNA for PRL in the human prostate. The identification of the product as PRL message was confirmed by Southern blotting and hybridization with a specific human PRL cDNA probe (77). The concentration of PRL in the culture medium of human prostate is 45 n*M;* this level of PRL is able to elicit biologic responses in prostate tissue in organ culture (64,71).

11.5. CONCLUSION

PRL is an endocrine hormone produced by the pituitary that acts on peripheral tissues. As shown in Figure 11.1, extrapituitary PRL may also act in an endocrine manner if it enters the circulation and affects other tissues. By definition, an autocrine/paracrine loop exists for any given factor if the tissue synthesizes the factor; if specific receptors for the factor are present in the same tissue; and if a biologic response to the factor occurs. It is now clear that in mammary gland, endometrium, and prostate, PRL is an autocrine/paracrine factor. Its action as an exocrine or intracrine factor has yet to be established.

REFERENCES

1. Vonderhaar BK: Prolactin: Transport, function, and receptors in mammary gland development and differentiation. In Neville MC, Daniel CW (eds): The Mammary Gland. New York: Plenum, pp 383–438, 1987.
2. Lachelin GCL, Yen SSC, Alksne JFN: Hormonal changes following hypophysectomy in humans. Obstet Gynecol 50:333–339, 1977.
3. Anderson E, Ferguson JE, Morten H, Shalet SM, Robinson EL, Howell A: Serum immunoreactive and bioactive lactogenic hormones in advanced breast cancer patients treated with bromocriptine and ocetotide. Eur J Cancer 29A:209–217, 1993.
4. Sinha YN: Structural variants of prolactin: Occurrence and physiological significance. Endocrine Rev 16:354–369, 1995.
5. Ginsburg E, Vonderhaar BK: Prolactin synthesis and secretion by human breast cancer cells. Cancer Res 55:2591–2595, 1995.

6. Nagasawa H, Miur K, Niki K, Namiki H: Interrelationship between prolactin and progesterone in normal mammary gland growth in SHN virgin mice. Exp Clin Endocrinol 86:357–360, 1985.
7. Bhattacharya A, Vonderhaar BK: Thyroid hormone regulation of prolactin binding to mouse mammary glands. Biochem Biophys Res Commun 88:1405–1411, 1979.
8. Vonderhaar BK: A role of thyroid hormones in differentiation of mouse mammary gland *in vitro*. Biochem Biophys Res Commun 67:1219–1225, 1975.
9. Plaut K, Ikeda M, Vonderhaar BK: Role of growth hormone and insulin-like growth factor I in mammary development. Endocrinology 133:1843–1848, 1993.
10. Atwood CS, Ikeda M, Vonderhaar BK: Involution of mouse mammary glands in whole organ culture: A model for studying programmed cell death. Biochem Biophys Res Commun 207:860–867, 1995.
11. Tonelli QJ, Sorof S: Epidermal growth factor requirement for development of cultured mammary gland. Nature 285:250–252, 1980.
12. Codegone ML, DiCarlo R, Muccioli G, Bussolati G: Histology and cytometrics in human breast cancers assayed for the presence of prolactin receptors. Tumori 67:549–552, 1981.
13. Peyrat JP, DeWailly D, Djiane J, Kelly PA, Vandewalle B, Bonneterre J, LeFebvre J: Total prolactin binding sites in human breast cancer biopsies. Breast Cancer Res Treat 1:369–373, 1981.
14. Bonneterre J, Peyrat JP, Vandewalle B, Beuscart R, Vie MC, Cappelaere P: Prolactin receptors in human breast cancer. Eur J Cancer Clin Oncol 18:1157–1162, 1982.
15. L'Hermite-Baleriaux M, Casteels S, Vokaer A, Loriaus C, Noel G, L'Hermite M: Prolactin and prolactin receptors in human breast disease. Prog Cancer Res Ther 31:325–334, 1984.
16. Kelly PA, Djiane J, Postel-Vinay MC, Edery M: The prolactin/growth hormone receptor family. Endocrine Rev 12:235–251, 1991.
17. Clevenger CV, Chang WP, Ngo W, Pasha TLM, Montone KT, Tomaszewski JE: Expression of prolactin and prolactin receptor in human breast carcinoma. Am J Pathol 146:695–705, 1995.
18. Kelly PA, Ali S, Rozakis M, Goujon L, Nagano M, Pellegrini I, Gould D, Djiane J, Edery M, Finidori J, Postel-Vinay MC: The growth hormone/prolactin receptor family. Rec Prog Hormone Res 48:123–164, 1993.
19. Shirota M, Banville D, Ali S, Jolicoeur C, Boutin JM, Edery M, Djiane J, Kelly PA: Expression of two forms of prolactin receptors in rat ovary and liver. Mol Endocrinol 4:1136–1143, 1990.
20. Lesueur L, Edery M, Ali S, Paly J, Kelly PA, Djiane J: Comparison of long and short forms of the prolactin receptor on prolactin-induced milk protein gene transcription. Proc Natl Acad Sci USA 88:824–828, 1991.
21. O'Neal KD, Yu-Lee L-Y: Differential signal transduction of the short, Nb2, and long prolactin receptors. J Biol Chem 269:26076–26082, 1994.
22. Das R, Vonderhaar BK: Transduction of prolactin's growth signal through both the long and short forms of the prolactin receptor. Mol Endocrinol 9:1750–1759, 1995.
23. Steinmetz RW, Grant AL, Malven PV: Transcription of prolactin gene in milk secretory cells of the rat mammary gland. J Endocrinol 36:271–276, 1993.
24. Kurtz A, Bristol LA, Toth BE, Lazar-Wesley E, Takacs L, Kacsoh B: Mammary epithe-

lial cells of lactating rats express prolactin messenger ribonucleic acid. Biol Reprod 48:1095–1103, 1993.

25. LeProvost F, Leroux C, Martin P, Gaye P, Djiane J: Prolactin gene expression in ovine and caprine mammary gland. Neuroendocrinology 60:305–313, 1994.
26. Ginsburg E, Vonderhaar BK: Prolactin: An autocrine growth factor in the mammary gland. In Wilde CJ, Peaker M, Taylor E (eds): Biological Signalling in the Mammary Gland. Hannah Research Institute Ayr, Scotland: pp 47–58, 1997.
27. Bancroft FC, Tashjian AH: Growth in suspension culture of rat pituitary cells which produce growth hormone and prolactin. Exp Cell Res 64:125–128, 1971.
28. Tanaka T, Shiu RPC, Gout PW, Beer CT, Noble RL, Friesen HG: A new sensitive and specific bioassay for lactogenic hormones: Measurement of prolactin and growth hormone in human serum. J Clin Endocrinol Metab 51:1058–1063, 1980.
29. Mershon J, Sall W, Mitchner N, Ben-Jonathan N: Prolactin is a local growth factor in rat mammary tumors. Endocrinology 136:3619–3623, 1995.
30. Gabou L, Boisnard M, Gourdou I, Jammes H, Dulor J-P, Djiane J: Cloning of rabbit prolactin cDNA and prolactin gene expression in the rabbit mammary gland. J Mol Endocrinol 16:27–37, 1996.
31. Golander A, Hurley T, Barrett J, Hizi A, Handwerger S: Prolactin synthesis by human chorion-decidual tissue: A possible source of amniotic fluid prolactin. Science 202:311–313, 1978.
32. Tadokoro N, Koibuchi N, Ohtake H, Kawatsu T, Kato Y, Yamaoka S, Kumasaka T: Expression of prolactin gene in human decidua during pregnancy studied by in situ hybridization histochemistry. Endocrine J 42:537–543, 1995.
33. Handwerger S, Wilson S, Conn PM: Different subcellular storage sites for decidua- and pituitary-derived prolactin: Possible explanation for differences in regulation. Mol Cell Endocrinol 37:83–87, 1984.
34. Thrailkill KM, Golander A, Underwood LE, Richards RG, Handwerger S: Insulin stimulates the synthesis and release of prolactin from human decidual cells. Endocrinology 124:3010–3014, 1989.
35. Thrailkill KM, Golander A, Underwood LE, Handwerger S: Insulin-like growth factor I (IGF-I) stimulates the synthesis and release of prolactin from human decidual cells. Endocrinology 123:2930–2934, 1988.
36. Huang JR, Tseng L, Bischof P, Janne OA: Regulation of prolactin production by progestin, estrogen, and relaxin in human endometrial stromal cells. Endocrinology 121: 2011–2017, 1987.
37. Lane B, Oxberry W, Mazella J, Tseng L: Decidualization of human endometrial stromal cells in vitro: Effects of progestin and relaxin on the ultrastructure and production of decidual secretory proteins. Hum Prod 9:259–266, 1994.
38. Meuris S, Soumenkoff G, Malengreau A, Robyn C: Immunoenzymatic localization of prolactin-like immunoreactivity in decidual cells of the endometrium from pregnant and nonpregnant women. J Histochem Cytochem 28:1347–1350, 1980.
39. Daly D, Maslar I, Riddick D: Term decidua response to estradiol and progesterone. Am J Obstet Gynecol 145:679–683, 1983.
40. Wang JD, Zhu JB, Shi WL, Zhu PD: Immunocytochemical colocalization of progesterone receptor and prolactin in individual stromal cells of human decidua. J Clin Endocrinol Metab 79:293–297, 1994.

41. Frank GR, Brar AK, Cedars MI, Handwerger S: Prostaglandin E2 enhances human endometrial stromal cell differentiation. Endocrinology 134:258–263, 1994.
42. Brar AK, Frank GR, Richards RG, Meyer AJ, Kessler CA, Cedars MI, Klein DJ, Handwerger S: Laminin decreases PRL and IGFBP-1 expression during in vitro decidualization of human endometrial stromal cells. J Cell Physiol 163:30–37, 1995.
43. Richards RG, Brar AK, Frank GR, Hartman SM, Jikihara H: Fibroblast cells from term human decidua closely resemble endometrial stromal cells: Induction of prolactin and insulin-like growth factor binding protein-1 expression. Biol Reprod 52:609–615, 1995.
44. Lee DW, Markoff E: Synthesis and release of glycosylated prolactin by human decidua in vitro. J Clin Endocrinol Metab 62:990–994, 1986.
45. Cook NE, Coit D, Shine J, Baxter JD, Martial JA: Human prolactin: cDNA structural analysis and evolutionary comparisons. J Biol Chem 256:4007–4016, 1981.
46. Takahashi H, Nabeshima Y, Nabeshima YI, Ogata K, Takeuchi S: Molecular cloning and nucleotide sequence of DNA complementary to human decidual prolactin mRNA. J Biochem 95:1491–1499, 1984.
47. Gellersen B, DiMattia G, Friesen HG, Bohnet H: Prolactin (PRL) mRNA from human decidua differs from pituitary PRL mRNA but resembles IM-9-P3 lymphoblast PRL transcript. Mol Cell Endocrinol 64:127–130, 1989.
48. Hiroka Y, Tatsumi K, Shiozawa M: A placenta-specific 5′ noncoding exon of human prolactin. Mol Cell Endocrinol 75:71–80, 1991.
49. Josimovich JB, Weiss G, Hutchinson DL: Sources and distribution of pituitary prolactin in maternal circulation, amniotic fluid, fetus and placenta in the pregnant rhesus monkey. Endocrinology 94:1364–1371, 1974.
50. Maaskant RA, Bogic LV, Gilger S, Kelley PA, Bryant-Greenwood GD: The human prolactin receptor in the fetal membranes, decidua and placenta. J Clin Endocrinol Metab 81:396–405, 1996.
51. Herington AC, Graham J, Healy DL: The presence of lactogen receptors in human chorion laeve. J Clin Endocrinol Metab 51:1466–1468, 1980.
51. Ormandy CJ, Clarke CL, Sutherland RL: Solubilization and characterization of a lactogenic receptor from human placental chorion membranes. J Cell Biochem 43:1–15, 1990.
53. Mercado M, Bauman G: A growth hormone/prolactin-binding protein in human milk. J Clin Endocrinol Metab 79:1637–1641, 1994.
54. Josimovich JB, Merisko K, Boccella L: Amniotic prolactin control over amniotic and fetal extracellular fluid water and electrolytes in the rhesus monkey. Endocrinology 100:564–570, 1977.
55. Talamantes F, Ogren L, Markoff E, Woodward S, Madrid J: The phylogenic distribution, regulation of secretion and the prolactin-like effects of placental lactogens. Fed Proc 39:2582–2587, 1980.
56. Handwerger S, Richards RG, Markoff E: The physiology of decidual prolactin and other decidual protein hormones. Trends Endocrinol Metab 3:91–95, 1992.
57. Furui T, Imai A, Takagi H, Takahashi K, Ohno T, Tamaya T: Inhibition by prolactin of membrane associated phosphatidylinositol kinase of human endometrial fibroblast. Proc Soc Exp Biol Med 205:140–145, 1994.
58. Higuchi T, Kanzaki H, Nakayama H, Fujimoto M, Hatayama H, Kojima K, Iwai M,

Mori T, Fujita J: Induction of tissue inhibitor of metalloproteinase 3 gene expression during in vitro decidualization of human endometrial stromal cells. Endocrinology 136: 4973–4981, 1995.

59. Perez-Villamil B, Bordiu E, Puente-Cueva M: Involvement of physiological prolactin levels in growth and prolactin receptor content of prostate glands and testes in developing male rats. J Endocrinol 132:449–459, 1992.
60. Bartke A: Role of prolactin in reproduction in male mammals. Fed Proc 39:2577–2581, 1980.
61. Costello LC, Franklin RB: Effect of prolactin on the prostate. Prostate 24:162–166, 1994.
62. Reiter E, Lardinois S, Klug M, Sente B, Hennuy B, Bruyninx M, Closset J, Hennen G: Androgen-independent effects of prolactin on the different lobes of the immature rat prostate. Mol Cell Endocrinol 112:113–122, 1995.
63. Yamanaka H, Kirdani RY, Saroff J, Murphy GP, Sandberg AA: Effects of testosterone and prolactin on rat prostatic weight, 5-alpha-reductase, and arginase. Am J Physiol 229:1102–1109, 1975.
64. Franklin RB, Costello LC: Prolactin directly stimulates citrate production and mitochondrial aspartate aminotransferase of prostate epithelial cells. Prostate 17:13–18, 1990.
65. Rui H, Purvis K: Prolactin selectively stimulates ornithine decarboxylase in the lateral lobe of the rat prostate. Mol Cell Endocrinol 50:89–97, 1987.
66. Wilson JD: The pathogenesis of benign prostatic hyperplasia. Am J Med 68:745–756, 1980.
67. Sissom JF, Eigenbrodt ML, Porter JC: Anti-growth action on mouse mammary and prostate glands of a monoclonal antibody to prolactin receptor. Am J Pathol 133:589–595, 1988.
68. Nakamura A, Shirai T, Ogawa K, Wada S, Fujimoto NA, Ito A, Ito N: Promoting action of prolactin released from a graft transplantable pituitary tumor (MtT/F84) on rat prostate carcinogenesis. Cancer Lett 53:151–157, 1990.
69. Janssen T, Darro F, Petein M, Raviv G, Pasteels JL, Kiss R, Schulman CC: In vitro characterization of prolactin-induced effects on proliferation in the neoplastic LNCaP, DU145, and PC3 models of the human prostate. Cancer 77:144–149, 1996.
70. Nevalainen MJ, Harkonen PL, Valve EM, Ping W, Nurmi M, Martikainen PM: Hormone regulation of human prostate in organ culture. Cancer Res 53:5199–5207, 1993.
71. Nevalainen MT, Valve EM, Makela SI, Blauer M, Tuohimaa PJ, Harkonen PL: Estrogen and prolactin regulation of rat dorsal and lateral prostate in organ culture. Endocrinology 129:612–622, 1991.
72. Nevalainen MJ, Valve EM, Ingleton PM, Harkonen PL: Expression and hormone regulation of prolactin receptors in rat dorsal and lateral prostate. Endocrinology 137:3078–3088, 1996.
73. Martikainen P: Maintenance of adult rat ventral prostate in organ culture. Anat Rec 218:166–174, 1987.
74. Romero L, Munoz C, Lopez A, Vilches J: Effects of prolactin on explant cultures of rat ventral prostate: Morphological and immunohistochemical study. Prostate 22:1–10, 1993.
75. Price D: Comparative aspects of development and structure in the prostate. Natl Cancer Inst Mongr 12:351–369, 1963.

76. Nevalainen MJ, Valve EM, Ingleton PM, Nurmi M, Martikainen PM, Harkonen PL: Prolactin and prolactin receptors are expressed and functioning in human prostate. J Clin Invest 99:618–627, 1997.
77. Gubbins EJ, Maurer RA, Lagrimini M, Erwin CR, Donelson JE: Structure of the rat prolactin gene. J Biol Chem 255:8655–8662, 1980.
78. Das R, Vonderhaar BK: Prolactin as a mitogen in mammary cells. J Mammary Gland Biol Neoplasia 2:29–39, 1997.

12

GROWTH FACTORS AS MEDIATORS OF STROMAL–EPITHELIAL INTERACTIONS IN STEROID HORMONE TARGET ORGANS

GERAL R. CUNHA
SIMON W. HAYWARD
YUN KIT HOM
ANNEMARIE A. DONJACOUR
TAKESHI KURITA
Department of Anatomy, University of California, San Francisco, CA 94143-0452

PAUL S. COOKE
Department of Veterinary Biosciences, University of Illinois, Urbana, IL 61801

DENNIS B. LUBAHN
Departments of Biochemistry and Child Health, University of Missouri, Columbia, MO 65211

12.1. INTRODUCTION AND HISTORICAL PERSPECTIVE

The key finding concerning the mechanism of action of steroidal sex hormones was the demonstration of specific high-affinity, low-capacity estrogen receptors (1). This seminal work initiated intense research that led to the discovery of receptors for all of the sex steroids: estrogen, androgen, and progesterone receptors. These studies emphasized and affirmed the central dogma of endocrinology: that hormones elicit their effects in target cells via specific hormone receptors. The

Hormones and Growth Factors in Development and Neoplasia, Edited by Robert B. Dickson and David S. Salomon
ISBN 0-471-16899-8

application of this concept to hormone target organs is complicated, however. Confounding the issue is the fact that hormonal effects in hormone target organs are elicited by paracrine factors mediating cell–cell interactions between epithelial and stromal cells, each of which express steroid hormone receptors. For example, in the uterus, estrogen receptors are expressed in the epithelium, stroma, and myometrium (2–4). The initial detection of estrogen receptors in homogenates of whole uteri correlated with the biology of estrogen action in the uterus as a whole (5). Ovariectomy leading to estrogen deprivation, and unoccupied estrogen receptors elicited rapid atrophy and reduction in uterine size, weight, and DNA content. Administration of exogenous estrogen to ovariectomized females led to rapid restoration of these parameters. The interpretation that the uterotrophic effects of estrogen were mediated by estrogen receptors was compelling, reasonable, and readily accepted by the scientific community. Focusing specifically on uterine epithelium, the initial demonstration of estrogen receptors by steroid autoradiography (later confirmed by immunocytochemistry) provided an equally compelling correlation between estrogen induction of uterine epithelial DNA synthesis (increase in the epithelial ^{3}H-thymidine labeling index) (6) and epithelial estrogen receptors (4). These correlations between estrogenic response in uterine epithelial cells and the presence of epithelial estrogen receptors led to the tacit assumption that estrogenic effects on uterine epithelial cells were elicited via the epithelial estrogen receptors. For the male reproductive tract, similar correlations were made regarding androgenic effects on prostatic epithelial cells and the presence of epithelial androgen receptors. Thus, androgen- and estrogen-induced epithelial proliferation in male and female genital tracts, respectively, were suggested to be mediated, respectively, by androgen and estrogen receptors present in the target epithelial cells themselves. While this concept was completely reasonable and widely accepted, it has proven to be incorrect because of the failure to consider the role of paracrine stromal–epithelial interactions between epithelial cells and hormone-receptor-positive stromal cells in mediating the effects of steroid sex hormones.

12.2. STROMAL–EPITHELIAL INTERACTIONS

The field of mesenchymal–epithelial interactions in mammalian systems was initiated in the 1950s by the pioneering studies of Clifford Grobstein (7). In the 1950s and 1960s, an extensive literature emerged in developmental biology demonstrating that all organs containing an epithelial parenchyme developed as a result of mesenchymal–epithelial interactions in which mesenchyme induced epithelial development (8,9). This concept applied to all integumental structures (composed of ectodermal epithelium) (10,11) and to the digestive system and its glandular derivatives (composed of endodermal epithelium) (12). Subsequent studies demonstrated an essential role of mesenchymal–epithelial interactions in hormone-dependent development and differentiation of male and female urogenital systems (containing either endodermal or mesodermal epithelia) (13,14), as well as the mammary gland (an ectodermal derivative) (15,16).

While cell–cell interactions certainly play an important role in development, it has become evident that interactions between epithelium and connective tissue continue to play critical roles in the regulation of tissue growth and homeostasis in adulthood (17). The equivalent of the mesenchymal–epithelial interaction in adulthood is the so-called stromal–epithelial interaction. *Stroma* is an inexact term that signifies the nonepithelial component of an organ. For most internal organs the predominant cells in stroma are fibroblasts and smooth muscle cells.

Initial studies in the embryonic male genital tract demonstrated that androgen-dependent development of the prostate and seminal vesicle occurred when embryonic prostatic (urogenital sinus) epithelium and seminal vesicle epithelium were grown in association with their homologous mesenchymes (i.e., with mesenchyme from androgen target organs). By contrast, androgen-dependent prostatic and seminal vesicle epithelial development was abortive when epithelium of the embryonic urogenital sinus or seminal vesicle was grown in association with mesenchyme from nontarget organs (18,19). From these studies it was proposed "that sex hormones may act on epithelia via effects on mesenchyme" (19). While the data supporting this conclusion were meager, subsequent studies have verified this interpretation.

The interface between developmental biology (mesenchymal–epithelial interactions) and endocrinology (steroid hormone action) was established in 1976 in an analysis of tissue recombinants composed of epithelium and mesenchyme of embryonic mouse mammary glands derived from wild-type and Tfm (testicular feminization) mice (20). Mammary gland development is initiated when embryonic mammary mesenchyme induces mammary differentiation from the overlying epidermis (21). The embryonic mammary epithelial bud, in turn, induces the expression of androgen receptors in the surrounding mesenchymal cells (22). Once androgen receptors are expressed in mammary mesenchyme, embryonic male and female mouse mammary glands become responsive to androgens that elicit condensation of the mesenchyme around the epithelial bud and ultimately the destruction of the epithelial rudiment (23). Androgen-induced regression of the mammary gland occurs naturally in male mouse fetuses in response to fetal testicular androgens but can also be elicited in females with exogenous androgen (24). Androgen-induced regression of embryonic mammary epithelium was shown to be elicited not through the epithelial androgen receptors, but instead via androgen receptors in the surrounding mesenchymal cells. This was determined in tissue recombinants composed of epithelium and mesenchyme of normal wild-type and androgen-insensitive Tfm mice. Androgen-insensitive Tfm epithelium (lacking androgen receptors) underwent androgen-induced regression when associated with wild-type mammary mesenchyme known to express androgen receptors. Conversely, wild-type embryonic mammary epithelium failed to regress when associated with androgen receptor-negative Tfm mammary mesenchyme. These findings demonstrated that mesenchyme was the actual target of androgen action in the developing mammary gland and that epithelial regression was elicited by androgens via mesenchymal androgen receptors. While Kratochwil's finding was a landmark for endocrinology, its impact was muted because androgens elicited destruction of the embryonic mouse mammary rudiment, thus precluding further analysis.

Building on the pioneering work of Kratochwil, we applied the paradigm of Tfm/wild-type tissue recombinants to the developing prostate, in which the effects of androgens were ductal morphogenesis, epithelial proliferation, and epithelial differentiation. These positive effects of androgens on epithelial development were shown to be elicited via mesenchymal androgen receptors (Fig. 12.1). Moreover, it was shown that androgen-insensitive Tfm epithelial cells lacking androgen receptors could be induced with androgens to undergo ductal branching morphogenesis, proliferate, and differentiate into columnar epithelium when grown in association with androgen receptor-positive mesenchymal cells (25–27). These findings provided compelling evidence for the existence of paracrine pathways mediating androgenic effects in the developing prostate, a finding recently extended to the seminal vesicle (28). Moreover, these findings emphasized the fallacy of assigning mechanistic significance based solely on correlation of the hormone response in epithelial cells with epithelial hormone receptors. If hormone receptors are also expressed by the stromal cells associated with the epithelium, it is always possible

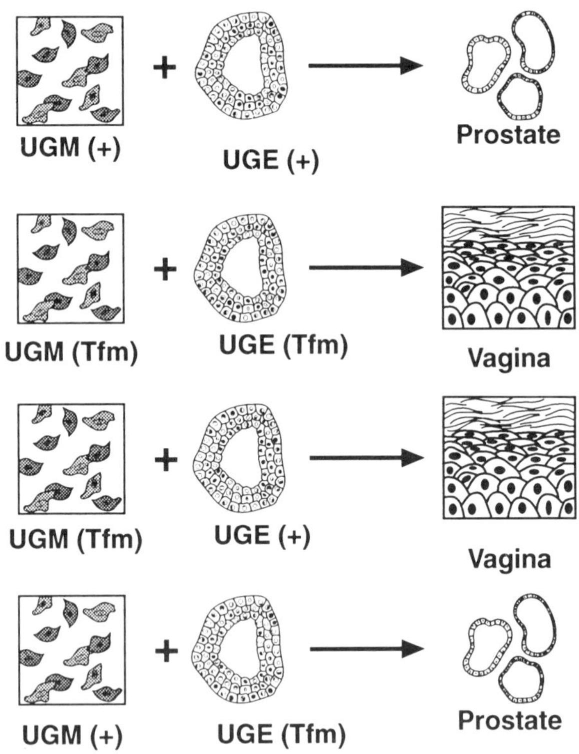

Fig. 12.1. A summary of tissue recombination experiments between urogenital sinus mesenchyme and epithelium from Tfm and wild-type embryos. A positive androgenic response (prostatic morphogenesis) occurs when wild-type mesenchyme is grown in association with either wild-type or Tfm epithelium. Conversely, vagina-like differentiation occurs when either wild-type or Tfm epithelium is grown in association with Tfm mesenchyme.

that certain hormonal effects on epithelial cells may, in fact, be elicited through stromal hormone receptors via paracrine interactions.

During the course of our analysis of androgenic effects in Tfm/wild-type tissue recombinants in the male genital tract, we also considered the possibility that estrogenic effects on epithelial cells of the female genital tract may likewise be elicited by paracrine mechanisms via estrogen receptors in the stromal cells. Several observations emerged throughout the 1980s that supported this idea. Given that hormones elicit their effects in target cells via specific hormone receptors, the initial view was that effects of estradiol on estrogen target epithelium were mediated directly through epithelial estrogen receptors. However, analysis of estrogen receptor expression and estradiol responsiveness in the developing mouse uterus demonstrated some intriguing inconsistencies with this central dogma of endocrinology. Using steroid autoradiography on neonatal Balb/c mice, estrogen receptors were detected in uterine mesenchyme (UtM) but not in uterine epithelium of the neonatal mouse uterus (29). Despite the apparent lack of uterine epithelial estrogen receptors, the injection of diethylstilbestrol (DES) into neonatal mice caused a doubling in the rate of uterine epithelial proliferation (30). One explanation for these results could be that DES induced epithelial estrogen receptors and then acted through these newly expressed estrogen receptors to induce epithelial mitogenesis. However, estrogen receptors remained undetectable in the uterine epithelium even after DES stimulation. These results indicated that even though estrogen receptors were undetectable in the neonatal mouse uterine epithelium, the uterine epithelium could respond mitogenically to estradiol, which suggested that the mitogenic effects of estrogen on neonatal uterine epithelium were mediated indirectly by estrogen receptor-positive mesenchymal cells.

Subsequent immunohistochemical studies reported a low level of estrogen receptor expression in uterine epithelium of neonatal CD-1 mice, raising the possibility that the presumed greater sensitivity of immunohistochemistry allowed detection of low levels of epithelial estrogen receptors not resolved by autoradiography (31). However, the apparent differences between the study of Yamashita et al. and the earlier work of Bigsby and Cunha were later found to reflect strain differences in the ontogeny of uterine estrogen receptors. CD-1 mice (used by Yamashita et al.) showed precocious uterine epithelial estrogen receptor expression compared to Balb/c mice (used by the Cunha group). Consistent with the idea that estradiol may not act through epithelial estrogen receptors to induce epithelial proliferation in the developing uterus was a later report (32) of a study using double labeling to examine simultaneously estrogen receptor expression and estrogen-induced DNA synthesis in neonatal mouse uterine epithelium. At 4 days postpartum, uterine epithelium of CD-1 mice contained both estrogen receptor-positive and estrogen receptor-negative cells. Significantly, DES-induced uterine epithelial proliferation was almost identical in both estrogen receptor-negative and estrogen receptor-positive uterine epithelial cells, again suggesting that estrogen stimulated epithelial proliferation indirectly via the mesenchyme/stroma. This interpretation was confirmed in a study on estrogen-induced uterine epithelial DNA synthesis in the monkey uterus (33). *In vitro* data were also consistent with the idea that the mitogenic effect of estrogen on female reproductive epithelia might involve indirect paracrine mechanisms. Many studies have shown that estradiol is not mitogenic for isolated uterine, vaginal, or mammary epithelial cells *in vitro* (34–38). However, when cultured vaginal or uterine epithelia were recombined with its respective stroma and grafted *in vivo*, estradiol was again mitogenic for the epithelium (39). Furthermore, estradiol increased epithelial DNA content in cocultures of uterine stroma and epithelium, an effect not observed with estradiol treatment of pure epithelial cultures (40). This further suggested that the normal mitogenic response to estrogen might be mediated indirectly via the stroma. However, despite these suggestive findings, alternative explanations existed and incontrovertible experiments were not possible. A definitive answer to the question of whether estrogens elicit epithelial proliferation via epithelial versus stromal estrogen receptors would require a Tfm equivalent for the estrogen receptor, an estrogen receptor-deficient mouse.

In the absence of such a transgenic mouse model, tissue recombination studies were performed using epithelium and mesenchyme from an estrogen target organ, the vagina (in which both epithelium and stroma express estrogen receptors), and a nontarget organ, the urinary bladder (in which estrogen receptors are undetectable). For this purpose, neonatal vaginae were separated into vaginal epithelium (VE) and vaginal stroma (VS), and the bladder was separated into stroma (BLS) and epithelium (BLE). The following tissue recombinants were constructed: VS + VE, VS + BLE, BLS + VE, and BLS + BLE. After 1 month of growth of all tissue recombinants in female hosts, the hosts were ovariectomized and 10 days later were injected with oil vehicle or estradiol. Epithelial proliferation was determined as the ^{3}H-thymidine labeling index. Estrogen-induced epithelial proliferation occurred in tissue recombinants prepared with vaginal stroma (VS + VE and VS + BLE) but not in tissue recombinants prepared with bladder stroma (BLS + VE or BLS + BLE). Moreover, when vaginal epithelium was retrieved from BLS+ VE tissue recombinants and recombined with fresh vaginal stroma, the proliferative response of the VE to estrogen was reestablished (41). These findings, in conjunction with the studies described above, strongly supported a paracrine mechanism of estrogen action for estrogen-induced epithelial proliferation.

A critical advance for the field has been the development of a transgenic estrogen receptor knockout (ERKO) mouse. These estrogen-insensitive animals, in which the estrogen receptor gene has been rendered nonfunctional by gene targeting (42,43), provide an opportunity to use tissue separation and recombination techniques to produce uterine tissue recombinants that lack estrogen receptors in their epithelium, stroma, or both. Indeed, the recent availability of the ERKO mouse has finally allowed analysis of uterine tissue recombinants composed of ERKO and wild-type tissues. For this purpose, tissue recombinants were prepared with uterine epithelium (UtE) and uterine stroma (UtS) from estrogen receptor-positive wild-type (wt) and ERKO mice (Fig. 12.2). All tissue recombinants were grown as subrenal capsular grafts in intact female nude mice for 4 weeks, at which time the hosts were ovariectomized. Ten days after ovariectomy, the hosts were injected with oil or estradiol. Sixteen hours later, the hosts were injected with ^{3}H-thymidine and sacrificed 2 hrs later. The epithelial labeling index was determined in tissue sections. In

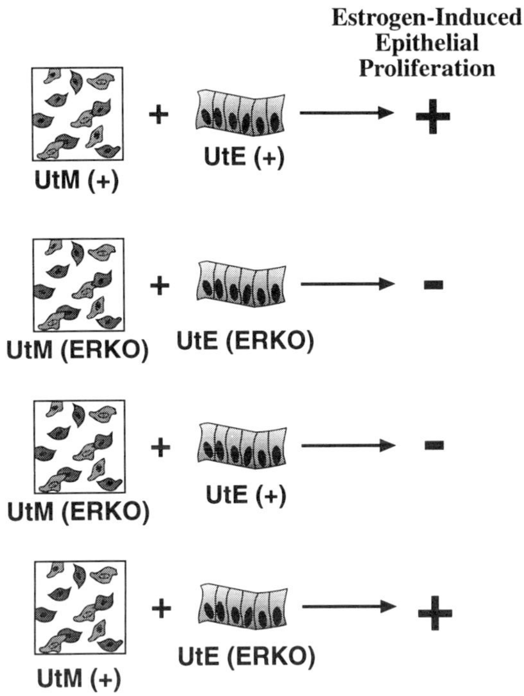

Fig. 12.2. Estradiol-induced epithelial proliferation in uterine tissue recombinants composed of epithelium and mesenchyme from ERKO and wt (+) mice.

tissue recombinants prepared with wt-UtS (wt-UtS + wt-UtE, wt-UtS + ERKO-UtE), the epithelial labeling index was increased severalfold by estradiol over oil-treated controls (Fig. 12.3). In contrast, in uterine tissue recombinants prepared with ERKO-UtS (ERKO-UtS + ERKO-UtE and ERKO-UtS + wt-UtE), the epithelial labeling index was low in both the estradiol- and oil-treated specimens. Similar results have been obtained with comparable vaginal tissue recombinants. These data clearly demonstrated that estrogen-induced epithelial proliferation is elicited via stromal estrogen receptors and that epithelial estrogen receptors are neither necessary nor sufficient for estrogen-induced epithelial proliferation (44). Thus, paracrine mechanisms play a key role in epithelial proliferation in estrogen target organs. The indirect effect of estrogen on epithelial proliferation in uterine and vaginal epithelia is consistent with the earlier finding that androgen-induced prostatic epithelial proliferation is mediated indirectly through mesenchymal/stromal androgen receptors (45). Thus, the apparent lack of a role for epithelial androgen receptors and estrogen receptors in regulating epithelial proliferation in male and female reproductive organs, respectively, suggests that indirect mediation of epithelial proliferation through sex steroid receptors in mesenchyme/stroma may be a general phenomenon for sex steroids in male and female genital systems. This

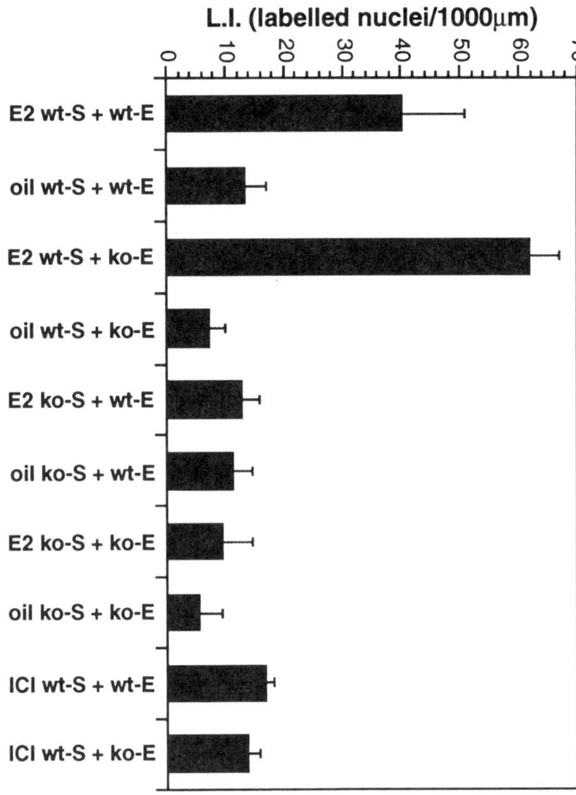

Fig. 12.3. Labeling index of epithelium in uterine tissue recombinants (wt-S + wt-E, wt-S + ko-E, ko-S + wt-E and ko-S + ko-E, as described in the text). Grafts were grown for 1 month in female nude mouse hosts, and then the hosts were ovariectomized. One week later, the hosts were injected with either 100 ng of E2 or vehicle alone. One group of hosts receiving E2 also were treated with the anti-estrogen ICI 182,780 (1 mg/kg) on days 5–7 post-ovariectomy. Epithelial LI in wt-S + wt-E and wt-S+ ko-E tissue recombinants from E2-treated hosts were significantly greater than that from oil-treated hosts ($p \leq 0.05$), but epithelial LI in E2-treated hosts was not different between wt-S + wt-E and wt-S + ko-E tissue recombinants. Epithelial LI values in all tissue recombinants prepared with ko-S and in all tissue recombinants from hosts that received E2 plus ICI 182,780 were not significantly different from each other or from the value of LI seen in oil-treated wt-S + wt-E tissue recombinants ($p \geq 0.05$).

generalization is further supported by examination of the regulation of uterine epithelial proliferation by progesterone.

One of the effects of progesterone is the inhibition of estrogen-induced uterine epithelial proliferation (6). Since progesterone receptors are present in both the epithelium and stroma of the uterus (4), progesterone inhibition of estrogen-induced uterine epithelial proliferation could be elicited by direct action of progesterone on the epithelium mediated via epithelial progesterone receptors. Alternatively, progesterone inhibition of estrogen-induced uterine epithelial proliferation could be elicited indirectly via stromal progesterone receptors. To determine whether progesterone inhibits estrogen-induced uterine epithelial proliferation via epithelial versus stromal progesterone receptors, we have used the progesterone receptor knockout mouse (PRKO) (46) to prepare the following tissue recombinants composed of UtE and UtS from progesterone receptor-positive wt and progesterone receptor-negative PRKO mice: wt-UtS + wt-UtE, wt-UtS + PRKO-UtE, PRKO-UtS + PRKO-UtE, and PRKO-UtS + wt-UtE (Fig. 12.4). All tissue recombinants were grown as subrenal capsular grafts in intact female nude mice for 4 weeks, at which time the hosts were ovariectomized. Ten days after ovariectomy, the hosts were treated with oil, estradiol, or estradiol + progesterone. The hosts were injected with ^{3}H-thymidine 2 hrs before sacrifice. The epithelial labeling index was determined in tissue sections. While the analysis of these uterine tissue recombinants is still underway, it is evident that inhibition of estrogen-induced uterine epithelial proliferation by progesterone is elicited via the stromal progesterone receptors (T. Kurita, G.R. Cunha, J.P. Lydon, and B.W. O'Malley, unpublished). Thus, mitogenic effects of androgens, estrogens, and progesterone on epithelial cells in all hormone target organs so far examined are mediated through paracrine mechanisms via stromal hormone re-

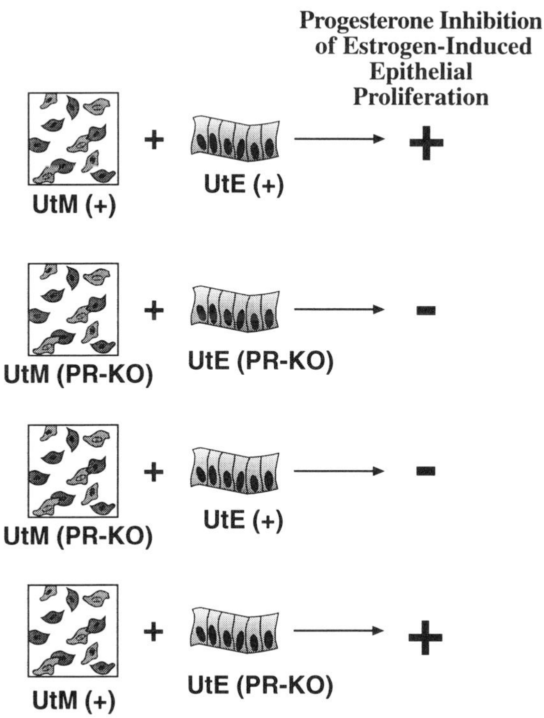

Fig. 12.4. Summary of the effects of progesterone on estradiol-induced epithelial proliferation in uterine tissue recombinants composed of epithelium and mesenchyme from PRKO and wt (+) mice.

ceptors. It must be recognized, however, that these concepts may not be relevant in all respects to the mammary gland, as progesterone has been shown to be a direct mitogen for mammary epithelial cells *in vitro* (37). Studies of mammary growth and development employing tissue recombinants composed of ERKO/wt and PRKO/wt epithelium and stroma should clarify this issue.

While the paracrine role of stroma has been emphasized in eliciting hormonal effects in epithelium via stromal hormone receptors, the role of epithelial hormone receptors can also be defined through analysis of tissue recombinants. For example, to assess the role of epithelial androgen receptors, the expression of androgen-dependent seminal vesicle secretory proteins was examined in tissue recombinants composed of rat seminal vesicle mesenchyme (SVM) plus ureter epithelium (URE) from either wt or Tfm mice (SVM + wt URE and SVM + Tfm URE). Following growth in male hosts, both wt and Tfm ureter epithelium was induced by SVM to differentiate into a simple columnar epithelium exhibiting the complex folded morphology characteristic of the seminal vesicle (28,47). In SVM + wt mouse URE tissue recombinants, epithelial androgen receptors were induced in the URE, and the epithelium expressed the full spectrum of seminal vesicle secretory proteins. By contrast, in SVM + Tfm mouse URE tissue recombinants, the Tfm epithelium is genetically incapable of producing functional androgen receptors and failed to produce seminal vesicle secretory proteins (28). These data demonstrate *in vivo* that the induction of seminal vesicle secretory proteins by androgens is an event requiring intraepithelial androgen receptors, and interpretation verified by comparable Tfm/wt experiments in the prostate (48). Thus, while androgen-induced epithelial morphogenesis and growth are elicited via stromal androgen receptors, epithelial androgen receptors play an essential role in expression of epithelial secretory proteins. Presumably, this concept applies to the role of epithelial estrogen receptors in the female genital tract. Indeed, recent ERKO/wt uterine tissue recombinants support this possibility (P.S. Cooke, D.B. Lubahn, and G.R. Cunha, unpublished).

An interesting finding in the human prostate illustrates the complexity of hormonal regulation of epithelial proliferation in which both epithelial and stromal cells are involved in the mitogenic response to insulin-like growth factor-II (IGF-II) (49). Stromal cells isolated from human prostatic tissue produce and secrete IGF-II. Prostatic epithelial cells express the type I IGF receptor and are stimulated via this receptor to grow in response to insulin, IGF-I, and IGF-II. IGF binding proteins (IGFBPs) modulate the effects of the IGFs by complexing with IGFs. Prostatic stromal cells are known to produce IGFBP-2, -3, and -4. IGFPB-3 has been shown to inhibit the mitogenic effects of both IGF-I and IGF-II on prostatic epithelial cells. IGFPBs are known to be cleaved by proteases, and one of the major epithelium-derived proteins in the human prostate is prostate specific antigen (PSA), a kallikrein-type serine protease. Recent studies have shown that IGFBP-3 inhibition of IGF-mediated growth of prostatic epithelial cells was abolished when PSA was added to the medium in conjunction with IGFBP-3 and IGF-I or IGF-II. Taken together these data suggest the following scenario. Prostatic stromal cells produce IGF-II and IGFPB-3. When bound to IGFPB-3, IGF-II is inactive. Prostatic epithelial cells under the direct influence of androgens produce PSA, which can cleave

IGFPB-3 to liberate IGFs, thus eliciting epithelial proliferation via the type I IGF receptors (49). The role of PSA in "activating" the IGF system in the prostate may not be operative in the normal human prostate, as PSA is normally secreted from the apical epithelial membrane into prostatic ducts to contribute to seminal fluid and thus may not have access to the basal-lateral cell compartment where growth factor receptors are expressed. However, in patients with prostatic carcinoma PSA is found in the serum, presumably due to aberrant secretion of PSA by cells having a disrupted basement membrane. Thus, PSA (whose expression is controlled by direct androgen action mediated by epithelial androgen receptors) may act in prostatic carcinomas as an androgen-dependent indirect epithelial mitogen.

12.3. GROWTH FACTORS AS PARACRINE/AUTOCRINE MEDIATORS OF HORMONAL RESPONSE

The demonstration that androgenic effects on prostatic epithelium are mediated by stromal androgen receptors via paracrine mechanisms in combination with the inability to demonstrate direct mitogenic effects of estrogens or androgens on epithelial cells of estrogen or androgen target organs, respectively, has led to the search for candidate paracrine mediators of steroid action in the prostate, uterus, vagina, and mammary gland. Growth factors are natural candidates as potential mediators of cell–cell interactions in hormone target organs. Such work has led to an enormous literature on the roles of many divergent growth factor families in the biology of both male and female genital tracts and the mammary gland.

As was the case previously for sex steroid action and hormone receptor expression, it is important to recognize that currently the role of growth factors in mediating hormone action is supported by an enormous body of compelling correlative data that in almost all cases are not bolstered by definitive genetic data employing mutant mice deficient in specific growth factor pathways. The use of transgenic mice in which specific growth factors or growth factor receptors have been knocked out will be particularly edifying and may provide definitive data on the role of a particular growth factor as a possible mediator of hormone action.

12.3.1. Role of the Epidermal Growth Factor Family as Autocrine/Paracrine Mediators of Hormone Action

In the uterus, vagina, and mammary gland, members of the epidermal growth factor (EGF) family (EGF, amphiregulin, heparin-binding EGF, and transforming growth factor-α [TGF-α]) have been implicated in the action of estrogen and/or progesterone. In the female genital tract, estradiol increases the expression of EGF, TGFα, and the EGF receptor (50–52). Moreover, many *in vivo* effects of estradiol in the female genital tract can be elicited in ovariectomized mice by exogenous EGF (53). Given this background, transgenic mice lacking a functional EGF receptor (54) or lacking TGFα (55) should exhibit abnormalities of growth and development, as well as impaired responsiveness to estradiol in the female genital tract. Recent studies show that this is not, in fact, the case. Uteri of TGFα-KO mice appear normal, and female TGFα-KO mice are fertile. Indeed, the only described phenotype of TGFα-KO mice is an abnormality of hair growth (55).

Epidermal growth factor receptor knockout (EGFR-KO) mice have been produced recently (54) but usually die shortly after birth. Morphologic examination of these neonatal EGFR-KO mice has revealed that female genital tracts are normal (Y.H. Hom and G.R. Cunha, unpublished). Since EGFR-KO mice usually die in the neonatal period, the interplay between estrogen action and EGFR signaling was examined in grafts of neonatal uteri of wt and EGFR-KO mice grown under the renal capsule of intact female nude mice. Four weeks after grafting, the hosts were spayed. In addition, the following uterine tissue recombinants were prepared with UtE and UtS from wt and EGFR-KO mice: wt-Ut S+ wt-UtE, EGFR-KO-UtS + EGFR-KO-UtE, wt-UtS + EGFR-KO-UtE, and EGFR-KO-UtS + wt-UtE. As above for uterine grafts, all tissue recombinants were grown under the renal capsule of intact female nude mice for 4 weeks, at which time the hosts were spayed. Ten days later, hosts bearing intact uterine grafts or uterine tissue recombinants were injected with oil or estradiol, and the ^{3}H-thymidine labeling index was determined for the epithelium in tissue sections of the uterine tissue recombinants and grafted uteri. The epithelial labeling index was induced severalfold by estrogen in epithelium of the hosts' uteri, in grafts of wt and EGFR-KO uteri, and in all uterine tissue recombinants irrespective of EGFR status of the constituent tissues (74). This suggests that EGFR signaling *is not essential* for estrogen-induced proliferation of uterine epithelium. Thus, while EGF ligands may play a role in the epithelial response to estrogens in wt mice, complete elimination of the EGFR signaling pathway by gene knockout clearly indicates that this pathway is, in fact, not required in the estrogenic response of estrogen target epithelial cells. Clearly, other potentially more important signaling pathways can compensate for the absence of the EGFR pathway. Future analysis of other types of knockout mice should identify critical versus dispensable growth factor pathways in hormone response.

12.3.2. Role of Keratinocyte Growth Factor in the Developing Male Urogenital Tract

Keratinocyte growth factor (KGF) is a member of the fibroblast growth factor (FGF) family (56). The receptor for KGF (KGF-R) is a splice variant of the bek gene (FGFR2) that is expressed in epithelial cells (57–59). The potential role of KGF in androgen-dependent development of the male genital tract stems from the postulated role of KGF as a paracrine mediator of mesenchymal–epithelial interactions (60,61). In this regard, KGF and KGF-R are also expressed during ductal branching morphogenesis in mesenchyme of the developing salivary gland, pancreas, thyroid, mammary gland, lung, and many other structures composed of epithelium and mesenchyme (56,62,63). KGF has also been identified in the adult rat and human prostate (64,65) and plays a role in androgen-dependent growth and development of the neonatal mouse seminal vesicle (66).

A possible role of KGF in the developing prostate is suggested by the observation that KGF transcripts were detected in the embryonic mouse urogenital sinus mesenchyme by *in situ* hybridization (62), in the neonatal rat ventral prostate, and in mesenchymal cells isolated from neonatal rat ventral prostate by reverse transcription - polymerase chain reaction (RT-PCR). Correspondingly, KGF-R transcripts were detected in embryonic mouse urogenital sinus epithelium and in the epithelial ducts of the neonatal rat ventral prostate by *in situ* hybridization (62,67) and in neonatal rat ventral prostate by RT-PCR (67). The functional role of KGF in prostatic development was assessed using a serum-free organ culture system. As shown in Figure 12.5a, the newborn rat ventral prostate is composed of four main ducts, each having a few branches surrounded by mesenchyme. Rat ventral prostates cultured in the presence of testosterone underwent considerable ductal branching morphogenesis in a dose-dependent fashion during 6 days of culture (Fig. 12.5b–d). Maximal ductal branching morphogenesis was attained with testosterone at 10^{-8} *M*. Growth (DNA content) of ventral prostates cultured with testosterone also increased in a dose-dependent fashion, with the maximal response observed at 10^{-8} *M* testosterone (67). Ductal branching morphogenesis was markedly reduced in ventral prostates cultured without testosterone (Fig. 12.5e). The minimal

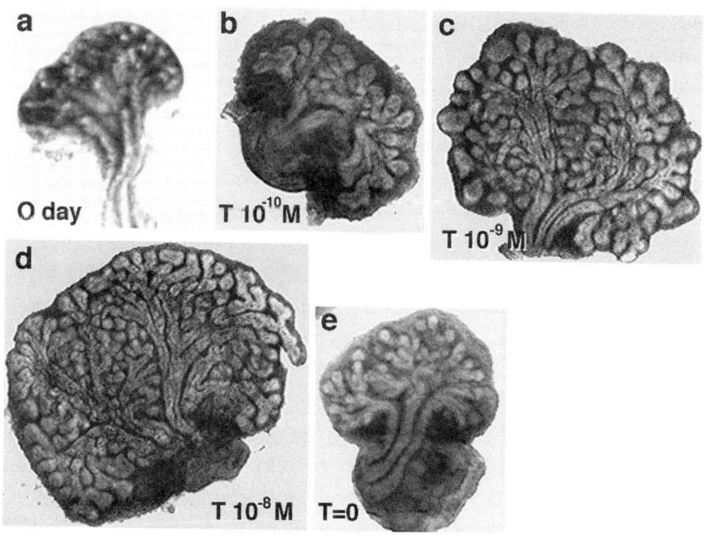

Fig. 12.5. (a) Rat ventral prostate (VP) prior to culture. Note that each main duct has three to six distal tips and that epithelial ducts are surrounded by mesenchyme. Neonatal rat VPs cultured for 6 days *in vitro* under serum-free conditions in the presence of testosterone at (b) 10^{-10} *M*, (c) 10^{-9} *M*, and (d) 10^{-8} *M*. Note the increased ductal growth and ductal branching morphogenesis at higher testosterone concentrations. (e) Neonatal rat VP cultured for 6 days *in vitro* under serum-free conditions in the absence of testosterone. While development of glands grown in the absence of testosterone has proceeded somewhat relative to glands before culture (compare with Fig. 12.1a), ductal growth and ductal branching morphogenesis in the absence of testosterone are far less than those elicited by an optimal dose of testosterone. 32× magnification.

growth observed in the absence of testosterone was attributed to the carryover of endogenous testosterone from the previous *in vivo* period of development.

KGF (50 or 100 ng/ml) by itself was able to elicit extensive ductal growth and branching morphogenesis in the absence of androgens. The pattern and magnitude of ductal branching induced by KGF were comparable to those elicited by testosterone (Fig. 12.6). DNA content of ventral prostates cultured with 100 ng/ml KGF alone versus testosterone at 10^{-8} M was equivalent: 1.80 ± 0.46 μg versus 1.88 ± 0.09 μg, respectively.

To further assess the role of KGF in prostatic development, newborn rat ventral prostates were cultured as explants for 6 days in the presence of testosterone (10^{-9} M) plus either a monoclonal antibody (Mab 1G4) to KGF (68) or peptide 412, a soluble portion of the KGF binding pocket of the KGF-R that competes with active receptors for endogenous KGF (68). Both Mab 1G4 and peptide 412 inhibited testosterone-induced ductal branching morphogenesis (Fig. 12.7) during 6 days of culture. The inhibitory effects of Mab 1G4 and peptide 412 were also manifested in a reduction in (a) the number of ductal tips per explant, (b) epithelial area per explant, and (c) DNA content per explant (Fig. 12.8) (67). Taken together these findings suggest that the effects of androgens on prostatic epithelial development are mediated, at least in part, by KGF, which is produced by the mesenchyme and, in turn, regulates epithelial development via cell surface receptors. These findings in the prostate extend the results of earlier studies in the developing mouse seminal vesicle demonstrating that KGF is present during the time of seminal vesicle morphogenesis and that KGF of mesenchymal origin functions as a paracrine mediator of androgen-stimulated seminal vesicle epithelial growth both *in vitro* (66) and *in vivo* (69). While KGF is clearly an important mesenchyme-derived paracrine regulator of androgen-induced growth and ductal branching morphogenesis in the male genital tract of wt rats and mice, the KGF signaling pathway may not be an essential, indispensable pathway since the only defects reported in KGF KO mice involve hair development (70), although it should be noted that the prostates of these KGF knockout mice have not yet been examined. Taking an entirely different approach, several investigators have constructed transgenic mice with dominant negative fibroblast growth factor receptors designed to impair the KGF signaling pathway in epithelial cells. When targeted to the lung or skin, such transgenic mice exhibit impaired pulmonary branching morphogenesis or impaired epidermal wound healing (71–73). Thus, KGF is clearly an important paracrine signaling molecule that is involved in androgen-induced epithelial growth and ductal branching morphogenesis of the prostate and seminal vesicle via the KGF-R on the epithelial cell surface. Clearly, other compensatory growth factor pathways (the IGF family, the EGF family, and hepatocyte growth factor [MGF]) are likely to also play important roles in the paracrine cell–cell interactions that mediate steroid hormone action.

12.4. CONCLUSIONS

While the steroid hormone response is mediated via specific high-affinity receptors, it has become apparent that hormone target organs contain a multitude of cell types

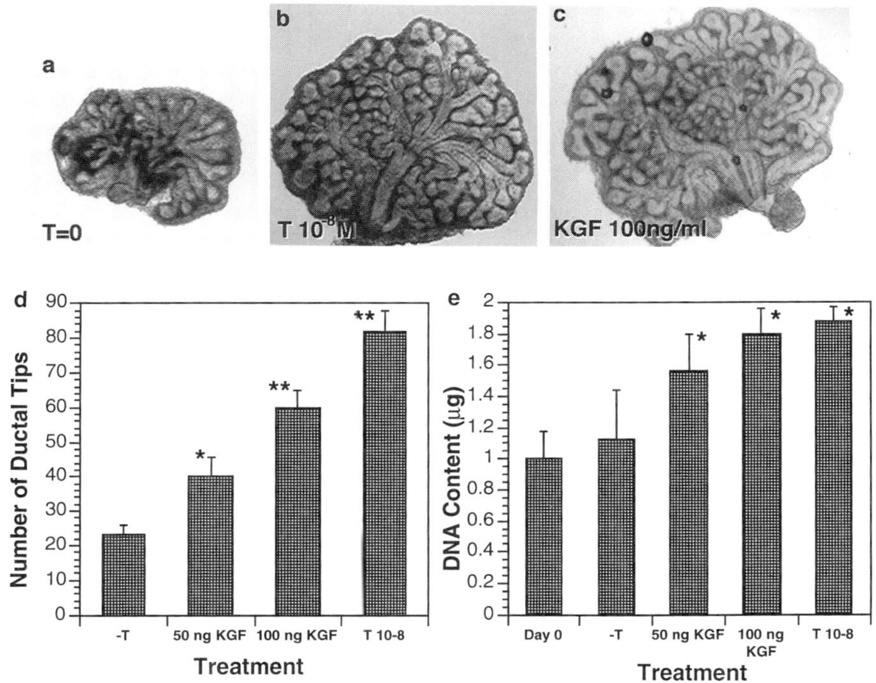

Fig. 12.6. Representative images of rat VPs cultured for 6 days in serum-free medium (a) without testosterone, (b) with testosterone (10^{-8} *M*), (c) without testosterone + KGF (100 ng/ml). Note that the ductal growth and branching morphogenesis elicited by KGF are comparable to those elicited by testosterone. 32× magnification. (d) Number of ductal tips of VPs cultured for 6 days in serum-free medium with KGF or testosterone at the concentrations indicated. (e) DNA contents of VPs cultured for 6 days in serum-free medium at different concentrations of KGF or testosterone at the concentrations indicated. Data are based on the analysis of four to six specimens per point ($*P < 0.05$, $**P < 0.001$).

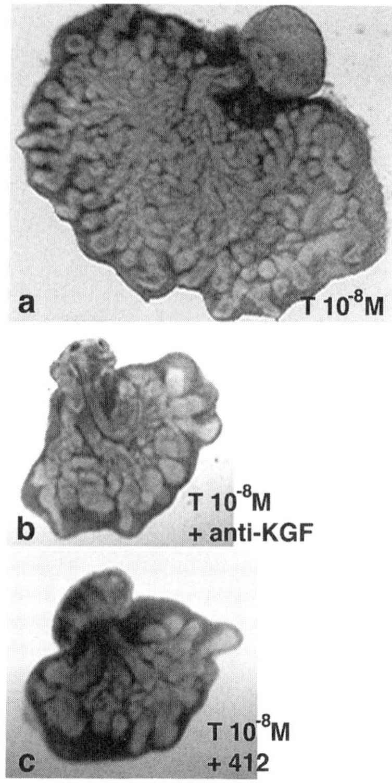

Fig. 12.7. Rat VPs incubated for 6 days in (a) the presence of testosterone (10^{-8} *M*) plus mouse immunoglobulin G (IgG) (60 μg/ml) or (b) testosterone in combination with anti-KGF Mab (60 μg/ml) or (c) peptide 412 (60 mg/ml). Note the dramatic inhibition of growth and ductal branching morphogenesis elicited by anti-KGF and peptide 412. 32× magnification.

(epithelium, fibroblasts, and visceral smooth muscle) that may all express one or more steroid receptor types. For example, in the female genital tract, evidence suggests that estrogen and progesterone receptors are coexpressed in both epithelial cells and the surrounding fibroblasts of the endometrial stroma. In addition, androgen receptors are also present in the uterus. Given the importance of morphogenetic cell–cell interactions between the epithelium and mesenchyme of hormone target organs and the expression of hormone receptors in mesenchyme, epithelium, or both, tissue recombination studies have clearly established the importance of paracrine mechanisms of hormone action on epithelial cells in both androgen and estrogen/progesterone target organs. Through the use of growth factor/growth factor receptor knockout mice, neutralizing antibodies, antisense oligonucleotides, and transgenic mice with dominant negative growth factor receptors, it should be possible to elucidate the complex cellular and molecular interactions involved in the paracrine cell–cell interactions that regulate hormone-induced growth and development of the male and female genital tract and the mammary gland. Growth regula-

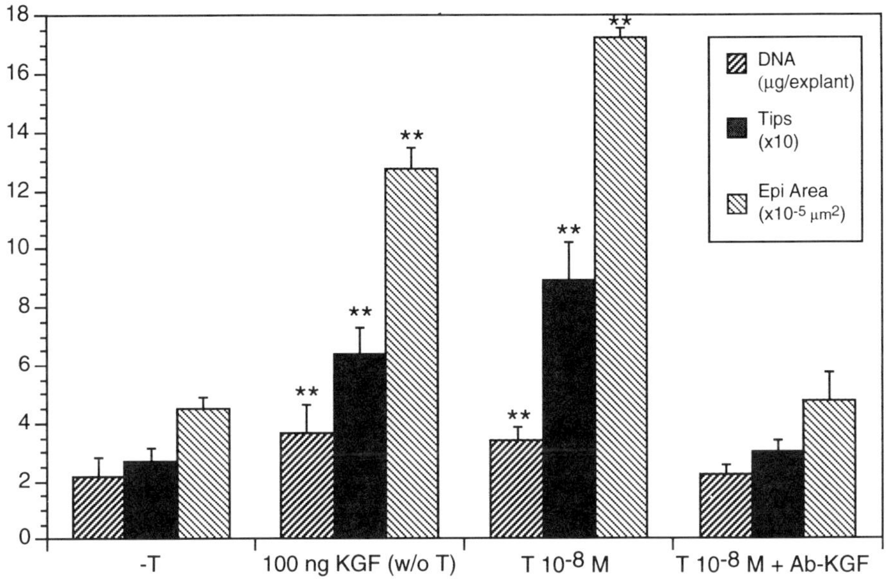

Fig. 12.8. Effect of testosterone, KGF, and anti-KGF on DNA content per rat VP (μg per explant), the number of ductal tips per rat VP, and the epithelial area (μm^2) per rat VP following 6 days of incubation in serum-free medium. Data are based on the analysis of four to six specimens per point ($**P < 0.001$).

tion will surely not be a simple process due to the levels of pathway redundancy in biologic systems, and elucidation of the fundamental mechanisms will require careful analysis in order to determine which signals are necessary and which are sufficient for a particular hormone response.

ACKNOWLEDGMENTS

The authors wish to thank Joel Brody for the preparation of the figures. Supported by National Institutes of Health grants DK 32157, CA 59831, DK 47517, DK 45861, CA 64872, DK 52708, and AG 13784.

REFERENCES

1. Jensen VE, Desombre ER, Hurst DJ, Kawashima T, Jungblut PW: Estrogen–receptor interactions in target tissues. Arch Anat Microsc Morphol Exp 56:547–569, 1967.
2. McClellan MC, West NB, Tacha DE, Greene GL, Brenner RM: Immunocytochemical localization of estrogen receptors in the macaque reproductive tract with monoclonal antiestrophilins. Endocrinology 114:2002–2014, 1984.

3. McCormack SA, Glasser SR: Differential response of individual uterine cell types from immature rats treated with estradiol. Endocrinology 106:1634–1649, 1980.
4. Stumpf W, Sar M: Autoradiographic localization of estrogen, androgen, progestin, and glucocorticosteroid in "target tissues" and "non-target tissues." In Pasqualini J (ed): Receptors and Mechanism of Action of Steroid Hormones. New York: Marcel Dekker, 1976, pp 41–84.
5. Clark J H, Peck E J: Female Sex Steroids: Receptors and Function. New York: Springer-Verlag, 1979.
6. Martin L: Estrogens, anti-estrogen and the regulation of cell proliferation in the female reproductive tract in vivo. In McLachlan J A (ed): Estrogens in the Environment. New York: Elsevier/North-Holland, 1980, pp 103–130.
7. Grobstein C: Tissue interactions in the morphogenesis of mouse embryonic rudiments in vitro. In Rudnick D (ed): Aspects of Synthesis and Order in Growth. Princeton, NJ: Princeton University Press, 1954, pp 233–256.
8. Grobstein C: Mechanisms of organogenetic tissue interaction. J Natl Cancer Inst Monogr 26:279–299, 1967.
9. Wessells NK: Tissue Interactions and Development. Menlo Park, CA: WA Benjamin, 1977.
10. Kollar EJ: The development of the integument: Spatial, temporal and phylogenetic factors. Am Zool 12:125–135, 1972.
11. Sengel P: The organogenesis and arrangement of cutaneous appendages in birds. Adv Morph 9:181–230, 1971.
12. Haffen K, Kedinger M, Simon-Assmann P: Cell contact dependent regulation of enterocytic differentiation. In Lebenthal E (ed): Human Gastrointestinal Development. New York: Raven Press, 1989, pp 19–35.
13. Cunha GR: Epithelial–stromal interactions in development of the urogenital tract. Int Rev Cytol 47:137–194, 1976.
14. Cunha GR, Chung LWK, Shannon JM, Taguchi O, Fujii H: Hormone-induced morphogenesis and growth: Role of mesenchymal–epithelial interactions. Recent Prog Horm Res 39:559–598, 1983.
15. Kratochwil K: Tissue combination and organ culture studies in the development of the embryonic mammary gland. In Gwatkin RBL (ed): Developmental Biology: A Comprehensive Synthesis. New York: Plenum, 1987, pp 315–334.
16. Sakakura T: Mammary embryogensis. In Neville CW, Daniel MC (eds): The Mammary Gland: Development, Regulation and Function. New York: Plenum, 1987, pp 37–66.
17. Cunha GR, Bigsby RM, Cooke PS, Sugimura Y: Stromal–epithelial interactions in adult organs. Cell Diff 17:137–148, 1985.
18. Cunha GR: Epithelio–mesenchymal interactions in primordial gland structures which become responsive to androgenic stimulation. Anat Rec 172:179–196, 1972.
19. Cunha GR: Tissue interactions between epithelium and mesenchyme of urogenital and integumental origin. Anat Rec 172:529–542, 1972.
20. Kratochwil K, Schwartz P: Tissue interaction in androgen response of embryonic mammary rudiment of mouse: Identification of target tissue of testosterone. Proc Natl Acad Sci USA 73:4041–4044, 1976.
21. Cunha GR, Hom YH: Role of mesenchymal–epithelial interactions in mammary gland development. J Mammary Gland Biol Neoplasia 1:21–35, 1996.

22. Heuberger B, Fitzka I, Wasner G, Kratochwil K: Induction of androgen receptor formation by epithelium–mesenchyme interaction in embryonic mouse mammary gland. Proc Natl Acad Sci USA 79:2957–2961, 1982.

23. Kratochwil K: Development and loss of androgen responsiveness in the embryonic rudiment of the mouse mammary gland. Dev Biol 61:358–365, 1977.

24. Kratochwil K: In vitro analysis of the hormonal basis for the sexual dimorphism in the embryonic development of the mouse mammary gland. J Embryol Exp Morphol 25:141–153, 1971.

25. Cunha GR, Lung B: The possible influences of temporal factors in androgenic responsiveness of urogenital tissue recombinants from wild-type and androgen-insensitive (Tfm) mice. J Exp Zool 205:181–194, 1978.

26. Shannon JM, Cunha GR: Characterization of androgen binding and deoxyribonucleic acid synthesis in prostate-like structures induced in testicular feminized (Tfm/Y) mice. Biol Reprod 31:175–183, 1984.

27. Sugimura Y, Cunha GR, Bigsby RM: Androgenic induction of deoxyribonucleic acid synthesis in prostatic glands induced in the urothelium of testicular feminized (Tfm/y) mice. Prostate 9:217–225, 1986.

28. Cunha GR, Young P: Inability of Tfm (testicular feminization) epithelial cells to express androgen-dependent seminal vesicle secretory proteins in chimeric tissue recombinants. Endocrinology 128:3293–3298, 1991.

29. Cunha GR, Shannon JM, Vanderslice KD, Sekkingstad M, Robboy SJ: Autoradiographic analysis of nuclear estrogen binding sites during postnatal development of the genital tract of female mice. J Steroid Biochem 17:281–286, 1982.

30. Bigsby RM, Cunha GR: Estrogen stimulation of deoxyribonucleic acid synthesis in uterine epithelial cells which lack estrogen receptors. Endocrinology 119:390–396, 1986.

31. Yamashita S, Newbold RR, McLachlan JA, Korach KS: Developmental pattern of estrogen receptor expression in female mouse genital tracts. Endocrinology 125:2888–2896, 1989.

32. Yamashita S, Newbold RR, McLachlan JA, Korach KS: The role of estrogen receptor in uterine epithelial proliferation and cytodifferentiation in neonatal mice. Endocrinology 127:2456–2463, 1990.

33. McClellan MC, Rankin S, West NB, Brenner RM: Estrogen receptors, progesterone receptors and DNA synthesis in the macaque endometrium during the luteal-follicular transition J Steroid Biochem Mol Biol 37:631–641, 1990.

34. Casimiri V, Rath NC, Parvez H, Psychoyos A: Effect of sex steroids on rat endometrial epithelium and stroma cultured separately. J Steroid Biochem 12:293–298, 1980.

35. Iguchi T, Uchima F -D A, Ostrander PL, Hamamoto ST, Bern HA: Proliferation of normal mouse uterine luminal epithelial cells in serum-free collagen gel culture. Proc Jpn Acad 61:292–295, 1985.

36. Iguchi T, Uchima FDA, Ostrander PL, Bern HA: Growth of normal mouse vaginal epithelial cells in and on collagen gels. Proc Natl Acad Sci USA 80:3743–3747, 1983.

37. Imagawa W, Bandyopadhyay GK, Nandi S: Regulation of mammary epithelial cell growth in mice and rats. Endocrine Rev 11:494–523, 1990.

38. Julian J, Carson DD, Glasser SR: Polarized rat uterine epithelium in vitro: Responses to estrogen in defined medium. Endocrinology 130:68–78, 1992.

39. Cooke PS, Uchima F -DA, Fujii DK, Bern HA, Cunha GR: Restoration of normal

morphology and estrogen responsiveness in cultured vaginal and uterine epithelia transplanted with stroma. Proc Natl Acad Sci USA 83:2109–2113, 1986.

40. Inaba T, Wiest WG, Strickler RC, Mori J: Augmentation of the response of mouse uterine epithelial cells to estradiol by uterine stroma. Endocrinology 123:1253–1258, 1988.

41. Cunha GR, Young P: Role of stroma in oestrogen-induced epithelial proliferation. Epithelial Cell Biol 1:18–31, 1992.

42. Korach K: Insights from the study of animals lacking functional estrogen receptor. Science 266:1524–1527, 1994.

43. Lubahn DB, Moyer JS, Golding TS, Couse JF, Korach KS, Smithies O: Alteration of reproductive function but not prenatal sexual development after insertional disruption of the mouse estrogen receptor gene. Proc Natl Acad Sci USA 90:11162–11166, 1993.

44. Cooke P, Buchanan D, Young P, Setiawan T, Brody J, Korach K, Taylor J, Lubahn D, Cunha G: Stromal estrogen receptors (ER) mediate mitogenic effects of estradiol on uterine epithelium. Proc Natl Acad Sci USA 94:6535–6540.

45. Cunha GR, Donjacour AA, Cooke PS, Mee S, Bigsby RM, Higgins SJ, Sugimura Y: The endocrinology and developmental biology of the prostate. Endocrine Rev 8:338–363, 1987.

46. Lydon JP, DeMayo FJ, Funk CR, Mani SK, Hughes AR, Montgomery CA, Shyamala G, Conneely OM, O'Malley BW: Mice lacking progesterone receptor exhibit pleiotropic reproductive abnormalities. Genes Dev 9:2266–2278, 1995.

47. Cunha GR, Young P, Higgins SJ, Cooke PS: Neonatal seminal vesicle mesenchyme induces a new morphological and functional phenotype in the epithelia of adult ureter and ductus deferens. Development 111:145–158, 1991.

48. Donjacour AA, Cunha GR: Assessment of prostatic protein secretion in tissue recombinants made of urogenital sinus mesenchyme and urothelium from normal or androgen-insensitive mice. Endocrinology 131:2342–2350, 1993.

49. Peehl DM, Cohen P, Rosenfeld RG: The insulin-like growth factor system in the prostate. World J Urol 13:306–311, 1995.

50. Mukku VR, Stancel GM: Regulation of epidermal growth factor receptor by estrogen. J Biol Chem 260:9820–9824, 1985.

51. Nelson KG, Takahashi T, Lee DC, Luetteke NC, Bossert NL, Ross K, Eitzman BE, McLachlan JA: Transforming growth factor-alpha is a potential mediator of estrogen action in the mouse uterus. Endocrinology 131:1657–1664, 1992.

52. Sakai Y, Nelson KG, Snedeker S, Bossert NL, Walker MP, McLachlan J, DiAugustine RP: Expression of epidermal growth factor in suprabasal cells of stratified squamous epithelia: Implication for a role in differentiation. Cell Growth Differ 5:527–535, 1994.

53. Nelson KG, Takahashi T, Bossert NL, Walmer DK, McLachlan JA: Epidermal growth factor replaces estrogen in the stimulation of female genital-tract growth and differentiation. Proc Natl Acad Sci USA 88:21–25, 1991.

54. Miettinen P, Berger J, Meneses J, Phung Y, Pedersen R, Werb Z, Derynck R: Epithelial immaturity and multi-organ failure in mice lacking the epidermal growth factor receptor. Nature 376:337–341, 1995.

55. Luetteke NC, Qiu TH, Peiffer RL, Oliver P, Smithies O, Lee DC: TGF alpha deficiency results in hair follicle and eye abnormalities in targeted and waved-1 mice. Cell 73:263–278, 1993.

56. Mason IJ, Pace FF, Smith R, Dickson C: FGF-7 (keratinocyte growth factor) expression

during mouse development suggests roles in myogenesis, forebrain regionalization and epithelial–mesenchymal interactions. Mech Dev 15:15–30, 1994.

57. Bottaro DP, Rubin JS, Ron D, Finch PW, Florio C, Aaronson SA: Characterization of the receptor for keratinocyte growth factor. Evidence for multiple fibroblast growth factor receptors. J Biol Chem 265:12767–12770, 1990.
58. Miki T, Bottaro DP, Fleming TP, Smith CL, Burgess WH, Chan AM-L, Aaronson SA: Determination of ligand-binding specificity by alternative splicing: Two distinct growth factor receptors encoded by a single gene Proc Natl Acad Sci USA 89:246–251, 1992.
59. Miki T, Fleming TP, Bottaro DP, Rubin JS, Ron D, Aaronson SA: Expression cDNA cloning of the KGF receptor by creation of a transforming autocrine loop. Science 251:72–75, 1991.
60. Aaronson SA, Bottaro DP, Miki T, Ron D, Finch PW, Fleming TP, Ahn J, Taylor WG, Rubin JS: Keratinocyte growth factor: A fibroblast growth factor family member with unusual target cell specificity. Ann NY Acad Sci 638:62–77, 1991.
61. Aaronson SA, Rubin JS, Finch PW, Wong J, Marchese C, Falco J, Taylor WG, Kraus MH: Growth factor-regulated pathways in epithelial cell proliferation. Am Rev Respir Dis 142:S7–S10, 1990.
62. Finch PW, Cunha GR, Rubin JS, Wong J, Ron D: Pattern of KGF and KGFR expression during mouse fetal development suggests a role in mediating morphogenetic mesenchymal–epithelial interactions. Dev Dynamics 203:223–240, 1995.
63. Orr-Urteger A, Bedford MT, Burakova T, Arman E, Zimmer Y, Yayon A, Givol D, Lonai P: Developmental localization of the splicing alternatives of fibroblast growth factor receptor-2 (FGFR2). Dev Biol 158:475–486, 1993.
64. Peehl D, Rubin J: Keratinocyte growth factor: An androgen-regulated mediator of stromal–epithelial interactions in the prostate. World J Urol 13:312–317, 1995.
65. Yan G, Fukabori Y, Nikolaropoulos S, Wang F, McKeehan WL: Heparin-binding keratinocyte growth factor is a candidate stromal to epithelial cell andromedin. Mol Endocrinol 6:2123–2128, 1992.
66. Alarid ET, Rubin JS, Young P, Chedid M, Ron D, Aaronson SA, Cunha GR: Keratinocyte growth factor functions in epithelial induction during seminal vesicle development. Proc Natl Acad Sci USA 91:1074–1078, 1994.
67. Sugimura Y, Foster BA, Hom YK, Rubin JS, Finch PW, Aaronson SA, Hayashi N, Kawamura J, Cunha GR: Keratinocyte growth factor (KGF) can replace testosterone in the ductal branching morphogenesis of the rat ventral prostate. Int J Dev Biol 40:941–951, 1996.
68. Bottaro DP, Fortney E, Rubin JS, Aaronson SA: A keratinocyte growth factor receptor-derived peptide antagonist identifies part of the ligand binding site. J Biol Chem 268:9180–9183, 1993.
69. Cunha GR, Foster BA, Sugimura Y, Hom YK: Keratinocyte growth factor as mediator of mesenchymal-epithelial interactions in the development of androgen target organs. Semin Cell Dev Biol 7:203–210, 1996.
70. Guo L, Degenstein L, Fuchs E: Keratinocyte growth factor is required for hair development but not for wound healing. Genes Dev 10:165–175, 1996.
71. Peters K, Werner S, Liao X, Wert S, Whitsett J, Williams L: Targeted expression of a dominant negative FGF receptor blocks branching morphogenesis and epithelial differentiation of the mouse lung. EMBO J 12:973–986, 1994.

72. Werner S, Smola H, Liao X, Longaker M, Krieg T, Hofschneider P, Williams LT: The function of KGF in morphogenesis of epithelium and reepithelialization of wounds. Science 266:819–822, 1994.

73. Werner S, Weinberg W, Liao X, Peters KG, Blessing M, Yuspa SH, Weiner RL, Williams LT: Targeted expression of a dominant-negative FGF receptor mutant in the epidermis of transgenic mice reveals a role of FGF in keratinocyte organization and differentiation. EMBO J 12:2635–2643, 1993.

74. Hom YK, Young P, Wiesen JF, Miettinen PJ, Derynck R, Werb Z, Cunha GR, 1997: Uterine and vaginal organ growth requires epidermal growth factor receptor (EGFR) signaling from stroma. Endocrinology in press.

13

HORMONES, INSULIN-LIKE GROWTH FACTORS, AND THEIR BINDING PROTEINS IN THE FEMALE REPRODUCTIVE TRACT

LIAM J. MURPHY

Departments of Internal Medicine and Physiology, University of Manitoba, Winnipeg R3E OW3, Canada

13.1. INTRODUCTION

Expression of various components of the insulin-like growth factor system has been demonstrated in the female reproductive tract from a variety of species, including the rodent and humans. The insulin-like growth factors (IGFs) function as mediators or amplifiers of both steroid and polypeptide hormones that have trophic effects in various cells and tissues of the female reproductive system. The IGFs appear to amplify hormone signals by activating the expression of other growth factor and cytokine systems and, in turn, may be activated by these and other growth factors. For example, epidermal growth factor (EGF), which as been shown to mimic many of the effects of estradiol in the uterus, also activates expression of IGF-I in this tissue. The IGFs also function as paracrine signals that allow for the interaction of the stromal and epithelial cell compartments. This interaction has been shown to be critically important for the development and maintenance of hormone responsiveness.

In this chapter, I will review the hormonal regulation of the IGFs in the female reproductive tract, the role of the IGFs in mediating hormone responsiveness, their role as mediators of stromal–epithelial interactions, and their interaction with other growth factor systems, particularly the EGF system.

Hormones and Growth Factors in Development and Neoplasia, Edited by Robert B. Dickson and David S. Salomon
ISBN 0-471-16899-8

Although IGF-I is only one of a long list of growth factors and cytokines that are expressed in the female reproductive tract, there is compelling evidence that they are important in the development and differentiation of reproductive tissues and in the hormonal responsiveness of reproductive tissue. Unlike other growth factor systems, the IGF system is quite complex, with multiple ligands, receptors, and binding proteins allowing for many potential levels of regulation. Considerable progress has been made in understanding the various components of the IGF system and how they interact.

13.2. THE INSULIN-LIKE GROWTH FACTOR SYSTEM

The IGFs were originally discovered during a search for circulating mediators of growth hormone action (1). To date, two IGFs have been identified, IGF-I and IGF-II. Only the former is growth hormone dependent and has been found to have a much more diverse role than as a simple mediator of growth hormone action (2). The IGFs are both structurally and functionally similar to insulin (2). However, unlike insulin, which is expressed at significant levels only in the endocrine pancreas, the IGFs are expressed in virtually all tissues, at least at some stages of development (3). The IGFs interact with at least three types of receptors. They interact with the insulin receptor with a considerably lower affinity than insulin itself, but binding affinity is sufficient that interaction of the IGFs with the insulin receptor is probably important physiologically, at least under some circumstances. In addition they interact with the type I or IGF-I receptor, with which insulin can also interact. In this case, insulin has a lower binding affinity than the IGFs for this receptor. Like their ligands, the insulin and IGF-I receptors are structurally similar. IGF-II can also interact with a distinct monomer receptor, the mannose-6-phosphate/IGF-II receptor. The exact function of this latter receptor is unclear, but it may serve as a scavenger system to limit the actions of IGF-II. In addition to the receptors, IGFs can interact with a group of structurally related high-affinity binding proteins, the insulin-like growth factor binding proteins (IGFBPs). To date, six IGFBPs, ranging in size from approximately 17 to 43 kDa, have been identified, and their cognate cDNAs have been cloned and sequenced (4). These binding proteins have been shown to be released by a variety of cell types and to be differentially regulated. Under various *in vitro* assay conditions they may either augment or, more commonly, inhibit IGF actions. The function of the IGFBPs has not yet been clearly elucidated, but they may serve to sequestrate IGFs or transport them to their cell surface receptors. Interestingly, proteolytic degradation of IGFBPs appears to be an important regulated mechanism whereby IGF action is modulated.

The importance of the IGF system in the female reproductive system is apparent from the observations in IGF-I receptor knockout mice (5). Although these animals die shortly after birth, the ovary and uterus are atretic and reduced in size. The myometrium of these mice, the site of expression of IGF-I and IGF-II, is markedly reduced in thickness.

13.3. REGULATION OF INSULIN-LIKE GROWTH FACTOR EXPRESSION AND ACTION IN THE OVARY

In the human ovary, IGF-II mRNA is localized to follicular blood vessels and the granulosa cells. Very little IGF-I expression is apparent in dominant follicles; however, IGF-I expression is apparent in thecal cells of atretic follicles (6). In the rat ovary, there is very little expression of IGF-II after the early postnatal period. In rat ovarian tissue, IGF-I expression is upregulated by both estradiol and gonadotropins (7). Furthermore, the IGFs synergize with gonadotropins in stimulating ovarian steroidogenesis (8,9). In thecal cells the IGFs enhance the effects of luteinizing hormone on androgen production, whereas they enhance the effects of follicle stimulating hormone (FSH) on estradiol and progesterone production from granulosa cells (6). In both the rodent and human ovary, the IGFs are thought to function as enhancers of FSH-stimulated aromatase activity and appear to have an important role in the development of follicular dominance (6).

At least five of the six known IGFBPs are expressed in ovarian tissue. In the human ovary, IGFBP-1, -2, and -3 are expressed in granulosa cells, and both IGFBP-1 and IGFBP-3 can inhibit IGF-I-stimulated E2 and progesterone production by these cells (10, 11). In the porcine ovary, expression of IGFBP-2 is correlated with follicular atresia, whereas IGFBP-3 expression is increased as follicles undergo preovulatory maturation (12). Similarly, in the human ovary, high levels of IGFBP-2 and IGFBP-4 are present in follicular fluid from androgen-dominant atretic follicles, whereas these binding proteins are present in very low levels in follicular fluid from estrogen-dominant follicles (6, 13). In these estrogen-dominant follicles, IGF-II levels are higher. In contrast, there are no differences in the levels of IGF-I, IGFBP-1, or IGFBP-3 in follicular fluid from estrogen-dominant and androgen-dominant follicles (6). Clearly, the ovarian IGF-I system has a role in the maturation of dominant follicles, and there are many potential steps whereby the action of these growth factors can be regulated. Specific protease for some of the binding proteins have been demonstrated in follicular fluid, and these would also appear to be regulated. The importance of the ovarian IGF system has been confirmed by recent observations in transgenic mice. Transgenic mice that overexpress IGFBP-1 in the ovary have reduced fecundity, smaller litter size, and a decreased ovulatory response to gonadotropins (14).

13.4. REGULATION OF INSULIN-LIKE GROWTH FACTOR EXPRESSION AND ACTION IN THE UTERUS

Estrogen-induced uterine proliferation is associated with induction of expression of a variety of growth factors and cytokines that are thought to function in an autocrine or paracrine fashion to amplify the estrogen-induced proliferative signal (15). Although these growth factors may mediate other actions in other tissues, they function as local estromedins in the uterus. In this regard, the best studied are IGF-I and EGF. The role of these growth factors in estrogen-induced uterine proliferation will be reviewed in detail below. Limitation of space does not allow for an extensive review of the growing number of potential estromedins that have been reported in recent years.

Although the IGFs are expressed in all tissues, relatively high levels of expression of both IGF-I and IGF-II are detectable in the female reproductive tract (3). The abundance of the mRNAs for both of these growth factors are upregulated in the uterus by estrogen administration to ovariectomized rodents (16,17). Similar data are available in other species, and there is evidence that this may also be true in the human reproductive tract. In the mature ovariectomized rat, a single injection of 17β estradiol results in a 10-fold increase in uterine IGF-I mRNA abundance (17). The increase in IGF-I mRNA reaches a maximum at 6 hr after injection and then gradually declines toward basal levels. A qualitatively and quantitatively similar response of uterine IGF-I to estradiol is demonstrable in the immature rat. This estrogen response is pituitary independent since a response of greater magnitude is seen in hypopituitary ovariectomized animals (16). Interestingly, although growth hormone can also increase uterine IGF-I levels, it appears to antagonize the estrogen response by mechanisms that have yet to be elucidated (17, 18). Repeated injections of estrogen result in a response in IGF-I mRNA as rapid as that seen after the initial injection, although there is a gradual reduction in the maximal response with subsequent injections and a gradual rise in the basal levels of IGF-I mRNA detected immediately prior to each injection (19). This may, in part, explain the relatively small variation in uterine IGF-I mRNA levels during the estrous cycle (19). In terms of total uterine RNA, the highest IGF-I mRNA levels are seen in late diestrus and proestrus. In the human reproductive tract, IGF-I transcripts are most abundant in human endometrium from the late proliferative stage of the menstrual cycle (20).

In situ hybridization localizes the IGF-I mRNA to the stromal and myometrial cells, with little expression in the luminal epithelium of the uterus (21). In the stroma, IGF-I mRNA appears to be particularly abundant in the basal, perimyometrial, or outer layer and in the periglandular stroma.

IGF-I receptors are present in uterine tissue and also appear to be under estrogen regulation. In both immature and ovariectomized rats, administration of estradiol increases the number of uterine IGF-I receptors, with no significant effect on receptor affinity (22). In addition to the IGF-I receptor, affinity cross-linking and ligand blotting experiments have demonstrated that at least four distinct binding proteins are present in uterine extracts and luminal fluid (15). Analysis of uterine RNA indicates that all of the six characterized IGFBPs are expressed to some extent in uterine tissue (23). Of these, IGFBP-1 and IGFBP-3 appear to be estrogen regulated. In the rodent, both of these binding proteins are downregulated by estrogen (23). The changes in IGFBP-1 levels during the estrous cycle are the inverse of those seen in uterine IGF-I receptors. The levels of IGFBP-1 mRNA are highest in diestrus and lowest in proestrus. There is evidence, predominantly from cell culture experiments (24,25) and more recently from transgenic animal experiments (26), to suggest that both IGFBP-1 and IGFBP-3 inhibit IGF-I action in uterine tissue. Thus, the mechanisms underlying estrogen-induced proliferation involve not only enhanced expression of growth factors but also attenuated expression of inhibitors of these growth factors.

In organ culture, IGF-I amplifies estrogen-induced DNA synthesis in rat uterine sections (21). In this system, IGF-I alone has little effect but enhances the estrogen response (21). Recently, we developed a transgenic mouse model in which IGFBP-1 is overexpressed under the phosphoglycerate kinase promoter. In these transgenic mice there is a high level of IGFBP-1 expression in the uterus, and the uterine response to estradiol is markedly attenuated. Both uterine DNA synthesis and the uterine weight gain in response to estradiol is reduced (ref. 26 and Fig. 13.1). Interestingly, in this transgenic mouse model, the effects of EGF on uterine DNA synthesis are also attenuated (Fig. 13.2).

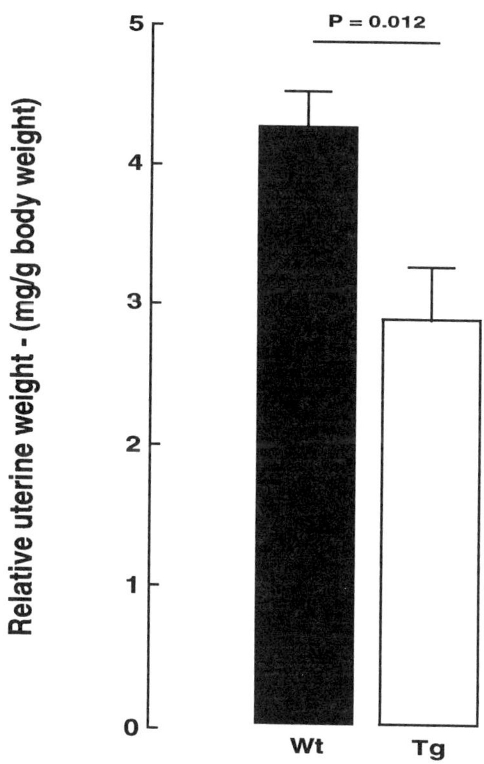

Fig. 13.1. Uterine weight gain in response to estradiol in wild-type (Wt) and transgenic (Tg) mice that overexpress IGFBP-1 in the uterus. Ovariectomized mice received three daily injections of 17β estradiol, 5 μg/100 g body weight. Uterine weight relative to body weight has been expressed as the mean ± SEM. (Modified, with permission, from data published in ref. 26).

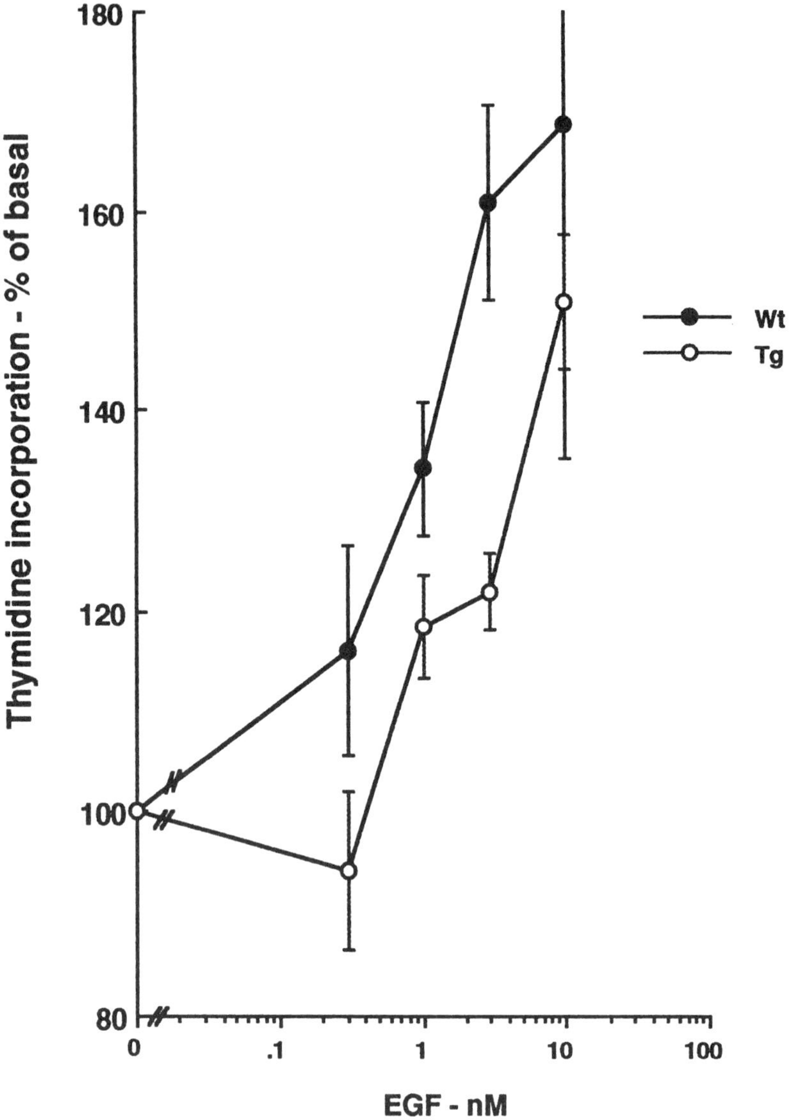

Fig. 13.2. The effect of EGF on uterine DNA synthesis in transgenic mice that overexpress IGFBP-1. Uterine tissue from wild-type (Wt) and transgenic (Tg) mice was cultured *in vitro* in the presence of various concentrations of EGF. The data represent the mean ± SEM. DNA synthesis has been expressed as a percentage of the basal DNA synthesis seen in control uterine tissue that was not significantly different between Tg and Wt mice. (Reproduced with permission from ref. 26).

13.5. INTERACTION OF INSULIN-LIKE GROWTH FACTOR-I AND EPIDERMAL GROWTH FACTOR IN THE UTERUS

Uterine expression of both EGF and the EGF receptor are enhanced by E2 (27,28). However, unlike uterine IGF-I expression, which is rapidly upregulated following estrogen administration to immature rodents, the increase in EGF expression is quite slow (26). It is possible to demonstrate uterine growth and morphologic changes in the uterus consistent with estrogenization by implantation of slow-release EGF pellets in ovariectomized mice (29). Furthermore, immunoneutralization of EGF is able to partially inhibit estrogen-induced uterine growth in mice (29). These latter studies indicate that EGF alone may be sufficient to mimic many if not all of the effects of estrogen in the uterus and provide compelling evidence that EGF functions as an estromedin in this tissue. Using a sensitive reverse transcription polymerase chain reaction assay and intraluminal administration of growth factors, we demonstrated the interdependence of EGF and IGF-I expression in the uterus (30).

As with estradiol, unilateral intraluminal administration of EGF to ovariectomized mice significantly increased the abundance of IGF-I mRNA compared to the contralateral saline-injected uterine horn (30). The antiestrogen ICI 164 384 partially blocked the EGF-induced increase in IGF-I mRNA (Fig. 13.3), consistent with previous reports that antiestrogen treatment inhibits EGF-induced uterine proliferation. Furthermore, intraluminal administration of sheep antibodies to EGF receptor significantly decreased the amount of IGF-I mRNA, indicating that endogenous EGF or an EGF-like ligand was involved in the maintenance of uterine IGF-I expression (30). Interestingly, intraluminal administration of IGF-I significantly increased the amount of EGF mRNA in the treated uterine horn, whereas it had no effect on IGF-I mRNA abundance. Similarly, intraluminal administration of EGF had no significant effect on uterine EGF mRNA levels. These data indicate that there is a complex interaction between these two growth factor systems and led us to propose that activation of IGF-I expression is a common downstream event in both estrogen- and EGF-induced uterine cellular proliferation. Further support for this hypothesis consists of the *in vitro* observations using a variety of cell lines in which EGF has been shown to enhance IGF-I secretion (31) or in which the mitogenic effect of EGF can be blocked by antisense oligonucleotides to the IGF-I receptor (32). In addition, as stated above, the effects of EGF on uterine DNA synthesis are reduced in transgenic mice that overexpress IGFBP-1.

The interaction between stromal and epithelial components in the uterus is important in the modulation of gene expression, in differentiation, and in the development of steroid hormone-responsive uterine epithelium (33). In addition, it may be important in estrogen-induced uterine proliferation. The distinct localization of expression of IGF-I and EGF in the uterus suggests that these growth factors may facilitate stromal–epithelial cell interaction. Paracrine actions of IGF-I, which is expressed predominantly in the stromal cells (21) on EGF expression in epithelial cells (27) and vice versa, may account for the more dramatic estrogen-induced proliferative response of uterine cells observed *in vivo* compared to *in vitro* dispersed cultures.

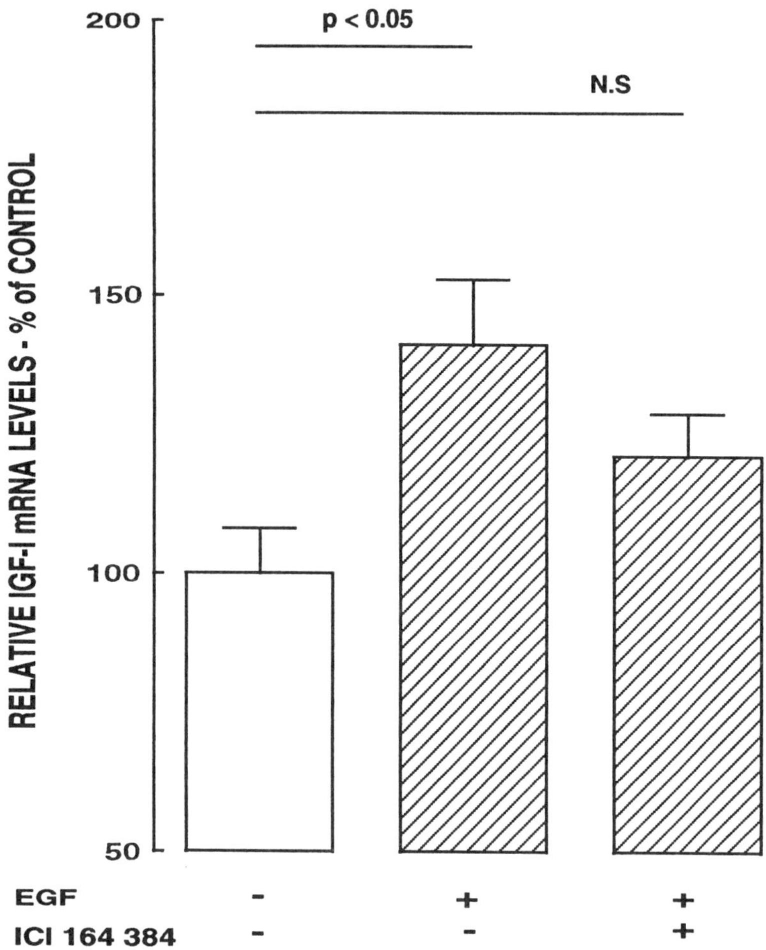

Fig. 13.3. The effects of intraluminal EFG on uterine IGF-I mRNA levels in ovariectomized mice. The data represent the mean ± SEM for seven animals per group killed 6 hr after intraluminal administration of EGF and have been expressed as a percentage of the mRNA level in the contralateral control uterine horns. One group of mice were pretreated with the antiestrogen ICI 164 384. (Modified from data published in ref. 30 and reproduced with permission).

13.6. CONCLUSIONS

Compelling data accumulated from experiments in a variety of species indicate that the IGF-I autocrine/paracrine system is important in ovarian folliculogenesis and in estrogen action in the uterus. The IGFs not only function as enhancers of steroid and peptide hormones that have trophic effects in the female reproductive tract but also are paracrine mediators of stromal–epithelial interaction. The development of transgenic mouse models in which various components of the IGF system have been overexpressed or deleted will provide a more detailed understanding of the role of this growth factor system in female reproduction.

REFERENCES

1. Salmon WD, Daughaday WH: A hormonally controlled serum factor which stimulates sulfate incorporation by cartilage in vitro. J Lab Clin Med 49:825–836, 1957.
2. Cohick WS, Clemmons DR: The insulin-like growth factors. Annu Rev Physiol 55:131–153, 1993.
3. Murphy LJ, Bell GI, Friesen HG: Tissue distribution of insulin-like growth factor I and II messenger ribonucleic acid in the adult rat. Endocrinology 120:1279–1282, 1987.
4. Rechler MM: Insulin-like growth factor binding proteins. Vit Horm 47:1–114, 1993.
5. Liu JP, Baker J, Perkins AS, Robertson EJ, Efstratiadis A: Mice carrying null mutations of the genes encoding insulin-like growth factor I (Igf-I) and type I IGF receptor. Cell 75:59072–59079, 1993.
6. Guidice LC, van Dessel HJ, Cataldo NA, Chandrasekher YA, Yap WS, Fauser BC: Circulating and ovarian IGF binding proteins: Potential roles in normo-ovulatory cycles and in polycystic ovarian syndrome. Prog Growth Factor Res 6:397–409, 1996.
7. Hernandez ER, Roberts CT, LeRoith D, Adashi EY: Rat ovarian insulin-like growth factor-I (IGF-I) gene expression is granulosa cell selective: 5′-untranslated mRNA variant representation and hormonal regulation. Endocrinology 89:572–574, 1989.
8. Adashi EY, Resnick CE, Hernandez ER, Hurwitz A, Roberts CT, LeRoith D, Rosenfeld R: Frontiers in human reproduction. Ann NY Acad Sci 626:161–168, 1991.
9. Cara JF, Rosenfeld RL: Insulin-like growth factor-I and insulin potentiates luteinizing hormone induced androgen synthesis by rat ovarian theca-interstitial cells. Endocrinology 122:1603–1612, 1988.
10. Guidice LC, Milki AA, Milkowski DA, Danasouri IE: Human granulosa cells contain messenger ribonucleic acids encoding insulin-like growth factor-binding proteins (IGF-BPs) and secrete IGFBPs in culture. Fertil Steril 56:475–480, 1991.
11. Mason HD, Willis D, Holly JMP, Cwyfan-Hughes SC, Seppala M, Franks S: Inhibitory effects of insulin-like growth factor-binding proteins on steroidogenesis by human granulosa cells in culture. Mol Cell Endocrinol 89:R1–R4, 1992.
12. Grimes RW, Guthrie HD, Hammond JM: Insulin-like growth factor-binding protein-2 and -3 are correlated with atresia and preovulatory maturation in the porcine ovary. Endocrinology 135:1996–2000, 1994.
13. Erickson GF, Nakatani A, Ling N, Shimasaki S: Cyclic changes in insulin-like growth factor binding protein-4 messenger ribonucleic acid in the rat ovary. Endocrinology 130:625–636, 1992.
14. Huang H, Rajkumar K, Murphy LJ: Reduced fecundity in insulin-like growth factor binding protein-1 transgenic mice. Biol Reprod 56:284–290, 1997.
15. Murphy LJ, Ghahary A: Uterine insulin-like growth factor-I: Regulation of expression and its role in estrogen-induced uterine proliferation. Endocrinol Rev 11:443–453, 1990.
16. Murphy LJ, Murphy LC, Friesen HG: A role for the insulin-like growth factors as estromedins in the rat uterus. Trans Assoc Am Physicians 99:204–214, 1988.

17. Murphy LJ, Murphy LC, Friesen HG: Estrogen induces insulin-like growth factor-I expression in the rat uterus. Mol Endocrinol 1:445–450, 1988.

18. Murphy LJ, Friesen HG: Differential effects of estrogen and growth hormone on uterine and hepatic insulin-like growth factor-I gene expression in the ovariectomized, hypophysectomized rat. Endocrinology 122:325–332, 1988.

19. Murphy LJ, Friesen HG: Regulation of insulin-like growth factor-I gene expression by estrogen. In LeRoith D, Raizada MK (eds): Molecular and Cellular Biology of Insulin-like Growth Factors and Their Receptor. New York: Plenum, 1989, pp 153–169.

20. Boehm KD, Daimon M, Gorodeski IG, Shelan LA, Utian WH, Ilan J: Expression of the insulin-like and platelet-derived growth factor genes in human uterine tissues. Mol Reprod Dev 27:93–97, 1990.

21. Ghahary A, Murphy LJ: In situ localization of the sites of synthesis and action of insulin-like growth factor-I in the rat uterus. Mol Endocrinol 4:191–195, 1990.

22. Ghahary A, Murphy LJ: Regulation of uterine insulin-like growth factor receptors by estrogen and variation throughout the estrous cycle. Endocrinology 125:597–604, 1989.

23. Molnar P, Murphy LJ: Estrogen effects on rat uterine insulin-like growth factor binding protein expression J Mol Endocrinol 13:59–67, 1994.

24. De Mellow JSM, Baxter RC: Growth hormone dependent insulin-like growth factor binding protein both inhibits and potentiates IGF-I stimulated DNA synthesis in human skin fibroblasts. Biochem Biophys Res Commun 156:199–204, 1988.

25. Elgin RG, Busby WH, Clemmons DL: An insulin-like growth factor (IGF) binding protein enhances the biological response to IGF-I. Proc Natl Acad Sci USA 84:3254–3259, 1987.

26. Rajkumar K, Dheen S, Krsek M, Murphy LJ: Impaired estrogen action in the uterus in insulin-like growth factor binding protein-1 transgenic mice. Endocrinology 137:1258–1264, 1996.

27. DiAugustine RP, Petrusz P, Bell GI, Brown CF, Korach KS, McLachlan JA, Tebg CT: Influence of estrogens on mouse uterine epidermal growth factor precursor protein and messenger ribonucleic acid. Endocrinology 122:2355–2363, 1988.

28. Mukku VR, Stancel GM: Regulation of epidermal growth factor receptor by estrogen. J Biol Chem 260:9820–9824, 1985.

29. Nelson KG, Takahashi T, Bossert NL, Walmer DK, McLachlan JA: Epidermal growth factor replaces estrogen in the stimulation of female genital-tract growth and differentiation. Proc Natl Acad Sci USA 88:21–25, 1991.

30. Hana V, Murphy LJ: Interdependence of epidermal growth factor and insulin like growth factor-I expression in the mouse uterus. Endocrinology 135:107–112, 1994.

31. Clemmons DR: Multiple hormones stimulate the production of somatomedin by cultured human fibroblasts. J Clin Endocrinol Metab 58:850–856, 1984.

32. Pietrzkowski Z, Sell C, Lammers R, Ullrich A, Baserga R: Roles of insulinlike growth factor I (IGF-I) and the IGF-I receptor in epidermal growth stimulated growth of 3T3 cells. Mol Cell Biol 12:3881–3889, 1992.

33. Cooke PS, Uchima FDA, Fujii DK, Bern HA, Cunha GR: Restoration of normal morphology and estrogen responsiveness in cultured vaginal and uterine epithelia transplanted with stroma. Proc Natl Acad Sci USA 83:2109–2113, 1986.

14

SIGNAL NETWORKS IN THE MAMMARY GLAND: LESSONS FROM ANIMAL MODELS

LOTHAR HENNIGHAUSEN

Laboratory of Biochemistry and Metabolism, National Institute of Diabetes, Digestive and Kidney Diseases, National Institutes of Health, Bethesda, MD 20892–1812

14.1. INTRODUCTION

The mammary gland is a recent acquisition on the phylogenetic scale of organ evolution. This gland is unique among the organs, as its development occurs predominantly after birth, in particular during pregnancy. During lactation this gland produces large amounts of milk, the sole source of nutrients for the newborn offspring. The cyclic development of the mammary gland reflects the fact that this organ is needed only during well-defined periods in the life of a mammal. Like the phoenix rising from the ashes, the secretory compartment arises from stem cells during each pregnancy, produces milk during lactation, and is fully remodeled after weaning of the young. This remodeling is accompanied by the loss of the entire secretory epithelium. Instead of designing new proteins, evolution picked old hands to guide and accomplish the extraordinary task of developing this highly specialized tissue. Gene deletion experiments in mice have demonstrated that development and differentiation of the mammary gland are critically dependent on individual genes whose absence in other organs has minor or no physiologic consequences. Receptors for estrogen, progesterone, and prolactin, the Stat5 (Stat: signal transducer and activator of transcription) transcription factors, cyclin D1, and the family of activins and inhibins are required to establish functional mammary tissue. The mammary gland

Hormones and Growth Factors in Development and Neoplasia, Edited by Robert B. Dickson and David S. Salomon
ISBN 0-471-16899-8

thus provides a unique opportunity to evaluate the role of specific proteins in the formation and function of a transient but highly specialized organ.

14.2. MAMMOPOIESIS AND LACTOGENESIS

Functional differentiation of the mammary gland is a crucial step in the reproductive cycle of mammals (1). The development of this organ proceeds in distinct phases. The rudimentary ductal system in newborn mice grows slowly until the onset of puberty. Pronounced ductal growth and branching occurs during puberty and early pregnancy. Development of the ducts continues in cycling virgins, leading to the formation of a ductal tree that fills the entire mammary fat pad. Terminal differentiation of alveolar epithelial cells is completed at the end of pregnancy, and milk secretion commences at parturition. Distinct steps of cellular differentiation are defined by the sequential activation of milk protein genes (2). Stricker and Grueter demonstrated in 1928 that pituitary extract injected into castrated virgin rabbits induced milk secretion (3)—a harbinger that prolactin (PRL) is essential for mammary function. However, the presence of lactogenic hormones is not sufficient to explain the complex development and temporal regulation of gene expression observed in the emerging mammary gland during pregnancy. Estrogen and progesterone are required for ductal outgrowth (4) and alveolar proliferation (5), respectively. In addition, growth regulators and cell cycle progression are obligatory for alveolar proliferation and differentiation.

In a meritorious effort, H. de Rothschild compiled in 1899 all publications on mammary gland development and milk—a total of 8375. In the preface to this publication, Duclaux states: "Such a discrepancy between the tremendous efforts and the paltriness of the results: hundreds of scientists and thousands of research years, just to create 200 or 300 pages of truth." Since then, thousands of researchers have added tens of thousands of publications. A milestone in the scientific community's efforts to understand the mechanisms of organogenesis has been reached with the generation of mice in which specific genes have been deleted. The signaling pathways that have been disrupted through the deletion of genes are outlined in Figure 14.1. Important lessons about the physiology of the estrogen, progesterone, and PRL pathways in distinct stages of mammary gland development and physiology have been learned.

14.3. PROLACTIN SIGNALING

In the cast of characters, PRL is a central player. Alveolar proliferation and the induction of terminal differentiation, as indicated by the transcriptional activation of milk protein genes, require the presence of PRL (1,6–8). Our appreciation that specific aspects of PRL signaling, in particular the JAK-STAT pathway, are critical for mammopoiesis was fostered by studies on the transcriptional regulation of milk protein genes.

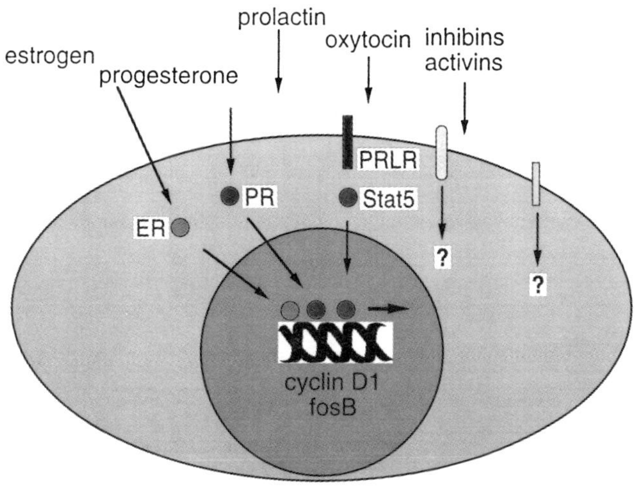

Fig. 14.1. Signaling molecules and pathways that are mandatory for mammary gland development. ER, estrogen receptor; PR, progesterone receptor; PRLP, prolactin receptor.

14.3.1. Prolactin and Its Receptor

It has now been over 60 years since Riddle, Bates, and Dykshorn purified a hormone from the anterior pituitary gland that stimulated milk secretion in the mammary gland of virgin rabbits (9). They named it *prolactin (PRL)*. Since then, the synergistic approaches of biochemistry, physiology, molecular biology, and cell biology have unveiled several molecular switches in the PRL signaling cascade (Fig. 14.2). Loss-of-function studies in the mouse have now provided clear insight into the biology of two components of the PRL pathway. A mandatory role for the prolactin receptor (PRLR) and the signal transducer and activator of transcription (Stat) 5 in mammopoiesis and lactogenesis was established (10,11). Although Stat5a and Stat5b are in the line of fire of many signals, such as PRL, growth hormone (GH), and several cytokines, their absence *in vivo* reveals an unexpected level of specificity.

PRL is a 23-kDa peptide that is synthesized mainly in lactotrophic cells of the anterior pituitary of vertebrates. Almost 100 effects have been attributed to this ancient hormone. PRL regulates gonadal functions (12) and behaviors such as nest building and pup retrieval (13), and it exerts multiple effects on the immune system (14). The best-characterized roles of PRL, however, are its ability to induce lobuloalveolar growth in the mammary gland (15) and to stimulate postpartum lactogenesis. It is known that these properties are mediated through the JAK-STAT pathway, which itself induces transcription of genes involved in growth control and differentiation (Fig. 14.2).

The cloning of the PRLR in 1988 (16) unleashed efforts to elucidate the cascade of molecular switches linking the receptor to target genes. The PRLR is a single-chain transmembrane protein that belongs to the cytokine superfamily and is expressed in a wide variety of tissues. Alternative splice products of the primary transcript yield PRLRs with short and long cytoplasmic tails. Both forms of the PRLR

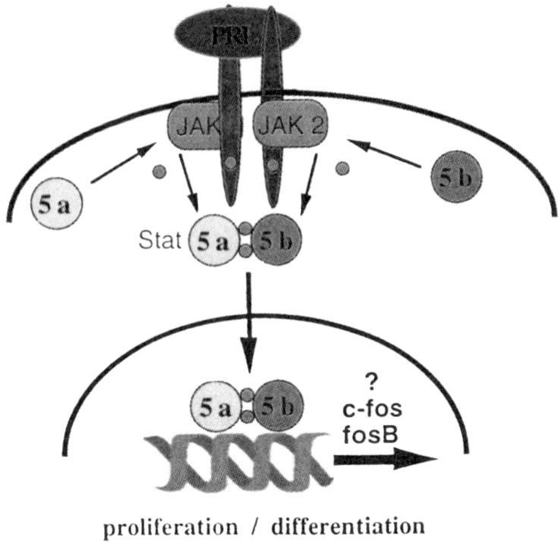

Fig. 14.2. Outline of our current understanding of the PRL signaling pathway in the mammary gland. PRL, prolactin; JAK, Janus kinase; STAT, signal transducer and activator of transcription.

can dimerize on ligand binding and activate the Janus kinase 2 (JAK2), Fyn, and the mitogen-activated protein (MAP) kinase to promote cell growth (17,18). However, only the long form can activate transcription of the β-casein promoter in transfected tissue culture cells (19). JAK2, in turn, phosphorylates Stat1, 3, and 5 on specific tyrosine residues (17,18). These activated Stat proteins bind to and induce transcription of genes containing γ-interferon activation sites (GAS) (TTCNNNGAA) (18).

14.3.2. Prolactin-Regulated Gene Expression

In a landmark experiment in 1965, Topper and colleagues demonstrated that synergistic signaling by insulin, hydrocortisone, and PRL is required to produce casein in mammary organ explants (20). Targeted efforts in the late 1970s and early 1980s led to the cloning of milk protein genes whose expression is controlled by PRL (21–24). Mammary-derived cell lines and studies in transgenic animals were instrumental in the identification, characterization, and biologic verification of PRL response elements in mammary gene promoters (6,25–29). By 1988 it was clear that promoter sequences from two mammary-specific genes, the whey acidic protein *(WAP)* and β-lactoglobulin *(BLG)* genes, contain sufficient genetic information to target transcription exclusively to mammary tissue and to respond to PRL signals (30,31). In the quest to identify PRL response elements, a sequence in the β-casein gene promoter was identified that was specifically recognized by a phosphorylated nuclear protein from mammary tissue. This protein was named *mammary gland factor (MGF)* (32). Transfection experiments of mutated β-casein gene promoter fragments into HC11 cells led to the demonstration that the MGF binding site (TTCNNNGAA) conveys PRL responsiveness (32). Experiments in transgenic mice demonstrated that MGF binding sites in the *WAP* (25) and *BLG* (28,29) gene promoters are critical for maximal gene activity and PRL response *in vivo*.

Our knowledge that the presence of Stat5 is essential for mammary gland development and differentiation was provided by mechanistic studies of the *BLG* and *WAP* genes. It is now clear that the primary function of Stat5 is not the transcriptional activation of milk protein genes but the induction of secretory cell differentiation itself. The ovine *BLG* gene promoter has been characterized in transgenic mice. Expression is induced during pregnancy, peaks during lactation, and declines at the onset of involution. In mammary gland explants and primary cultures, insulin, cortisol, and PRL are required to stimulate *BLG* expression. A 410-bp promoter fragment is sufficient to activate high-level expression of BLG in mammary tissue of transgenic mice (33). Deletion analyses demonstrated that the essential lactogenic hormone response elements reside between -150 and -410 in the sheep *BLG* promoter. The *BLG* promoter contains three binding sites for the transcription factor Stat5 (28,29). The role of individual transcription factors in regulating transcription was evaluated through their inactivation using site-directed mutagenesis. Abolishing binding to a single Stat5 site had no effect on the induction of *BLG* expression in response to the lactogenic hormones PRL, dexamethasone, and insulin (28). However, the disruption of two Stat5 sites abrogated induction, and mutation of all three elements completely abolished the hormone response (28). Clearly, the binding of a Stat factor, most likely Stat5, is dispensable for the expression of *BLG* in the mammary gland but mediates the hormonal response. Stat5a and Stat5b form heterodimers that have different affinities for the three *BLG* motifs (34). Since at least two Stat sites are required for maximal *BLG* expression, it is likely that proteins binding to these sites synergize with each other.

The *WAP* gene has also been studied extensively (for a review, see ref. 35), and several transcription factors are critical for temporal and hormonal gene control (25,36,37). These include a promoter proximal Ets binding site (37) and promoter distal Stat5 and NF1 sites (25). The Stat5 binding site by itself, however, is not sufficient for optimal gene activity; it may cooperate with neighboring transcription elements, such as NF1 (25).

14.3.3. The JAK-STAT Pathway

PRL, GH, epidermal growth factor (EGF), erythropoietin (EPO), and many cytokines use STAT, (signal transducer and activator of transcription) proteins to regulate the transcription of specific genes through the JAK-STAT pathway (18). JAK proteins contain a kinase domain that phosphorylates the respective receptors and Stat molecules. The ability of individual receptors (cytokines, GH, PRL) to activate overlapping but distinct sets of homo- and heterodimerizing STAT, (signal transduces and activator of transcription) proteins is thought to contribute to their signal specificity. For example, interferon α activates Stats 1, 2, and 3 and exerts a growth-retarding effect. PRL, in contrast, activates Stats 1, 3, and 5 in many cell lines. It induces transcription from the β-casein promoter in mammary epithelial cells and proliferation in Nb2 lymphoid cells. Stat5a and Stat5b are expressed in most if not all tissues, and they can be activated in tissue culture cells by PRL, GH, EGF, and many cyctokines (17). This suggests that Stat5 transcription factors are components of different signaling pathways leading to cell growth and differentiation.

Stat5 was originally identified as an MGF that binds to promoter sequences of milk protein genes and activates their transcription. Stat5-MGF was cloned in 1994 from sheep mammary tissue and recognized as a new member of the family of STAT proteins (38). In the mouse, Stat5 exists as two isoforms (5a and 5b) with 96% similarity (39). Early studies attempting to link Stat5 to terminal differentiation of mammary epithelial cells and the transcription of milk protein genes yielded inconclusive results. While the level of milk protein mRNAs in mammary tissue from immature virgins was at the threshold of detection, expression during pregnancy increased several thousandfold and remains high throughout lactation (6). In contrast, the level of Stat5a and Stat5b mRNA increased only slightly during pregnancy. The peak mRNA levels coincided with the onset of the milk protein gene expression, and a decrease in Stat5a and Stat5b mRNA was observed during lactation (39). Similarly, at the protein level, Stat5 showed only a slight increase during pregnancy and again a slight decrease after lactation. Most important, Stat5 is already present in mammary tissue from virgins when little milk protein gene transcription is detected (39).

Tyrosine phosphorylation of STATs is essential for their dimerization, translocation to the nucleus, binding to the GAS sites, and gene activation. Phosphorylation of Stat5 at Tyr 694 has been shown to be important for DNA binding and β-casein gene activation in cell culture studies. The presence of tyrosine phosphorylated Stat5 proteins may therefore provide a more physiologic indicator of the activity of these proteins and a mechanistic explanation for the detected DNA binding activities in the gel shift assay. Using isoform-specific antibodies against Stat5a or Stat5b, we have demonstrated that both isoforms are present in similar amounts at the protein level in mammary tissues of virgin, pregnant, lactating, and involuting mice (40). However, Stat5 phosphorylation is very low in immature virgins, rises sharply during late pregnancy, and declines rapidly during involution. On phosphorylation, Stat5a and Stat5b form homo- and heterodimers. The induction of Stat5 phosphorylation during late pregnancy correlates with the transcriptional activation of milk protein genes. The hypothesis that Stat5 is directly involved in mammary cell differentiation was tested in estrous cycle and transgenic mice with impaired mammary development. Transient differentiation of mammary alveolar cells and milk protein gene expression during estrus in virgin female mice coincide with transient Stat5 phosphorylation (40).

14.3.4. Genetic Disruptions of the Prolactin Pathway

Evidence that PRL signaling induces mammopoiesis and lactogenesis comes largely from studies with organ cultures and cell lines. Similarly, our knowledge of the roles of Stat5a and Stat5b in PRL, GH, and cytokine signaling comes mainly from tissue culture cells. Nonetheless, these observations fall short of uncovering the roles of PRL, Stat5a, and Stat5b in physiologic processes *in vivo*. The establishment of mice from which the genes encoding the PRLR (11), Stat5a (10), and

Stat5b (J. Ihle, personal communication) have been inactivated has now provided a clear insight into their biology. Since the PRLR is expressed in many organs of the developing fetus and since Stat5a and Stat5b are in the line of fire of multiple signaling pathways, late fetal or neonatal lethality could have been expected in the absence of these proteins. In contrast to this anticipation, mice deficient in the PRLR or Stat5a or Stat5b were born and survived until adulthood. Defects were confined to a few tissues and specific physiologic conditions.

Females with only one intact PRLR allele failed to lactate after their first pregnancy due to greatly reduced mammary development (11). Clearly, epithelial cell proliferation during pregnancy depends on a threshold of PRLR that cannot be obtained with just one functional allele. However, mammary gland development after the second pregnancy was sufficient for successful lactation, demonstrating that continued hormonal stimuli will eventually lead to functional development. It is currently not clear whether other growth pathways will compensate for the reduced number of PRLRs. In contrast, the progesterone (5) and estrogen (4) receptors, although essential for mammopoiesis, mediate functional development in the presence of only one functional allele.

The only noticeable phenotype of Stat5a-deficient mice (10) is their inability to lactate due to a failure of the gland to develop and differentiate during pregnancy (Fig. 14.3). The superimposable expression patterns of Stat5a and Stat5b during pregnancy and lactation (40), in conjunction with the 96% sequence similarity between the two isoforms (39), was a harbinger that Stat5b might be able to compensate for the loss of Stat5a. Unexpectedly, Stat5b protein levels, and even more pronounced the extent of phosphorylation, were greatly reduced in Stat5a-deficient

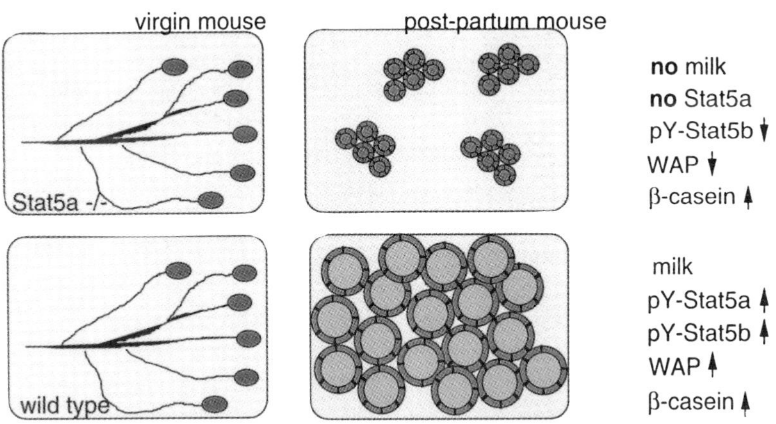

Fig. 14.3. An outline of the roles of Stat5a in mammary gland development and lactation. In the absence of Stat5a alveolar cell proliferation during pregnancy is impaired and functional differentiation at parturition does not occur. The amount of phosphorglated Stat5b and WAP is greatly reduced, but levels of β-casein are only slightly altered. No milk is produced in the absence of Stat5a.

mammary tissue (10). This demonstrates that efficient phosphorylation of Stat5b requires the presence of Stat5a and suggests that a docking of both molecules to JAK2 is critical for maximum activation. The importance of Stat5a-Stat5b heterodimerization for mammary cell differentiation had been suggested previously (40). There are two plausible explanations for the lack of compensation by Stat5b. First, Stat5b has a unique biochemical activity distinct from that of Stat5a. Alternatively, a certain threshold level of phosphorylated Stat5 (regardless of a or b) may be required for functional mammary development, similar to that seen in PRLR-deficient mice. Clearly, Stat5a and Stat5b may have unique nonoverlapping activities, as suggested by their different C termini (39), which encompass a transcriptional activator sequence (41). Whether Stat5a and Stat5b have unique functions will also be addressed by the derivation of Stat5b-deficient mice. Two groups have recently obtained such mice (44).The phenotype of the Stat5b-deficient mice is distinct from that of the Stat5a-deficient mice. These mice exhibit reduced growth of males and severely compromised fertility in females. Interestingly, the reproductive lesions in Stat5b-deficient mice are similar to those observed in PRLR-deficient mice. This phenotype would establish PRL signaling through Stat5b as a key regulator of mammalian reproduction (42). Although both the WAP and β-casein genes contain GAS sites, only WAP gene expression was reduced in Stat5a-deficient mice, suggesting that β-casein gene expression is controlled by other transcription factors.

It is intriguing to speculate why widespread and general signaling cascades involving transcription factors, cyclins, hormones, and growth factors are so critical for mammopoiesis and lactogenesis, while their requirements for the development of other tissues appear to be far less stringent. Since the mammary gland arrived late, if not last, in the phylogenetic scale of organ evolution, it may well be possible that alternative compensatory signaling pathways have not been developed in the mammary gland. There is experimental evidence to support such a view. Many cytokines, including EGF, GH, EPO, granulocyte-macrophage colony-stimulating factor (GM-CSF), and PRL, can activate Stat5a and Stat5b, as well as Stat1 and Stat3 (for recent references, see ref. 40). While PRL activates Stat5 and Stat3 to a similar extent in some lymphoid and breast cancer cells (43), Stat1 and Stat3 activities are barely detectable in the lactating mammary gland and do not correlate with PRL levels (40). Thus, PRL-mediated pathways activating Stat1 or Stat3 were perhaps never established in the mammary gland and therefore cannot compensate for the loss of Stat5a.

The past 60 years of PRL research were characterized by a synergy of biochemistry, genetics, and physiology that culminated in the identification of molecular switches in the PRL signaling pathway. Future research will focus on how these switches control the physiology of the mammary gland. Progress in understanding the cell- and ligand-specific effects of signaling molecules with a widespread distribution, such as the PRLR and the transcription factors Stat5a and Stat5b, will depend on our ability to further manipulate the genetic components of the mouse. And certainly I have to agree with Cato's statement made 2000 years ago: O sacculum, o literae iuvat vivere!

REFERENCES

1. Topper YJ, Freeman CS: Multiple hormone interactions in the developmental biology of the mammary gland. Physiol Rev 60:1049–1056, 1980.
2. Robinson GW, McKnight RA, Smith GH, Hennighausen L: Mammary epithelial cells undergo secretory differentiation in cycling virgins but require pregnancy for the establishment of terminal differentiation. Development 121:2079–2090, 1995.
3. Stricker P, Grueter R: Action du lobe anterieur de l'hypophyse sur la montee laiteuse. Compt Rend Soc Biol 99:1978–1980, 1928.
4. Korach KS: Insights from the study of animals lacking functional estrogen receptor. Science 266:1524–1527, 1994.
5. Lydon JP, DeMayo FJ, Funk CR, Mani SK, Hughes AR, Montgomery CA, Shyamala G, Conneely OM, O'Malley BW: Mice lacking progesterone receptor exhibit pleiotropic reproductive abnormalities. Genes Dev 9:2266–2278, 1995.
6. Burdon T, Sankaran L, Wall RJ, Spencer M, Hennighausen L: Expression of a whey acidic protein transgene during mammary development: Evidence for different mechanisms of regulation during pregnancy and lactation. J Biol Chem 266:6909–6914, 1991.
7. Pittius CW, Sankaran S, Topper Y, Hennighausen L: Comparison of the regulation of the whey acidic protein gene to a hybrid gene containing the whey acidic protein gene promoter in transgenic mice. Mol Endocrinol 2:1027–1032, 1988.
8. Vonderhaar BK: Prolactin: Transport, function, and receptors in mammary gland development and differentiation. In Neville MC, Daniel CW (eds): The Mammary Gland. New York: Plenum, 1987, pp 383–438.
9. Riddle O, Bates RW, Dykshorn SW: The preparation, identification and assay of prolactin—a hormone of the anterior pituitary. Am J Physiol 105:191–216, 1933.
10. Liu X, Robinson GW, Wagner K-U, Garrett L, Wynshaw-Boris A, Hennighausen L: Stat5a is mandatory for adult mammary gland development and lactogenesis. Genes Dev 11:179–186, 1997.
11. Ormandy CJ, Camus A, Barra J, Damotte D, Lucas B, Buteau H, Edery M, Brousse N, Babinet C, Binart N, Kelly PA: Null mutation of the prolactin receptor gene produces multiple reproductive defects in the mouse. Genes Dev 11(2):167–177, 1997.
12. Kelly PA, Djiane J, Postel-Vinay MC, Edery M: The prolactin/growth hormone receptor family. Endocrinol Rev 12:235–251, 1991.
13. Bridges RS: The role of lactogenic hormones in maternal behavior in female rats. Acta Paediatr Suppl 397:33–39, 1994.
14. Gala RR: Prolactin and growth hormone in the regulation of the immune system. Proc Soc Exp Biol Med 198:513–527, 1991.
15. Lyons WR: Hormonal synergism in mammary growth. Proc R Soc Lond 149:303–325, 1958.
16. Boutin JM, Jolicoeur C, Okamura H, Gagnon J, Edery M, Shirota M, Banville D, Dusanter-Fourt I, Djiane J, Kelly PA: Cloning and expression of the rat prolactin receptor, a member of the growth hormone/prolactin receptor gene family. Cell 53:69–77, 1988.
17. Ihle JN, Kerr IM: Jaks and Stats in signaling by the cytokine receptor superfamily. Trends Genet 11:69–74, 1995.
18. Schindler C, Darnell JE: Transcriptional responses to polypeptide ligands: the Jak-Stat pathway. Annu Rev Biochem 64:621–651, 1995.

19. Das R, Vonderhaar BK: Transduction of prolactin's (PRL) growth signal through both long and short forms of the PRL receptor. Mol Endocrinol 9:1750–1759, 1995.

20. Juergens WG, Stockdale FE, Topper YJ, Elias JJ: Hormone-dependent differentiation of mammary gland *in vitro*. Proc Natl Acad Sci USA 54:629–634, 1965.

21. Campbell SM, Rosen JM, Hennighausen L, Strech-Jurk U, Sippel AE: Comparison of the whey acidic protein genes of the rat and mouse. Nucleic Acids Res 12:8685–8697, 1984.

22. Hobbs AA, Richards DA, Kessler DJ, Rosen JM: Complex hormonal regulation of rat casein gene expression. J Biol Chem 3598–3605, 1982.

23. Jones WK, Yu-Lee L-Y, Clift SM, Brown LT, Rosen JM: The rat casein multigene family. J Biol Chem 260: 7042–7050, 1985.

24. Hennighausen LG, Sippel AE: Characterization and cloning of the mRNAs specific for the lactating mouse mammary gland. Eur J Biochem 125:131–141, 1982.

25. Li S, Rosen JM: Nuclear factor I and mammary gland factor (STAT5) play a critical role in regulating rat whey acidic protein gene expression in transgenic mice. Mol Cell Biol 15:2063–2070, 1995.

26. Li S, Rosen JM: Distal regulatory elements required for rat whey acidic protein gene expression in transgenic mice. J Biol Chem 269:14235–14243, 1994.

27. Li S, Rosen JM: Glucocorticoid regulation of rat whey acidic protein gene expression involves hormone-induced alterations of chromatin structure in the distal promoter region. Mol Endocrinol 8:1328–1335, 1994.

28. Burdon TG, Maitland KA, Clark AJ, Wallace R, Watson CJ: Regulation of the sheep beta-lactoglobulin gene by lactogenic hormones is mediated by a transcription factor that binds an interferon-gamma activation site-related element. Mol Endocrinol 8:1528–1536, 1994.

29. Burdon TG, Demmer J, Clark AJ, Watson CJ: The mammary factor MPBF is a prolactin-induced transcriptional regulator which binds to STAT factor recognition sites. FEBS Lett 350:177–182, 1994.

30. Andres A-C, Schönenberger C-A, Groner B, Hennighausen L, LeMeur M, Gerlinger P: Ha-ras oncogene expression directed by a milk protein gene promoter: Tissue specificity, hormonal regulation, and tumor induction in transgenic mice. Proc Natl Sci USA 84:1299–1303, 1987.

31. Pittius CW, Hennighausen L, Lee E, Westphal H, Nichols E, Vitale J, Gordon K: A milk protein gene promoter directs the expression of human tissue plasminogen activator cDNA to the mammary gland in transgenic mice. Proc Natl Acad Sci USA 85: 5874–5878, 1988.

32. Schmitt-Ney M, Doppler W, Ball RK, Groner B: Beta-casein gene promoter activity is regulated by the hormone mediated relief of transcriptional repression and a mammary gland specific factor. Mol Cell Biol 11:3745–3755, 1991.

33. Watson CJ, Gordon KE, Robertson M, Clark AJ: Interaction of DNA-binding proteins with a milk protein gene promoter in vitro: identification of a mammary gland-specific factor. Nucleic Acids Res 19:6603–6610, 1991.

34. Philp JAC, Burdon TG, Watson CJ: Differential activation of STATs 3 and 5 during mammary gland development. FEBS Lett 396:77–80, 1996.

35. McKnight RA, Burdon T, Pursel VG, Shamay A, Wall RJ, Hennighausen L: The whey acidic protein. In Dickson RB, Lippman ME (eds): Genes, Oncogenes, and Hormones. Boston: Kluwer Academic, 1992, pp. 399–412.

36. Lubon H, Hennighausen L: Nuclear proteins from lactating mammary glands bind to the promoter of a milk protein gene. Nucleic Acids Res 15:2103–2121, 1987.

37. McKnight RA, Spencer M, Dittmer J, Brady J, Wall RJ, Hennighausen L: An Ets site in the whey acidic protein gene promoter mediates transcriptional activation in the mammary gland of pregnant mice but is dispensable during lactation. Mol Endocrinol 9:717–724, 1995.

38. Wakao H, Gouilleux F, Groner B: Mammary gland factor (MGF) is a novel member of the cytokine regulated transcription factor gene family and confers the prolactin response. EMBO J 13:2182–2191, 1994.

39. Liu X, Robinson GW, Gouilleux F, Groner B, Hennighausen L: Cloning and expression of Stat5 and an additional homologue (Stat5b) involved in prolactin signal transduction in mouse mammary tissue. Proc Natl Acad Sci USA 92:8831–8835, 1995.

40. Liu X, Robinson GW, Hennighausen L: Activation of Stat5 and Stat5b by tyrosine phosphorylation is tightly linked to mammary gland differentiation. Mol Endocrinol 10:1496–1506, 1996.

41. Moriggl R, Gouilleux-Gruart V, Jaehne R, Berchtold S, Gartmann C, Liu X, Hennighausen L, Sotiropoulos A, Groner B, Gouilleux F: Deletion of the carboxyl-terminal transactivation domain of MGF-Stat5 results in sustained DNA binding and a dominant negative phenotype. Mol Cell Biol 16:5691–5700, 1996.

42. Russell DL, Norman RL, Dajee M, Liu X, Hennighausen L, Richards JS: Prolactin-induced activation and binding of Stat proteins to the IL-6RE of the α2-macroglobulin promoter: Relation to the expression of α2M in the rat ovary. Biol Reprod 55:1029–1038, 1996.

43. DaSilva L, Rui H, Erwin RA, Howard OMZ, Kirken RA, Malabarba MG, Hackett RH, Larner AC, Farrar WL: Prolactin recruits Stat1, Stat3 and Stat5 independent of conserved receptor tyrosines TYR402, TYR479, TYR515 and TYR580. Mol Cell Endocrinol 117:131–140, 1996.

44. Udy GB, Towers RP, Snell RG, Wilkins RJ, Park S-H, Ram PA, Waxman DJ, Davey HW: Requirement of Stat5b for sexual dimorphism of body growth rates and liver gene expression. Proc Natl Acad Sci USA 94:7239–7244, 1997.

PART IV

GROWTH FACTOR-HORMONAL INTERACTIONS IN TUMORIGENESIS AND MALIGNANT PROGRESSION

15

SEX STEROIDS AND CANCER

BRADLEY D. FENSTER, ELLY-GERALD STOICA, JIANMING LU, MARK DANIELSEN, AND MARY BETH MARTIN

Department of Biochemistry and Molecular Biology, Lombardi Cancer Center, Georgetown University, Washington, DC 20007

15.1. BREAST CANCER

It has been estimated that approximately 20% of all deaths in the United States are due to cancer. Cancer in hormone target tissues such as breast, uterus, and ovary in women and prostate in men accounts for about 8% and 5% of total mortality and about 30% and 11% of cancer mortality in women and men, respectively (1). Among women, breast cancer is the most common malignancy and accounts for approximately 27% of cancers (2). Although lung cancer has been the leading cause of cancer deaths among women for the past decade, the number of deaths due to breast cancer is still staggering. Statistics from the National Cancer Institute reveal that 18% of female cancer deaths are due to breast cancer, and breast cancer is the leading cause of death for women in the United States between the ages of 40 and 55. Annually, approximately 200,000 new cases of breast cancer are diagnosed and approximately 50,000 women die from the disease. The lifetime risk of developing breast cancer is one in nine for women in the United States, and 4% of patients will die from the disease.

15.1.1. Risk Factors for Breast Cancer

Family history is one risk factor for the development of breast cancer. Recently, two genes, BRCA1 (3) and BRCA2 (4), have been identified that influence

Hormones and Growth Factors in Development and Neoplasia, Edited by Robert B. Dickson and David S. Salomon
ISBN 0-471-16899-8

susceptibility to breast cancer. However, only 5% of cases of breast cancers are believed to be truly hereditary (3). Other factors that influence the risk of developing breast cancer include early menarche, nulliparity or late first pregnancy, and late menopause (5). All of these risk factors result in increased exposure of breast tissue to estrogens relative to the general population. However, there are no definitive studies showing an increased risk of breast cancer with the use of oral contraceptives or postmenopausal estrogen replacement therapy. Additionally, a high-fat diet and increased alcohol consumption have been suggested to increase the risk of developing breast cancer, although the data are not completely in agreement. No interventions, other than the drastic approach of bilateral prophylactic mastectomy, have been shown to decrease the breast cancer risk (2).

15.1.2. Biology of the Mammary Gland

The mammary gland consists of a tubuloalveolar structure similar to a tree that radiates from the nipple. It contains approximately 15 to 25 lobes subdivided into lobules. From each lobule, lactiferous ducts emerge. By the fifth week of gestation the mammary gland is formed, but it does not develop further until puberty. With the onset of puberty, there is significant growth of the gland. Estrogens are primarily responsible for stimulating ductal and stromal proliferation, whereas progestins stimulate alveolobular development. The role of other factors, such as growth factors, in the normal development of the gland is not well understood.

15.1.3. Role of the Estrogen Receptor in Breast Cancer

Estrogens are believed by many investigators to be the dominant etiologic factor in the development of human breast cancer (4). The prevailing theory of breast cancer development begins with estrogen-responsive cells that subsequently progress to estrogen independence (6). Many of the epidemiologic risk factors described above point to this model of progression. In addition, estrogens have been shown to promote breast tumors in rodents (7). Estrogens have a mitogenic effect on many established breast cancer cell lines, while antiestrogens are a well-established treatment regimen for breast cancer (8). Expression of many growth factors and their receptors has been shown to be induced by estrogens in breast cancer cell lines (9).

In nonmalignant breast epithelium, it has been estimated that approximately 7% of cells express the estrogen receptor at any given time (10), while estimates for breast cancer cells range from 40% to 70% (11). Thus, high estrogen receptor levels appear to be linked to neoplastic transformation. Following transformation, estrogen receptor expression by tumor cells is a positive prognostic factor for a more favorable course of the disease (12). Compared to estrogen receptor-negative tumors, estrogen receptor-positive tumors are more likely to be diploid and well differentiated. The size of the tumor is also related to estrogen receptor status. There is an inverse relationship between estrogen receptor status and size (13). Estrogen receptor-positive tumors are more common in postmenopausal women and are usually less aggressive than estrogen receptor-negative tumors (12). Estrogen receptor-negative tumors have been shown to express higher levels of growth factors and to have an increased rate of genetic alterations and oncogene expression (6).

Perhaps most important, the estrogen receptor status of a tumor is used to predict the response to endocrine therapy, one common component of breast cancer treatment. The most common agent used in endocrine therapy is the antiestrogen tamoxifen >. The response to tamoxifen requires a functional estrogen receptor signaling pathway. To test for a functional estrogen receptor pathway, the presence of both the estrogen receptor and the progesterone receptor is measured in breast tumors. Expression of the progesterone receptor in breast cancer is generally under the control of estrogens; the presence of the progesterone receptor is strongly suggestive of an intact estrogen receptor signal system (4). Clinically, 50–60% of estrogen receptor-positive tumors respond favorably to tamoxifen, whereas 80% of tumors that are positive for estrogen receptor and progesterone receptor respond to treatment. In contrast, fewer than 10% of receptor-negative tumors respond to endocrine therapy, as discussed below (14).

15.1.4. Hormone-Resistant Breast Cancer

Tamoxifen was first used in the treatment of breast cancer in 1978 (15). Reviews of numerous clinical trials and studies has led to the acceptance of adjuvant tamoxifen treatment for all stages of breast cancer, especially in postmenopausal women (2,16). Meta-analysis of 133 randomized trials showed that adjuvant tamoxifen treatment produces a significant reduction in both 5- and 10-year mortality rates (17). Tamoxifen has relatively few side effects, so it is generally better tolerated than conventional chemotherapy (8). Currently, clinical trials are underway to test the use of tamoxifen as a prophylactic agent in the prevention of breast cancer in high-risk patients (18). Although approximately 60% of patients with estrogen receptor tumors will respond to tamoxifen treatment, most of these initial responders will become hormone resistant. The majority of these women will relapse (19). Thus, tamoxifen treatment is generally considered to be palliative rather than curative (19). Tamoxifen resistance has also been demonstrated in breast cancer cells grown in culture (20).

The reasons for the discrepancy between the level of estrogen receptor measured in the tumor and its ability to predict the response to hormonal treatment remain unclear but may be attributed to laboratory error in the measurement of estrogen receptor and progesterone receptor or to heterogeneity of estrogen receptor content in the tumor (21,22). Although the majority of human tumors are believed to be clonal in origin, originating from a hypothetical stem cell (23), most tumors ultimately exhibit a significant degree of heterogeneity (21,22). The appearance of these phenotypically diverse subpopulations within the tumor probably represents subpopulations with different growth advantages. The factors responsible for generating tumor heterogeneity are varied and complex (23). In common with many other cancers, breast tumors are markedly heterogeneous. Within tumors that express high levels of estrogen receptor and progesterone receptor, there are areas that appear devoid of these receptors (21,22). The presence of subpopulations of cells that are estrogen receptor negative prior to endocrine treatment may be one reason for treatment failure. The selective pressures applied by hormonal treatment would be expected to remove the sensitive populations and facilitate the emergence of tumors composed of predominantly hormone resistant cells.

A considerable amount of research effort has been directed toward elucidating the mechanisms underlying the development of tamoxifen resistance. At least five possible mechanisms have been proposed (24,25). In the first mechanism, it has been suggested that loss of hormone responsiveness is due to the loss of estrogen receptor expression. This is the simplest model for antiestrogen resistance. If a cell no longer expresses the estrogen receptor, then it will no longer respond to compounds that block or alter estrogen receptor activity. In effect, these cells have bypassed the need for estrogen for growth. However, this explanation is unlikely for a large number of cases. Although loss of estrogen receptor expression after prolonged tamoxifen treatment has been reported (26), more than 50% of tamoxifen-resistant tumors continue to express the estrogen receptor (27). The second mechanism that has been proposed includes mutant or variant estrogen receptor proteins. A number of point mutants, alternately spliced forms, and truncated estrogen receptor variants have been identified in human breast tumors (28,29). Some of these variants may respond to tamoxifen differently than wild-type estrogen receptors, allowing tamoxifen to act as an agonist (30). In addition, some of these mutant receptors are constitutively active; for example, deletion of the hormone binding domain has been shown to result in a constitutively active estrogen receptor (31). Insertion of two amphipathic helices in the hinge regions of the *Xenopus laevis* estrogen receptor was shown to increase the affinity of the receptor for DNA and to increase the constitutive activity (32). The role of these estrogen receptor mutants in progression and resistance of breast cancer remains to be seen. A third possible mechanism of tamoxifen resistance involves changes in the pharmacology and metabolism of the antiestrogen. Possible alterations include deficiency in uptake or retention of the compound, as well as increased metabolism to by-products with increased estrogenic potency or decreased antiestrogenic potency (33). The fourth mechanism, activation of the estrogen receptor by alternative pathways, involves two basic mechanisms. Activation of other cellular signaling pathways has been shown to result in a potentiation of agonist activity and a blunting of antagonist activity by mixed steroid agonists/antagonists such as tamoxifen. In this model, tamoxifen would act as a stimulant to growth. Some of the same signaling pathways have also been shown to activate the estrogen receptor in a ligand-independent manner. Finally, the fifth mechanism involves defects downstream of estrogen receptor activation and is related to the first mechanism. Activation of the estrogen receptor has been shown to increase the expression of some growth factors and proto-oncogenes. Constitutive or alternative activation of these downstream factors would lead to an uncoupling of cell growth and division from the requirement for estrogens (34,35). Paracrine effects can also be included in this category. Alteration in growth factor production by surrounding tissue may be a possible cause of tamoxifen resistance (15). From the preceding discussion, it is clear that, the term *tamoxifen resistant* is often used to denote two separate types of tumors: tumors that no longer respond to the inhibitory effects of tamoxifen and tumors that may be stimulated by tamoxifen (20,36). The latter group of tumors would be expected to respond to withdrawal of the antiestrogen with decreased growth. This *antagonist withdrawal syndrome* has been observed for breast cancer (8) and for recurrent prostate cancer in patients treated with the antiandrogen flutamide (37).

15.1.5. Mechanisms of Action of Tamoxifen

Tamoxifen (brand name Nolvadex) is a nonsteroidal antiestrogen that is the most widely used estrogen antagonist in the treatment of breast cancer (19). However, tamoxifen has complex pharmacologic effects ranging from agonist to full antagonist activity (38). These effects have been shown to be tissue, cell, and gene specific. Thus, tamoxifen is generally considered to be a mixed agonist/antagonist, depending on the context (39); for example, in the endometrium, tamoxifen acts as an agonist. Tamoxifen and a large series of related compounds have been shown to bind to the hormone binding domain of the estrogen receptor (15). Binding of tamoxifen was shown to induce a different receptor conformation than binding of estradiol (40), and binding of tamoxifen to the estrogen receptor exposes a unique epitope (not exposed when bound to estradiol) that is recognized by the H222 antiestrogen receptor antibody (41). Studies have suggested that different regions of the estrogen receptor hormone binding domain are responsible for the binding of tamoxifen than for the binding of estradiol (42,43). Mutagenesis studies have identified the region surrounding cysteine 530 as a major determinant in the discrimination between estrogens and antiestrogens (44). Mutations within the region between amino acids 538 and 552 have been shown to decrease estradiol-dependent transactivation without affecting hormone or DNA binding. In these same mutants, tamoxifen was shown to be an estrogen agonist (45).

It has been shown that when tamoxifen binds to the estrogen receptor several steps in the activation of receptor occur, including nuclear binding and DNA binding (46,47). The main difference between the effects of estradiol and tamoxifen appears to lie in the activation of gene transcription. The model that has been proposed for the differential agonist and antagonist effects of tamoxifen suggests that the agonist activity results from the activation of the N-terminal transactivation function (TAF-1), while the antagonist activity results from inhibition of the C-terminal transactivation function (TAF-2), which is located in the hormone binding domain (48,49). Thus, for estrogen receptor effects that are dependent on TAF-1 activity, tamoxifen would be expected to act as an agonist. Numerous examples of mixed agonist/antagonist activity of tamoxifen exist (50–53). In the treatment of breast cancer the mixed activity of tamoxifen may be beneficial, as the antagonist activity is the predominant function in the inhibition of tumor growth, while the agonist activity is the predominant function in the maintenance of the cardiovascular system and bone (54). Although all the actions of tamoxifen were believed at first to be mediated by the estrogen receptor, there is considerable evidence for estrogen receptor-independent mechanisms of action of tamoxifen including inhibition of protein kinase C (55), inhibition of calmodulin-dependent cAMP phosphodiesterase (56), and membrane effects (57).

15.2. ENDOMETRIAL CANCER

Uterine cancer is the fourth most common malignancy among women in the United States (58). In 1996, it was estimated that 34,000 new cases of uterine cancer were diagnosed and approximately 6000 women died from the disease. It is the most frequent and also the most curable of the gynecologic cancers, perhaps due to the early diagnosis of the disease. Ninety-seven percent of all uterine cancers are endometrial carcinomas.

15.2.1. Biology of the Uterus and the Development of Endometrial Cancer

The lining of the cavity of the uterus, the endometrium, consists of a layer of columnar epithelial cells overlying a layer of stromal cells. During the proliferative phase of the menstrual cycle, estrogens stimulate proliferation of both the epithelial cells and the stromal cells to generate the functional layer of the uterus. During the secretory phase, progesterone inhibits mitosis of the estrogen-stimulated epithelial and stromal cells, stimulates the secretory activity of the epithelial cells, and increases the vasculature in the stromal layer. As the levels of estrogens and progestins diminish, necrosis of the blood vessels in the stromal layer occurs, followed by exfoliation of the epithelium.

There are generally two classes of endometrial cancers, the adenocarcinomas and the serous carcinomas (59). The adenocarcinomas are the most frequent malignancy in the uterus and develop from atypical endometrial hyperplasia. They are believed to develop in the presence of estrogen without cyclic exposure to progesterone, i.e., unopposed exposure to estrogen. In contrast to the adenocarcinomas, the serous carcinomas are endometrial tumors that develop in older women who have endometrial atrophy due to the absence of hormones. Serous carcinomas are very rare and comprise only a small proportion of endometrial cancers. They appear to develop from surface lesions and are very aggressive tumors. Serous carcinomas lack the risk factors associated with unopposed estrogen exposure.

15.2.2. Risk Factors for Endometrial Cancer

Many of the risk factors for endometrial cancer are associated with exposure to estrogen and are similar to the risk factors for breast cancer including early menarche, nulliparity, late menopause, and obesity (60,61). Early menarche and late menopause increase the length of exposure of the endometrium to estrogen. During pregnancy, progesterone concentrations increase and protect the endometrium from the continuous stimulation by estrogen. In obesity, androstenedione is converted to estrone by aromatase in adipose tissue, thereby increasing the exposure to estrogens. It is believed that estrogens have a promoting rather than an initiating role in the development of the disease. Tamoxifen therapy has also been shown to increase the risk of endometrial carcinomas (62). Although tamoxifen has antiestrogenic properties in the breast and is used in the treatment of breast cancer, it has estrogen-like activities in the endometrium.

15.2.3. Role of the Estrogen Receptor in Endometrial Cancer

The amounts of estrogen receptor and progesterone receptor appear to be important prognostic factors in endometrial cancer. Patients with receptor-positive disease have a significantly better disease-free survival than patients with receptor-negative disease (63). High amounts of estrogen receptor and progesterone receptor correlate with greater tumor differentiation, less myometrial invasion, and a lower incidence of lymph node metastases (64,65). The identification of several estrogen receptor variants, including alternately spliced forms of the estrogen receptor in human breast cancer, has prompted the investigation of the role of receptor variants in the initiation and/or progression of endometrial carcinomas. However, both normal and neoplastic uterine tissue contain several estrogen receptor splice variants, including variants that lack exons 4, 5, and 7 (66–68). The presence of these variants in both normal and neoplastic tissue does not appear to support a role for these variants in uterine tumor formation at the present time.

15.2.4. Role of Tamoxifen in the Development of Endometrial Cancer

Tamoxifen is a mixed-function agonist/antagonist, depending on the target tissue. It functions as an antiestrogen in the breast and is used in the adjuvant treatment of breast cancer. In the cardiovascular system, bone, the endometrium, and possibly the liver, tamoxifen acts as an estrogen. The principal effect of tamoxifen in the uterus is an increased incidence of proliferative abnormalities of the endometrium ranging from simple hyperplasia to invasive malignancies (69–71). In post-menopausal breast cancer patients, there is evidence that treatment with tamoxifen gives rise to an increase in endocervical and/or endometrial polyps. However, in premenopausal breast cancer patients, there is no increase in polyps associated with tamoxifen therapy (72). It is estimated that tamoxifen treatment only doubles the risk of developing endometrial cancer in breast cancer patients (62). The increased risk of endometrial cancer needs to be weighed against the 40% decreased incidence of contralateral breast cancer in women treated with tamoxifen.

15.3. PROSTATE CANCER

In the United States, cancer of the prostate is the most common malignancy among men. The estimated probability of developing a clinically apparent prostatic carcinoma by 50 years of age is 9.5% and 11.4% for white and black American men, respectively. Before the age of 50, prostate cancer is rare. However, the incidence of prostatic carcinomas increases with increasing age. Prostate cancer is curable when diagnosed early, but due to the great difficulties in early detection, the mortality rate from prostate cancer is high. In addition to digital rectal examination and transrectal ultrasonography, the level of prostate specific antigen (PSA) in serum is used to detect early prostate cancer. PSA is a serine protease that is synthesized in the ductal and acinar epithelium of the prostate gland (73,74). The expression of PSA is regulated by androgenic steroids, growth factors, and extracellular matrix. In an androgen-independent prostate cancer cell line, PSA is expressed at high levels (75).

15.3.1. Risk Factors for Prostate Cancer

The causes of prostate cancer have not been determined; however, age, a family history of prostate cancer, and exposure to androgens appear to be important risk factors (76,77). Androgens play a key role in the structure and function of the prostate gland, leading to the hypothesis that the effects of the hormone are an important component in the development of the disease. Although it has been suggested that androgens are involved in the etiology of the disease, most studies have been difficult to interpret. A recent prospective study demonstrates that high levels of circulating testosterone and low levels of serum hormone binding globulin are associated with an increased risk of prostate cancer (78). This study also suggests that low levels of circulating estradiol may be an additional risk factor for prostate cancer. A number of environmental factors have also been proposed to affect the risk of prostate cancer including occupational exposures to chemicals, dietary fat intake, alcohol consumption, and cigarette smoking (79–81). Studies of these potential risk factors have been inconsistent.

15.3.2. Biology of the Prostate Gland and Development of Prostate Cancer

The prostate is one of the major accessory glands of the male reproductive system. It has a branching structure that consists of tubular or sack-like glands that secrete prostatic fluid into the urethra and ejaculatory ducts. The ducts of the individual glands are lined with epithelial cells. These cells consist of columnar cells, which synthesize the prostatic secretions, and basal cells, which are the progenitor cells of the columnar cells. Surrounding the secretory portions of the gland is connective tissue composed of basal and stromal cells, a thick network of elastic fibers, and blood vessels. The growth, differentiation, and function of the prostate are dependent on testicular androgens. Both epithelial cells and stromal cells contain androgen receptor and are responsive to androgens. Starting at puberty, the prostate increases in weight until it reaches its maximum size at the age of 20. Between the ages of 40 and 60, there is atrophy of the prostatic smooth muscle and an increase in the fibrous tissue due in part to the decrease in circulating androgen levels. After the age of 60, the organ is largely replaced by fibrous tissue. The dependence of the prostate gland on androgens is demonstrated by the involution of the gland following the reduction in the level of androgen.

Prostate cancer arises from the epithelium lining the prostatic acini and small ducts. The majority of prostatic cancers are adenocarcinomas that tend to infiltrate the stroma and eventually invade the capsule and metastasize to the rectum, the base of the bladder, and more distant sites by dissemination via lymphatic or blood vessels. Bone is a common site of prostate metastases.

15.3.3. Role of the Androgen Receptor in Prostate Cancer

Prostate cancer is rarely observed in men who have undergone orchiectomy, and it regresses following castration. Since testicular hormones regulate the growth and development of the prostate and stimulate the growth of prostate cancer, one method for the treatment of prostate cancer is hormonal manipulation (82). In the prostate, the most potent androgen is dihydrotestosterone formed from the less active precursors, testosterone from the testis, and dehydroepiandrosterone from the adrenal. Testosterone is the principal circulating androgen and is converted to dihydrotestosterone by 5-alpha-reductase. In the case of dehydroepiandrosterone, conversion to dihydrotestosterone involves 3-beta-hydroxysteroid dehydrogenase and 17-beta-hydroxysteroid dehydrogenase, as well as 5-alpha-reductase. The genes for 3-beta-hydroxysteroid dehydrogenase and 17-beta-hydroxysteroid dehydrogenase are expressed in the prostate and may explain the failure of castration in the treatment of prostate cancer since dehydroepiandrosterone can be readily converted to dihydrotestosterone in the gland. Following castration, the level of dihydrotestosterone is approximately 40% of control levels. Alternatively, antiandrogen therapy may be employed. At the present time, flutamide is the only nonsteroidal antiandrogen approved for the treatment of prostate cancer in the United States. It is generally well tolerated in prostate cancer patients. Although the rate of response to antiandrogen treatment is high, many patients relapse. Clinical studies have shown that antiandrogen withdrawal is therapeutically effective for approximately 20% of hormone-refractory prostate cancer patients (83). This observation has led to the suggestion that androgen independence does not necessarily mean that a tumor is resistant to further hormonal manipulations (see below) (84).

15.3.4. Hormone-Resistant Prostate Cancer

The mechanism by which the originally androgen-dependent prostate cancer cells no longer require androgen for their growth are largely unknown. However, much attention has been focused on the androgen receptor. Attempts to correlate the amount of androgen receptor with prostate disease have been unsuccessful (85). In the case of prostate cancer, loss of androgen responsiveness is not associated with loss of androgen receptor expression. Metastases of prostate cancer usually retain the expression of the androgen receptor. Indeed, in many cases, the amount of androgen receptor is higher than in the primary tumor (86), which may result in hypersensitivity to hormone. In approximately 30% of recurrent tumors the androgen receptor gene is amplified, a possible response to androgen deprivation. Mutations in the androgen receptor gene are rare in prostate cancer, and most reported mutations are found in the hormone binding domain of the receptor. Some of these mutations show altered responsiveness to estrogens, progestins, and antiandrogens, suggesting a role for these mutations in the resistance to androgen therapy. The first androgen receptor mutation found in prostate cancer was from a patient with stage B disease. This mutation results in a valine to methionine change at codon 730 (87). This mutation was a somatic mutation since the DNA from the patient's peripheral blood lymphocytes contained a normal androgen receptor gene. A second mutation at codon 877 results in the substitution of alanine for threonine and loss of hormone specificity (88–90). It is also activated by flutamide and appears to occur frequently in metastatic prostate cancer (91,92). A possible association between the length of the CAG (glutamine) and GGC (glycine) microsatellites in exon 1 of the androgen receptor has also been implicated in prostate cancer. In one prostate cancer patient, the glutamine repeat number decreased from 24 to 18 (93). Other studies also show that shorter glutamine repeats in the androgen receptor are associated with a higher incidence of prostate cancer (94). Shorter glutamine repeats increase the transcriptional activity of the androgen receptor. African Americans have, on average, shorter glutamine repeats, which may be one explanation for the increased incidence of prostatic cancer in this racial group.

REFERENCES

1. Rao GN: Influence of diet on tumors of hormonal tissue. Prog Clin Biol Res 394:41–56, 1996.
2. Harris JR, Morrow M, Bonadonna G: Cancer of the breast. In DeVita VT Jr, Hellman S, Rosenberg SA (eds): Cancer: Principles and Practice of Oncology. Philadelphia: JB Lippincott, 1993, pp 1264–1332.
3. Miki Y, Swensen J, Shattuck-Eidens D, Futreal PA, Harshman K, Tavtigian S, Liu Q, Cochran C, Bennett LM, Ding W, et al: A strong candidate for the breast and ovarian cancer susceptibility gene BRCA1. Science 266:66–71, 1994.
4. King RJ, William L: McGuire Memorial Symposium. Estrogen and progestin effects in human breast carcinogenesis. Breast Cancer Res Treat 27:3–15, 1993.
5. Lippman ME: Endocrine responsive cancers of man. In Williams RH (ed): Textbook of Endocrinology. Philadelphia: WB Saunders, 1985, pp 1309–1326.
6. Dickson RB, Johnson MD, Bano M, Shi E, Kurebayashi J, Ziff B, Martinez-Lacaci I, Amundadottir LT, Lippman ME: Growth factors in breast cancer: Mitogenesis to transformation. J Steroid Biochem Mol Biol 43:69–78, 1992.
7. Henderson BE, Ross R, Bernstein L: Estrogens as a cause of human cancer: The Richard and Hinda Rosenthal Foundation award lecture. Cancer Res 48:246–253, 1988.
8. Santen RJ, Manni A, Harvey H, Redmond C: Endocrine treatment of breast cancer in women. Endocrinol Rev 11:221–265, 1990.
9. Dickson RB, Lippman ME: Growth factors in breast cancer. Endocrinol Rev 16:559–589, 1995.
10. Peterson OW, Hoyer PE, van Deurs B: Frequency and distribution of estrogen receptor positive cells in normal, nonlactating human breast tissue. Cancer Res 47:5748–5751, 1987.
11. Petrangeli E, Lubrano C, Ortolani F, Ravenna L, Vacca A, Sciacchitano S, Frati L, Gulino A: Estrogen receptors: New perspectives in breast cancer management. J Steroid Biochem Mol Biol 49:327–331, 1994.
12. Sheikh MS, Garcia M, Pujol P, Fontana JA, Rochefort H: Why are estrogen receptor-negative breast cancers more aggressive than the estrogen receptor-positive breast cancers? Invasion Metastasis 14:329–336, 1994.
13. Thorpe SM, Rose C, Pedersen BV, Rasmussen BB: Estrogen and progesterone receptor profile in primary breast cancer. Breast Cancer Res Treat 3:103–110, 1983.

14. Osborne CK, Yochmowitz MG, Knight WA III, McGuire WL: The value of estrogen and progesterone receptors in the treatment of breast cancer. Cancer 46:2884–2888, 1980.

15. Jordan VC, Murphy CS: Endocrine pharmacology of antiestrogens as antitumor agents. Endocrinol Rev 11:578–610, 1990.

16. Early Breast Cancer Trialists' Collaborative Group: Effects of adjuvant tamoxifen and of cytotoxic therapy on mortality in early breast cancer. An overview of 61 randomized trials among 28,896 women. N Engl J Med 319:1681–1692, 1988.

17. Early Breast Cancer Trialists' Collaborative Group: Systemic treatment of early breast cancer by hormonal, cytotoxic, or immune therapy. 133 randomized trials involving 31,000 recurrences and 24,000 deaths among 75,000 women. Lancet 339:71–85, 1992.

18. Henderson BE, Ross RK, Pike MC: Hormonal chemoprevention of cancer in women. Science 259:633–638, 1993.

19. Wakeling AE: The future of new pure antiestrogens in clinical breast cancer. Breast Cancer Res Treat 25:1–9, 1993.

20. Clarke R, Thompson EW, Leonessa F, Lippman J, McGravey M, Frandsen TL, Brunner N: Hormone resistance, invasiveness, and metastatic potential in breast cancer. Breast Cancer Res Treat 24:227–239, 1993.

21. Dexter DL, Calabresi P: Intraneoplastic diversity. Biochim Biophys Acta 695:97–112, 1982.

22. VanNetten DA, Greene GL, Lippman ME: Estrogen receptor distribution in the peripheral, intermediate, and central regions of breast cancers. Eur J Cancer Clin Oncol 24:1885–1889, 1988.

23. Nowell PC: The clonal evolution of tumor cell populations. Science 194:23–28, 1976.

24. Katzenellenbogen BS: Antiestrogen resistance: Mechanisms by which breast cancer cells undermine the effectiveness of endocrine therapy. J Natl Cancer Inst 83: 1434 1435, 1991.

25. Horwitz KB: When tamoxifen turns bad. Endocrinology 136:821–823, 1995.

26. Melchor JC, Rodriquez-Eschdero FJ, Lujan S, Corcostegui B: Variation of estrogen and progesterone receptor status in breast cancer after tamoxifen therapy. Oncology 47:467–470, 1990.

27. Encarnacion CA, Ciocca DR, McGuire WL, Clark GM, Fuqua SA, Osborne CK: Measurement of steroid hormone receptors in breast cancer patients on tamoxifen. Breast Cancer Res Treat 26:237–246, 1993.

28. Fuqua S, Chamness GC, McGuire WL: Estrogen receptor mutations in breast cancer. J Cell Biochem 51:135–139, 1993.

29. Dotzlaw H, Alkhalaf M, Murphy LC: Characterization of estrogen receptor variant mRNAs from human breast cancers. Mol Endocrinol 6:773–785, 1992.

30. McDonnell DP, Dana SL, Hoener PA, Lieberman BA, Imhof MO, Stein RB: Cellular mechanisms which distinguish between hormone- and antihormone-activated estrogen receptor. Ann NY Acad Sci 761:121–137, 1995.

31. Chambraud B, Berry M, Redeuilh G, Chambon P, Baulieu EE: Several regions of human estrogen receptor are involved in the formation of receptor-heat shock protein 90 complexes. J Biol Chem 265:20686–20691, 1990.

32. Xing H, Shapiro DJ: An estrogen receptor mutant exhibiting hormone-independent transactivation and enhanced affinity for the estrogen receptor element. J Biol Chem 268:23227–23233, 1993.

33. Osborne CK, Coronado E, Allred DC, Wiebe V, DeGregorio M: Acquired tamoxifen resistance: Correlation with reduced breast tumor levels of tamoxifen and isomerization of trans-4-hydroxytamoxifen. J Natl Cancer Inst 83:1477–1482, 1991.
34. Koga M, Musgrove EA, Sutherland RL: Differential effects of phorbol ester on epidermal growth factor receptors in estrogen receptor-positive and -negative breast cancer cell lines. Cancer Res 50:4849–4855, 1990.
35. Koga M, Musgrove EA, Sutherland RL: Modulation of the growth-inhibitory effects of progestins and the antiestrogen hydroxyclomiphene on human breast cancer cells by epidermal growth factor and insulin. Cancer Res 49:112–116, 1989.
36. Gottardis MM, Jordan VC: Development of tamoxifen-stimulated growth of MCF-7 tumors in athymic mice after long-term antiestrogen administration. Cancer Res 48:5183–5187, 1988.
37. Scher HI, Kelly WK: Flutamide withdrawal syndrome: Its impact on clinical trials in hormone-refractory prostate cancer. J Clin Oncol 11:1566–1572, 1993.
38. Nardulli AM, Katzenellenbogen BS: Progesterone receptor regulation in T47D human breast cancer cells: Analysis by density labeling of progesterone receptor synthesis and degradation and their modulation by progestin. Endocrinology 122:1532–1540, 1988.
39. Parker MG, Arbuckle N, Dauvois S, Danielian P, White R: Structure and function of the estrogen receptor. Ann NY Acad Sci 684:119–126, 1993.
40. Garcia M, Greene G, Rochefort H, Jensen EV: Effect of antibodies to estrogen receptor on the binding of 3H-labeled antiestrogens and androstanediol in the uterus. Endocrinology 110:1355–1361, 1982.
41. Martin PM, Bertois Y, Jensen EV: Binding of antiestrogens exposes an occult antigenic determinant in the human estrogen receptor. Proc Natl Acad Sci USA 85:2533–2537, 1988.
42. Wrenn CK, Katzenellenbogen BS: Structure-function analysis of the hormone binding domain of the human estrogen receptor by region-specific mutagenesis and phenotypic screening in yeast. J Biol Chem 268:24089–24098, 1993.
43. Danielian PS, White R, Hoare SA, Fawell SE, Parker MG: Identification of residues in the estrogen receptor that confer differential sensitivity to estrogen and hydroxytamoxifen. Mol Endocrinol 7:232–240, 1993.
44. Katzenellenbogen BS, Fang H, Ince BA, Pakdel F, Reese JC, Wooge CH, Wrenn CK: William L. McGuire Memorial Symposium. Estrogen receptors: ligand discrimination and antiestrogen action. Breast Cancer Res Treat 27:17–26, 1993.
45. Mahfoudi A, Roulet E, Dauvois S, Parker MG, Wahli W: Specific mutations in the estrogen receptor change the properties of antiestrogens to full agonists. Proc Natl Acad Sci USA 92:4206–4210, 1995.
46. Reese JC, Katzenellenbogen BS: Examination of the DNA-binding ability of estrogen receptor in whole cells: Implications for hormone-independent transactivation and the actions of antiestrogens. Mol Cell Biol 12:4531–4538, 1992.
47. Pham TA, Elliston JF, Nawaz Z, McDonnel DP, Tasi M, O'Malley BW: Antiestrogen can establish nonproductive receptor complexes and alter chromatin structure at target enhancers. Proc Natl Acad Sci USA 88:3125–3129, 1991.
48. Berry M, Metzger D, Chambon P: Role of the two activating domains of the oestrogen receptor in the cell-type and promoter-context dependent agonistic activity of the anti-oestrogen 4-hydroxytamoxifen. EMBO J 9:2811–2818, 1990.

49. Tzuderman MT, Esty A, Snatiso-Mere D, Danielian P, Parker MG, Stein RB, Pike JW, McDonnell DP: Human estrogen receptor transactivational capacity is determined by both cellular and promoter context and mediated by two functionally distinct intramolecular regions. Mol Endocrinol 8:21–30, 1994.

50. Maudelonde T, Escot C, Pujol P, Rouanet P, Defrenne A, Brouillet JP, Rochefort H: In vivo stimulation by tamoxifen of cathepsin D RNA level in breast cancer. Eur J Cancer 30A:2049–2053, 1994.

51. Ramkumar T, Adler S: Differential positive and negative transcriptional regulation by tamoxifen. Endocrinology 136:536–542, 1995.

52. Shull JD, Beams FE, Baldwin TM, Gilchrist CA, Hurbek MJ: The estrogenic and antiestrogenic properties of tamoxifen in GH4C1 pituitary tumor cells are gene specific. Mol Endocrinol 6:529–535, 1992.

53. Webb P, Lopez GN, Uht RM, Kushner PJ: Tamoxifen activation of the estrogen receptor/AP-1 pathway: Potential origin for the cell-specific estrogen-like effects of antiestrogens. Mol Endocrinol 9:443–456, 1995.

54. Love RR, Mazess RB, Barden HS, Epstein S, Newcomb PA, Jordan VC, Carbone PP, De Mets DL: Effects of tamoxifen on bone mineral density in postmenopausal women with breast cancer. N Engl J Med 326:852–856, 1992.

55. Antila E, Mussalo-Rauhamaa H, Kantola M, Atroshi F, Westermarck T: Association of cadmium with human breast cancer. Sci Total Environ (Netherlands) 186:251–256, 1996.

56. Fanidi A, Pageaux JF, Courion C, Fayard JM, Laugier C: Growth inhibition and the regulation of cyclic AMP by the triphenylethylene anti-estrogen tamoxifen in the quail oviduct. Biol Cell 72:181–186, 1991.

57. Wiseman H: Tamoxifen: New membrane-mediated mechanisms of action and therapeutic advances. Trends Pharmacol Sci 15:83–89, 1994.

58. Rose PG: Endometrial carcinoma. N Engl J Med 335:640–649, 1996.

59. Platz CE, Benda JA: Female genital tract cancer. Cancer 75:270–294, 1995.

60. Barker VW: Potential criteria for cohort selection in chemoprevention trials of uterine adenocarcinoma. J Cell Biochem 23:184–188, 1995.

61. Poteschman N, Hoover RN, Brinton LA, Siiteri P, Dorgan JF, Swanson CA, Berman ML, Mortel R, Twiggs LB, Barrett RJ, Wilbanks GD, Persky V, Lurain JR: Case-control study of endogenous steroid hormones and endometrial cancer. J Natl Cancer Inst 88:1127–1135, 1996.

62. Jordan VC: Third Annual William L. McGuire Memorial Lecture. "Studies on the estrogen receptor in breast cancer"—20 years as a target for the treatment and prevention of cancer. Breast Cancer Res Treat 36:267–285, 1995.

63. Creasman WT: Prognostic significance of hormone receptors in endometrial cancer. Cancer 71:1467–1470, 1993.

64. Zaino RJ, Satyaswaroop PG, Mortel R: The relationship of histologic and histochemical parameters to progesterone receptor status in endometrial adenocarcinoma. Gynecol Oncol 16:196–208, 1983.

65. Creasman WT: Prognostic significance of hormone receptors in endometrial cancer. Cancer 71:467–470, 1993.

66. Daffada AA, Dowsett M: Tissue-dependent expression of a novel splice variant of the human estrogen receptor. J Steroid Biochem Mol Biol 55:413–421, 1995.

67. Hirata S, Yamada-Mouri N, Nara M, Takizawa M, Ito H, Kato J: Presence of alternatively spliced-estrogen receptor mRNA variants in normal uterine endometrium and endometrial cancer. Endocrinol J 42:289–293, 1995.

68. Hu C, Hyder SM, Needleman DS, Baker VV: Expression of estrogen receptor variants in normal and neoplastic uterus. Mol Cell Endocrinol 118:173–179, 1996.

69. Ross D, Whitehead M: Hormonal manipulation and gynaecological cancer: The tamoxifen dilemma. Curr Opin Obstet Gynecol 7:63–68, 1995.

70. Ismail SM: The effects of tamoxifen on the uterus. Curr Opin Obstet Gynecol 8:27–31, 1996.

71. Crabbe WW: The tamoxifen controversy. Oncol Nurs Forum 23:761–766, 1996.

72. McGonigle KF, Lantry SA, Odom-Maryon TL, Vasilev SA, Simpson JF: Histopathologic effects of tamoxifen on the uterine epithelium of breast cancer patients—analysis by menopausal status. Cancer Lett 101:59–66, 1996.

73. Papsidero L, Kuriyama M, Wang ML, Horoszewicz JS, Leong SS, Valenzuela L, Murphy GP, Chu TM: Prostate antigen: A marker for human prostate epithelial cells. J Natl Cancer Inst 66:37–42, 1981.

74. Watt KWK, Lee P, M'Timkulu T, Chan W, Loor R: Human prostate-specific antigen: Structural and functional similarity with serine proteases. Proc Natl Acad Sci USA 83:3166–3170, 1986.

75. Hsieh JT, Wu HC, Martin EG, von Eschenbach AC, Chung LWK: Autocrine regulation of prostate-specific antigen gene expression in human prostatic cancer (LNCaP) subline. Cancer Res 53:2852–2857, 1993.

76. Wynder EL, Mabuchi K, Whitemore WJ: Epidemiology of cancer of the prostate. Cancer 28:344–366, 1971.

77. Ernster VL, Selvin S, Hinkelstein M Jr: Cohort mortality for prostatic cancer among United States non-whites. Science 200: 1165–1166, 1978.

78. Gann PH, Hennekens CH, Ma J, Longcope C, Stampfer MJ: Prospective study of sex hormone levels and risk of prostate cancer. J Natl Cancer Inst 88:1118–1126, 1996.

79. Hayes RB, Brown LM, Schoenberg JB, Greenberg RS, Silverman DT, Schwartz AG, Swanson GM, Benichou J, Liff JM, Hoover RN, Pottern LM: Alcohol use and prostate cancer risk in US blacks and whites. Am J Epidemiol 143:692–697, 1996.

80. Meikle AW, Smith JA Jr: Epidemiology of prostate cancer. Urol Clin North Am 17:709–718, 1990.

81. Hsing AW, McLaughlin JK, Schuman LM, Bjelke E, Gridley G, Wacholder S, Chien HT, Blot WJ: Diet, tobacco use, and fatal prostate cancer: Results from the Lutheran Brotherhood Cohort Study. Cancer Res 51:6836–6840, 1991.

82. Trachtenberg J: Hormonal therapy in metastatic prostatic cancer. In Bruce AW, Trachtenberg J (eds): Adenocarcinoma of the Prostate. New York: Springer-Verlag, 1987, pp 173–184.

83. Naito S: New development of endocrine therapy in prostate cancer. Gan To Kagaku Ryoho 23:684–688, 1996.

84. Scher HI, Zhang ZF, Nanus D, Kelly WK: Hormone and antihormone withdrawal: Implications for the management of androgen-independent prostate cancer. Urology 47:61–69, 1996.

85. Brown T: Provocative aspects of androgen genetics. Prostate 6:9–12, 1996.

86. Visakorpi T, Hyytinen E, Koivisto P, Tanner M, Keinanen R, Palmberg C, Palotie A, Tammela T, Isola J, Kallioniemei OP: In vivo amplification of the androgen receptor and progression of human prostate cancer. Nat Genet 9:401–406, 1995.

87. Newmark JR, Hardy DO, Tonb DC, Carter BS, Epstein JI, Isaacs WB, Brown TR, Barrack ER: Androgen receptor gene mutations in human prostate cancer. Proc Natl Acad Sci USA 89:6319–6323, 1992.

88. Gaddipati JP, McLeod DG, Heinenberg HB, Sesterhenn IA, Finger MJ, Moul JW, Srivastava S: Frequent detection of codon 877 mutation in the androgen receptor gene in advanced prostate cancers. Cancer Res 54:2861–2864, 1994.

89. Veldsholte J, Ris SC, Kuiper GG, Jenster G, Berrevoets C, van Claassen E, Trapman J, Brinkman AO, Mulder E: A mutation in the ligand binding domain of the androgen receptor of human LNCaP cells affects steroid binding characteristics and response to anti-androgens. Biochem Biophys Res Commun 173:534–540, 1990.

90. Kokontis J, Ito K, Hiipakka RA, Liao S: Expression and function of normal and LNCaP androgen receptors in androgen-insensitive human prostatic cancer cells. Altered hormone and antihormone specificity in gene transactivation. Receptor 1:271–279, 1991.

91. Suzuke H, Akakura K, Komiya A, Aida S, Akimoto S, Shimazaki J: Codon 877 mutation in the androgen receptor gene in advanced prostate cancer: Relation to antiandrogen withdrawal syndrome. Prostate 29:153–158, 1996.

92. Taplin ME, Bubley GJ, Shuster TD, Frantz ME, Spooner AE, Ogata GK, Keer HN, Balk SP: Mutation of the androgen receptor gene in metastatic androgen-independent prostate cancer. N Engl J Med 322:1393–1398, 1995.

93. Schoneberg MP, Hakimi JM, Wang S, Bova GSR, Epstein JI, Fischbeck KH Isaacs WB, Walsh PC, Barrack ER: Microsatellite mutation (CAG24-18) in the androgen receptor gene in human prostate cancer. Biochem Biophys Res Commun 198:74–80, 1994.

94. Irvine RA, Yu MC, Ross RK, Coetzee GA: The CAG and GGC microsatellites of the androgen receptor gene are in linkage disequilibrium in men with prostate cancer. Cancer Res 55:1937–1940, 1995.

16

THE IGF-I RECEPTOR IN NORMAL AND ABNORMAL GROWTH

RENATO BASERGA
Kimmel Cancer Institute, Thomas Jefferson University, Philadelphia, PA 19107-5541

16.1. INTRODUCTION

The insulin-like growth factor-I receptor (IGF-IR) and its ligands play an important role in normal and abnormal growth. A tyrosine kinase receptor with extensive homologies to the insulin receptor (IR), the activated IGF-IR is necessary for the optimal development and growth of the animal body. It regulates cell proliferation, protects cells from apoptosis, and is crucially important in the establishment and maintenance of the transformed phenotype and therefore in tumorigenesis. For these reasons, targeting of the IGF-IR offers some intriguing possibilities for inhibition of abnormal growth.

16.1.1 The IGF System

A simplified representation of the IGF system (Fig. 16.1) includes three receptors, three ligands and six IGF binding proteins (IGFBPs), but there are other players in the system whose role has not yet been clarified. The following brief summary is based on a recent review by Rubin and Baserga (1), and the reader is referred to that review for appropriate references. The three ligands are the mature IGF-I (70 amino acids), the mature IGF-II (67 amino acids), and insulin. The receptors are the IGF-IR, the IR, and the IGF-IIR. Of the three receptors, the IGF-IR is the most im-

Hormones and Growth Factors in Development and Neoplasia, Edited by Robert B. Dickson and David S. Salomon
ISBN 0-471-16899-8

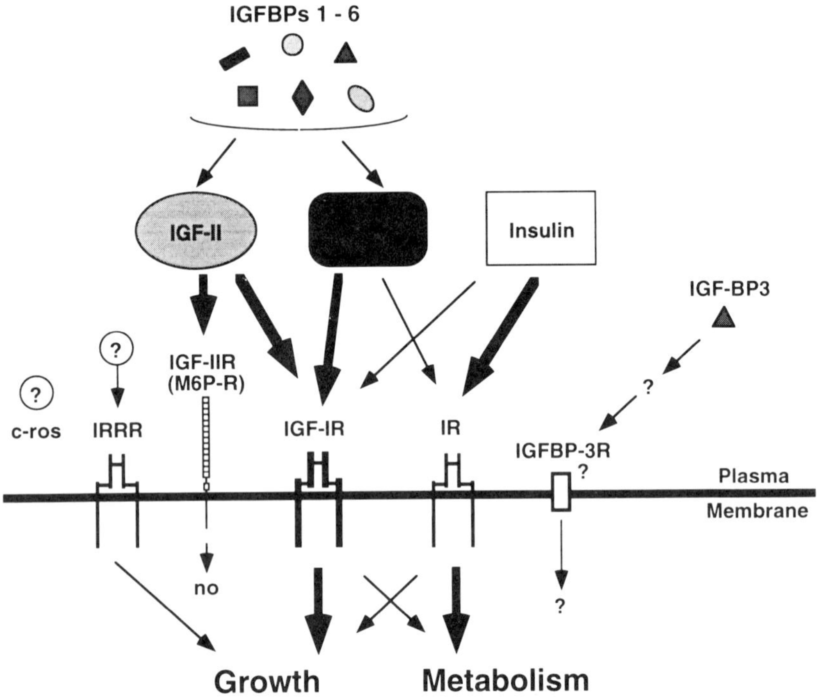

Fig. 16.1. The IGF system. Three growth factors interact with three major receptors, but only the IGF-I and insulin receptors transmit a mitogenic signal. The IGF-II receptor and the IGF binding proteins regulate the availability of IGF-I and IGF-II. Thr , other receptors could be part of the IGF system, but their roles have not yet been clarifie

portant for stimulation of cell proliferation (see below): It is activated by all three ligands, and although it has several features in common with the IR (see review in ref. 2), the β subunit of the IGF-IR is 10 times more mitogenic than the β subunit of the IR (3). Insulin activates the IGF-IR only at supraphysiologic concentrations, several orders of magnitude above the plasma concentration of insulin. An important point to remember is that, in the laboratory, investigators growing cells in culture in defined media use insulin at microgram concentrations: at these concentrations, insulin exerts its mitogenic stimulation through the IGF-IR (4). The IR is also activated by the three ligands, but the concentration required for activation by IGF-I or IGF-II is on the order of nanograms, i.e., three orders of magnitude above the optimal concentration for activation by insulin. Under appropriate conditions, the IR is modestly mitogenic.

It is now generally agreed that the IGF-IIR has no growth stimulatory properties: Indeed, deletion of the IGF-IIR genes in mice causes an increase in body size (see below). The IGF-IIR should be rightfully considered as a mechanism for the downregulation of IGF-II, i.e., a way for the organism and the cells to regulate the availability of IGF-II. The availability of IGF-I and IGF-II is also regulated by the six IGFBPs, which can serve as a storage site for these ligands (insulin does not bind to IGFBPs). However, there is evidence that at least the IGFBP-3 may inhibit cellular proliferation in its own right, i.e., independently of its capacity to bind IGF-I and IGF-II (5).

As mentioned above, there are other players in the system. There is an insulin receptor-related receptor (IRRR), with extensive homologies to the IGF-IR and the IR (6). Its tyrosine kinase domain is mitogenic, but neither insulin nor the mature IGFs can activate the IRRR, and its specific ligand is still unknown (7). There is also a c-ros (8), with homologies to the other receptors, whose function has not yet been clarified. And there are several other ligands—e.g., an IGF-I lacking the first 3 amino acids, which is especially important in brain function (9), and several IGF-IIs (propeptides), with molecular weights of 10–15 kDa, that are known to be mitogenic (10–12).

16.1.2. The Receptor

The IGF-IR belongs to the family of tyrosine kinase receptors (13), and its amino acid sequence has a 70% homology to that of the IR (14). it is synthesized as a single precursor polypeptide of 1367 amino acids, including a 30-amino acid signal peptide. The organization of the preproreceptor is NH_2-signal peptide, α subunit β subunit-COOH. After removal of the signal peptide, the proreceptor is cleaved after residue 706 to form the α and β subunits, linked by disulfide bonds (15). The α subunit is entirely extracellular, is involved in ligand binding (16), and forms a dimer with another subunit; indeed, the active form of the IGF-IR is a dimer. The transmembrane β subunit has a tyrosine kinase domain, which is intracellular. The highest homology between insulin and IGF-IRs is in the tyrosine kinase domain (84%), and the lowest homology is in the cysteine-rich regions of the α subunit (48%) and in the C terminus (44%). Following Ullrich et al. (14), in this review the amino acid residues are numbered from the first amino acid of the mature peptide (after removal of the signal peptide) up to 1337. The IGF-IR, like the IR, has several functions, including anabolic functions, but, in terms of growth, it has three properties that are attenuated in the IR: It is mitogenic, it is required for the establishment and maintenance of the transformed phenotype, at least in many cell types, and it protects cells from apoptosis.

16.1.3. The Signal Transducing Pathway

When activated by its ligands, the IGF-IR transmits a signal to its major substrates, IRS-1, IRS-2, and Shc proteins (17–21). A fourth substrate of the IGF-IR, Grb10, has recently been described (22). Several reports have appeared clearly indicating that IRS-1 plays an essential role in the mitogenicity of both insulin and IGF-I in cells in culture (23–26). The roles of the other substrates, Shc proteins (25), IRS-2 and Grb10, in IGF-IR-mediated mitogenesis and transformation are less well defined. From these substrates originate signal transducing pathways. One of these pathways, for both the IR and the IGF-IR, is the *ras* pathway, which has been reviewed extensively in recent articles and involves, among others, PI-3 kinase, Grb2, Sos, and other transducing molecules (27). Blenis (28) gives a list of the proteins involved in the signal transduction cascade, from ras to c-raf to MEK kinase, MEK, ERKK, p92, and, finally, transcription factors (see also ref. 27). This pathway is used by the three growth factors—platelet-derived growth factor (PDGF), epidermal growth factor (EGF), and IGF-I—that together sustain the growth of wild-type mouse embryo cells in serum-free medium, and although the evidence is overwhelming that *ras* activation is required for optimal cell proliferation and transformation, other findings indicate that it is not sufficient since a constitutively activated *ras* fails to induce growth of R^- cells (29) in serum-free medium supplemented with purified growth factors (R^- cells, to be described later, are mouse embryo fibroblasts with a targeted disruption of the IGF-IR genes). Ras-independent pathways for growth (and transformation; see below) have also been suggested independently by others (30,31). Recently, Barone and Courtneidge (32) have reported that the oncogene v-*src* may bypass the *ras* pathway in PDGF-induced activation of c-myc expression. This pathway is a good candidate for the ras-independent pathway originating from the IGF-IR, especially in view of the fact that v-src can tyrosyl phosphorylate both the IGF-IR and IRS-1 (33).

16.2. THE INSULIN-LIKE GROWTH FACTOR SYSTEM IN THE GROWTH AND DEVELOPMENT OF THE ANIMAL BODY

Pediatricians have known for a long time that, in children and adolescents, there is a good correlation between body height and IGF-I plasma levels (34). By contrast, low IGF-I plasma levels are found in pygmies and in Laron-type dwarfs (reviewed in refs. 1 and 35). The correlation is better than with growth hormone levels, and it extends also to dogs and mice. In dogs, differences in body size parallel IGF-I levels; thus, large standard poodles have six times the plasma IGF-I concentration of toy poodles, and the Newfoundland, a giant breed, exhibits significantly higher IGF-I plasma levels than the German shepherd, a large breed (36). Indeed, in transgenic mice, a modest 84% increase in IGF-I plasma level is sufficient to cause a significant increase in body weight (37). The importance of the activated IGF-IR in murine development has been rigorously demonstrated *in vivo* by Efstratiadis and coworkers, who showed that mouse embryos with a targeted disruption of the *IGF-IR* and *IGF-II* genes (*Igfr-/-*/mice) have a size at birth that is only 30% of the size of their wild-type littermates (38,39). Subsequent experiments by the same group have further elucidated the role of the IGF system in murine embryo growth. Mouse mutants inheriting maternally a targeted disruption of the imprinted *Igf2r* gene have increased serum and tissue levels of IGF-II and exhibit overgrowth of 135% compared to their wild-type littermates. These mutants usually die perinatally, the explanation being that, in the absence of IGF-IIR-mediated turnover, the IGF-IR is overstimulated by the excess IGF-II (40). Consistent with this hypothesis is the finding that Igf2r mutants are completely rescued when they carry a second mutation, eliminating either IGF-II or the IGF-IR. Triple mutants lacking IGF-IR, IGF-IIR, and IGF-II are nonviable dwarfs (40) 30% of the size of their wild-type littermates, like the double mutants lacking IGF-II and IGF-IR (38,39). This suggests that IGF-II may also signal through an unknown or unidentified receptor, for the moment designated XR.

16.3. THE INSULIN-LIKE GROWTH FACTOR-I RECEPTOR IN CELL GROWTH

For many years, the IGF-IR activated by its ligands has been known to be mitogenic to cells in culture; however, in growth-regulated cells (like mouse 3T3 cells and human diploid fibroblasts), IGF-I by itself cannot sustain growth of cells in serum-free medium but requires the cooperation of other growth factors, for instance PDGF and/or EGF, which by themselves also fail to induce a mitogenic response (41). 3T3-like cells derived from mouse embryos with a targeted disruption of the *IGF-IR* genes and designated R^- cells (see Section 16.2) grow in 10% fetal bovine serum, albeit at a rate that is roughly 40% of W cells, which are derived from mouse embryo wild-type littermates to the *Igfr-/-* mice. In R^- cells, all phases of the cell cycle are equally elongated (29), indicating that, at variance with established wisdom, the activation of the IGF-IR is important in maintaining the flow of cells through all phases of the cell cycle. R^- cells do not grow at all in serum-free medium supplemented by a variety of growth factors, which can sustain the growth of cells derived from wild-type littermate embryos (W cells) and other 3T3 cells (29). The important role of the IGF-IR in cell cycle progression has been demonstrated in a variety of cell types; a list of cells requiring IGF-I for optimal growth can be found in Baserga and Rubin (42). However, an important point to remember is that the activated IGF-IR is not essential for growth, as demonstrated by R^- cells that do grow in 10% serum (29). A correct statement would be that the activated IGF-IR is required for *optimal* growth, both *in vivo* and *in vitro* (in 10% serum), but it is not an absolute requirement. It becomes an absolute requirement only in serum-free medium supplemented with most of the known growth factors; the corollary is that there must be an unidentified growth factor in the serum (and in the animal) and/or an unidentified growth factor receptor that can at least partially overcome the IGF-I requirement for cell growth.

Another important observation made possible by R^- cells is that overexpression of the EGF (43), PDGF (44), and insulin (45) receptors is not sufficient for ligand-dependent growth unless a functional IGF-IR is present. By contrast, an overexpressed IGF-IR renders cells capable of growing in IGF-I only (46), without any activation of the PDGF and EGF receptors. This would suggest a central role for the IGF-IR in the mitogenic action of other growth factors.

16.4. INSULIN-LIKE GROWTH FACTOR-I RECEPTOR IN APOPTOSIS

There have been sporadic reports that IGF-I may protect cells, either *in vivo* or *in vitro,* from cell death in general and apoptosis in particular. For instance, Gluckman et al. (47) reported a protective effect of IGF-I in ischemic injuries of the central nervous system, and D'Mello et al. (48) found that IGF-I inhibited low potassium-induced apoptosis of cerebellar granule neurons. There are also intriguing reports that IGF-I inhibits apoptosis caused by interleukin-3 (IL-3) withdrawal in IL-3-dependent hemopoietic cells (49) and in other cell types. While these reports dealt with the protective effect of the ligands, more recent experiments have shown that the crucial component in the protection against apoptosis is the receptor. Thus, an overexpressed IGF-IR protects cells *in vitro* from apoptosis induced by IL-3 withdrawal (50), by etoposide (51), or by tumor necrosis factor (52). The results are even more dramatic when the ability of the IGF-IR to protect tumor cells from apoptosis is tested *in vivo:* A decrease in IGF-IR levels below wild-type levels (using antisense strategies) causes massive apoptosis of several types of tumor cells *in vivo* (53,54). Similar results were obtained when tumor cells were stably transfected with a dominant negative of the IGF-IR (55). Under the same conditions, growth in monolayer is only partially inhibited, with no increase in the percentage of apoptotic cells. These findings confirm that the role of the IGF-IR in monolayer growth is modest; its role becomes quasi-obligatory in anchorage-independent growth and in protection from apoptosis. The role of the IGF-IR in apoptosis, and the possibility that it may discriminate between normal and tumor cells, have been discussed in two recent reviews by Baserga (56,57).

There is another consideration in terms of the antiapoptotic effect of the IGF-IR: Neither IRS-1 nor Shc can protect 32D cells from apoptosis caused by IL-3 withdrawal, while the IGF-IR can (Fig. 16.2). IRS-1 and Shc are the two major substrates of the IGF-IR (see above), and both bind to tyrosine residue 950 of the receptor. A mutation of the IGF-IR at Y950 has no effect on its ability to protect cells from apoptosis induced by IL-3 withdrawal (50), thus confirming that these two substrates are not involved in the antiapoptotic signaling of the receptor. The unavoidable conclusion is that protection by the IGF-IR must involve another, unidentified signaling pathway. In this respect, a clue to this possible pathway comes from the experiments of Jung et al. (58), who found that IGF-I (but not EGF or fibroblast growth factor [bFGF]) protected cells from (ICE)-induced apoptosis.

16.5. THE INSULIN-LIKE GROWTH FACTOR-I RECEPTOR IN TRANSFORMATION

Overexpression and/or constitutive activation of IGF-IR in a variety of cell types leads to ligand-dependent growth in serum-free medium and to the establishment of a transformed phenotype; i.e., the ability to form colonies in soft agar or to produce tumors in nude mice (29,43,46,49,59–61). This, by itself, is not remarkable because

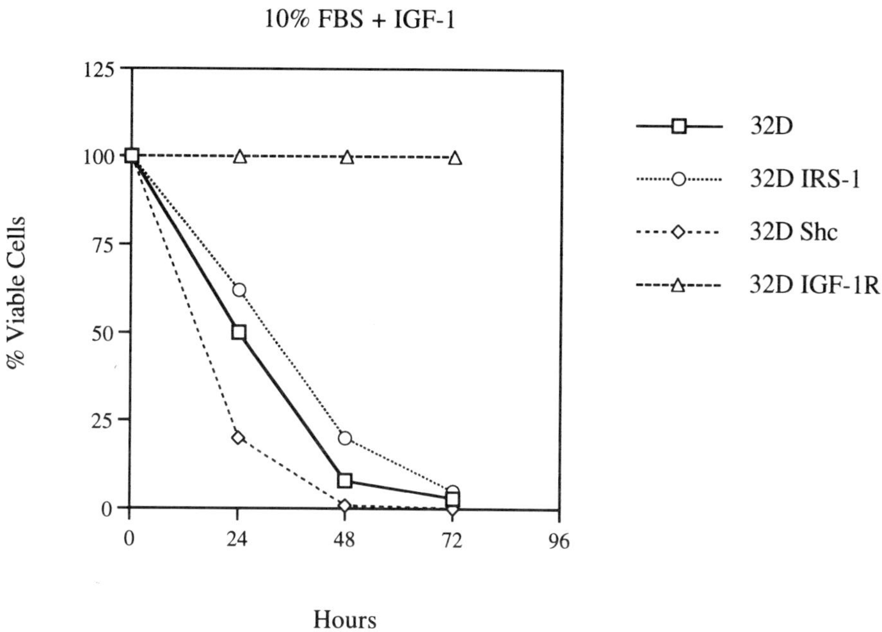

Fig. 16.2. The IGF-I receptor protects cells from apoptosis. 32D cells undergo apoptosis when IL-3 is withdrawn from the medium. We have generated lines of 32D cells (squares) overexpressing either the IGF-I receptor (triangles), IRS-1 (circles), or Shc proteins (diamonds) and tested them for resistance to apoptosis after IL-3 withdrawal. Only the IGF-I receptor conferred resistance to apoptosis; indeed, overexpression of Shc may actually accelerate apoptosis.

a large variety of overexpressed gene products can transform cells, including a number of growth factor receptors (for a list of these receptors, see ref. 43). What sets the IGF-IR apart from other growth factor receptors is that R^- cells are refractory to transformation by SV40 large T antigen, by an activated ras or a combination of both (29,62), by the bovine papilloma virus E5 protein (63), and by overexpressed growth factor receptors, such as the EGF receptor (43), the PDGF β receptor (44), and the IR (45). All these conditions readily transform mouse embryo fibroblasts with a normal number of IGF-IRs. It would be exaggerated to claim that R^- cells are untransformable; we have been growing these cells in our laboratory for over 4 years, and we have observed spontaneous transformation at a rate roughly three logarithms lower than that of mouse cells with a physiologic number of IGF-IRs. 3T3-like cells have such a strong propensity toward chromosomal rearrangements and spontaneous transformation that it is easy to visualize how a gain-of-function mutation in a signal-transducing molecule downstream of the IGF-IR may bypass the IGF-IR requirement for transformation.

If deletion of the IGF-IR genes renders cells refractory to transformation, it is not surprising that a reversal of the transformed phenotype can be obtained in a variety of tumor cells by decreasing the number of IGF-IRs or by interfering with their function. For instance, antisense expression plasmids or antisense oligodeoxynucleotides against either IGF-II (64), IGF-I (65), or the IGF-IR (29,62,66–71), antibodies to the IGF-IR (72,73), and dominant negative mutants of the IGF-IR (55,74–76) can all reverse the transformed phenotype, inhibit tumorigenesis, and induce loss of the metastatic phenotype (77).

In theory, and possibly also in practice, targeting the IGF-IR may be more efficient than targeting its ligands. The adult rat has negligible plasma levels of IGF-II, but other adult animals, including humans, have substantial circulating concentrations of both IGFs (78). Interference with one of the ligands could leave the other one free to activate the IGF-IR.

16.6. FUNCTIONAL DOMAINS OF THE INSULIN-LIKE GROWTH FACTOR-I RECEPTOR

Mutations in the various domains of the IGF-IR can affect its various functions: For instance, a single point mutation at lysine 1003 (the ATP-binding site) results in an IGF-IR that has essentially lost all its functions (43). A triple mutation of the three Ys (1131, 1135, 1136) in the tyrosine kinase domain results in a markedly reduced level of autophosphorylation and in the abolition of both mitogenicity and transforming activity (75,79). These mutations affect both the mitogenic and transforming activities of the IGF-IR. More interesting are those mutant receptors in which only selective functions are deleted. For instance, a human IGF-IR mutated at Y950 is neither mitogenic nor transforming, but it does protect cells from apoptosis induced by IL-3 withdrawal (50). Although Liu et al. (60) had reported that truncations of the C terminus of the IGF-IR affected its transforming potential, it was Surmacz et al. (80) who first demonstrated that the mitogenic and transforming activities of the IGF-IR could be dissociated. Deletion of the C-terminal 108 amino acids from the human IGF-IR resulted in a receptor that had conserved its ability to transmit an IGF-I-mediated mitogenic signal but was no longer capable of transforming activity. Subsequently, Miura et al. (45) showed that a single point mutation at Y1251 (to phenylalanine) also greatly reduced the transforming activity of the IGF-IR without affecting its mitogenicity. Interestingly, a Y1250F mutation had no effect on the receptor, which behaved like a wild-type IGF-IR (45). In these experiments, mutant receptors were transfected into R^- cells to generate stable cell lines with no other IGF-IR but the transfected one. Subsequently, other mutants of the IGF-IR (81,82) were generated and tested; the results are summarized in Table 16.1.

The conclusions from these experiments can be summarized as follows: 1) mitogenicity and transforming activity of the IGF-IR can be localized on separate domains (45,80,81,82; see also Table 16.1), clearly indicating that there is at least one pathway for transformation that is additional to and distinct from the mitogenic pathways and is very likely an unidentified ras-independent pathway (29); 2) the transforming domain of the IGF-IR can be localized between residues 1245 and 1310 (81; see also Table 16.1); and 3) Y1251 and the serine quartet at 1280–1283

TABLE 16.1. Summary of the Mitogenic, Transforming, and Antiapoptotic Activities of the IGF-I Receptor and Its Mutants

Receptor	Mitogenicity	Transformation	Antiapoptosis
wt	+	+	+
d 1229	+	−	+
d1245	+	−	+
d1270	ND	+	+
d1293	+	+?	+
d1310	+	+	+
K 1003	−	−	−
Y950F	−	−	+
Y3F	−	−	−
Y1250F	+	+	+
Y1251F	+	−	−
S1280/1283A	+	−	?
1293/1294(FL)	+	+	−
Y1316F	+	+	+

Note: These data are summarized from refs. 45, 49, 59, 60, 75, and 79–82 for mitogenicity and transforming activity and from ref. 50 for protection from apoptosis. Mitogenicity is defined as the ability to make cells grow in SFM supplemented solely with IGF-I; transforming activity as the ability to make cells form colonies in soft agar; and protection from apoptosis as protection of FL5.12 cells from apoptosis induced by IL-3 withdrawal. ND, not done; d, truncation of the receptor at the indicated residues; Y, tyrosine; F, phenylalanine; S, serine; K, lysine; A, alanine; Y3F, mutations at Y 1131, 1135, and 1136.

are very important for transformation (45,82; see also Table 16.1), with the region around 1293 (81) playing some role. When these residues are mutated, the transforming activity of the IGF-IR is severely impaired, whereas its mitogenic activity in response to IGF-I is perfectly normal.

Table 16.1 also shows that the domain(s) for protection from apoptosis can be separated from the mitogenic and transforming domains. The C-terminal truncated receptors are mitogenic, nontransforming, and fully protect cells from apoptosis, while the mitogenic and transforming 1293/1294 mutant does not protect cells from apoptosis. One must therefore conclude that the three functions of the IGF-IR that relate to cell proliferation are spatially separated on the receptor, presumably signaling through distinct pathways. It should not be surprising that the transforming pathway is additional to and distinct from the mitogenic pathways since transformed cells are anchorage independent and this property is not required by cells growing in monolayers. Less expected is the finding that transforming activity and protection from apoptosis can also be separated from each other. This finding, though, is quite intriguing. The fact that the three pathways originate from different domains of a single growth factor receptor suggests the possibility of interfering with one pathway while sparing the others.

Table 16.2 summarizes the mitogenic, transforming, and antiapoptotic properties of the IGF-IR and its major substrates. It is significant that the wild-type

TABLE 16.2. Effect of Overexpression of the IGF-I Receptor or Its Major Substrates on Mitogenesis, Transformation, and Protection from Apoptosis

Overexpression	Mitogenesis	Transformation	Apoptosis
IGF-I receptor	+++	+++	+++
IRS-1	+++	+	−
Shc	+++	−	−
IRS-1 + Shc	+++	+	−

Note: Mitogenesis is the ability to grow in serum-free medium, in monolayer, in response to IGF-I. Transformation is the ability to form colonies in soft agar in 10% serum. Apoptosis means protection from apoptosis. IRS-1 (and IRS-1 + Shc) is transforming only in cells that have a normal number of wild-type IGF-IRs; in R^- cells (with no IGF-IRs), IRS-1 is nontransforming (26).

IGF-IR by itself is fully transforming (R^+ and p6 cells), as already reported by several laboratories (see above), while its two major substrates, IRS-1 and Shc, singly or in combination, cannot transform R^- cells. IRS-1 is weakly transforming, though, in cells with a normal number of IGF-IRs. In agreement with these findings, both Surmacz et al. (80) and Miura et al. (45) reported increased tyrosyl phosphorylation of IRS-1 and Shc in cells expressing the mutant IGF-IRs that are mitogenic but nontransforming. This indicates that the additional transforming pathway of the IGF-IRC terminus is independent of IRS-1 and Shc. Indeed, all the available evidence suggests that IRS-1 and Shc proteins are very important for IGF-I- or insulin-mediated mitogenesis (23,24,83), but play only a permissive role in transformation.

16.7. THE INSULIN-LIKE GROWTH FACTOR-I RECEPTOR IN HUMAN TUMORS

There are abnormalities in the expression or amplification of the genes for the IGF-IR and its ligands in several human tumors (84), but they are sporadic and do not seem to follow a consistent pattern. There are also intriguing reports of an increased incidence of malignancies in acromegalics, in whom IGF-I levels are elevated (85). However, the important point that emerges from the experiments in rodents is that it is not the overexpression or the gain of function of the IGF-IR that is crucial, but rather its absence or loss of function. True, overexpression of the IGF-IR can lead to transformation, but the most striking observation is that transformation does not occur, or is even reversed, when IGF-IR signaling is impaired, either because the number of receptors falls below normal levels or because of interference with its function (dominant negatives). It is, therefore, not surprising that deletions or loss of function of the IGF-IR are not found in human tumors since many types of tumors could not develop in the absence of a certain level of functional IGF-IRs.

16.8. THE INSULIN-LIKE GROWTH FACTOR-I RECEPTOR AS A TARGET FOR INHIBITION OF ABNORMAL GROWTH

The fact that interference with the function of the IGF-IR results in tumor cell death, inhibition of tumorigenesis (see above), and prevention of metastases (77) in itself is not especially remarkable since many agents and many modalities can do the same. But there is something unique about the IGF-IR; these unique aspects are summarized in Table 16.3. We have previously hypothesized (56,57) that a decrease or a loss of function of the IGF-IR will cause massive apoptosis of tumor cells while being much less effective on normal cells, which can take refuge in the G_0 state. There are three separate pieces of evidence that normal cells are more resistant to apoptosis than tumor cells. First is the finding that, in mouse embryos with a targeted disruption of the IGF-IR genes (see above), there is no evidence of increased apoptosis. Similarly, the cells derived from these mouse embryos, R^- cells (see above), grow more slowly but do not seem more prone to apoptosis by serum deprivation than the cells generated from their wild-type littermates. Second, we have found that WI-38 human diploid fibroblasts do not undergo apoptosis *in vivo* even when treated with high concentrations of antisense oligodeoxynucleotides to the IGF-IR RNA (unpublished data). And third, Wang (86) reported that human diploid fibroblasts are more resistant to apoptosis induced by serum deprivation than are 3T3 cells. The latter, although often considered normal cells, are in fact immortalized aneuploid cells prone to spontaneous transformation; human diploid fibroblasts, by contrast, do not transform spontaneously and have a finite life span.

In addition, targeting the IGF-IR seems to induce a host response that gives an additional means to control tumor growth. This host response is a peculiar one and merits some discussion. As shown in Table 16.4, rats previously injected with, or carrying in a diffusion chamber for 24 hr, tumor cells with a targeted IGF-IR (antisense strategies or dominant negatives) become totally resistant to a subsequent challenge with wild-type C6 tumor cells (67). Untreated wild-type cells do not confer this resistance. This would not be particularly surprising except that the animals become resistant to challenge with wild-type C6 rat glioblastoma cells even when they are pretreated with unrelated tumor cells—indeed, even tumor cells of different species (Table 16.4). In these experiments, the "immunizing" tumor cells are either injected subcutaneously or loaded in a diffusion chamber, which is then inserted into the subcutaneous tissue of a rat for 24 hr. The chamber is then removed,

TABLE 16.3. Uniqueness of the IGF-I Receptor as a Target for Cancer Therapy

Targeting of the IGF-I receptor by antisense strategies, antibodies, or use of dominant negatives:

- causes massive apoptosis of tumor cells *in vivo*
- inhibits tumor growth and metastases
- elicits a host response that leads to the elimination of surviving tumor cells
- has effects on normal cells that are seemingly limited to partial growth inhibition, without apoptosis

TABLE 16.4. Host Response Elicited in Rats by Cells with Impaired IGF-I Receptor Function

Cell Type	Strategy	Tumor Development (C6)
C6 wild type	SC, CH	+
C6 sense plasmid or random ODN	SC, CH	+
C6, antisense plasmid	SC, CH	−
C6, antisense ODN	SC, CH	−
C6 with dominant negative	CH	−
Human melanoma cells	CH	+
do, with antisense	CH	−
Rat rhabdomyosarcoma (antisense)	CH	−
Balb/c 3T3	CH	+
R^- (no IGF-I receptors)	CH	−
R^+ (overexpressing the IGF-IR)	CH	+
Human ovarian carcinoma	CH	+
do, with dominant negative	CH	−
WI-38 human diploid fibroblasts		
+ random ODN	CH	+
+ antisense ODN	CH	−

Note: The rats were treated with the types of cells indicated in column 1, which also includes information on how the IGF-I receptor function was targeted in these cells, i.e., the use of antisense or random or sense oligodeoxynucleotides (ODN) or expression plasmids. Plasmid means that the cells were stably transfected with a plasmid expressing either a sense or an antisense RNA to the IGF-IR RNA (53). Column 2 gives the strategy used to induce a host response, i.e., whether the cells were injected subcutaneously (SC) or placed in the diffusion chamber (CH), left in the subcutaneous tissue of the animal for 24 hr. In every case, the pretreated rats were subsequently challenged with 10^7 wild-type C6 cells injected subcutaneously 1 to 30 days after the first treatment. The appearance of tumors in these rats is given as +; protection is given as −, no tumor.

and from 1 week to 6 weeks later, the rats are challenged with wild-type C6 cells. Provided that the immunizing cells had a decrease in the number of or a functional impairment of the IGF-IR, they invariably elicited a response that made the rats resistant to subsequent challenge with C6 cells.

We realize that this observation is somewhat unusual, but we have observed this puzzling phenomenon for several years and there is no question that it is reproducible, whether the immunizing cells express an antisense RNA to the IGF-IR RNA (67), or are treated with antisense oligodeoxynucleotides against the IGF-IR RNA (54), or are expressing a dominant negative of the IGF-IR (55). The reported association of the IGF-IR and IR with major histocompatibility complex (MHC) (87–94) offers a possible lead to the explanation of the peculiar phenomenon we have observed.

16.9. CONCLUSIONS

Targeting the IGF-IR by a variety of strategies (antisense, antibodies, dominant negatives) has been shown to be very effective in preventing tumorigenesis and even in causing tumor regression in rodents. Successful eradication of transplantable tumors in rats and mice may or may not be extrapolated to human tumors. But at least we can say that the trials on animals have been remarkably effective.

What makes the IGF-IR an attractive target for chemotherapeutic interventions is its uniqueness. First, it seems that tumor cells are much more sensitive than normal cells to apoptosis induced by interference with the receptor, which should give it a favorable therapeutic index. Second, the IGF-IR is seemingly downstream from other growth factor receptors (see review in ref. 57), which makes it a more generalized target. Third, the role of the IGF-IR in protecting tumor cells from apoptosis is even more striking *in vivo* than *in vitro*. And finally, these strategies use a double-edged sword: On one side, they induce apoptosis of tumor cells; on the other, they provoke some kind of immune response that mops up residual surviving cells.

These considerations justify plans for therapeutic applications of strategies that impair the function of the IGF-IR. Such therapy should be especially useful in eradicating residual tumors or small metastases after surgical ablation of primary tumors. The problems with targeting the IGF-IR are the usual ones with any anticancer biologic approach: specificity of action and delivery. As mentioned above, specificity may not be a difficult problem to solve if we can confirm in humans that normal cells and tumor cells differ in their requirement for a functional IGF-IR. Delivery will be the biggest developmental problem, especially if one wishes to use gene therapy with expression plasmids; antisense oligodeoxynucleotides will avoid the delivery problem, only to incur specificity pitfalls, until better and more specific antisense probes are developed.

REFERENCES

1. Rubin R, Baserga R: Biology of disease. Insulin-like growth factor I receptor. Lab Invest 73:311–331, 1995.
2. Tavare JM, Siddle K: Mutational analysis of insulin receptor function: Consensus and controversy. Biochim Biophys Acta 1178:21–39, 1993.
3. Lammers R, Gray A, Schlessinger J, Ullrich A: Differential signaling potential of insulin- and IGF-I receptor cytoplasmic domain. EMBO J 8:1369–1375, 1989.
4. Flier JS, Usher P, Moses AC: Monoclonal antibody to the type 1 insulin-like growth factor (IGF-I) receptor blocks IGF-I receptor-mediated DNA synthesis: Clarification of the mitogenic mechanisms of IGF-I and insulin in human skin fibroblasts. Proc Natl Acad Sci USA 83:664–668, 1986.
5. Valentinis B, Bhala A, DeAngelis T, Baserga R, Cohen P: The human insulin-like growth factor (IGF) binding protein-3 inhibits the growth of fibroblasts with a targeted disruption of the IGF-I receptor gene. Mol Endocrinol 9:361–367, 1995.
6. Shier P, Watt VM: Primary structure of a putative receptor for a ligand of the insulin family. J Biol Chem 264:14605–14608, 1989.

7. Zhang B, Roth RA: The insulin receptor-related receptor. J Biol Chem 267:18320–18328, 1992.

8. Chen J, Hanafusa T, Wang LH: Ala Gly mutation in the putative catalytic loop confers temperature sensitivity on ros, insulin receptor and insulin-like growth factor I receptor protein-tyrosine kinases. Proc Natl Acad Sci USA 91:321–325, 1994.

9. Carlsson-Skwirut C, Lake M, Hartmani M, Hall M, Sara VR: A comparison of the biological activities of recombinant intact and truncated insulin-like growth factor I (IGF-I). Biochim Biophys Acta 1011:192–197, 1989.

10. Gowan LK, Hampton B, Hill DJ, Schlueter RJ, Perdue JF: Purification and characterization of a unique high molecular weight form of insulin-like growth factor II. Endocrinology 121:449–458, 1987.

11. Hylka VW, Straus DS: The E-domain peptide of rat proinsulin-like growth factor II (proIGF-II): Properties of the peptide in serum and production by rat cell lines. Biochim Biophys Acta 1051:6–13, 1990.

12. Hylka VW, Teplow DB, Kent SBH, Straus DS: Identification of a peptide fragment from the carboxyl-terminal extension region (E-domain) of rat proinsulin-like growth factor II. J Biol Chem 260:14417–14420, 1985.

13. Ullrich A, Schlessinger J: Signal transduction by receptors with tyrosine kinase activity. Cell 61:203–212, 1990.

14. Ullrich A, Gray A, Tam AW, Yang-Feng T, Tsubokawa M, Collins C, Henzel W, Le Bon T, Kahuria S, Chen E, Jakobs S, Francke U, Ramachandran J, Fujita-Yamaguchi Y: Insulin-like growth factor I receptor primary structure: Comparison with insulin receptor suggests structural determinants that define functional specificity. EMBO J 5: 2503–2512, 1986.

15. Werner H, Woloschak M, Stannard B, Shen-Orr Z, Roberts CT, LeRoith D: The insulin-like growth factor receptor. In LeRoith D (ed): Insulin-like Growth Factors: Molecular and Cellular Aspects. Boca Raton, FL: CRC Press, 1991, pp 17–47.

16. Schumacher R, Soos MA, Schlessinger J, Brandenburg D, Siddle K, Ullrich A: Signaling-competent receptor chimeras allow mapping of major insulin receptor binding domain determinants. J Biol Chem 268:1087–1094, 1993.

17. White MF, Kahn CR: The insulin signaling system. J Biol Chem 269:1–4, 1994.

18. Myers MJ, Sun XJ, Cheatham B, Jachna BR, Glasheen EM, Backer JM, White MF: IRS-1 is a common element in insulin and insulin-like growth factor-I signaling to the phosphatidylinositol 3′-kinase. Endocrinology 132:1421–1430, 1993.

19. Keller S, Lienhard G: Insulin signaling: The role of insulin receptor substrate 1. Trends Cell Biol 4:115–119, 1994.

20. O'Neill TJ, Craparo A, Gustafson TA: Characterization of an interaction between insulin receptor substrate-1 and the insulin receptor by using the two-hybrid system. Mol Cell Biol 14:6433–6442, 1994.

21. Araki E, Lipes MA, Patti ME, Bruning JC, Haag B III, Johnson RS, Kahn CR: Alternative pathway of insulin signaling in mice with targeted disruption of the IRS-1 gene. Nature 372:186–190, 1995.

22. Morrione A, Valentinis B, Li S, Ooi JYT, Margolis B, Baserga R: Grb10: A new substrate of the insulin-like growth factor I receptor. Cancer Res 56:3165–3167, 1996.

23. Waters SB, Yamauchi K, Pessin JE: Functional expression of insulin receptor substrate-1 is required for insulin-stimulated mitogenic signaling. J Biol Chem 268:22231–22234, 1993.

24. Rose DW, Saltiel AR, Majumdar M, Decker SJ, Olefsky JM: Insulin receptor substrate 1 is required for insulin-mediated mitogenic signal transduction. Proc Natl Acad Sci USA 91:797–801, 1994.

25. Yamauchi K, Pessin JE: Enhancement or inhibition of insulin signaling by insulin receptor substrate 1 is cell context dependent. Mol Cell Biol 14:4427–4434, 1994.

26. D'Ambrosio C, Keller SR, Morrione A, Lienhard GE, Baserga R, Surmacz E: Transforming potential of the insulin receptor substrate 1. Cell Growth Diff 6:557–562, 1995.

27. Hill CS, Treisman R: Transcriptional regulation by extracellular signals: Mechanisms and specificity. Cell 80:199–211, 1995.

28. Blenis J: Signal transduction via the MAP kinases: Proceed at your own RSK. Proc Natl Acad Sci USA 90:5889–5892, 1993.

29. Sell C, Dumenil G, Deveaud C, Miura M, Coppola D, DeAngelis T, Rubin R, Efstratiadis A, Baserga R: Effect of a null mutation of the type 1 IGF receptor gene on growth and transformation of mouse embryo fibroblasts. Mol Cell Biol 14:3604–3612, 1994.

30. Silvennoinen O, Schindler C, Schlessinger J, Levy DE: Ras-independent growth factor signaling by transcription factor tyrosine phosphorylation. Science 261:1736–1739, 1993.

31. Buscher D, Hipskind RA, Krautwald S, Reimann T, Baccarini M: Ras-dependent and -independent pathways target the mitogen activated protein kinase network in macrophages. Mol Cell Biol 15:466–475, 1995.

32. Barone MV, Courtneidge SA: Myc but not Fos rescue of PDGF signaling block caused by kinase-inactive Src. Nature 378:509–512, 1995.

33. Peterson JE, Jelinek T, Kaleko M, Siddle K, Weber MJ: Phosphorylation and activation of the IGF-I receptor in src-transformed cells. J Biol Chem 269:27315–27321, 1994.

34. Gourmelen M, LeBouc Y, Girard F, Binoux M: Serum levels of insulin-like growth factor (IGF) and IGF-binding proteins in constitutionally tall children and adolescents. J Clin Endocrinol Metab 59:1197–1203, 1984.

35. LeRoith D, Adamo M, Werner H, Roberts CT Jr: Insulin-like growth factors and their receptors as growth regulators in normal physiology and pathologic states. Trends Endocrinol Metab 2:134–139, 1991.

36. Eigenman JE, Amador A, Patterson DF: Insulin-like growth factor I levels in proportionate dogs, chondrodystrophic dogs and in giant dogs. Acta Endocrinol 118:105–108, 1988.

37. Reiss K, Cheng W, Ferber A, Kajstura J, Li P, Li B, Olivetti G, Homcy CJ, Baserga R, Anversa P: Overexpression of insulin-like growth factor I in the heart is coupled with myocyte proliferation in transgenic mice. Proc Natl Acad Sci USA 93:8630–8635, 1996.

38. Liu J-P, Baker J, Perkins AS, Robertson EJ, Efstratiadis A: Mice carrying null mutations of the genes encoding insulin-like growth factor I (igf-1) and type 1 IGF receptor (igf1r). Cell 75:59–72, 1993.

39. Baker J, Liu J-P, Robertson EJ, Efstratiadis A: Role of insulin-like growth factors in embryonic and postnatal growth. Cell 73:73–82, 1993.

40. Ludwig T, Eggenschwiler J, Fisher P, D'Ercole JP, Davenport ML, Efstratiadis A: Mouse mutants lacking the type 2 IGF receptor (IGF2R) are rescued from perinatal lethality in Igf2 and Igf1r null backgrounds. Dev Biol 177:517–535, 1996.

41. Stiles CD, Capone GT, Scher CD, Antoniades NH, Van Wyk JJ, Pledger WJ: Dual control of cell growth by somatomedins and platelet derived growth factor. Proc Natl Acad Sci USA 76:1279–1283, 1979.

42. Baserga R, Rubin R: Cell cycle and growth control. Crit Rev Eukar Gene Express 3:47–61, 1993.
43. Coppola D, Ferber A, Miura M, Sell C, D'Ambrosio C, Rubin R, Baserga R: A functional insulin-like growth factor I receptor is required for the mitogenic and transforming activities of the epidermal growth factor receptor. Mol Cell Biol 14:4588–4595, 1994.
44. DeAngelis T, Ferber A, Baserga R: The insulin-like growth factor I receptor is required for the mitogenic and transforming activities of the platelet-derived growth factor receptor. J Cell Physiol 164:214–221, 1995.
45. Miura M, Surmacz E, Burgaud JL, Baserga R: Different effects on mitogenesis and transformation of a mutation at tyrosine 1251 of the insulin-like growth factor I receptor. J Biol Chem 270:22639–22644, 1995.
46. Pietrzkowski Z, Sell C, Lammers R, Ullrich A, Baserga R: Role of insulin-like growth factor I (IGF-I) and the IGF-I receptor in epidermal growth factor-stimulated growth of 3T3 cells. Mol Cell Biol 12:3883–3889, 1992.
47. Gluckman P, Klempt N, Guan J, Mallard C, Sirimanne E, Dragunow M, Klempt M, Singh K, Williams C, Nikolics K: A role for IGF-1 in the rescue of CNS neurons following hypoxic-ischemic injury. Biochem Biophys Res Commun 182:593–599, 1992.
48. D'Mello SR, Galli C, Ciotti T, Calissano P: Induction of apoptosis in cerebellar granule neurons by low potassium: Inhibition of death by insulin-like growth factor I and cAMP. Proc Natl Acad Sci USA 90:10989–10993, 1993.
49. McCubrey JA, Stillman LS, Mayhew MW, Algate PA, Dellow RA, Kaleko M: Growth promoting effects of insulin-like growth factor 1 (IGF-1) on hematopoietic cells. Overexpression of introduced IGF-1 receptor abrogates interleukin-3 dependency of murine factor dependent cells by ligand dependent mechanism. Blood 78:921–929, 1991.
50. O'Connor R, Kauffmann-Zeh A, Liu Y, Lehar S, Evan GI, Baserga R, Blattler WA: The IGF-I receptor domains for protection from apoptosis are distinct from those required for proliferation and transformation. Mol Cell Biol 17:427–435, 1997.
51. Sell C, Baserga R, Rubin R: Insulin-like growth factor I (IGF-I) and the IGF-I receptor prevent etoposide-induced apoptosis. Cancer Res 55:303–306, 1995.
52. Wu Y, Tewari M, Cui S, Rubin R: Activation of the insulin-like growth factor I receptor inhibits tumor necrosis factor-induced cell death. J Cell Physiol 168:499–509, 1996.
53. Resnicoff M, Abraham D, Yutanawiboonchai W, Rotman H, Kajstura J, Rubin R, Zoltick P, Baserga R: The insulin-like growth factor I receptor protects tumor cells from apoptosis *in vivo*. Cancer Res 55:2463–2469, 1995.
54. Resnicoff M, Burgaud J-L, Rotman HL, Abraham D, Baserga R: Correlation between apoptosis, tumorigenesis and levels of insulin-like growth factor I receptors. Cancer Res 55:3739–3741, 1995.
55. D'Ambrosio C, Ferber A, Resnicoff M, Baserga R: A soluble insulin-like growth factor I receptor that induces apoptosis of tumor cells in vivo and inhibits tumorigenesis. Cancer Res 56:4013–4020, 1996.
56. Baserga R: Oncogenes and the strategy of growth factors. Cell 79:927–930, 1994.
57. Baserga R: The insulin-like growth factor 1 receptor: A key to tumor growth? Cancer Res 55:249–252, 1995.
58. Jung YK, Miura M, Yuan J: Suppression of interleukin-1 beta converting enzyme-mediated cell death by insulin-like growth factor. J Biol Chem 271:5112–5117, 1996.
59. Kaleko M, Rutter WG, Miller AD: Overexpression of the human insulin-like growth fac-

tor 1 receptor promotes ligand-dependent neoplastic transformation. Mol Cell Biol 10:464–473, 1990.

60. Liu D, Zong C, Wang LH: Distinctive effects of the carboxyl-terminal sequence of the insulin-like growth factor I receptor on its signaling functions. J Virol 67:6835–6840, 1993.

61. Rogler CE, Yang D, Rossetti L, Donohoe J, Alt E, Chang CJ, Rosenfeld R, Neely K, Hintz R: Altered body composition and increased frequency of diverse malignancies in insulin-like growth factor II transgenic mice. J Biol Chem 269:13779–13784, 1994.

62. Sell C, Rubini M, Rubin R, Liu J-P, Efstratiadis A, Baserga R: Simian virus 40 large tumor antigen is unable to transform mouse embryonic fibroblasts lacking type-1 IGF receptor. Proc Natl Acad Sci USA 90:11217–11221, 1993.

63. Morrione A, DeAngelis T, Baserga R: Failure of the bovine papilloma virus to transform mouse embryo fibroblasts with a targeted disruption of the insulin-like growth factor I receptor gene. J Virol 69:5300–5303, 1995.

64. Christophori G, Naik P, Hanahan D: A second signal supplied by insulin-like growth factor II in oncogene-induced tumorigenesis. Nature 369:414–418, 1994.

65. Trojan J, Johnson TR, Rudin SD, Ilan J, Tykocinski ML, Ilan I: Treatment and prevention of rat glioblastoma by immunogenic C6 cells expressing antisense insulin-like growth factor I RNA. Science 256:94–97, 1993.

66. Resnicoff M, Coppola D, Sell C, Rubin R, Ferrone S, Baserga R: Growth inhibition of human melanoma cells in nude mice by antisense strategies to the type 1 insulin-like growth factor receptor. Cancer Res 54:4848–4850, 1994.

67. Resnicoff M, Sell C, Rubini M, Coppola D, Ambrose D, Baserga R, Rubin R: Rat glioblastoma cells expressing an antisense RNA to the insulin-like growth factor I (IGF-I) receptor are nontumorigenic and induce regression of wild-type tumors. Cancer Res 54:2218–2222, 1994.

68. Shapiro DN, Jones BG, Shapiro LH, Dias P, Houghton PJ: Antisense-mediated reduction in insulin-like growth factor 1 receptor expression suppresses the malignant phenotype of a human alveolar rhabdomyosarcoma. J Clin Invest 94:1235–1242, 1994.

69. Neuenschwander S, Roberts CT Jr, LeRoith D: Growth inhibition of MCF-7 breast cancer cells by stable expression of an insulin-like growth factor 1 receptor antisense ribonucleic acid. Endocrinology 136:4298–4303, 1995.

70. Lee CT, Wu S, Gabrivolich D, Chen H, Nadaf-Rahrov S, Ciernick IF, Carbone DP: Antitumor effect of an adenovirus expressing antisense insulin-like growth factor I receptor on human lung cancer cell lines. Cancer Res 56:3038–3041, 1996.

71. Pass HI, Mew DJY, Carbone M, Matthews WA, Donington JS, Baserga R, Walker CL, Resnicoff M, Steinberg SM: Inhibition of hamster mesothelioma tumorigenesis by an antisense expression plasmid to the insulin-like growth factor I receptor. Cancer Res 56:4044–4048, 1996.

72. Arteaga CL: Interference of the IGF system as a stratgy to inhibit breast cancer growth. Breast Cancer Res Treat 22:101–106, 1992.

73. Kalebic T, Tsokos M, Helman LJ: In vivo treatment with antibody against the IGF-I receptor suppresses growth of human rhabdomyosarcoma and down regulates p34/cdc2. Cancer Res 54:5531–5534, 1994.

74. Prager D, Li HL, Asa S, Melmed S: Dominant negative inhibition of tumorigenesis in vivo by human insulin-like growth factor receptor mutant. Proc Natl Acad Sci USA 91:2181–2185, 1994.

75. Li S, Ferber A, Miura M, Baserga R: Mitogenicity and transforming activity of the insulin-like growth factor I receptor with mutations in the tyrosine kinase domain. J Biol Chem 269:32558–32564, 1994.

76. Burgaud J-L, Resnicoff M, Baserga R: Mutant IGF-I receptors as dominant negatives for growth and transformation. Biophys Biochem Res Commun 214:475–481, 1995.

77. Long L, Rubin R, Baserga R, Brodt P: Loss of the metastatic phenotype in murine carcinoma cells expressing an antisense RNA to the insulin-like growth factor I receptor. Cancer Res 55:1006–1009, 1995.

78. Conlon MA, Tomas FM, Owens PC, Wallace JC, Howarth GS, Ballard FJ: Long R/3 insulin-like growth factor I (IGF-I) infusion stimulates organ growth but reduces plasma IGF-I, IGF-II and IGF binding protein concentrations in the guinea pig. J Endocrinol 146:247–253, 1995.

79. Gronborg M, Wulff BS, Rasmussen JS, Kjeldsen T, Gammeltoft S: Structure–function relationship of the insulin-like growth factor I receptor tyrosine kinase. J Biol Chem 268:23435–23440, 1993.

80. Surmacz E, Sell C, Swantek J, Kato H, Roberts CT Jr, LeRoith D, Baserga R: Dissociation of mitogenesis and transforming activity by C-terminal truncation of the insulin-like growth factor I receptor. Exp Cell Res 218:370–380, 1995.

81. Hongo A, D'Ambrosio C, Miura M, Morrione A, Baserga R: Mutational analysis of the mitogenic and transforming activities of the insulin-like growth factor I receptor. Oncogene 12:1231–1238, 1996.

82. Li S, Resnicoff M, Baserga R: Effect of mutations at serines 1280–1283 on the mitogenic and transforming activities of the insulin-like growth factor I receptor. J Biol Chem 271:12254–12260, 1996.

83. Pronk GJ, McGlade J, Pelicci G, Pawson T, Bos JL: Insulin-induced phosphorylation of the 46- and 52-kDa Shc proteins. J Biol Chem 268:5748–5753, 1993.

84. Macaulay VM: Insulin-like growth factors and cancer. Br J Cancer 65:311–320, 1992.

85. Bengtsson BA: Acromegaly and neoplasia. J Pediatr Endocrinol 6:73–78, 1993.

86. Wang E: Senescent human fibroblasts resist programmed cell death, and failure to suppress bcl-2 is involved. Cancer Res 55:2284–2292, 1995.

87. Due C, Simonsen M, Olsson L: The major histocompatibility complex class 1 heavy chain as a structural subunit of the human cell membrane insulin receptor: Implications for the range of biological functions of histocompatibility antigens. Proc Natl Acad Sci USA 83:6007–6011, 1986.

88. Fehlmann M, Peyron JF, Samson M, Van Obberghen E, Brandenburg D, Brossette N: Molecular association between major histocompatibility complex class 1 antigens and insulin receptors in mouse liver membranes. Proc Natl Acad Sci USA 82:8364–8367, 1985.

89. Phillips ML, Moule ML, Delovitch DL, Yip CC: Class 1 histocompatibility antigens and insulin receptors: Evidence for interactions. Proc Natl Acad Sci USA 83:3474–3478, 1986.

90. Hansen T, Stagsted J, Pedersen L, Roth RA, Goldstein A, Olsson L: Inhibition of insulin receptor phosphorylation by peptides derived from major histocompatibility complex class 1 antigens. Proc Natl Acad Sci USA 86:3123–3126, 1989.

91. Hsu D, Olefsky JM: Effect of a major histocompatibility complex class I peptide on insulin-like growth factor I receptor internalization and biological signaling. Endocrinology 133:1247–1251, 1993.

92. Olsson L, Goldstein A, Stagsted J: Regulation of receptor internalization by the major histocompatibility complex class 1 molecule. Proc Natl Acad Sci USA 91:9086–8090, 1994.

93. Stagsted J, Reaven GM, Hansen T, Goldstein A, Olsson L: Regulation of insulin receptor functions by a peptide derived from a major histocompatibility complex class 1 antigen. Cell 62:297–307, 1990.

94. Verland S, Simonsen M, Gammeltoft S, Allen H, Flavell RA, Olsson L: Specific molecular interaction between the insulin receptor and a D product of MHC class 1. J Immunol 43:945–951, 1989.

17

MUTATIONAL ACTIVATION OF RECEPTOR TYROSINE KINASES

MATTHIAS LOHMEYER AND WILLIAM J. GULLICK

ICRF Molecular Oncology Laboratory, Hammersmith Hospital, London, W12 0HS, United Kingdom

17.1. INTRODUCTION

It is well established that the great majority of transforming oncogenes are integral members of the cell signaling cascade, involving growth factor receptors, their ligands, or associated downstream kinases. This review focuses on the role of constitutively activated receptor tyrosine kinases (RTKs) associated with a number of pathologic conditions. Constitutive receptor activation may be caused by a number of different mechanisms, the most prominent of which are receptor overexpression, autocrine stimulation, and receptor mutation.

Overexpression of RTKs is a major factor in cellular transformation and can often be correlated directly with therapeutic response and survival in the clinical setting. Moreover, tumor cells often coexpress growth factors as well as their specific receptors, thus subverting normal growth control through unregulated autocrine stimulation. These two important mechanisms of constitutive receptor activation have been reviewed previously (1) and are not the subject of this chapter. Instead, we focus exclusively on activating mutations, some of which are involved in developmental aberrations or malignant transformation. To date, however, transforming RTK mutations have been identified in only a minority of human cancers, but it is possible that this represents an underestimate of their actual involvement.

Hormones and Growth Factors in Development and Neoplasia, Edited by Robert B. Dickson and David S. Salomon

ISBN 0-471-16899-8

In this review, we will focus on activating mutations, as RTK mutations leading to loss of function, such as the insulin receptor (InsR) mutations responsible for non-insulin-dependent diabetes mellitus (NIDDM) or the c-kit mutations leading to the white spotting (rodent) or piebald (human) phenotypes, have been considered in detail elsewhere (2–4).

RTKs fall into about 10 distinct classes based on their broad structural similarities, the nature of their ligands, and the degree of sequence identity in their catalytic domains (5). Essentially all of them, however, have a common layout in which the extracellular ligand binding domain is important for directing ligand-induced oligomerization; the single transmembrane domain serves as the membrane anchor and participates in dimer formation; and the juxtamembrane region is an important determinant of receptor transmodulation. The kinase domains are generally very highly conserved, whereas the carboxy-terminal residues are the most divergent part, even within receptor families (5). The latter region contains many of the tyrosine residues and second-messenger docking sites important for signal transduction.

Activating point mutations and deletions have been identified in all of these different receptor domains. Mechanistically, these mutations fall into four broad classes: 1) mutations that induce ligand-independent dimerization or stabilize the dimeric complex, 2) those that enhance RTK activity apparently independently of the dimerization state, 3) changes that alter the substrate specificity or quality of downstream signaling, and 4) mutations affecting the longevity of the receptor on the cell surface.

Interestingly, receptor activation by different mechanisms is nonequivalent. There are a number of examples where different types of mutations activate the receptors to different extents, causing qualitatively different clinical presentations. These will be explored in more detail below.

The identification and subsequent molecular characterization of naturally occurring activating mutations has provided important insights into receptor function and, together with powerful site-directed mutagenesis analysis, has greatly enhanced our understanding of oncogenic transformation. Moreover, some of the mutations described below, such as a large deletion in the extracellular domain of the epidermal growth factor receptor (EGFR), have formed the basis of new avenues of diagnosis and targeted therapy (6).

17.2 MUTATIONS INDUCING OR STABILIZING RECEPTOR DIMERIZATION

Formation of a receptor dimer following ligand binding is the initial activating step in nearly all RTKs, the exception being the type II (Ins-R) family, which is naturally dimeric. Mutations or truncations that permit dimerization in the absence of ligand or stabilize the dimeric conformation are clearly potential mechanisms for generating a transforming oncogene. In fact, such a mechanism of action is one model for the transforming potency of overexpressed wild-type receptors. This model assumes that unligated receptors are in a dynamic equilibrium between monomer and dimer and that receptor overexpression leads to increased numbers of dimers, thus stimulating cell proliferation even in the absence of ligand (7). As the extracellular domain and the transmembrane domain are the principal determinants of receptor dimerization, many of the mutations affecting the dimerization status have been localized to these two regions.

17.2.1. Activation by Truncation

The most extreme activating change found is the complete deletion of the extracellular domain. This is most commonly seen in virally captured DNA sequences encoding a host RTK. The first such extracellular domain deletion was seen in the transforming v-*erbB* oncogene, where almost the complete extracellular domain of the acquired chicken epidermal growth factor (EGF) receptor was missing (8). This virally captured gene invariably also shows C-terminal deletions, and the coding sequence is often found fused with parts of the viral *gag* or *env* proteins. Other mutations are also present in the majority of viral isolates, although not all are implicated in the potent transforming ability of v-*erbB* (9). It is apparent, however, that deletion of the extracellular domain alone, when caused by viral integration, is sufficient to create a transforming oncogene but that the C-terminal mutations found in v-*erbB* greatly enhance the transformed phenotype (10,11).

Subsequent laboratory studies with truncated human EGFR constructs have also shown that deletion of the extracellular domain of the protein resulted in constitutive activation, ligand-independent tyrosine kinase (TK) activity, and greatly enhanced transforming efficiency (10,12). As with v-*erbB*, however, the ability of truncated EGFR to transform cells was further enhanced by additional C-terminal deletions (10).

The importance of an intact extracellular domain for maintaining negative control over activation has also been suggested with InsR. Although the receptor is a naturally disulfide-bonded dimer, tryptic digestion of the extracellular α subunit resulted in constitutive activation of the β subunit TK (13). Similarly, v-*ros*, a potent transforming oncogene most closely related to the *Drosophila sevenless (sev)* RTK (14,15), and the *sea* proto-oncogene, a relative of the met/hepatocyte growth factor/scatter factor receptor (16), can both be activated by the loss of substantial parts of their extracellular domain. In fact, *sea* was first isolated as an env-sea fusion protein from the avian erythroblastosis virus S13 in which only the cytoplasmic domain of c-*sea* was captured and fused in-frame to the viral *env* gene (17). It was subsequently shown that the env domains are not required for cell transformation, as a myristoylated v-sea protein without any of the *env* sequences retained its transforming ability (18). In this case, fusion with *env* probably provided the signal sequence and membrane anchor required for membrane-spanning proteins without actively directing dimerization.

Given these detailed studies in different RTKs, it appears that the extracellular domain of these receptors exerts some negative control over the intracellular kinase and that deletion of the extracellular domain or its replacement by fusion with an otherwise irrelevant protein can act to activate at least some RTKs.

17.2.2. Activation by Genomic Rearrangement

Activation by genomic rearrangement generally results in N- or C-terminal truncation of the RTK and fusion with a heterologous gene. Moreover, such genomic rearrangements may also place the RTK gene under the control of a different promoter. The oncogenic met receptor was first identified as the result of a genomic rearrangement that generated a hybrid protein composed of parts of the *tpr* gene, coding for a protein of unknown function, and the met-encoded kinase domain (19). This chimera was transforming with a constitutively active kinase domain. In this particular case, the *tpr* sequence was vital for oncogenic transformation, as it contains a leucine zipper motif that directed dimerization of the chimera, thus activating the receptor kinase domain (19).

Somatic rearrangement and fusion with heterologous genes has also been shown to be constitutively activating for both the *ret* and *trk* proto-oncogenes (20–22). Rearrangement of *ret* has been reported in 11–33% of papillary thyroid carcinomas in adult Caucasians (23), as well as in approximately 60% of papillary carcinomas in Russian children exposed to radiation as a result of the Chernobyl reactor accident (reviewed in ref 24). However, although the causative link appears to be established, it is not clear whether ret fusion proteins are transforming due to active dimerization, as in the *met-tpr* chimera, or as a result of diminished control over the constitutive kinase activity.

Activating rearrangements of the platelet-derived growth factor receptor (PDGFR) β have also been reported in a sizable proportion of chronic myelomonocytic leukemia (CMML) patients expected to develop acute myelogenous leukemia (AML) at a later stage (25). In this population, a translocation of chromosomes 5 and 13 generates a chimeric protein composed of the novel transcription factor-like gene *tel* and the transmembrane and tyrosine kinase domains of the PDGFR (25). This fusion protein may be involved in the pathogenesis of CMML, but it is not yet established whether the chimera actually functions as a constitutively activated tyrosine kinase. Moreover, the PDGFR family is conspicuous for its lack of transforming mutations identified either *in vitro* or *in vivo,* so that this translocation may not be the dominant activating lesion in CMML.

17.2.3. Activation by Internal Deletion or Mutation Within the Extracellular Domain

Receptor TKs are also vulnerable to activation by internal deletions and point mutations in the extracellular domain. A case in point is the prominent 801-amino acid deletion within the extracellular domain of EGFR (the type III mutation), identified initially in a large number of gliomas but also apparently found in breast, ovary, and non-small cell lung cell (NSCLC) cells (26,27). This deletion, a result of the elimination of exons 2–7 from the EGFR gene, produces constitutively phosphorylated receptors that enhance the tumorigenic potency of cells when injected into nude mice (6,28–30). However, the type III deletion has been observed only in cells already amplifying the EGFR gene, suggesting that it must precede and indeed may arise as part of the gene amplification process (31). The type III mutant possesses only a partially activated kinase; thus, subsequent amplification and overexpression probably represent a further evolutionary step toward increasing oncogenic potential (31). Other, less extensive tumor-associated deletions within the extracellular domain of EGFR have also been described (types I and II) (32). These in-frame deletions were located in either of the two cysteine-rich domains in the extracytoplasmic domain. The potential receptor-activating mechanism of such deletions is discussed below.

Activating deletions within the extracellular part of PDGFR-α have also been found. In a human glioma, PDGFR-α with an 81-amino acid deletion within the extracellular immunoglobulin-like domain was overexpressed (33).

Extracellular domain control over receptor dimerization can also be subverted by single point mutations. One of the primary examples is the substitution of leucine at codon 301 of the CSF-1/MCSF/fms receptor found in some patients with myelodysplastic syndromes. This mutation may even be induced by some forms of anticancer chemotherapy (34–35), but its relevance to the pathogenesis of acute myeloid leukemia in the majority of patients has been questioned (36). This single mutation confers ligand-independent transforming activity in NIH-3T3 cells and is further enhanced *in vitro* and *in vivo* by a concomitant mutation at codon 969 in the C-terminal tail of the receptor (37–39).

The deletion of Asn281 in the extracellular domain of InsR is also highly pathogenic *in vivo*. This mutation was seen in two cases of familial leprechaunism and resulted in severe insulin desensitization (40). Codon 281 is located in the cys-rich region of the α subunit extracellular domain, and deletion resulted in constitutive activation of the InsR and corresponding cellular desensitization to insulin (40). Therefore, it is likely that Asn281 plays a critical role in the inhibitory constraint of the α over the cytoplasmic β subunits of InsR.

A further example from the *Drosophila* torso RTK illustrates the frequent occurrence of activating point mutations. Torso function is required for the determination of anterior and posterior terminal structures in the embryo, and a number of gain-of-function mutations have been described. The Tyr327Cys, Trp317Arg, and His242Leu point mutations in the extracellular domain all cause ligand-independent activation, resulting in segment defects and enlarged terminal structures in the developing embryo (41).

It is not yet clear how, on a molecular level, the point mutations in fms, InsR, and torso activate their receptors. It can only be hypothesized that mutations in critical residues change the overall conformation of the receptor extracellular domain to one similar to that of the ligand-bound receptor, thus facilitating ligand-independent dimerization. However, there are certain mutations in the extracellular domain in which receptor dimerization can be attributed to a novel activating mechanism. These mutations invariably involve the generation of a free cysteine residue, often reasonably close to the transmembrane domain. This free cysteine may then be in a position to form a disulfide bond with its homolog in a neighboring mutant receptor. In fact, the Tyr327Cys mutation in *Drosophila* torso may be activating the receptor tyrosine kinase by this mechanism.

In practice, cysteine-mediated covalent stabilization of RTK dimers probably represents the mechanism of activation for a significant proportion of the extracellular domain mutations and deletions observed *in vivo*. Clearly, functional free cysteines can be generated either by point mutation, resulting in the expression of a new cysteine residue, or by a suitable deletion/mutation that removes one of a cysteine pair, leaving the other available for intermolecular bonding.

Early work had introduced the concept of disulfide-mediated receptor activation using the erythropoietin receptor as well as the erbB2 receptor TK (42,43). The Arg129Cys mutation identified in the extracellular domain of the erythropoietin receptor was shown to result in a constitutively active and oncogenic receptor (43). Similarly, artificial introduction of an additional cysteine proximal to the transmembrane domain of erbB2 greatly increased the formation of stable receptor dimers, although this did not translate into transforming potency (42). The *ret* oncogene was the first RTK in which an extracellular domain cysteine mutation was detected *in vivo* (44). More than 97% of patients with multiple endocrine neoplasia type 2A (MEN2A) carry the Cys634Arg mutation in the *ret* PTK (44,45). Mutation of a different cysteine residue (Cys618) has also been reported in a number of pheochromocytomas (46). Both of these cysteine mutations in the extracellular domain result in ligand-independent receptor dimerization and correspondingly increased TK activity. Interestingly, the loss-of-function Hirschsprung mutation in the kinase domain of *ret* abolishes the tumorigenicity of Cys634Arg-*ret*. In such MEN2A/Hirschsprung double mutants, levels of covalent receptor dimerization are still high but TK activity is minimal (47,48). Clearly, TK activity, and not dimerization per se, is required for transformation.

Cysteine mutations analogous to those observed in *ret* have also been identified in the FGF-2 receptor, where the Cys332Tyr mutation is responsible for the autosomal dominant Crouzon syndrome or craniosynostosis (49). Again, results indicate that disulfide bond formation and ligand-independent activation of the mutated FGF-2 receptor are important contributors to the etiology of this syndrome (49). The *Caenorhabditis elegans* LET-23 receptor, an EGFR homolog, can also be constitutively activated by point mutation of a conserved cysteine residue in the extracellular domain (50). This was the first such mutation discovered in the EGFR family of growth factor receptors. Work in our own laboratory has shown that a point mutation near the transmembrane domain can also activate the human EGF receptor (N. Tuzi and W. Gullick, unpublished results). It appears that the disruption of certain strategically placed disulfide bonds or mutational introduction of free cysteines can lead to receptor dimerization. In these cases, intermolecular disulfide bond formation produces covalently bonded receptor dimers akin to the insulin receptor family. Such dimers are generally constitutively active.

17.2.4. Activation by Mutations in the Transmembrane Domain

A further mechanism of stabilizing dimer formation involves interactions within the transmembrane domain. The archetypal mutation of this type is that found in the rat homolog of the erbB-2 receptor, known as $p185^{neu}$ or neu. Exposure of rats to chemical carcinogens causes activation of the *neu* gene by a single point mutation within the transmembrane domain. The Val664Glu mutation invariably found in the resulting tumors greatly increases the oncogenic potency of rat *neu* in model systems (reviewed in ref. 51). Stereochemical modeling has suggested that the packing of the transmembrane α helices in the receptor dimer of mutant *neu* is stabilized by additional hydrogen bonds and that this additional stabilization may be responsible for the spontaneous and efficient aggregation of mutant *neu* (52). Subsequent mutational analyses (53) and biophysical experiments (54,55) have provided direct experimental evidence for this model. However, other models have also been proposed (56).

This mechanism of RTK activation observed in rat *neu* is applicable to *neu* homolog from many different species. An identical activating point mutation has also been found in the *neu* gene cloned from nitrosourea-induced tumors in the Syrian hamster (57). Moreover, placing an aspartic acid residue in the equivalent position of the human *neu* equivalent, erbB2, also increased the oncogenic potency of this RTK (58). However, this activating mutation is not seen in human tumors (59), probably because the valine-glutamic acid mutation in erbB2 requires the coordinate mutation of two nucleotides, while in rat *neu* a single point mutation is sufficient.

Introducing a similar mutation into the transmembrane domain of EGFR did not produce a potent transforming oncogene (60,61), but NIH-3T3 cells expressing the mutant showed some characteristics of the transformed phenotype (62). Moreover, moderate ligand-independent activation of the receptor kinase and constitutive activation of the ras pathway were noted (62).

A superficially different transmembrane domain mutation in FGFR-3 is found to be the causative lesion in achondroplasia, characterized by short-limbed dwarfism and macrocephaly (63). Clinically, achondroplasia is essentially homogeneous, with over 98% of all patients carrying a Gly380Arg missense mutation in the transmembrane region of FGFR-3 (64,65). Site-directed mutagenesis studies have shown that the hydrogen bonding potential of the substituting residue is important for the constitutive activation of FGFR-3 (66). This clearly suggests that the mechanism of receptor activation is likely to be similar to that seen with mutant *neu*. Another mutation, the Gly375Cys conversion at the very N-terminal edge of the transmembrane domain in FGFR-3, also leads to achondroplasia (67). In this case, disulfide bonding is probably the most likely mechanism of dimer stabilization.

Curiously, yet another activating FGFR-3 transmembrane domain mutation (Ala391Glu) was not associated with achondroplasia. Instead, the mutation was identified in three unrelated cases of familial Crouzon syndrome (68). Unlike achondroplasia, Crouzon syndrome is a non-dwarfing, non-skeletal condition and is normally associated with constitutively activating mutations of FGFR-2 (49). Clearly, these mutations may affect not only the quantity of the FGFR signal, but also the quality of the transduced signal. This highlights the biologic complexity of growth factor signaling within the FGFR family and, no doubt, equally applies to the other growth factor receptor families.

In yet another system, a screen of SCLC cell lines for activating mutations of the c-*kit* proto-oncogene has also identified a Met541Leu substitution in the transmembrane domain (69). However, whether this is a functional activating mutation or a random event has yet to be ascertained, as this mutation was present in only one of the cell lines tested.

Finally, in experimental models, introducing a Val to Glu mutation at codon 922 into the transmembrane domain of the insulin-like growth factor-I (IGF-I) receptor also resulted in receptors with a constitutively active TK (70), even though the IGF-I receptor is naturally dimeric. Similarly, replacement of the transmembrane domains of InsR with that of the oncogenic rat *neu* sequence produced a constitutively activated receptor (71). The fact that even the naturally dimeric InsR family can be constitutively activated by transmembrane domain mutations suggests that α-helix association and dissociation may be important for signal transduction even in the type II receptor family, which is naturally dimeric.

17.3. MUTATIONS AFFECTING TYROSINE KINASE ACTIVITY, SUBSTRATE SPECIFICITY, AND RECEPTOR TURNOVER

Although the promotion of receptor dimerization appears to be the predominant mechanism of constitutive receptor activation, a large number of mutations have been identified that directly upregulate TK activity and downstream signaling. Dimerization aside, the signaling activity of RTKs is regulated by a number of different mechanisms. These include control of the rate of kinase activity, substrate specificity, receptor transmodulation, and control of receptor availability on the cell surface.

Phosphorylation of specific serine, threonine, and tyrosine residues has been shown to modulate the catalytic rate of the TK, as well as directing the exact substrate specificities toward second messengers such as Grb-2, PI-3-kinase, or PLC-γ, to name but a few. These phosphorylation sites are predominantly located either in the juxtamembrane domain, the kinase insert, or the C-terminal tail of the receptors. The activity of individual RTKs is also subject to built-in restraints, responsive to transmodulation by other growth factor receptors. For example, when PDGF or bombesin receptors are activated by stimulation with the appropriate ligand, cells are not only desensitized to further stimulation by PDGF or bombesin but also lose the ability to respond to exogenous EGF (72). This is achieved rapidly by downregulation of high-affinity EGF-receptor binding sites, but exactly how this is achieved on a molecular level remains to be elucidated. Clearly, given the complex regulatory interactions between the cytoplasmic domains of RTKs and their intracellular substrates, there is much scope for activating mutations. In fact, the juxtamembrane domain, the carboxy-terminal tail, and the kinase domain itself have all yielded different activating mutations that may be contributing to the neoplastic phenotype.

17.3.1. Activation Through Increased Kinase Activity

Mutations enhancing the catalytic rate of the TK have been found in a number of different receptor families. Probably the best-characterized mutation, responsible for the 2B type of multiple endocrine neoplasia (MEN2B), is the Met918Thr conversion in *ret*. This mutation is located in the TK domain itself and has been shown to increase TK domain enzymatic activity, as well as altering the substrate specificity of the TK (73,74). Thus, both the quantity and the quality of the signal are affected. However, it appears that homodimerization of *ret* is still required for the MEN2B-associated mutation to unleash its full transforming ability (75,76).

One of the most informative receptors with respect to activating mutations in the cytoplasmic part of the receptor is the c-*kit* proto-oncogene. Sequencing of c-*kit* cDNA from the human mast cell leukemia cell line HMC-1 identified two point mutations that constitutively activated the receptor even in the absence of its ligand, stem cell factor (77). The Asp816Val mutation is located within the TK domain in close proximity to a major tyrosine phosphorylation site. Corresponding Asp mutations in codons 814 and 817 have also been identified in the mouse P-815 and rat RBL-2H3 mastocytoma cell lines, respectively (78,79). A number of different studies have shown that mutation of this particular Asp residue in c-*kit* induced ligand-independent mast cell growth *in vitro,* as well as mast cell differentiation and oncogenic transformation *in vivo* (77,80). Curiously, mutations of Asp814 to a number of different residues all resulted in constitutively active murine *kit* receptors, while deletion of this residue abolished TK activation and receptor phosphorylation completely (81).

The Asp816Val mutation has also recently been identified in mast cells from human patients suffering from aggressive, potentially neoplastic, mastocytosis (82). This mutation was found exclusively in tumor cells and in only one of the two c-*kit* expressing alleles, suggesting that the lesion arose spontaneously. It is interesting to note that mutation of only one c-*kit* allele was sufficient to promote clonal expansion of the affected cell, resulting in extensive mastocytosis and hyperpigmentation (urticaria pigmentosa), consistent with c-*kit*'s behaving as a dominant oncogene (82). This demonstrates clearly that the specific TK domain mutation identified in c-*kit* from rat, mouse, and human cell lines is relevant to the clinical development of mast cell neoplasms.

A second mutation identified in c-*kit* from the human HMC-1 cell line replaced Val560 with glycine (77). Introducing the corresponding Val559Gly mutation into murine c-*kit* rendered it constitutively active (83,84). Residue 559/560 is located in the juxtamembrane domain, and analysis has shown that mutant c-*kit* with this single mutation dimerized even in the absence of stem cell factor (83). In contrast, c-*kit* bearing the Val814 mutation in their kinase domain did not dimerize spontaneously (84). Clearly, these two mutations both activate c-*kit* constitutively by two distinct mechanisms.

When tumor cells bearing both the Gly559 and Val814 (murine) c-*kit* mutations were injected into nude mice, all animals rapidly succumbed to leukemia and died (84,85). Moreover, transgenic mice expressing the *kit* Val814 mutation alone developed early acute leukemia or malignant lymphomas (80,85). Thus, the corresponding mutations, as already identified in human mast cell carcinomas, may well contribute to the development of human neoplasias.

Like the HMC-1, P-815, and RBL-2H3 cell lines discussed earlier, the mouse mastocytoma cell line FMA-3 also expresses a constitutively active *kit* receptor. However, in this case, sequencing of the c-*kit* coding region revealed a 21-base pair in-frame deletion, removing seven amino acids from the juxtamembrane region (codons 573–579) (86). When c-*kit* bearing this deletion was expressed in cultured murine mast cells, the receptor dimerized independently of SCF binding and was constitutively activated and phosphorylated (86). As this deletion and the Gly559 mutation illustrate, the juxtamembrane domain, in addition to the extracellular domain or the transmembrane region, is also involved in the control of receptor dimerization. However, although it is conceptually clear how free cysteines or increased hydrogen-bonding interactions may induce dimerization, the important molecular interactions between two juxtamembrane domains remain to be elucidated.

Probably the most dramatic enhancement of TK activity was reported for the Lys650Glu mutant of FGFR-3. This mutation is associated *in vivo* with thanatophoric dysplasia type II (TDII), a skeletal disorder characterized by neonatal lethality and very profound dwarfism (64). TDII is a genetically homogeneous disease, with essentially 100% of affected individuals carrying an identical mutation (64). The Lys650Glu substitution is located in the activating loop of the TK domain of FGFR-3 and results in constitutive activation of the kinase to more than 100-fold above that of the wild-type receptor (87).

A much less profound phenotype is seen with the Asn540Lys mutation in the first TK domain of FGFR-3. This Asn540Lys conversion was detected in only 60% of individuals with hypochondroplasia (88). FGFR-3 activity was still upregulated, but was much less pronounced than in either achondroplasia or TDII. Moreover, in 40% of cases, FGFR-3 is wild type and FGFR-3-unrelated lesions are responsible for the hypochondroplasia phenotype. This indicates that the Asn540Lys TK domain mutation probably affects the quality of the receptor signal rather than the level of kinase hyperactivation (reviewed in ref. 64).

Direct TK activation is also seen with the Asp802Val mutation in the CSF-1/MCSF/fms receptor (89). This substitution in the TK domain is located in a region important for triggering kinase activation. Replacement of leucine 301, as already discussed, effects spontaneous receptor dimerization, and transforming activity can be enhanced by the Tyr969Phe substitution near the carboxyl terminus at a site shown to be involved in negative regulatory activity (90). Mutation at this locus was reported in 14/110 (12.7%) patients with human myeloid malignancies (91). Two further activating mutations were detected within the C-terminal tail of the fms receptor. In the FDC-P1 hematopoietic cell line, the substitutions Ser929Gly or Leu926Pro both constitutively activated fms (89). These additional observations reinforce the concept that in c-fms, residues in the C-terminal tail exert important negative regulatory control over receptor kinase activity.

Similar conclusions have also been reached regarding the EGFR family. Here the C termini also appear to modulate the intrinsic TK activities rather than affect substrate specificity per se (92). Interestingly, when the carboxyl terminal of EGFR was replaced with that of the highly transforming erbB2 receptor, the TK activity and transforming ability of EGFR were much enhanced (92). The specificity for mitogenic signaling, on the other hand, appeared to reside in the juxtamembrane domain and the amino-terminal half of the TK domain of EGFR and ErbB2 (93,94).

Clearly, EGFR and ErbB2 differ not only in their relative TK activities, but also in their efficiency of signal transduction to different substrate molecules (95). For example, within the EGFR family only EGFR itself interacts with the cbl protein (96). It is likely that a combination of these two factors accounts for the comparatively lower transforming potency of EGFR—even in its activated state.

In a study comparing the relative kinase activities of chicken EGFR and three different avian ErbB proteins, the latter showed a more than threefold higher apparent V_{max} for autophosphorylation, while the apparent K_m values for ATP were broadly similar (97). Moreover, the three different ErbB proteins all had distinctive TK activities, suggesting that the different mutations and deletions in their cytoplasmic domains directly affected the kinetics of the TK (97). Whether any of the mutations or the C-terminal deletions affected the substrate specificity of the erbB proteins was not investigated, but these may also be important for imparting the distinct disease potentials observed (97).

17.3.2. Activation Through Altered Substrate Specificity

The juxtamembrane, the kinase insert, and the C-terminal tail of RTKs are populated with many of the tyrosine phosphorylation sites vital for signal transduction. In addition, regulatory sites involving phosphorylated serine or threonine and docking sites for various second messenger molecules are located in these domains. Naturally, deletions, mutations, or truncations in any of these domains may affect not only substrate phosphorylation kinetics (by removal of negative kinase control residues) but also substrate specificity (by elimination of docking/interaction sites) (98). Truncations of the C termini are particularly prevalent in virally captured activated oncogenes, which may account for some of the altered substrate specificities observed. However, it is unclear whether changes in substrate specificity alone can confer the transformed phenotype without the support of a genuinely kinase-activating mutation.

This contention is borne out in studies examining the oncogenic potential of v-ErbB (9). Sequence analyses of v-*erbB* genes isolated from a chicken infected with rous associated virus 1 (RAV-1) revealed five different isoforms, all with identical amino and carboxy-terminal truncations, originating from the same capture event. One such v-*erbB* gene had apparently not undergone any further mutation and lacked a detectable capacity to transform either fibroblasts or erythroblasts. Two other isolates, one with a mutation within the kinase domain, the other with a frameshift termination just after the kinase domain, differed in their potential to induce fibroblast and erythroblast transformation (9). Others have noted that mutant forms of v-ErbB with varying deletion and mutation patterns differ in their ability to transform erythroid cells, whereas their ability to transform fibroblasts appear unaffected (99). Clearly, the capacity to dissociate these two transforming abilities on the basis of mutations affecting phosphotyrosine residues is a clear indicator that substrate specificity is an important determinant of transformation potential as well as tumor type.

Examination of the transforming oncogene encoded by the Hardy-Zuckermann-4 strain of the feline sarcoma virus showed that this virus has captured part of the feline c-*kit* sequence. Both the extracellular domain and the transmembrane domain are absent, and the remaining cytoplasmic part of c-*kit* also contains a number of mutations (100). It has been proposed that the greatly enhanced v-*kit* oncogenicity is a result of the two-codon deletion (Tyr569 + Val570) in the juxtamembrane domain. This mutation deletes an important tyrosine residue and repositions the adjacent tyr571 into a context where its phosphorylation can promote transformation (100).

An example of a change in receptor substrate specificity by a single point mutation is the Met918Thr substitution in the TK domain of *ret*. As mentioned above, this mutation causes MEN2A and results in upregulated receptor kinase activity. However, this mutation also results in a shift in peptide substrate specificity that may well be responsible for conferring the particular type 2B phenotype (98).

Changes in substrate specificity have also been reported for a C-terminal deletion of 69 amino acids in InsR. This truncation enhanced InsR ligand-dependent activation of mitogen-activated protein (MAP) kinase activity while not affecting the phosphorylation kinetics of the 90-kDa S6 kinase in CHO cells (101).

Receptor dimerization and activation of tyrosine kinases are comparatively easy to ascertain, whereas the intrinsic complexity of downstream signaling is slowing progress in areas relating to the quality of transforming signals. Mutational studies have shown how a number of transforming mutations, especially those involving tyrosine phosphorylation residues or residues in SH2/SH3 consensus binding sites, will affect downstream signaling. However, very little information is available on mutations relevant to the clinical manifestations of disease. Clearly, the nature of the transforming signal is dependent not only on the nature and mutational status of the receptor-oncogene, but also on the predisposition of the cell type involved.

17.3.3. Inhibition of Receptor Internalization

Irrespective of activating mutations that enhance dimerization or TK activity, the efficacy of receptor signaling is also determined by the receptors' intracellular processing and the speed with which receptors are turned over at the cell surface. The rate of receptor endocytosis appears to be controlled by the phosphorylation of key residues located in the juxtamembrane domain or the C-terminal tail of the molecule. As summarized above, many receptors, including EGFR, InsR, and mutant *neu,* are constitutively phosphorylated as a result of activating mutations, and this has been shown to enhance their turnover (88,102,103). In the case of ErbB2, it has been shown that the activated oncogene has a half-life of only about 1.5 hr, compared to over 7 hr for wild-type ErbB2 (104). It has also been postulated that the different transforming potential of mutant fms in rat-2 fibroblasts and FDC-P1 cells was due to greatly enhanced receptor degradation in the rat-2 cells (89).

Clearly, endocytosis and the control of receptor degradation are important mechanisms for keeping even constitutively activated receptors in check. In some cases, however, deletions within the cytoplasmic domain or mutations involving critical tyrosine residues actually remove endocytotic control, resulting in a longer half-life of the mutated receptor. It is unlikely that such lesions alone would be sufficient to transform normal cells, but they may well compound the effect of other activating mutations.

Inhibition of receptor recycling has been noted in the mouse c-*kit* mutant, where seven amino acids (including Tyr577) were deleted from the juxtamembrane domain (86). A part of the juxtamembrane domain of fms has similarly been implicated in the control of receptor internalization where mutation of Tyr569 to alanine eliminated ligand-induced receptor endocytosis (105). Moreover, the four-amino acid sequence surrounding Tyr569 in c-*fms* facilitated endocytosis when introduced into the juxtamembrane region of a glycophorin A construct (105). Other detailed studies have shown that the receptor sites controlling the internalization and degradation signals are located in different parts of the receptor (106). Receptor degradation of *fms* required an intact kinase insert region and intrinsic receptor TK activity, whereas receptor internalization did not. In fact, the authors suggest that the signal for *fms* degradation is the phosphorylation of a second protein that binds to the kinase insert region of the receptor (106).

In EGFR, the carboxyl tail, and serine residues 1046 and 1047 in particular, may control downregulation of the receptor. In EGFR mutants where these CaMKII phosphorylation sites are removed, TK activity and downstream signaling are potentiated on EGF binding (107). Thus serine phosphorylation in the carboxy terminal region of EGFR appears to be important for directing receptor downregulation after EGF binding (107). Studies with EGFR/ErbB2 chimeras have also identified the importance of the C-terminal domain for the regulation of ligand-induced internalization (108).

17.4. SUMMARY

In summary, constitutive activation of various growth factor receptors can result in many different clinical pathologies, such as cancer, altered sensitivity to metabolic hormones, dwarfism, and craniosynostosis. Clearly, the identity of the activated receptor is an important determinant of the resulting phenotype, but even different types of activating mutation in the same receptor can often be distinguished by their different clinical manifestations. A case in point is the *ret* proto-oncogene, in which a constitutively dimerizing mutation is responsible for >97% of MEN2A papillary thyroid tumors and an apparently similarly activating TK-domain mutation is associated almost exclusively with MEN2B. The latter mutation still requires ligand-dependent dimerization for its full oncogenic potential, and this may provide the molecular explanation of why MEN2B tumors are not associated with parathyroid disease and are often less aggressive than those of the MEN2A type (44).

A similar gradation of clinical severity has been found with FGFR-3 mutations responsible for various degrees of dwarfism. Chondrodysplasias are characterized by short-limbed dwarfism and macrocephaly. The least severe form of this condition, hypochondroplasia, is genetically heterogeneous where about 60% of cases appear to be associated with an activating mutation (Asn540Lys) in the first part of the TK domain of FGFR-3 (88). In other cases, the FGFR-3 locus appears normal. Achondroplasia, the most common form of dwarfism, produces a more severe but nonlethal phenotype. Over 97% of patients with achondroplasia carry a characteristic Gly380Arg mutation in one allele of FGFR-3 (63). Where the condition is homozygous, the Gly380Arg mutation produces the same phenotype as thanatophoric dysplasia (TD). TD is neonatal lethal and produces very profound dwarfism. A powerful kinase-activating mutation in FGFR-3 is responsible for essentially 100% of TD type II cases (64,87). TD type I is associated with a number of different activating mutations, predominantly the Arg248Cys substitution in the extracellular domain of FGFR-3 (109).

The *ret* and FGFR-3 mutations provide elegant molecular explanations of how different types of activating mutation may be involved in the generation of distinct clinical phenotypes. However, there is much still to be discovered about the molecular functioning of growth factor receptors and, importantly, about how these functionalities are disturbed in activated oncogenes. This information will be of great use in the future design of specifically targeted antioncogene therapeutics. In theory at least, effective neutralization of the principal transforming lesion should prove to be more efficacious than current approaches that target downstream proliferation. Until this can be achieved, mutations such as the type III deletion in the extracellular domain of EGFR are providing new opportunities for targeting antiproliferative therapy to tumor cells. This mutant receptor features a novel and unique epitope, which has been used successfully in antibody targeting. As laboratory research and genome sequencing efforts combine to reveal more and more about how receptors interact with downstream signaling molecules, the rational design of selective inhibitors of activated receptor protein kinases becomes ever more attainable.

REFERENCES

1. Salomon DS, Brandt R, Ciardiello F, Normanno N: Epidermal growth factor-related peptides and their receptors in human malignancies. Crit Rev Oncol Hematol 19: 183–232, 1995.
2. Tan JC, Nocka K, Ray P, Traktman P, Besmer P: The dominant W42 spotting phenotype results from a missense mutation in the c-kit receptor kinase. Science 247:209–212, 1990.
3. Fleischmann RA: Human piebald trait resulting from a dominant negative mutant allele of the c-kit membrane receptor gene. J Clin Invest 89:1713–1717, 1992.
4. Kahn CR, Vicent D, Doria A: Genetics of non-insulin-dependent (type-II) diabetes mellitus. Annu Rev Med 47:509–531, 1996.
5. Heldin C-H: Protein tyrosine kinase receptors. In Tooze J, Parker PJ, TP (eds): Cancer Surveys: Cell Signalling. Plainview, NY: Cold Spring Harbour Laboratory Press, 1996, pp. 7–24.
6. Hills D, Rowlinson-Busza G, Gullick WJ: Specific targeting of a mutant, activated EGF receptor found in glioblastoma using a monoclonal antibody. Int J Cancer 63:537–543, 1995.

7. Heldin C-H: Dimerization of cell surface receptors in signal transduction. Cell 80:213–223, 1995.
8. Downward J, Parker P, Waterfield MD: Autophosphorylation sites on the epidermal growth factor receptor. Nature 311:483–485, 1984.
9. Vennstrom B, Raynoscheck C, Jansson L, Doederlein G, Lhotak V, Johnsson A, Beug H: Retroviral capture of c-erbB proto-oncogene sequences: Rapid evolution of distinct viral genomes carrying mutant v-erbB genes with different transforming capacities. Oncogene 9:1307–1320, 1994.
10. Haley JD, Hsuan JJ, Waterfield MD: Analysis of mammalian fibroblast transformation by normal and mutated human EGF receptors. Oncogene 4:273–283, 1989.
11. Massoglia S, Gray A, Dull TJ, Munemitsu S, Kun HJ, Schlessinger J, Ullrich A: Epidermal growth factor receptor cytoplasmic domain mutations trigger ligand-independent transformation. Mol Cell Biol 10:3048–3055, 1990.
12. Chantry A: The kinase domain and membrane localization determine intracellular interactions between epidermal growth factor receptors. J Biol Chem 270:3068–3073, 1995.
13. Shoelson SE, White MF, Kahn CR: Tryptic activation of the insulin receptor. Proteolytic truncation of the alpha-subunit releases the beta-subunit from inhibitory control. J Biol Chem 263:4852–4860, 1988.
14. Birchmeier C, Birnbaum D, Waitches G, Fasano O, Wigler M: Characterization of an activated human ros gene. Mol Cell Biol 6:3109–3016, 1986.
15. Zong CS, Poon B, Chen J, Wang LH: Molecular and biochemical bases for activation of the transforming potential of the proto-oncogene c-ros. J Virol 67:6453–6462, 1993.
16. Huff JL, Jelinek MA, Borgman CA, Lansing TJ, Parsons JT: The protooncogene c-sea encodes a transmembrane protein-tyrosine kinase related to the Met/hepatocyte growth factor/scatter factor receptor. Proc Natl Acad Sci USA 90:6140–6144, 1993.
17. Smith DR,Vogt PK, Hayman MJ: The v-sea oncogene of avian erythroblastosis retrovirus S13: Another member of the protein-tyrosine kinase gene family. Proc Natl Acad Sci USA 86:5291–5295, 1989.
18. Crowe AJ, Hayman MJ: A myristylated form of the sea oncoprotein can transform chicken embryo fibroblasts. J Virol 65:2533–2538, 1991.
19. Rodrigues GA, Park M: Dimerization mediated through a leucine zipper activates the oncogenic potential of the met receptor tyrosine kinase. Mol Cell Biol 13:6711–6722, 1993.
20. Borrello MG, Pelicci G, Arighi E, De-Filippis L, Greco A, Bongarzone I, Rizzetti M, Pelicci PG, Pierotti MA: The oncogenic versions of the Ret and Trk tyrosine kinases bind Shc and Grb2 adaptor proteins. Oncogene 9:1661–1668, 1994.
21. Santoro M, Grieco M, Melillo RM, Fusco A, Vecchio G: Molecular defects in thyroid carcinomas: Role of the RET oncogene in thyroid neoplastic transformation. Eur J Endocrinol 133:513–522, 1995.
22. Herrmann ME: Proto-RET is rearranged in the new human papillary thyroid cancer cell line PTC-1113A. Cancer Genet Cytogenet 89:14–20, 1996.
23. Santoro M, Carlomagno F, Hay ID, Herrmann MA, Grieco M, Melillo R, Pierotti MA, Bongarzone I, Della-Porta G, Berger N: Ret oncogene activation in human thyroid neoplasms is restricted to the papillary cancer subtype. J Clin Invest 89:1517–1522, 1992.
24. Takahashi M: Oncogenic activation of the ret protooncogene in thyroid cancer. Crit Rev Oncog 6:35–46, 1995.

25. Golub TR, Barker GF, Lovett M, Gilliland DG: Fusion of PDGF receptor β to a novel *ets*-like gene, *tel,* in chronic myelomonocytic leukemia with t(5;12) chromosomal translocation. Cell 77:307–316, 1994.
26. Garcia-de-Palazzo IE, Adams GP, Sundareshan P, Wong AJ, Testa JR, Bigner DD, Weiner LM: Expression of mutated epidermal growth factor receptor by non-small cell lung carcinomas. Cancer Res 53:3217–3220, 1993.
27. Wikstrand CJ, Hale LP, Batra SK, Hill ML, Humphrey PA, Kurpad SN, McLendon RE, Moscatello D, Pegram CN, Reist CJ, Traweek ST, Wong AJ, Zalotsky MR, Bigner DD: Monoclonal antibodies against EGFRvIII are tumor specific and react with breast and lung carcinomas and malignant gliomas. Cancer Res 55:3140–3148, 1995.
28. Nishikawa R, Ji XD, Harmon RC, Lazar CS, Gill GN, Cavenee WK, Huang HJ: A mutant epidermal growth factor receptor common in human glioma confers enhanced tumorigenicity. Proc Natl Acad Sci USA 91:7727–7731, 1994.
29. Montgomery RB, Moscatello DK, Wong AJ, Cooper JA, Stahl WL: Differential modulation of mitogen-activated protein (MAP) kinase/extracellular signal-related kinase kinase and MAP kinase activities by a mutant epidermal growth factor receptor. J Biol Chem 270:30562–30566, 1995.
30. Moscatello DK, Montgomery RB, Sundareshan P, McDanel H, Wong MY, Wong AJ: Transformation and altered signal transduction by a naturally occurring mutant EGF receptor. Oncogene 13:85–96, 1996.
31. Ekstrand AJ, Sugawa N, James CD, Collins VP: Amplified and rearranged epidermal growth factor receptor genes in human glioblastomas reveal deletions of sequences encoding portions of the N- and/or C-terminal tails. Proc Natl Acad Sci USA 89:4309–4313, 1992.
32. Humphrey PA, Gangarosa LM, Wong AJ, Archer GE, Lund-Johansen M, Bjerkvig R, Laerum OD, Friedman HS, Bigner DD: Deletion-mutant epidermal growth factor receptor in human gliomas: Effects of type II mutation on receptor function. Biochem Biophys Res Commun 178:1413–1420, 1991.
33. Kumabe T, Sohma Y, Kayama T, Yoshimoto T, Yamamoto T: Amplification of alpha-platelet-derived growth factor receptor gene lacking an exon coding for a portion of the extracellular region in a primary brain tumor of glial origin. Oncogene 7:627–633, 1992.
34. Tobal K, Pagliuca A, Bhatt B, Bailey N, Layton DM, Mufti GJ: Mutation of the human FMS gene (M-CSF receptor) in myelodysplastic syndromes and acute myeloid leukemia. Leukemia 4:486–489, 1990.
35. Stephenson J, Mufti GJ, Yoshida Y: Myelodysplastic syndromes: From morphology to molecular biology. Part II. The molecular genetics of myelodysplasia. Int J Hematol 57:99–112, 1993.
36. Springall F, O'Mara S, Shounan Y, Todd A, Ford D, Iland H: c-fms point mutations in acute myeloid leukemia: fact or fiction? Leukemia 7:978–985, 1993.
37. Roussel MF, Dull TJ, Rettenmier CW, Ralph P, Ullrich A, Sherr CJ: Transforming potential of the c-fms proto-oncogene (CSF-1 receptor). Nature 325:549–552, 1987.
38. Roussel MF, Downing JR, Ashmun RA, Rettenmier CW, Sherr CJ: Colony-stimulating factor 1-mediated regulation of a chimeric c-fms/v-fms receptor containing the v-fms-encoded tyrosine kinase domain. Proc Natl Acad Sci USA 85:5903–5907, 1988.
39. Woolford J, McAuliffe A, Rohrschneider LR: Activation of the feline c-fms proto-oncogene: Multiple alterations are required to generate a fully transformed phenotype. Cell 55:965–977, 1988.
40. Debois-Mouthon C, Sert-Langeron C, Magre J, Oreal E, Blivet MJ, Flori E, Besmond

C, Capeau J, Caron M: Deletion of Asn281 in the alpha-subunit of the human insulin receptor causes constitutive activation of the receptor and insulin desensitization. J Clin Endocrinol Metab 81:719–727, 1996.

41. Sprenger F, Nüsslein-Volhard C: Torso receptor activity is regulated by a diffusible ligand produced at the extracellular terminal regions of the Drosophila egg. Cell 71:987–1001, 1992.
42. Cao HN, Bangalore L, Dompe C, Bormann BJ, Stern DF: An extra cysteine proximal to the transmembrane domain induces differential cross-linking of p185neu and p185neu. J Biol Chem 267:20489–20492, 1991.
43. Watowich SS, Yoshimura A, Longmore GD, Hilton DJ, Yoshimura Y, Lodish HF: Homodimerization and constitutive activation of the erythropoietin receptor. Proc Natl Acad Sci USA 89:2140–2144, 1992.
44. Mulligan LM, Eng C, Healey CS, Clayton D, Kwok JB, Gardner E, Ponder MA, Frilling A, Jackson CE, Lehnert H, Neumann HPH, Thibodeau SN, Ponder BAJ: Specific mutations of the Ret proto-oncogene are related to disease phenotype in MEN 2A and FMTC. Nat Genet 6:70–74, 1994.
45. Asai N, Iwashita T, Matsuyama M, Takahashi M: Mechanism of activation of the ret proto-oncogene by multiple endocrine neoplasia 2A mutations. Mol Cell Biol 15:1613–1619, 1995.
46. Wada M, Asai N, Tsuzuki T, Maruyama S, Ohiwa M, Imai T, Funahashi H, Takagi H, Takahashi M: Detection of ret homodimers in MEN 2A-associated pheochromocytomas. Biochem Biophys Res Commun 218:606–609, 1996.
47. Carlomagno F, De-Vita G, Berlingieri MT, de-Franciscis V, Melillo RM, Colantuoni V, Kraus MH, Di-Fiore PP, Fusco A, Santoro M: Molecular heterogeneity of RET loss of function in Hirschsprung's disease. EMBO J 15:2717–2725, 1996.
48. Cosma MP, Panariello L, Quadro L, Dathan NA, Fattoruso O, Colantuoni V: A mutation in the RET proto-oncogene in Hirschsprung's disease affects the tyrosine kinase activity associated with multiple endocrine neoplasia type 2A and 2B. Biochem J 314:397–400, 1996.
49. Neilson KM, Friesel RE: Constitutive activation of fibroblast growth factor receptor-2 by a point mutation associated with Crouzon syndrome. J Biol Chem 270:26037–26040, 1995.
50. Katz WS, Lesa GM, Yannoukakos D, Clandinin TR, Schlessinger J, Sternberg PW: A point mutation in the extracellular domain activates LET-23, the Caenorhabditis elegans epidermal growth factor receptor homolog. Mol Cell Biol 16:529–537, 1996.
51. Dougall WC, Qian X, Peterson NC, Miller MJ, Samanta A, Greene MI: The neu-oncogene: Signal transduction pathways, transformation mechanisms and evolving therapies. Oncogene 9:2109–2123, 1994.
52. Sternberg MJE, Gullick WJ: Neu receptor dimerisation. Nature 339:587, 1989.
53. Cao H, Bangalore L, Bormann BJ, Stern DF: A subdomain in the transmembrane domain is necessary for p185neu* activation. EMBO J 11:923–932, 1992.
54. Gullick WJ, Bottomley AC, Lofts FJ, Doak DG, Mulvey D, Newman R, Crumpton MJ, Sternberg MJE, Campbell ID: Three-dimensional structure of the transmembrane region of the proto-oncogenic and oncogenic forms of the neu protein. EMBO J 11:43–48, 1992.
55. Smith SO, Smith CS, Bormann BJ: Strong hydrogen bonding interactions involving a buried glutamic acid in the transmembrane domain sequence of the neu/erbB-2 receptor. Nature Struct Biol 3:252–258, 1996.

56. Brandt-Rauf PW, Rackovsky S, Pincus MR: Correlation of the structure of the transmembrane domain of the neu oncogene-encoded p185 protein with its function. Proc Natl Acad Sci USA 87:8660–8664, 1990.
57. Nakamura T, Ushijima T, Ishizaka Y, Nagao M, Arai M, Yamazaki Y, Ishikawa T: Cloning and activation of the Syrian hamster neu proto-oncogene. Gene 140:251–255, 1994.
58. Segatto O, King CR, Pierce JH, Di Fiore PP, Aaronson SA: Different structural alterations upregulate in vitro tyrosine kinase activity and transforming potency of the erbB-2 gene. Mol Cell Biol 8:5570–5574, 1988.
59. Lemoine NR, Staddon S, Dickson C, Barnes DM, Gullick WJ: Absence of activating transmembrane mutations in the c-erbB-2 proto-oncogene in human breast cancer. Oncogene 5:237–239, 1990.
60. Kashles O, Szapary D, Bellot F, Ullrich A, Schlessinger J, Schmidt A: Ligand-induced stimulation of epidermal growth factor receptor mutants with altered transmembrane regions. Proc Natl Acad Sci USA 85:9567–9571, 1988.
61. Carpenter CD, Ingraham HA, Cochet C, Walton GM, Lazar CS, Sowadski JM, Rosenfeld MG, Gill GN: Structural analysis of the transmembrane domain of the epidermal growth factor receptor. J Biol Chem 266:5750–5755, 1991.
62. Miloso M, Mazzotti M, Vass WC, Beguinot L: SHC and GRB-2 are constitutively activated by an epidermal growth factor receptor with a point mutation in the transmembrane domain. J Biol Chem 270:19557–19562, 1995.
63. Rousseau F, Bonaventure J, Legeai-Mallet L, Pelet A, Rozet JM, Maroteaux P, Le-Merrer M, Munnich A: Mutations in the gene encoding fibroblast growth factor receptor-3 in achondroplasia. Nature 371:252–254, 1994.
64. Bonaventure J, Rousseau F, Legeai-Mallet L, Le-Merrer M, Munnich A, Maroteaux P: Common mutations in the fibroblast growth factor receptor 3 (FGFR 3) gene account for achondroplasia, hypochondroplasia, and thanatophoric dwarfism. Am J Med Genet 63:148–154, 1996.
65. Niu DM, Hsiao KJ, Wang NH, Chin LS, Chen CH: Chinese achondroplasia is also defined by recurrent G380R mutations of the fibroblast growth factor receptor-3 gene. Hum Genet 98:65–67, 1996.
66. Webster MK, Donoghue DJ: Constitutive activation of fibroblast growth factor receptor 3 by the transmembrane domain point mutation found in achondroplasia, EMBO J 15:520–527, 1996.
67. Ikegawa S, Fukushima Y, Isomura M, Takada F, Nakamura Y: Mutations of the fibroblast growth factor receptor-3 gene in one familial and six sporadic cases of achondroplasia in Japanese patients. Hum Genet 96:309–311, 1995.
68. Meyers GA, Orlow SJ, Munro IR, Przylepa KA, Jabs EW: Fibroblast growth factor receptor 3 (FGFR3) transmembrane mutation in Crouzon syndrome with acanthosis nigricans. Nat Genet 11:462–464, 1995.
69. Sekido Y, Takahashi T, Ueda R, Takahashi M, Suzuki H, Nishida K, Tsukamoto T, Hida T, Shimokata K, Zsebo KM, Recombinant human stem cell factor mediates chemotaxis of small-cell lung cancer cell lines aberrantly expressing the c-kit protooncogene. Cancer Res 53:1709–1714, 1993.
70. Takahashi K, Yonezawa K, Nishimoto I: Insulin-like growth factor I receptor activated by a transmembrane mutation. J Biol Chem 270:19041–19045, 1995.
71. Yamada K, Goncalves E, Kahn CR, Shoelson SE: Substitution of the insulin receptor

transmembrane domain with the c-neu/erbB2 transmembrane domain constitutively activates the insulin receptor kinase in vitro. J Biol Chem 267:12452–12461, 1992.

72. Collins MK, Sinnett-Smith JW, Rozengurt E: Platelet-derived growth factor treatment decreases the affinity of the epidermal growth factor receptors of Swiss 3T3 cells. J Biol Chem 258:11689–11693, 1983.

73. Santoro M, Carlomagno F, Romano A, Bottaro DP, Dathan NA, Grieco M, Fusco A, Vecchio G, Matoskova B, Kraus MH, Di Fiore PP: Activation of RET as a dominant transforming gene by germline mutations of MEN2A and MEN2B. Science 267:381–383, 1995.

74. Pandit SD, Donis-Keller H, Iwamoto T, Tomich JM, Pike LJ: The multiple endocrine neoplasia type 2B point mutation alters long-term regulation and enhances the transforming capacity of the epidermal growth factor receptor. J Biol Chem 271:5850–5858, 1996.

75. Borrello MG, Smith DP, Pasini B, Bongarzone I, Greco A, Lorenzo MJ, Arighi E, Miranda C, Eng C, Alberti L, Bocciardi R, Mondellini P, Scopsi L, Romeo G, Ponder BAJ, Pierotti MA: RET activation by germline MEN2A and MEN2B mutations. Oncogene 11:2419–2427, 1995.

76. Xing S, Smanik PA, Oglesbee MJ, Trosko JE, Mazzaferri EL, Jhiang SM: Characterization of ret oncogenic activation in MEN2 inherited cancer syndromes. Endocrinology 137:1512–1519, 1996.

77. Kanakura Y, Furitsu T, Tsujimura T, Butterfield JH, Ashman LK, Ikeda H, Kitayama H, Kanayama Y, Matsuzawa Y, Kitamura Y: Activating mutations of the c-kit proto-oncogene in a human mast cell leukemia cell line. Leukemia 8:S18–S22, 1994.

78. Tsujimura T, Furitsu T, Morimoto M, Isozaki K, Nomura S, Matsuzawa Y, Kitamura Y, Kanakura Y: Ligand-independent activation of c-kit receptor tyrosine kinase in a murine mastocytoma cell line P-815 generated by a point mutation. Blood 83:2619–2626, 1994.

79. Tsujimura T, Furitsu T, Morimoto M, Kanayama Y, Nomura S, Matsuzawa Y, Kitamura Y, Kanakura Y: Substitution of an aspartic acid results in constitutive activation of c-kit receptor tyrosine kinase in a rat tumor mast cell line RBL-2H3. Int Arch Allergy Immunol 106:377–385, 1995.

80. Piao X, Bernstein A: A point mutation in the catalytic domain of c-kit induces growth factor independence, tumorigenicity, and differentiation of mast cells. Blood 87: 3117–3123, 1996.

81. Hallek M, Danhauser-Riedl S, Herbst R, Warmuth M, Winkler A, Kolb HJ, Druker B, Griffin JD, Emmerich B, Ullrich A: Interaction of the receptor tyrosine kinase p145c-kit with the p210bcr/abl kinase in myeloid cells. Br J Haematol 94:5–16, 1996.

82. Longley BJ, Tyrrell L, Lu SZ, Ma YS, Langley K, Ding TG, Duffy T, Jacobs P, Tang LH, Modlin I: Somatic c-KIT activating mutation in urticaria pigmentosa and aggressive mastocytosis: Establishment of clonality in a human mast cell neoplasm. Nat Genet 12:312–314, 1996.

83. Furitsu T, Tsujimura T, Tono T, Ikeda H, Kitayama H, Koshimizu U, Sugahara H, Butterfield JH, Ashman LK, Kanayama Y: Identification of mutations in the coding sequence of the proto-oncogene c-kit in a human mast cell leukemia cell line causing ligand-independent activation of c-kit product. J Clin Invest 92:1736–1744, 1993.

84. Kitayama H, Kanakura Y, Furitsu T, Tsujimura T, Oritani K, Ikeda H, Sugahara H, Mitsui H, Kanayama Y, Kitamura Y, Matsuzawa Y: Constitutively activating mutations of c-kit receptor tyrosine kinase confer factor-independent growth and tumorigenicity of factor-dependent hematopoietic cell lines. Blood 85:790–798, 1995.

85. Kitayama H, Tsujimura T, Matsumura I, Oritani K, Ikeda H, Ishikawa J, Okabe M, Suzuki M, Yamamura K, Matsuzawa Y, Kitamura Y, Kanakura Y: Neoplastic transformation of normal hematopoietic cells by constitutively activating mutations of c-kit receptor tyrosine kinase. Blood 88:995–1004, 1996.

86. Tsujimura T, Morimoto M, Hashimoto K, Moriyama Y, Kitayama H, Matsuzawa Y, Kitamura Y, Kanakura Y: Constitutive activation of c-kit in FMA3 murine mastocytoma cells caused by deletion of seven amino acids at the juxtamembrane domain. Blood 87:273–283, 1996.

87. Webster MK, D'Avis PY, Robertson SC, Donoghue DJ: Profound ligand-independent kinase activation of fibroblast growth factor receptor 3 by the activation loop mutation responsible for a lethal skeletal dysplasia, thanatophoric dysplasia type II. Mol Cell Biol 16:4081–4087, 1996.

88. Carpentier JL, Hamer I, Gilbert A, Paccaud JP: Molecular and cellular mechanisms governing the ligand-specific and non-specific steps of insulin receptor internalization. Z Gastroenterol 3:73–75, 1996.

89. Glover HR, Baker DA, Celetti A, Dibb NJ: Selection of activating mutations of c-fms in FDC-P1 cells. Oncogene 11:1347–1356, 1995.

90. Roussel MF, Downing JR, Sherr CJ: Transforming activities of human CSF-1 receptors with different point mutations at codon 301 in their extracellular domains. Oncogene 5:25–30, 1990.

91. Ridge SA, Worwood M, Oscier D, Jacobs A, Padua RA: FMS mutations in myelodysplastic, leukemic, and normal subjects. Proc Natl Acad Sci USA 87:1377–1380, 1990.

92. Di Fiore PP, Segatto O, Lonardo F, Fazioli F, Pierce JH, Aaronson SA: The carboxy-terminal domains of erbB-2 and epidermal growth factor receptor exert different regulatory effects on intrinsic receptor tyrosine kinase function and transforming activity. Mol Cell Biol 10:2749–2756, 1990.

93. Segatto O, Lonardo F, Wexler D, Fazioli F, Pierce JH, Bottaro DP, White MF, Di Fiore PP: The juxtamembrane regions of the epidermal growth factor receptor and gp185erbB-2 determine the specificity of signal transduction. Mol Cell Biol 11:3191–3202, 1991.

94. Di Fiore PP, Helin K, Kraus MH, Pierce JH, Artrip J, Segatto O, Bottaro DP: A single amino acid substitution is sufficient to modify the mitogenic properties of the epidermal growth factor receptor to resemble that of gp185erbB-2. EMBO J 11:3927–3933, 1992.

95. Di Fiore PP, Segatto O, Taylor WG, Aaronson SA, Pierce JH: EGF receptor and erbB-2 tyrosine kinase domains confer cell specificity for mitogenic signaling. Science 248:79–82, 1990.

96. Levkowitz G, Klapper LN, Tzahar E, Freywald A, Sela M, Yarden Y: Coupling of the c-Cbl protooncogene product to ErbB-1/EGF-receptor but not to other ErbB proteins. Oncogene 12:1117–1125, 1996.

97. Nair N, Davis RJ, Robinson HL: Protein tyrosine kinase activities of the epidermal growth factor receptor and ErbB proteins: Correlation of oncogenic activation with altered kinetics. Mol Cell Biol 12:2010–2016, 1992.

98. Songyang S, Carraway KL, Eck MJ, Harrison SC, Feldman RA, Mohammadi M, Schlessinger J, Hubbard SR, Smith DP, Eng C, Lorenzo MJ, Ponder BAJ, Mayer BJ, Cantley LC: Catalytic specificity of protein-tyrosine kinases is critical for selective signalling. Nature 373:536–539, 1995.

99. Meyer S, LaBudda K, McGlade J, Hayman MJ: Analysis of the role of the Shc and Grb2 proteins in signal transduction by the v-ErbB protein. Mol Cell Biol 14:3253–3262, 1994.
100. Herbst R, Munemitsu S, Ullrich A: Oncogenic activation of v-kit involves deletion of a putative tyrosine-substrate interaction site. Oncogene 10:369–379, 1995.
101. Dickens M, Chin JE, Roth RA, Ellis L, Denton RM, Tavare JM: Characterization of insulin-stimulated protein serine/threonine kinases in CHO cells expressing human insulin receptors with point and deletion mutations. Biochem J 287:201–209, 1992.
102. Gilboa L, Ben-Levy R, Yarden Y, Henis YI: Roles for a cytoplasmic tyrosine and tyrosine kinase activity in the interactions of Neu receptors with coated pits. J Biol Chem 270:7061–7067, 1995.
103. Lamaze C, Schmid SL: Recruitment of epidermal growth factor receptors into coated pits requires their activated tyrosine kinase. J Cell Biol 129:47–54, 1995.
104. Stern DF, Kamps MP, Cao H: Oncogenic activation of p185neu stimulates tyrosine phosphorylation in vivo. Mol Cell Biol 8:3969–3973, 1988.
105. Myles GM, Brandt CS,Carlberg K, Rohrschneider LR: Tyrosine 569 in the c-Fms juxtamembrane domain is essential for kinase activity and macrophage colony-stimulating factor-dependent internalization. Mol Cell Biol 14:4843–4854, 1994.
106. Carlberg K, Tapley P, Haystead C, Rohrschneider L: The role of kinase activity and the kinase insert region in ligand-induced internalization and degradation of the c-fms protein. EMBO J 10:877–883, 1991.
107. Theroux SJ, Latour DA, Stanley K, Raden DL, Davis RJ: Signal transduction by the epidermal growth factor receptor is attenuated by a COOH-terminal domain serine phosphorylation site. J Biol Chem 267:16620–16626, 1992.
108. Sorkin A, Di Fiore PP, Carpenter G: The carboxyl terminus of epidermal growth factor receptor/erbB-2 chimerae is internalization impaired. Oncogene 8:3021–3028, 1993.
109. Naski MC, Wang Q, Xu J, Ornitz DM: Graded activation of fibroblast growth factor receptor 3 by mutations causing achondroplasia and thanatophoric dysplasia. Nat Genet 13:233–237, 1996.

18

ESTROGEN, GROWTH FACTORS, AND CARCINOGENESIS OF THE REPRODUCTIVE TRACT

KAREN GRAY

Department of Obstetrics and Gynecology, Uniformed Services, University of Health Sciences, Bethesda, MD 20814, and Laboratory of Reproductive and Developmental Toxicology, National Institute of Environmental Health Sciences, Research Triangle Park, NC 27709

BENJAMIN EITZMAN, KIMBERLY RASZMANN WASHBURN

Laboratory of Reproductive and Developmental Toxicology, National Institute of Environmental Health Sciences, Research Triangle Park, NC 27709

TSUNEO TAKAHASHI

Laboratory of Reproductive and Developmental Toxicology, National Institute of Environmental Health Sciences, Research Triangle Park, NC 27709, and Department of Obstetrics and Gynecology, School of Medicine, Yokohama City University, Yokohama 236, Japan

YASUHIRO SAKAI

Laboratory of Reproductive and Developmental Toxicology, National Institute of Environmental Health Sciences, Research Triangle Park, NC 27709, and Department of Anatomy, School of Medicine, Kitasato University, Kanagawa 229, Japan

GLENN MERLINO

Laboratory of Molecular Biology, National Cancer Institute, National Institutes of Health, Bethesda, MD 20892

JOHN MCLACHLAN

Laboratory of Reproductive and Developmental Toxicology, National Institute of Environmental Health Sciences, Research Triangle Park, NC 27709, and, Center for Bioenvironmental Research, Tulane University, New Orleans, LA 70112

Hormones and Growth Factors in Development and Neoplasia, Edited by Robert B. Dickson and David S. Salomon
ISBN 0-471-16899-8

18.1. ESTROGEN ACTION, GROWTH FACTORS, LACTOFERRIN, AND WOUND HEALING

Steroid hormones elicit complex biologic responses in the reproductive tract at the level of gene transcription mediated by their specific receptors, which coordinate the expression of gene networks (1). A number of cell- and tissue-specific target genes for estrogens have been identified including proto-oncogenes, enzymes, cytokines, and peptide growth factors that participate at various points within growth pathways (2–12). Studies in our laboratory have demonstrated that the platelet-derived growth factors (PDGFs) and their receptors, the transforming growth factor betas (TGFβs), epidermal growth factor (EGF), transforming growth factor alpha (TGFα), insulin-like growth factor-1, and lactoferrin are estrogen-inducible genes in the reproductive tract of prepubescent CD-1 female mice (13–18). Figure 18.1A outlines the time periods of induction for the estrogen-regulated genes that we have investigated.

Treatment of immature (17–19 days old) mice with the potent estrogen diethylstilbestrol (DES) induces mRNA expression for the PDGF A and PDGF B chains and for the PDGF receptor α and β subunits in the uterus and vagina, localized predominantly to the epithelium (14). This regulation suggests that PDGF autocrine or paracrine pathways may play roles in cell-to-cell communication important in coordinating a synchronized, integrated tissue response to estrogen. Although PDGFs are documented mitogens for cells of mesenchymal origin, growth stimulation of some epithelial cell types has been reported. In fact, PDGF receptors are expressed by epithelial cells of the mammary gland, fetal lung, skin, olfactory glands, and a number of epithelial cancer cells. Another point to consider is that, unlike most epithelia, the epithelium of the uterus and upper vagina is derived from mesoderm that may program these epithelial cells to respond to the PDGFs in a more mesenchyme-like manner.

The expression of the three mammalian TGFβ isoforms, TGFβ1, TGFβ2, and TGFβ3, is also subject to estrogen regulation in both the uterus and the vagina of the prepubescent mouse (18). Immunohistochemical protein detection, as well as *in situ* hybridization and Northern blot RNA analyses, demonstrate overlapping but distinct time-dependent and site-specific induction of all three TGFβ genes by estrogen, most notably in the epithelium. Northern blot analysis clearly demonstrates that DES significantly but transiently upregulates TGFβ3 mRNA within 30 min, and TGFβ1 and TGFβ2 mRNAs by 3 hr, with decreases to or below control levels by 6 hr. The vagina also responds to DES with kinetics of induction for TGFβ2 and TGFβ3 mRNAs similar to those observed in the uterus; however, TGFβ1 mRNA levels increase gradually and peak at around 16 hr after treatment. Unlike the transient nature of TGFβ mRNA induction elicited by estrogen, immunohistochemistry demonstrates that estrogen treatment results in a more prolonged elevation of the

Fig. 18.1. Correlation of the induction of various estrogen-regulated genes (A) to the onset of DNA synthesis (C) as determined by ^{3}H-thymidine autoradiography (B).

(a)

TIME-DEPENDENT INDUCTION OF GROWTH FACTOR EXPRESSION IN MOUSE UTERUS AND VAGINA FOLLOWING ESTROGEN TREATMENT OF PREPUBESCENT CD-1 MICE

	TGFα	$TGF\beta_1$	$TGF\beta_2$	$TGF\beta_3$	PDGF-A	PDGF-B	PDGF-RI	PDGF-RII	IGF-1	PR	Lactoferrin	c-myc
UTERUS	peak 3hr down 48hr	peak 3hr down 6hr	peak 3hr down 6hr	peak 1hr down 6hr	peak 3hr down >72hr	peak 3hr down 24hr	peak 6hr down >72hr	peak 6hr down >72hr	peak 3hr down 48hr	peak 3hr down 24 hr	peak 24hr down 72hr	peak 30min down 3hr
VAGINA	peak 24hr down 48hr	peak 16hr down 24hr	peak 3hr down 6hr	peak 1hr down 6hr	peak 3 hr down 24hr	peak 3hr down 24hr	peak 3hr down 24hr	peak 3 hr down 24hr	peak 3hr down 48hr	peak 16hr down 24hr	peak 24 hr down 72hr	peak 15min down 30min

(b)

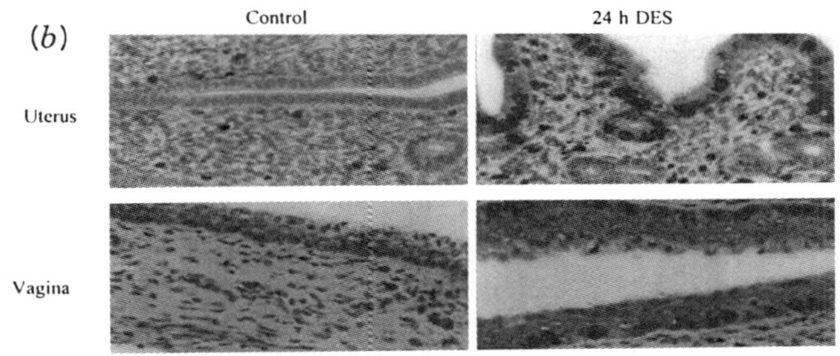

(c)

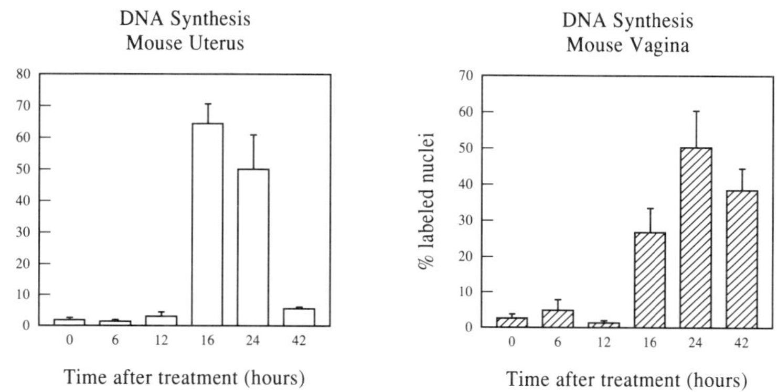

proteins for TGFβ1, TGFβ2, and TGFβ3 in the epithelium of both tissues. It is known that the TGFβs are bifunctional growth regulators having either positive or negative effects on cell growth, depending on the cell type and the extracellular environment. Most ectoderm derived-epithelial cells, endothelial cells, and lymphoid cells are growth inhibited, whereas cells of mesenchymal origin are generally stimulated to proliferate. Since the epithelial cells of the uterus and upper vagina are derived from mesoderm, these cells may be programmed to respond to the TGFβs in a more mesenchyme-like manner, with enhanced growth.

Estrogen regulation of the TGFα and EGF receptor pathways may also play an important role in normal and abnormal reproductive physiology (13,15–22). EGF and TGFα belong to a growing family of homologous growth factors, which elicit similar biologic effects by binding to and activating the tyrosine kinase of the EGF receptor. We have shown that estrogen coordinately upregulates TGFα mRNA and protein levels in both the uterine and vaginal epithelium of immature mice, starting as early as 3 hr after treatment; these levels remain elevated for at least 24 hr. Although extensive time-dependent expression studies have not been carried out for EGF, authentic EGF is also present in mouse uterus and vagina, and multiple estrogen treatments enhance expression of both protein and mRNA for EGF. Functional TGFα/EGF receptors are expressed by most uterine and vaginal cell types and are evident as early as day 13 of gestation. Specific neutralizing antibodies to these growth factors can partially inhibit estrogen-induced uterine and vaginal growth *in vivo*. Exogenous EGF can mimic the effects of estrogen *in vivo* by stimulating growth, which, like that seen with estrogen, is accompanied by enhanced phosphatidylinositol lipid turnover, modulation of the estrogen and progesterone receptors, and the induction of lactoferrin. *In vitro* studies have shown that EGF and TGFα transcription activate estrogen-responsive elements in the absence of estrogen, demonstrating that there is cross-talk between growth factor and steroid receptor pathways. Overall, these data suggest that an integral relationship exists between the EGF family and estrogen action.

To gain a better understanding of the sequential changes in gene expression in the reproductive tract on estrogen treatment, we have compared the kinetics of induction of the PDGFs, TGFβs, and TGFα to that of other genes documented to be susceptible to estrogen control, including c-*myc*, IGF-1, progesterone receptor (PR), and lactoferrin (Fig. 18.1A) (18). In general, the induction of the TGFβs and c-*myc* in mouse uterus and vagina are the most transient (decreased by 30 min to 6 hr) compared to the estrogen-regulated genes TGFα, PDGFs, IGF-, PR, and lactoferrin, which exhibit prolonged upregulation of mRNA expression ranging from 24 hr to more than 72 hr after estrogen treatment. The c-*myc* proto-oncogene is a short-lived transcription factor that governs the decision of a cell to either enter the cell cycle or undergo programmed cell death (23), and is proposed to act as a mediator of the action of diverse signaling molecules including estrogen. Similar to the induction of TGFβ3 by estrogen, c-*myc* mRNA levels are elevated within 15 min in both the uterus and vagina; however, c-*myc* induction in the vagina is actually more transient than that of TGFβ3, returning to baseline by 3 hr, whereas the uterine c-*myc* response is maintained for up to 3 hr. Of equal importance, because IGF-1 is proposed to play a pivotal role in the control of cell proliferation, is the upregulation of IGF-1 transcripts by estrogen in both the uterus and vagina at 30 min and 3 hr, respectively (24). Analysis of the progesterone receptor, the classical gene marker for estrogen action, demonstrates early, prolonged induction of PR mRNA in the mouse reproductive tract similar to the time-dependent pattern of TGFα and IGF-1 expression.

Identified as a biomarker of estrogen action, the induction of lactoferrin mRNA and protein expression in the epithelium of the mouse uterus and vagina correlates with the onset of DNA synthesis (25–29). Lactoferrin is a 70,000–80,000 molecular weight iron binding protein belonging to the transferrin family and is found in milk, most body fluids, and as the major constituent of secondary granules of neutrophils. A variety of biologic functions is proposed for lactoferrin, which links this protein to growth regulation including iron transport, modulation of various receptors involved in lipid transport, immune function, RNase activity, and the ability to act as a paracrine transcriptional regulator. In contrast to the regulation of other genes investigated, estrogen does not significantly upregulate gene expression for lactoferrin until 16 hr, as determined by Northern blot analysis, or until 12 hr, as determined by *in situ* hybridization (18). Even though the lactoferrin gene promoter contains an estrogen response element, the late induction, compared to that of the growth factor genes, classifies lactoferrin as an intermediate or late gene in estrogen action and suggests that other nuclear factors are required for the transactivation of lactoferrin gene transcription, such as growth factors and retinoid acid.

A time course study of estrogen stimulation of epithelial DNA synthesis, as measured by the percentage of [^{3}H]thymidine-labeled nuclei detected by autoradiography, demonstrates that maximal DNA synthesis occurs 16 to 24 hr in both the uterus and vagina of prepubescent CD-1 mice (Fig. 18.1B,C). The epithelial cells stimulated to grow by estrogen are the same cells that are induced to express increased mRNA levels for the TGFβs, TGFα, PDGFs, IGF1, c-*myc,* and lactoferrin (Fig. 18.1B). Comparison of the kinetics of estrogen induction of growth factor synthesis to the onset of DNA synthesis in the uterine and vaginal epithelium reveals that the upregulation of the TGFβs, TGFα, PDGFs, IGF1, and lactoferrin mRNAs occurs prior to the peak of DNA synthesis, implicating these factors in the regulation of cell cycle progression. Thus, studies in our laboratory and others have revealed the presence of a number of autocrine/paracrine loops in the reproductive tract that are upregulated. *In vitro* studies show that mitogenesis is a two-step process in cultured fibroblasts in which quiescent (G_0) cells proceed first into the preliminary (G_1) phase of the cell cycle under the influence of *competence* factors such as PDGF, where they subsequently become committed to DNA synthesis by the separate action of *progression* factors such as EGF/TGFα and IGF1 (24). Similarly, estrogen induction of the coexpression of both competence and progression growth factors and their respective receptors in the reproductive tract may provide conditions conducive for complete cell cycle transition and thus optimal growth.

The physiologic changes observed in the reproductive tract during the normal estrous cycle and pregnancy mimic some of the events that occur during wound repair (8,10,12,30–35). Similarities between steroid hormone action in the reproductive tract and a modified inflammatory response such as that seen in wound healing include marked edema, coordinated infiltration of inflammatory cells, neovascularization, modulation of the extracellular matrix, elevation of prostaglandins and leukotrienes, activation of proteases, induction of chemotactic factors, complement factors and numerous cytokines, enhanced expression of proto-oncogenes and other transcriptional factors, cell proliferation, and differentiation. The fact that estrogen induces a repertoire of growth factors and cytokines similar to that found in wounding supports the hypothesis that estrogen regulation of reproductive tissue growth mimics the regenerative repair associated with wounding. Similarly, excessive wounding or unopposed estrogen exposure can produce conditions that promote abnormal tissue growth, leading to tumorigenesis (18,36,37).

18.2. ESTROGEN-INDUCED REPRODUCTIVE TRACT CARCINOGENESIS

Diethylstilbestrol (DES) is a synthetic estrogen that was prescribed to millions of pregnant women to prevent miscarriages and other pregnancy complications. Women exposed *in utero* during the first trimester of pregnancy exhibited genital tract abnormalities such as vaginal adenosis, cervical ectropion, ridges, pseudopolyps, uterine hypoplasia, and occasional vaginal and cervical clear cell adenocarcinoma (38). Similarly, prenatal or neonatal exposure of mice and rats to high levels of estrogen induces reproductive tract lesions that range from the retention of Wolffian duct remnants to uterine squamous metaplasia, reduction of uterine gland formation, vaginal adenosis, hyperplastic vaginal and cervical epithelium with persistent cornification, and uterine and vaginal adenocarcinomas (21,39–41). Disruption of the uterine stroma and muscle also occurs, which may contribute indirectly to alterations in epithelial cell function due to perturbation of critical mesenchymal–epithelial interactions (42). Additionally, there is abnormal growth characterized by persistent proliferation, which is maintained in the absence of ovaries and is referred to as *ovarian-independent* growth. Figures 18.2 and 18.3 diagram events and lesions that are induced in mouse uterus and vagina, respectively. Because extensive reproductive tract development occurs neonatally in the rodent, which is the developmental equivalent to the end of the first trimester in the human fetus, many studies utilize neonatal treatment of rodents as models for the exploration of the mechanism of the developmental toxicity of estrogens.

During the first 5 days after birth in the mouse, the reproductive tract undergoes extensive growth, remodeling, and differentiation, which occur independently of steroid hormones. It is during this critical developmental period that the *imprinting* of the adult pattern of steroid hormone responsiveness is reported to occur. It is also during this period that the reproductive tract is most susceptible to the teratogenic and carcinogenic effects of exogenous estrogens. Moreover, little detectable estrogen receptor is expressed by the uterine epithelium during this period (43). Once the estrogen receptor is expressed, the uterine epithelium is no longer sensitive to the toxic effects of estrogen, suggesting that the appearance of the estrogen receptor repre-

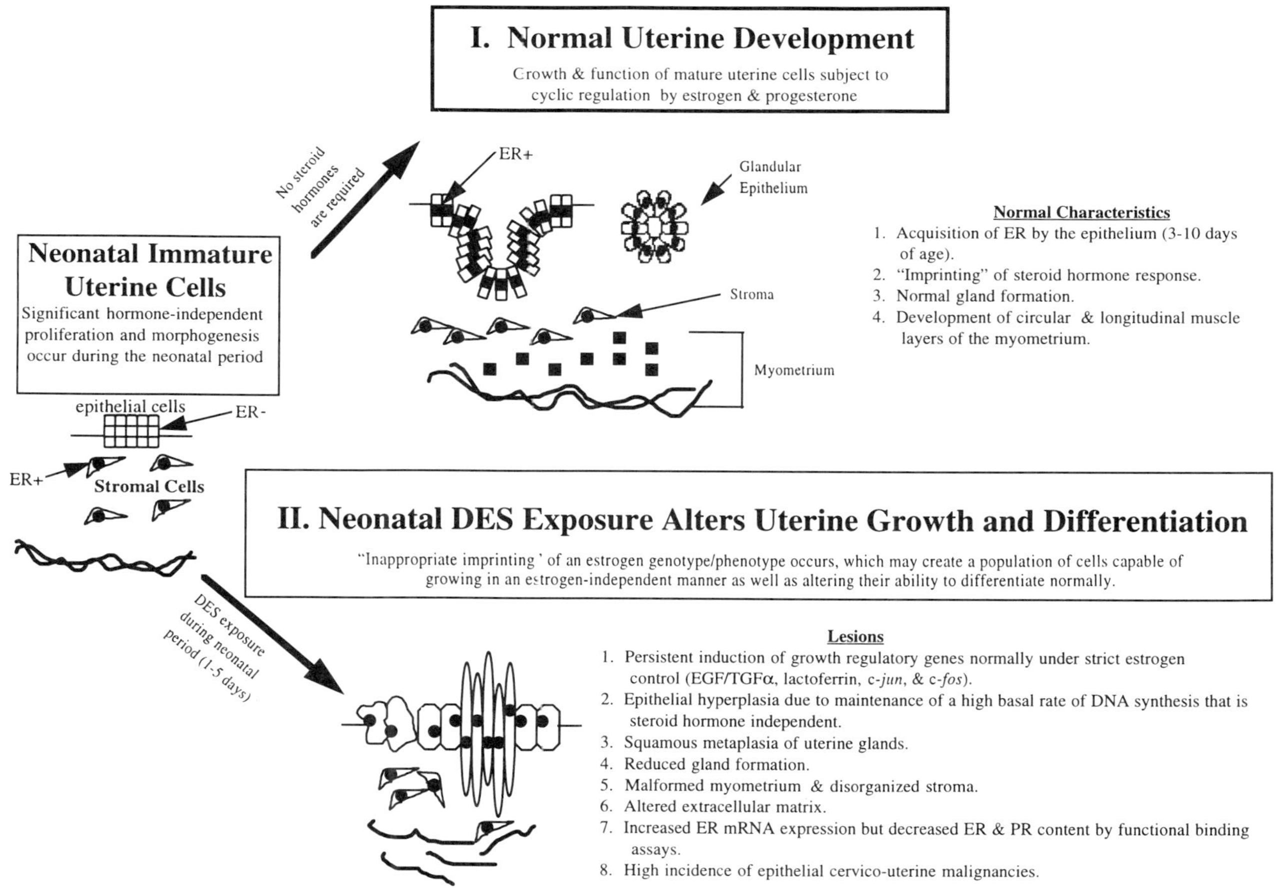

Fig. 18.2. Diagram of some of the development effects of DES treatment during the neonatal period on mouse uterine histology and gene expression.

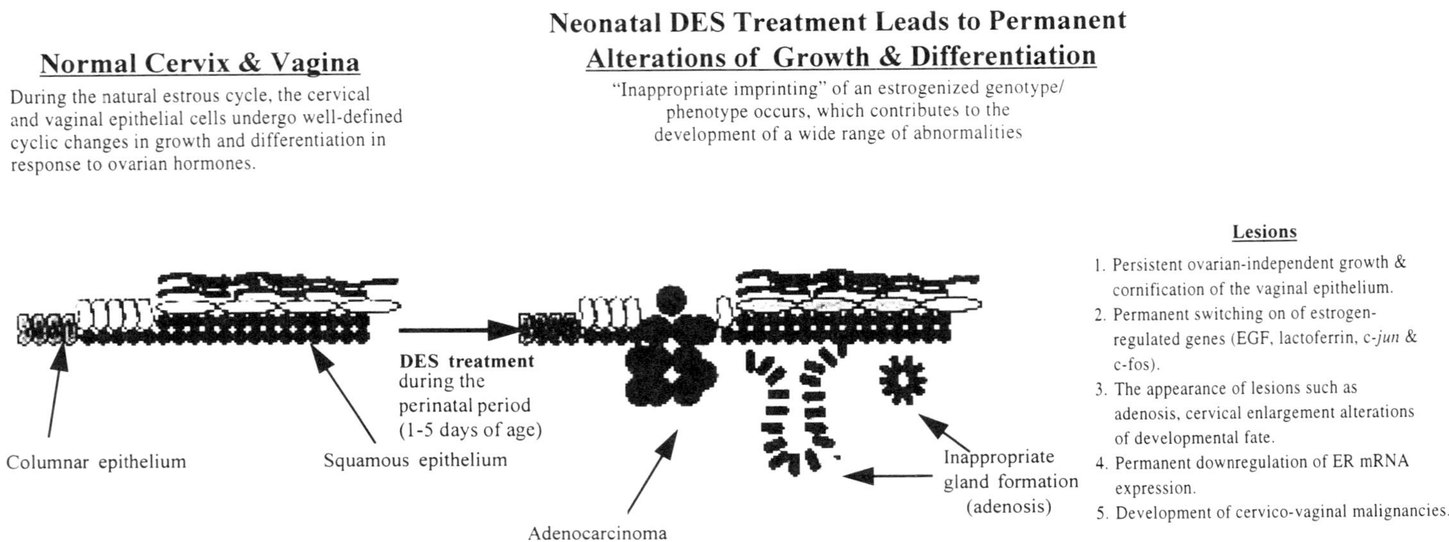

Fig. 18.3. Some of the cervical and vaginal lesions induced by neonatal estrogen exposure.

sents the final stage of differentiation or imprinting that imposes rigid steroid hormone-mediated control over the genes associated with the mature reproductive tract.

The actual mechanism for the toxicity of DES remains unknown, but both genotoxic and nongenotoxic pathways are proposed. Support for the idea that DES acts as an initiator of transformation in the genital tract is derived from studies showing that DES is metabolized to reactive intermediates in animals, interacts with microtubules, causes mitotic arrest, induces chromosomal abnormalities, elicits neoplastic transformation in ER-negative fibroblasts, and increases DNA ploidy of mouse vaginal epithelium (44–47). Epigenetic mechanisms are also hypothesized to involve the premature and permanent induction of genes normally under steroid hormone control in older animals, ultimately leading to disease (40,41,48,49).

In fact, a number of studies have documented that perinatal DES treatment induces persistent biochemical and molecular changes in the reproductive tracts of mice. Since dysregulation of autocrine/paracrine growth pathways such as that of the EGF receptor family is often a consistent alteration observed in many neoplastic cells, we are seeking to determine if early estrogen treatment leads to abnormal induction of these estrogen-regulated genes in reproductive tissues. Our initial studies indicate that neonatal exposure to DES does cause persistent ovarian-independent induction in both the uterus and vagina of the two estrogen-regulated genes studied, EGF and lactoferrin (49). Although the physiologic role of lactoferrin in the reproductive tract is not known, lactoferrin probably has multiple functions that, on overexpression, may facilitate malignant transformation. In fact, endometrial adenocarcinomas in women exhibit marked upregulation of lactoferrin expression, which further supports a role for lactoferrin in tumorigenesis of reproductive tissues (50). The hormone-independent expression of both EGF and lactoferrin strongly indicates that these genes may contribute to the establishment of the persistent "estrogenized" morphology and estrogen-independent growth exhibited by the genital tract following developmental estrogen treatment.

The role of members of the EGF family in estrogen carcinogenesis is further confirmed by our study using transgenic mice that demonstrated that the presence of the TGFα transgene promoted DES-induced vaginal adenosis, uterine polyps, hyperplasia, and benign ovarian cysts but, interestingly, did not lead to an elevation in the incidence of neoplasia (51). The fact that the TGFα-mediated elevation of hyperplasia and adenosis does not translate into increased tumorigenesis suggests that DES carcinogenesis involves more than simple alterations in growth potential elicited by dysregulation of a single growth factor such as TGFα.

Also reported to be among the persistent changes in gene expression induced by neonatal DES are increased levels of c-*fos* and c-*jun* mRNAs in both uterus and vagina that coincide with the higher mitotic activity exhibited by these tissues (51). c-*Jun* and c-*fos* proteins dimerize to form the potent transcription factor AP-1, which is involved in the regulation of gene batteries associated with cell cycle progression. The fact that deregulation of c-*jun* and c-*fos* can neoplastically transform rat fibroblasts has led to the hypothesis that estrogen-induced deregulation of c-*jun* and c-*fos* expression in the vagina and uterus may contribute to the evolution of growth and differentiation abnormalities. Also, since the promoter regions of several growth factor genes induced by estrogen (TGFα, TGFβ, PDGFs, IGFs, lactoferrin) contain putative AP-1 sites, it is reasonable to propose that elevation of c-*fos* and c-*jun* may lead to persistent enhancement of transcription of these factors in the absence of estrogen. Additionally, neonatal treatment causes permanent changes in the expression of the estrogen and progesterone receptors, key regulators of uterine and vaginal function, that are associated with changes in the responsiveness of reproductive tissues to subsequent hormone stimulation as adults (52).

Thus, neonatal estrogenization leads to permanent alterations in a number of genes such that cells of the genital tract bear these permanent epigenetic imprints into adulthood, which may contribute to the pathogenesis of reproductive tract abnormalities. The cascade model of steroid hormone action proposes that steroid receptors affect a multitude of events in target cells by the regulation of only a few primary regulatory genes, which in turn alter the expression of specific batteries of late-responding genes that bring about designated phenotypic changes. Based on this model, we propose that the upregulation of only a few genes due to aberrant imprinting, the transcriptional factors (c-*jun* and c-*fos*), as well as growth-related genes (EGF and lactoferrin) by neonatal estrogen treatment, may be sufficient to perpetuate a persistent ovarian-independent growth signal.

Although it is documented that exposure to estrogens neonatally converts the genital tract to an estrogenized phenotype, neoplastic transformation of the reproductive tract requires further challenge of the adult female with ovarian hormones. The results of our study and those of others allow us to propose that one mechanism by which secondary exposure to estrogen promotes neoplastic progression is that cells altered by DES not only exhibit an altered response to endogenous steroids, but also have a proliferative advantage due to deregulation of the expression of several key growth regulatory genes (c-*jun*, c-*fos*, EGF, lactoferrin). This amplifies the cellular growth response and creates conditions that favor the acquisition of secondary neoplastic lesions associated with progression. To conclude, there is now sufficient evidence that the teratogenic and carcinogenic effects of developmental estrogen exposure are due in part to the precocious, persistent induction of estrogen-regulated genes, an imprinting phenomenon that sets the stage for further neoplastic progression.

18.3. DISCUSSION

The function of the reproductive tract is exquisitely regulated by ovarian steroid hormones via complex interactions among multiple cell types that communicate, in part, by the synthesis of soluble mediators. Current theories of wound repair and of estrogen action propose that the production of endogenous growth factors is triggered in response to an initiating stimulus, and that these factors then serve to propagate, amplify, and maintain the mitogenic response (1). Interruption of these orderly processes can lead to aberrant gene expression and interference with normal growth and differentiation.

We and others have used the mouse as a model system to explore the molecular mechanism of developmental and hormonal toxicity and carcinogenicity of natural and synthetic estrogens. Neonatal estrogen exposure is found to result in the persistent constitutive expression of a number of estrogen-regulated genes (EGF, lactoferrin, c-*jun*, c-*fos*, ER, PR) that probably are instrumental in the development of abnormal tissue morphogenesis, function, and neoplasia in the female reproductive tract. Even with this progress, understanding the mechanism of estrogen carcinogenesis will not be easy.

Within the last decade, research tools have been developed that will allow us to begin to define the molecular basis that determines normal and abnormal physiology in female reproductive tissues. An important question for future research is: What is so special about the neonatal period that allows estrogen to act as a carcinogen? Studies need to be directed at elucidating the epigenetic environment of the uterus and vagina during this critical developmental period by identifying genes, expressed or not expressed, that determine the estrogen sensitivity of the neonatal reproductive tract. These investigations will provide greater insight into the mechanism of hormonal carcinogenesis (Fig. 18.4).

As found in the rodent models, evidence is beginning to suggest that altered patterns of gene expression, especially of growth factors and other growth regulatory molecules, are likely to be involved in the etiology of reproductive tract diseases in women, warranting further study (53). Approximately 150,000 new cases of endometrial cancer are diagnosed every year worldwide, which represents the fifth leading cancer in women. Endometrial cancer is more common in elderly women, in urban populations, in whites compared to blacks, and in Western societies. There is a well-documented association in women between endometrial carcinoma and unopposed estrogen exposure due to menopausal replacement treatment, obesity, and irregular menstrual cycles, while factors that decrease exposure to estrogens or increase progesterone levels, such as oral contraceptives or smoking, tend to be protective. However, these known risk factors account for only 30–60% of the cases of endometrial cancer, which suggests that other factors are probably responsible. In most cases, hormonal carcinogenesis may take decades to develop, suggesting that the initiating events leading to clinically manifested cancer may occur at a young age. Except for the well-documented association between prenatal DES use and genital tract cancer, it is not clear whether exposure of humans to other unsuspected

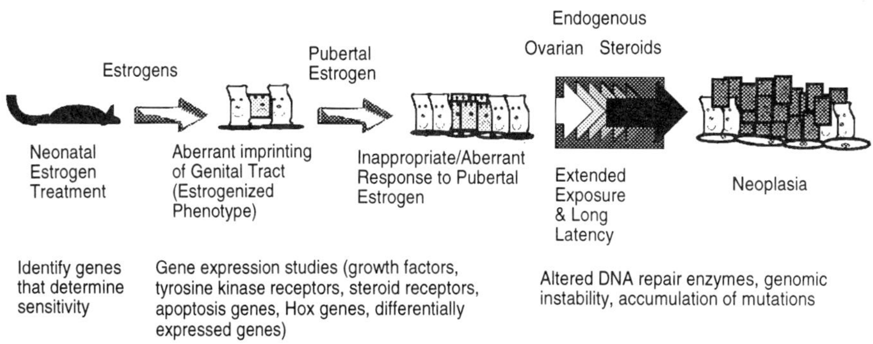

Fig. 18.4. Questions for future research.

exogenous or endogenous estrogens, either prenatally or at critical periods of development such as puberty, may induce subtle abnormalities that increase the susceptibility of these women later in life to the development of reproductive malignancies.

The prevalence of phytoestrogens and other industrial, estrogenically active compounds in human foods and environment has increased concern about their impact on human health. There is sufficient evidence that many structurally different environmental chemicals with estrogenic activity can disturb development of the endocrine system and of the target organs on exposure during prenatal and/or early postnatal life (54). It seems that the unborn animal or the neonate may be especially at risk from these chemicals because of rapid growth and development during these periods. Permanent and irreversible neurobehavioral, immunologic, endocrine, skeletal, and reproductive alterations can result on developmental exposure to estrogens. Furthermore, epidemiologic studies have shown that the incidence of some hormonal cancers (prostate and breast) increases in immigrants moving from low-risk to high-risk areas, indicating that the risk is modified by exposure of adults to environmental hormone mimics (55–57). Overall, there is sufficient evidence to propose that excessive estrogen exposure during development or as an adult may contribute to the progression of endometrial and other gynecologic tumors in women, which may account for those tumors that cannot be associated with known risk factors. Clearly, a better definition of the impact of dietary and industrial estrogen exposure in the pathogenesis of reproductive tract diseases is needed.

REFERENCES

1. Briehl MM, Flomerfelt FA, Wu X-P, Miesfeld RL: Transcriptional analyses of steroid-regulated gene networks. Mol Endocrinol 4:287–294, 1990.
2. Cicatiello L, Ambrosino C, Coletta B, Scalona M, Sica V, Bresciani F, Weisz A: Transcriptional activation of jun and actin genes by estrogen during mitogenic stimulation of rat uterine cells. J Steroid Biochem Mol Biol 41:523–528, 1992.
3. Guidice L: Growth factors and growth modulators in human uterine endometrium: their potential relevance to reproductive medicine. Fertility and Sterility 61:1–17, 1994.
4. Loose-Mitchell DS, Chiappetta C, Stancel GM: Estrogen regulation of c-fos messenger ribonucleic acid. Mol Endocrinol 2:946–951, 1988.
5. Murphy LJ, Murphy LC, Friesen HG: Estrogen induction of n-myc and c-myc proto-oncogene expression in the rat uterus. Endocrinology 120:1882–1888, 1987.
6. Murphy LJ, Murphy LC, Friesen HG: A role for the insulin-like growth factors as estromedins in the rat uterus. Trans Assoc Am Physicians 99:204–214, 1988.
7. Pollard JW: Regulation of polypeptide growth factor synthesis and growth factor-related gene expression in the rat and mouse uterus before and after implantation. J Reprod Fertil 88:721–731, 1990.
8. Pollard JW: Lymphohematopoietic cytokines in the female reproductive tract. Curr Opin Immunol 3:772–777, 1991.
9. Scrocchi LA, Jones LA: Alteration of proto-oncogene c-fos expression in neonatal estrogenized BALB/c female mice and murine cervicovaginal tumor LJ6195. Endocrinology 129:2251–2253, 1991.

10. Wegmann TG: The cytokine basis for crosstalk between the maternal immune and reproductive systems. Curr Opin Immunol 2:765–769, 1990.
11. Weisz A, Cicatiello L, Persico E, Scalona M, Bresciani F: Estrogen stimulates transcription of c-jun protooncogene. Mol Endocrinol 4:1041–1050, 1990.
12. Zhang Z, Funk C, Roy D, Glasser S, Mulholland J: Heparin-binding epidermal growth factor-like growth factor is differentially regulated by progesterone and estradiol in rat uterine epithelial and stromal cells. Endocrinology 134:1089–1094, 1994.
13. DiAugustine RP, Petrusz P, Bell GI, Brown CF, Korach KS, McLachlan JA, Teng CT: Influence of estrogens on mouse uterine epidermal growth factor precursor protein and messenger ribonucleic acid. Endocrinology 122:2355–2363, 1988.
14. Gray KD, Raszmann K, McLachlan J, Bidwell M: Coordinate regulation by diethylstilbestrol of the platelet-derived growth factor-A (PDGF-A) and -B chains and the PDGF receptor-α and β-subunits in the mouse uterus and vagina: Potential mediators of estrogen action. Endocrinology 136:2325–2340, 1995.
15. Gray Nelson K, Takahashi T, Bossert NL, Walmer DK, McLachlan JA: Epidermal growth factor replaces estrogen in the stimulation of female genital-tract growth and differentiation. Proc Natl Acad Sci USA 88:21–25, 1991.
16. Gray Nelson K, Takahashi T, Lee DC, Luetteke NC, Bossert NL, Ross K, Eitzman BE, McLachlan JA: Transforming growth factor-α is a potential mediator of estrogen action in the mouse uterus. Endocrinology 131:1657–1664, 1992.
17. Sakai Y, Gray Nelson K, Snedeker S, Bossert NL, Walker MP, McLachlan J, DiAugustine RP: Expression of epidermal growth factor in suprabasal cells of stratified squamous epithelia: Implications for a role in differentiation. Cell Growth Differ 5:527–535, 1994.
18. Takahashi T, Eitzman B, Bossert NL, Walmer DK, Sparrow K, Flanders KC, McLachlan JA, Gray Nelson K: Transforming growth factor beta$_{1,2\&3}$ mRNA and protein expression in mouse uterus and vagina during estrogen-induced growth: a comparison to other estrogen regulated genes. Cell Growth Differ 5:919–935, 1994.
19. Bossert NL, Nelson Gray K, Ross KA, Takashi T, McLachlan JA: Epidermal growth factor binding and receptor distribution in the mouse reproductive tract during development. Dev Biol 142:75–85, 1990.
20. Ignar-Trowbridge DM, Gray Nelson K, Bidwell MC, Curtis SW, Wasburn TF, McLachlan JA, Korach KS: Coupling of dual signaling pathways: EGF action involves the estrogen receptor. Proc Natl Acad Sci USA 89:4658–4662, 1992.
21. Halling A, Forsberg J-G: Acute and permanent growth effects in the mouse uterus after neonatal treatment with estrogens. Reprod Toxicol 7:137–153, 1993.
22. Huet-Hudson YM, Chakraaborty C, De SK, Suzaki Y, Andrews GK, Dey SK: Estrogen regulates synthesis of epidermal growth factor in mouse uterine epithelial cells. Mol Endocrinol 4:510–517, 1990.
23. Evan GI, Littlewood TD: The role of c-myc in cell growth. Curr Opin Genet Dev 3:44–49, 1993.
24. Baserga R, Rubin R: Cell cycle and growth control. Crit Rev Eukaryotic Gene Expression 3:47–61, 1993.
25. He J, Furmanski P: Sequence specificity and transcriptional activation in the binding of lactoferrin to DNA. Nature 373:721–724, 1995.
26. Hiesberger T, Hermann M, Jacobsen L, Novak S, Hodits RA, Bujo H, Meilinger M, Huttinger M, Scheider WJ, Nimpf J: The chicken oocyte receptor for yolk precursors as a

model for studying the action of receptor-associated protein and lactoferrin. J Biol Chem 270:18219–18226, 1995.

27. Lee MO, Liu Y, Zhang XK: A retinoic acid response element that overlaps an estrogen response element mediates multihormonal sensitivity in transcriptional activation of the lactoferrin gene. Mol Cell Biol 15:4194–4207, 1995.
28. Levay PF, Viljoen M: Lactoferrin: A general review. Haematologica 80:252–267, 1995.
29. Walmer DK, Wrona MA, Hughes CL, Gray Nelson K: Lactoferrin expression in the mouse reproductive tract during the natural estrous cycle: Correlation with circulating estradiol and progesterone. Endocrinology 131:1458–1466, 1992.
30. Davidson JM: Wound repair. In Gallin JI, Goldstein IM, Synderman R (eds): Inflammation: Basic Principles and Clinical Correlates. New York: Raven Press, 1992, pp 809–819.
31. Henrikson KP, Jazin EE, Greenwood JA, Dickerman HW: Prothrombin levels are increased in the estrogen-treated immature rat uterus. Endocrinology 126:167–175, 1990.
32. Hosgood G: Wound healing. The role of platelet-derived growth factor and transforming growth factor beta. Veterinary Surg 22(6):490–495, 1993.
33. Kachkache M, Acker GM, Chaouat G, Noun A, Garabedian M: Hormonal and local factors control the immunohistochemical distribution of immunocytes in the rat uterus before conceptus implantation: Effects of ovariectomy, fallopian tube section, and injection. Biol Reprod 45:860–868, 1991.
34. Kuivanen PC, Capulong RB, Harkins RN, DeSombre ER: The estrogen-responsive 110K and 74K rat uterine secretory proteins are structurally related to complement component C3. Biochem Biophys Res Commun 158:898–905, 1989.
35. McMaster MT, Newton RC, Dey SK, Andrews GK: Activation and distribution of inflammatory cells in the mouse uterus during the preimplantation period. J Immunol 148:1699–1505, 1992.
36. Cross M, Dexter TM: Growth factors in development, transformation, and tumorigenesis. Cell 64:271–280, 1991.
37. Fowlis DJ, Flanders KC, Duffie E, Balmain A, Akhurst RJ: Discordant transforming growth factor $\beta 1$ RNA and protein localization during chemical carcinogenesis of the skin. Cell Growth Differ 3:81–91, 1992.
38. Herbst AL: In Herbst AI, Bern HA (eds): Developmental Effects of Diethystilbestrol (DES) in Pregnancy. New York: Thieme Stratton, 1981, pp 203–223.
39. Newbold R, McLachlan JA: DES associated defects in murine gential tract development. In McLachlan JA (ed): Estrogens in the Environment. II. Influences on Development. New York: Elsevier, 1985, pp 288–318.
40. Iguchi T: Cellular effects of early exposure to sex hormones and antihormones. Int Rev Cytol 139:1–57, 1992.
41. Takasugi N: Development of permanently proliferated and cornified vaginal epithelium in mice treated neonatally with steroid hormones and the implication in tumorigenesis. Natl Cancer Inst Monogr 51:57–66, 1979.
42. Cunha GR: Role of epithelial–mesenchymal interactions in the development of hormonal responsiveness. In Hochberg RB, Naftolin F (eds): The New Biology of Steroid Hormones, Vol 74. New York: Raven Press, 1991, pp 301–310.
43. Yamashita S, Newbold RR, McLachlan JA, Korach KS: Developmental pattern of estrogen receptor expression in female mouse genital tracts. Endocrinology 125:2888–2896, 1989.
44. Hajek RA, Van NR, Johnston DA, Jones LA: In vivo induction of increased DNA ploidy

of mouse cervicovaginal epithelium by neonatal estrogen treatment. Biol Reprod 49:908–917, 1993.
45. McLachlan JA, Wong A, Degen GH, Barrett JC: Morphological and neoplastic transformation of Syrian hamster embryo fibroblasts by diethylstilbestrol and its analogs. Cancer Res 42:3040–3045, 1982.
46. Metzler M, McLachlan JA: Oxidative metabolites of diethylstilbestrol in the fetal, neonatal, and adult mouse. Biochem Pharmacol 27:1087–1094, 1977.
47. Wheeler WJ, Cherry LM, Downs T, Hsu TC: Mitotic inhibition and aneuploidy induction by naturally occurring and syntectic estrogens in Chinese hamster cells in vitro. Mutat Res 171:31–41, 1986.
48. Bern HA, Edery M, Mills KT, Kahrman AF, Mori T, Larson L: Long-term alterations in histology and steroid receptor levels of the genital tract and mammary gland following neonatal exposure of female BALB/cCrgl mice to various doses of diethystilbestrol. Cancer Res 47:4165–4172, 1987.
49. Gray Nelson K, Sakai Y, Eitzman B, Steed T, McLachlan J: Exposure to diethylstilbestrol during a critical developmental period of the mouse reproductive tract leads to persistent induction of two estrogen-regulated genes. Cell Growth Differ 5:595–606, 1994.
50. Walmer DK, Padin CJ, Wrona MA, Healy BE, Bentley RC, Tsao M-S Kohler MF, McLachlan JA, Gray KD: Malignant transformation of the human endometrium is associated with overexpression of lactoferrin messenger RNA and protein. Cancer Res 55:1168–1175, 1995.
51. Gray K, Bullock B, Dickson R, Raszmann K, Walmer D, McLachlan J, Merlino G: Potentiation of diethylstilbestrol-induced alterations in the female mouse reproductive tract by transforming growth factor-α transgene expression. Mol Carcin 17:163–173, 1996.
52. Kamiya K, Sato T, Nishimura N, Goto SL, Kano K, Iguchi T: Expression of estrogen receptor and proto-oncogene messenger ribonucleic acids in reproductive tracts of neonatally diethystilbestrol-exposed female mice with or without postpubertal estrogen administration. Exp Clin Endocrinol Diabetes 104:111–122, 1996.
53. Reinartz JJ: Expression of transforming growth factor alpha, epidermal growth factor receptor, and erbB-2 in endometrial carcinoma and correlation with survival and known predictors of survival. Hum Pathol 25:1075–1083, 1994.
54. Colborn T: Environmental estrogens: Health implications for humans and wildlife. Environ Health Perspectives 103:135–136, 1995.
55. Adlercreutz CHT, Goldin BR, Gorbach SL, Hockerstedt KA, Watanabe S, Hamalainen EK, Markkanen MH, Makela TH, Wahala KT, Adlercreutz T: Soybean phytoestrogen intake and cancer risk. J Nutr 125:757S–770S, 1995.
56. McLachlan JA, Korach KS: Symposium on Estrogens in the Environment III. Environmental Health Perspectives 103:3–4, 1995.
57. Sheehan DM: The case for expanded phytoestrogen research. Proc Soc Exp Biol Med 208:3–5, 1995.

19

ONCOGENES, GROWTH FACTORS, AND HORMONES IN PROSTATE CANCER

TIMOTHY C. THOMPSON

Scott Department of Urology and Departments of Cell Biology and Radiology, Baylor College of Medicine, Houston TX 77030

TERRY L. TIMME AND INDER SEHGAL

Scott Department of Urology, Bayler College of Medicine, Houston, TX 77030

19.1. INTRODUCTION

Prostate cancer is now the most common internal malignancy in American men (1), resulting in an estimated 41,000 deaths in 1996. Unacceptably high levels of age-adjusted prostate cancer mortality persist despite a dramatic increase in the detection and treatment of this disease in recent years (2). Prostate cancer is multifocal, with documented inter- and intrafocal heterogeneity of detectable chromosomal abnormalities (3,4). Overall, the molecular and cellular profile of primary prostate cancer suggests that there is only limited clonal expansion at the local site. This pattern could result from tumor growth suppressor activities. Evidence for the involvement of tumor suppressors comes from cytogenetic data indicating numerous chromosomal losses in localized prostate cancer tissues (5). Interestingly, both clinical tissue analysis and experimental results suggest that prostate cancer metastases may originate from clones within relatively small, nondominant tumors (3,4,6). An abrupt transition in clinical stage from localized to metastatic cancer may be the result of multiple levels of interrelated molecular pathways involved in metastatic progression.

Hormones and Growth Factors in Development and Neoplasia, Edited by Robert B. Dickson and David S. Salomon
ISBN 0-471-16899-8

19.1.1. Amplification of the Malignant Phenotype

Recent studies have clearly documented that genetic alterations underlie the stepwise progression of human carcinogenesis (7). This paradigm was initially established through the careful documentation of genetic alterations in the well-defined phenotypes presented in colorectal carcinoma specimens (8). Although the fundamental principles established by these landmark studies have served to anchor molecular modeling of human carcinogenesis, they may require modification to accommodate the addition of nonlinear molecular pathways that better describe the abrupt transitions seen in prostate cancer progression (9,10). We have described a variation on the incremental multistep model of carcinogenesis, the "big bang" model (11) (Fig. 19.1), in which the generation of specific mutations could lead to the establishment of a set of potentially complementing genetic alterations over relatively short periods of time that drive tumor progression. The clearest example of this type of molecular pathway is the recently documented occurrence of germ line and somatic mutations in DNA mismatch repair genes (12–16). Disabling mutations in this important set of genes appear to result in the propagation of genetic alterations through defective replication and repair mechanisms (12). A less well defined molecular pathway for the relatively rapid amplification of the malignant phenotype may involve transcription factors, which through mutation and/or aberrant expression could modulate multiple downstream targets and thus account for the hundreds of alterations in gene expression observed in malignancies (17–19). These examples suggest that either inappropriate expression or mutation of specific "control genes" that determine the fidelity of DNA replication and repair or that regulate the transcription of critical downstream target genes could lead to amplification of the malignant phenotype. Loss of control gene functions could therefore result in the establishment of sets of complementary genetic activities involving oncogenes, growth factors, and hormones and further manifest and expand specific characteristics of the malignant phenotype acquired previously. These types of molecular pathways may underlie the abrupt transitions in the clinical phenotype observed in prostate cancer progression (9,10).

This chapter will review and discuss the separate and interactive roles of oncogenes, tumor suppressor genes, growth factors, and hormones in prostate cancer. We will attempt to place these factors within the context of prostate cancer progression using a three-compartment model consisting of the local tumor compartment, the vascular compartment, and the metastatic site compartment. We will also further develop the notion that certain genes, e.g., p53, may function as control genes as defined above. Abrogation of the normal function of these control genes could trigger a cascade or multiplication of events that may underlie the abrupt transition in the local prostatic compartment leading to the amplification of the metastatic phenotype.

19.2. ONCOGENES: GENETIC COMPLEMENTATION IN MALIGNANT PROGRESSION

The notion that transforming genes cooperate to produce additive or synergistic changes during malignant progression, i.e., amplification of the malignant pheno-

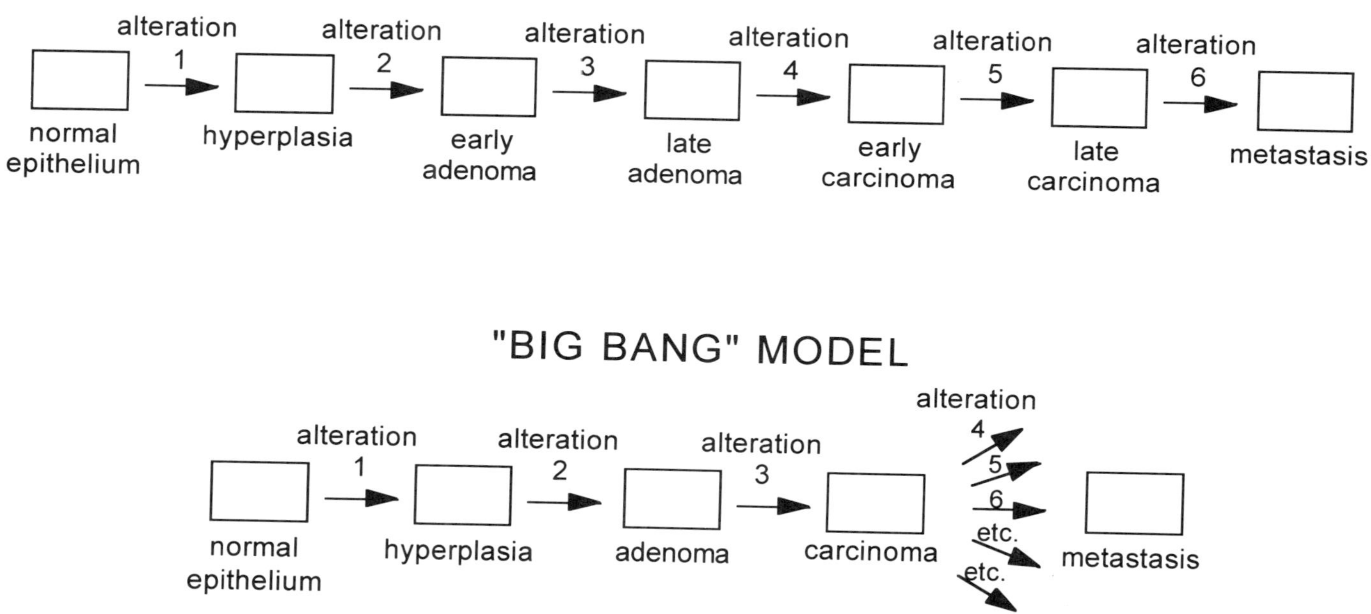

Fig. 19.1. Models for cancer progression. The upper panel depicts the incremental, multistep model proposed by Fearon and Vogelstein (8) for colorectal carcinogenesis. The lower panel proposes an alternative model, the "big bang" model for prostate cancer based on the concepts presented in this chapter and in ref. 11. Because of the uncertainties regarding the prem,alignant phenotype(s) in prostate cancer, we have used the term *adenoma* to describe the benign phenotype.

type, was advanced as the concept of *oncogene cooperativity* (20). One of the central tenets of this concept involved complementation between cytoplasmic and nuclear oncogenes, e.g., *ras* and *myc*. After many years, this paradigm has gained support from the results of numerous experimental and clinical studies (21). In addition, the accumulation of more comprehensive information regarding the *ras* pathway and the *myc* gene transactivation system, as well as the interaction of these important transformation pathways with tumor suppressor genes such as p53, has greatly increased our understanding of molecular oncology and provided a useful mechanistic framework for understanding prostate cancer.

19.2.1. Involvement of *RAS* and the *RAS* Pathway

A complex signal transduction pathway involving RAS (a guanine nucleotide-binding protein) could conceivably play a central role in the linkage of growth factor stimulation and the development of prostate malignancies. RAS mediates the transduction of some signals that are initiated by tyrosine kinase receptors such as the epidermal growth factor receptor (EGFR). Both membrane-bound receptor tyrosine kinase and soluble tyrosine kinase activities have been detected in prostate cancer (22). Although the biochemistry and molecular biology of tyrosine kinase activities in prostate cancer remain largely unexplored, aberrant expression of EGFR and related family members *erb*B-2 (HER-2 or *neu*) and *erb*B-3 have been reported in prostate cancer (23–26). In response to various extracellular stimuli including activated tyrosine kinase receptors, *ras* has been shown to activate the mitogen-activated protein kinase (MAPK) signaling cascade (27–33). Other oncogenes, including *raf* (34–36), *src* (37,38), and *mos* (39–41), also activate the MAPK signaling cascade. It has ben shown that the MAPK activities are essential for the transforming activity of these oncogenes (42–44) and that activation of the MAPK signaling cascade independently can lead to cell transformation (43). Interestingly, it was recently demonstrated that loss of wild-type p53 function can synergize with the *mos*/MAPK pathway to induce cell transformation and chromosome instability (45). This synergy is mediated through loss of $G_1 \rightarrow S$ and $G_2 \rightarrow M$ checkpoints, as well as through the abrogation of apoptosis.

The reported frequency of *ras* point mutations in localized prostate cancer is 6–25% (46–49); the roles of *raf, src,* and *mos* in prostate cancer have not been sufficiently investigated. Point mutations in the *ras* oncogene may represent only a fraction of the alterations that lead to aberrant activation of the MAPK signaling cascade. For example, additional proteins, such as GTPase activating proteins and guanine-nucleotide exchange factors that regulate the rate of GTP hydrolysis and the rate of guanine-nucleotide exchange on the RAS protein and other small GTP binding proteins, have been shown to harbor mutations involved in other growth disorders such as neurofibromatosis (50). The presence of gene mutations that indirectly affect *ras* or independently activate the MAPK signaling cascade in prostate cancer has not been sufficiently investigated.

Because of its critical function in the activation of the MAPK signaling cascade, the RAS protein represents a target for therapeutic approaches. Recently developed pharmacologic agents that block RAS farnesylation and thus prevent its normal attachment to the inner surface of the plasma membrane may prove useful in suppressing prostate cancer growth (51). In a preclinical evaluation of one of these agents, *ras*-induced mammary and salivary gland tumor regression was documented in mouse mammary tumor virus (MMTV)-*v*-Ha-*ras* transgenic mice following administration of a farnesyltransferase inhibitor (52).

Activation of the MAPK signaling cascade by *ras* or possibly other oncogene activities may complement a separate set of genetic alterations occurring in the molecular network that regulates the activities of the proto-oncogene c-*myc*.

19.2.2. The c-*myc* Proto-oncogene: Activities and Interactions

The c-*myc* gene product is a functionally diverse nuclear protein involved in the regulation of both proliferation and apoptosis (53,54). The c-MYC protein has a characteristic transcription factor structure, an amino-terminal transactivating domain, and a carboxy-terminal DNA binding domain with a basic helix-loop-helix leucine zipper motif (55,56). The identification of another protein, MAX, with a similar basic helix-loop-helix leucine zipper structure that can heterodimerize with c-MYC or homodimerize with itself has provided a structural model for transcriptional activation (57–59). Recently, the stoichiometric model for the MYC-MAX transcriptional unit has been expanded to include the MAX interacting proteins MAD and MXI1 (reviewed in ref. 60). Although the complete structure and functional significance of the network regulating c-MYC activities are not fully resolved, it is clear that c-MYC transactivation activities can be tightly regulated by the levels of these interacting proteins.

At this time, only a few transcriptional targets of MYC have been identified. In the absence of adequate growth factors c-MYC, regulated gene activities lead to apoptosis (61-64). Recently it was shown that cdc25, a cyclin-dependent kinase-activating phosphatase may be responsible for c-MYC-induced apoptosis in fibroblast cells (61). In addition to the transcriptional activation of target genes, c-MYC can suppress the transcriptional activity of specific genes (56,65) including major histocompatibility complex (MHC) type I genes (66).

The level of c-*myc* mRNA is increased in the normal rat ventral prostate following castration (67,68), and increased levels of c-*myc* RNA and c-MYC protein in prostate cancer have been described (69–71). Amplification of c-*myc* in prostate cancer tissues using fluorescent *in situ* hybridization has been reported (72). Mutations in and allelic loss of the *MXI1* gene in 5 of 10 cases of prostate cancer with karyotypically documented deletions of 10q24–25, the chromosomal location of *MXI1* (73), suggests that alterations in the molecular network that regulates c-*myc* activity may be involved in prostate cancer.

In regard to the concept of *ras* + *myc* cooperativity in transformation, it is of interest that both the oncogene-driven MAPK signaling cascade and c-*myc*-stimulated cdc25 can induce p53-dependent apoptosis (45,61). This point of convergence may be relevant for the concept of oncogene cooperativity as well as for prostate cancer.

19.2.3. Tumor Suppressor Genes

Numerous studies have documented chromosomal abnormalities in prostate cancer specimens (3–5,74,75). Specific chromosomal aberrations identified by loss of heterozygosity at informative markers have suggested the presence of prostatic tumor suppressor loci on chromosomes 7 (76), 8 (77–80), 10 (81–83), 16 (81,82), and 17 (84). Several specific tumor suppressor genes have been evaluated in prostate cancer including the retinoblastoma gene *(Rb)*, the *nm*23 gene, the *KAI1* gene, and the p53 gene.

The human prostate cancer cell line DU 145 contains a mutated *Rb* gene, and gene replacement reversed its tumorigenicity (85). *Rb* mutations or deletions have been reported in up to 35% of clinical prostate cancer specimens (82,86–89). The *nm*23 genes (*nm*23-H1 and *nm*23-H2) have been implicated as metastasis suppressors in both rodent and human tumors, and aberrant expression of the *nm*23 genes has been correlated with a poor prognosis (reviewed in 90). However, there have been conflicting results from immunohistochemical studies of *nm*23 expression in prostate cancer specimens (91,92). The *KAI1* gene was initially identified as capable of suppressing metastasis of a rat prostate cancer cell line (93). Subsequent work has revealed it to be homologous with a human lymphocyte cell surface marker, CD 82, which is expressed in prostatic hyperplasia and low-grade prostate cancer, but expression is reduced in high-grade cancer (94). This loss of expression does not appear to be related to mutation or allelic loss (95).

The p53 gene has been the most thoroughly studied tumor suppressor gene and will be the focus of this section. The p53 gene product inhibits progression of the cell cycle at the $G_1 \rightarrow S$ interface to allow DNA damage repair before replication (96). Inactivation of the p53 gene could result in fixation of mutations and lead to genetically unstable cells more likely to develop a malignant phenotype (97). Mutation of the p53 gene is probably the most common molecular alteration in human cancer, including colon, lung, and breast carcinoma (98–100). Detection of p53 mutations frequently relies on the association of mutations with stabilization of the half-life of the p53 protein leading to immunoreactivity with specific antibodies (101,102).

A comprehensive review of the literature indicates that p53 alterations in primary prostate cancer tissues are less frequent than the levels reported at metastatic sites. Although numerous studies of primary prostate tumors have detailed the association of p53 alterations with increased tumor grade and/or androgen insensitivity (23,103–116), Figure 19.2 summarizes other studies that used molecular as well as immunohistochemical analysis of p53 alterations in both primary and metastatic disease (117–127). In *every* study in which both primary and metastatic specimens were analyzed, there was increased detection of p53 alterations in the metastatic tissue relative to the primary tumors. The median of alterations detected in primary cancers is 9%, whereas in metastases the median is 23%.

The most compelling clinical evidence for the involvement of p53 alterations in the development of the metastatic phenotype comes from studies in which primary and metastatic samples from the same patient were analyzed. There are several reports of the same p53 mutation in a primary tumor and a metastasis in one or a few

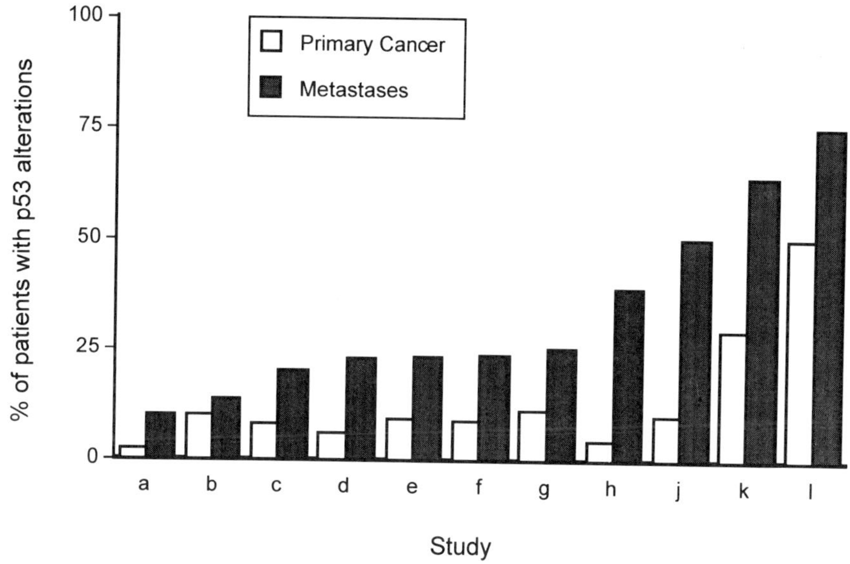

Fig. 19.2. Frequency of p53 alterations in prostate cancer. A summary of 11 studies that evaluated both primary prostate tumors (open boxes) and prostate tumor metastases (closed boxes). The frequency is based on immunohistochemistry results for studies a, b, and f–k, while studies c–e and l used sequence analysis to detect mutations. In the immunohistochemical studies, sequence analysis was used to confirm the presence of mutations in many samples. The studies depicted are from the references cited ( ), and the number of patient primary (1° =) and metastatic (2° =) tumor samples evaluated were: a = (121), (1° = 85), (2° = 10);b = (120), (1° = 20), (2° = 15); c = (126), (1° = 25), (2° = 20); d = (127), (1° = 51), (2° = 22); e = (124), (1° = 65), (2° = 13); f = (123), (1° = 47), (2° = 47); g = (122), (1° = 36), (2° = 4); h = (118), (1° = 45), (2° = 49); j = (117), (1° = 10), (2° = 2); k = (125), (1° = 38), (2° = 39); l = (119), (1° = 30), (2° = 4).

patients (117,118,122,127). We have conducted an extensive analysis of paired primary and metastatic patient samples (123,128). Although the frequency of patients with p53 alterations in primary tumor samples is lower than in metastatic samples, as noted in Figure 19.2, the primary tumor samples can have considerable molecular heterogeneity (124,128). We analyzed multiple cloned DNAs from microdissected cancer samples to estimate the frequency of specific p53 mutations and found multiple cases where a mutation detectable in the primary tumor at a low frequency was more abundant in the metastatic cancer, i.e., clonally expanded (128).

Although there are certainly other (non-p53) pathways involved in prostate cancer progression, the association of p53 malfunction with disease progression to hormone therapy resistance provides a compelling argument for understanding the complexities of the p53 pathway.

To study the biology of p53 function in prostate cancer, we developed an experimental *in vivo* mouse model. We use a single recombinant retrovirus that transduces the *ras* and *myc* oncogenes to produce an initiating event in epithelial cells that have been carefully dissected from fetal urogenital sinus tissue. The urogenital sinus differentiates into prostate under the influence of androgenic hormones and can be obtained from normal inbred mice or from mice that have been genetically altered (e.g., transgenic or knockout mice). The oncogene-initiated epithelial cells along with urogenital sinus-derived stromal cells are grafted under the renal capsule of adult male mice, where they develop into prostatic cells with a phenotype that varies from normal prostate through hyperplasia to adenocarcinoma, depending on the initiating event and the genetic background of the epithelial cells (6,129–131). The molecular and cellular phenotypes associated with this experimental progression model have been found to be relevant to human prostate cancer progression (e.g., accumulation of transforming growth factor-beta 1) (132,133). Using the p53 knockout mouse (134) as a source of urogenital sinus tissue, we demonstrated that loss of p53 function leads to rapid progression of prostate cancer with extensive metastatic activity (6). Metastasis was invariably associated with loss of p53 function, and we further demonstrated that metastases did not necessarily originate from the most abundant clone present in the primary tumor. Rapid clonal progression from relatively small tumor foci containing p53 mutations fits the observed experimental data and provides a molecular basis for the abrupt transition in the clinical phenotype observed in prostate cancer (9,10).

Based on this experimental and clinical data, we consider p53 to be a critical control gene in prostate cancer progression. A large and expanding list of genes under the transcriptional control of p53 has been reported. Activity of the p53 gene is regulated by its own gene product through transcriptional transactivation and inhibited through protein–protein interactions with MDM-2 (135,136). Interestingly, *mdm*-2 is under the transcriptional control of p53, thus forming a negative feedback loop (137). The most notable class of genes regulated by p53 is involved in growth suppression and/or cell cycle checkpoint controls at the $G_1 \rightarrow S$ transition and includes p21, GADD45, and cyclin G, (refs: 138–141 and reviewed in ref. 142). There are numerous biologic activities that p53 may regulate in addition to its role in controlling the cell cycle. Among these are differentiation (143) and apoptosis (144–149), as well as regulation of genes that encode cell type-specific polypeptide growth factors (150–156).

Loss of normal p53 function through dominant-negative mutations, gain-of-function mutations, or gene loss may lead to aberrant transcriptional regulation of the multiple downstream target genes under p53 control. Alterations in this molecular p53 network may amplify the malignant phenotype since loss of function for p53 could simultaneously affect multiple gene activities and thus result in the clonal expansion and metastatic progression of p53 mutant cells. It is also possible that gain-of-function mutations could activate cryptic target genes and provide additional mechanisms of malignant progression. Therefore, functional alterations in the p53 gene could lead to a wide spectrum of inappropriate transcriptional activities that contribute to metastatic progression. Furthermore, loss of p53-mediated apoptosis could potentiate the oncogenic activities of the MAPK signaling cascade (45) and c-*myc* (61,157). Although we are of the opinion that p53 fits our concept of a control gene and has particular relevance for prostate cancer progression, it is likely that other genes also function in this capacity. Such control genes may regulate DNA mismatch repair since there is evidence of a mutator phenotype in prostate cancer (158–160). However, alterations in a specific gene or genes that maintain control of this type of genetic stability have not been reported for prostate cancer.

19.3. POLYPEPTIDE GROWTH FACTORS: AUTOCRINE AND PARACRINE EFFECTORS OF ABERRANT GENE ACTIVITIES

Several polypeptide growth factors were originally identified through their oncogene-induced expression in retrovirus-transformed cells (161–164). It was subsequently reported that these growth factors (165–168) or their mutated receptors (169) can independently induce transformation through the establishment of autocrine and paracrine activities. More recently, it has been documented that in many cases the signal transduction pathways for these growth factors either directly involve oncogenes such as *ras* or converge with oncogene-mediated pathways, e.g., the MAPK signaling cascade (170,171). Below we discuss families of growth factors that are relevant to prostate cancer.

19.3.1. Transforming Growth Factor-β

Transforming growth factor-β's (TGF-β's) are a family of multifunctional polypeptides with several isoforms. The expression of each TGF-β isoform, as well as expression of other members of the TGF-β superfamily (such as activin, inhibin, bone morphogenic protein, and mullerian inhibiting substance), is temporally and spatially controlled during development and differentiation (172). TGF-β responses vary among cell types and depend on the cellular microenvironment since other growth factors may influence the cellular response to TGF-β (173). Although usually inhibitory to the growth of normal epithelial cells, TGF-β1 induces stromal cell proliferation (174). Many biologic activities demonstrated by TGF-β1, e.g., induction of angiogenesis and immune suppression, may enhance the progression of malignant cells (172,173). A role for TGF-β1 in malignant progression is also suggested since TGF-β1 can also induce type IV collagenase expression in primary cultures of human keratinocytes (175), an activity associated with invasion of the basement membrane. TGF-β1 is permissive for tumor development in Rous sarcoma virus-infected chickens (176) and increases the invasive and metastatic potential of a mammary adenocarcinoma cell line (177).

The cDNA for human TGF-β2 was originally cloned from PC-3 cells, a human prostate cancer cell line (178). Accumulation of TGF-β1 has been observed in surgically excised prostate tissue samples from patients with either prostate cancer or benign prostatic hyperplasia (BPH) (133,179). Overall accumulation encompassed a greater area in prostate cancer (59%) than in BPH (26%), and both had much greater staining than on adjacent areas of the slide with normal morphology. Studies using antiserum that specifically recognizes an intracellular form of TGF-β1 suggested that the TGF-β1 in BPH was derived from the stromal cells, whereas in prostate cancer the neoplastic epithelial cells are the predominant source. Blood levels of TGF-β1 in patients with prostate cancer have been reported to be elevated at least twofold over those of control patients and patients with BPH (180).

Increased TGF-β1 mRNA and protein have also been detected in experimental oncogene-induced mouse prostate cancer (130), with a pattern of accumulation that is remarkably similar to the one observed in human prostate cancer (132). Overexpression of TGF-β1 has been associated with a more aggressive phenotype in anaplastic rat prostate tumor cells (181). An increase in TGF-β1 expression accompanies castration-induced apoptosis in normal rodent prostate (182), as well as in rodent (183) and human (184) prostate cancers. Although these observations do not provide evidence of a direct causal relationship between expression of TGF-β1 and prostate cancer, they indicate that aberrant accumulation of TGF-β1 accompanies progression to the malignant phenotype.

The molecular and cellular mechanisms that lead to elevated TGF-β1 levels in prostate cancer are unclear, yet studies to determine the expression and functionality of the TGF-β receptors may produce important information regarding the TGF-β response. We have evaluated the expression of the type II TGF-β receptor by immunohistochemistry in radical prostatectomy specimens and observed that normal and benign epithelia stained intensely but that there was progressively diminished staining as the histologic grade of the cancer increased (185). Inactivation of the type II receptor has also been associated with other cancers (186,187) and mechanistically linked to genomic instability (187). The type I receptor has been reported to be defective in the TGF-β1-insensitive prostate cancer cell line LNCaP (188).

19.3.2. Fibroblast Growth Factors

The fibroblast growth factor (FGF) family consists of at least nine gene products. Substantial amounts of FGF-2 mRNA and protein are present in mature human prostate (189), and BPH tissue contains elevated FGF-2 mRNA relative to normal prostatic tissue (190). FGF-1 and FGF-2 demonstrate significant angiogenic activity *in vivo,* a prognostically significant feature for prostate cancer as well as other malignancies (90,191,192). Androgen-insensitive PC-3 and DU 145 cells produce significant levels of FGF-2 mRNA and protein, whereas the androgen-sensitive and generally less aggressive LNCaP cells do not express detectable levels of FGF-2 protein (193). In addition, the level of FGF-1 receptor mRNA is elevated in PC-3 and DU 145 cells relative to LNCaP cells (193).

Although the mechanisms of aberrant expression of FGFs and the FGF response have not been fully elucidated, several FGF activities and their responses may be involved in prostate cancer progression. For example, overexpression of FGF-3 (INT-2) in a transgenic mouse induces prostatic epithelial hyperplasia (194) and, in a rat prostate cancer model, a specific change in mRNA splicing for the FGF receptor 2 gene leads to loss of stromal cell dependency for growth in malignant epithelial cells (195). Some FGF activities have been reported to be regulated by androgens in normal and malignant prostate tissues. The production of FGF-7 in rat prostate stromal cells is under the control of androgens (196); therefore FGF-7 has been termed *andromedin,* a paracrine factor mediating epithelial cell growth. In general, stromal and epithelial cells respond differentially to some FGF family members: Stromal cells respond to FGF-1 and FGF-2 but not to FGF-7, whereas epithelial cells respond to FGF-1 and FGF-7 but not to FGF-2 (195). An androgen-induced growth factor termed FGF-8 appears to be expressed at the RNA level in moderately and poorly differentiated prostate cancer cells but not in prostatic stroma, BPH, or most normal epithelia except occasional basal cells (197).

19.3.3. Epithelial Growth Factors

The EGF growth factor family, including EGF, TGF-α, and amphiregulin, as well as their receptors, has been reported in normal and malignant prostatic tissues (198,199). EGF has been detected in human prostate cancer cell lines and rodent prostate. It has been shown to be mitogenic for both normal and malignant rat and human prostate cells *in vitro*. TGF-α also interacts with the EGF receptor and has been implicated as an autocrine growth factor in human prostate cancer. When over-expressed in transgenic mice, TGF-α induces significant prostatic epithelial hyperplasia and focally dysplastic changes resembling carcinoma *in situ* (200). Amphiregulin has been reported to be regulated by androgens in LNCaP cells (201). Several studies have demonstrated a decrease in EGFR expression in malignant versus normal prostate tissue (23–26), suggesting that aberrant EGFR expression could play a role in prostate cancer progression. There is also a report that EGFR and TGF-α expression is increased in advanced and metastatic prostate cancer (202). Loss of regulation of expression of growth factors and/or their receptors may lead to unchecked autocrine growth mechanisms. The expression of EGFR has been shown to be regulated by normal p53 (203), and thus loss of normal function for this tumor suppressor gene could result in loss of regulation in the expression of the EGFR.

19.3.4. Other Growth Factors

There are reports that specific binding proteins for the insulin-like growth factors are overexpressed in prostate cancer (204), and that these proteins may be cleaved by the serine protease prostate specific antigen (205,206). The complexities of the autocrine and paracrine interactions of insulin-like growth factors, their receptors, and their binding proteins are only now being revealed for prostate cancer (204). The role of platelet-derived growth factors and their receptors in prostate cancer progression has not been clearly established (198,199).

In addition to these polypeptide growth factors, neuropeptide hormones are often expressed in prostate cancers (reviewed in ref. 207). Foci of prostate cells with neuroendocrine differentiation markers secrete a variety of these small peptides including calcitonin, serotonin, somatostatin, bombesin/gastrin-releasing peptide, thyroid-stimulating hormone, and neurotensin (208,209). These clusters of neuroendocrine cells have been reported to be a constituent of most prostatic adenocarcinomas, with incidences ranging up to 100%, depending on the criteria used to define neuroendocrine differentiation, and the prevalence of neuroendocrine cells is correlated with a poor prognosis and high-grade disease (207). Although the origin of these cells is not known, in the normal prostate they have been suggested to originate from pluripotent basal stem cells (210). Interestingly, a recent study has suggested that some prostate adenocarcinoma cells can be induced to differentiate toward a neuronal morphology after pharmacologic elevation of cAMP (211). Neuroendocrine tumors are typically unresponsive to hormonal therapy (207), and prostate neuroendocrine cells are probably not androgen dependent (210). Androgen withdrawal often leads to the evolution of hormone-refractory disease. Therefore, it is possible that such withdrawal is accompanied by a selection for cells responsive to paracrine and autocrine factors secreted by surviving cell populations and that neuropeptides are a potential source of these factors. Androgen deprivation of the androgen-sensitive human prostate cancer line LNCaP leads to downregulation of a neuropeptide-degrading proteinase, facilitating the accumulation of at least one peptide, neurotensin, which stimulates growth in cultures of hormone-deprived but not hormone-stimulated cells (212). Although it is likely that these small peptides play several roles in prostate cancer, most studies have focused primarily on growth as the endpoint of neuropeptide stimulation. However, other mitogenic polypeptides also activate a multitude of biologic responses in normal and tumor cells, and it is therefore quite likely that neuropeptides induce responses other than proliferation that must still be carefully defined.

The nerve growth factor (NGF) family has been implicated in prostate cancer. Normal prostate stromal cells produce NGF-like proteins, and prostate epithelial and cancer cells contain receptors for these compounds (213), suggesting that a paracrine loop may exist between production of NGF-like proteins by prostatic stroma and the malignant epithelium. The NGF-like protein activity induces chemotaxis and chemokinesis of human prostate cancer cell lines (214).

19.4. STEROID HORMONES: MODULATION OF MALIGNANT PROGRESSION

Although a clear mechanistic basis for the role of androgenic hormones in prostate cancer remains to be established, the steroid hormones are likely involved in several critical aspects of prostate tumorigenesis and progression. Androgens are essential for normal prostate development and maintenance of homeostasis—the balance between cellular proliferation, growth arrest, and programmed cell death (apoptosis). Androgen ablation has been a mainstay of treatment for metastatic prostate cancer for over 50 years (215); however, this therapy is not curative since hormone-independent cells almost invariably emerge and result in disease progression. Although androgens have been shown to induce prostate cancer in rats (216), epidemiologic studies in humans have produced conflicting results regarding the role of serum steroid hormone levels in prostate cancer risk evaluation (217).

Alterations in androgen-responsive genes may facilitate prostate cancer progression and could occur at several different levels. One such alteration is mutation of the intracellular androgen receptor (AR). Some AR mutations have been documented in both primary and metastatic prostate cancer tissues (218–220). In general, the frequency of mutations appears to be relatively low and it has been difficult to assign functional significance to the mutant ARs identified. One mutation detected in an androgen-insensitive prostate cancer was activated by adrenal androgens and progesterone (220). However, AR mutations could define an important subset of advanced disease. Video image analysis of prostate biopsy specimens reacted with anti-peptide antibodies specific for either the carboxyl or amino terminus of AR has been used to predict the outcome of hormonal therapy in advanced disease (221). Interestingly, overexpression of AR in the androgen-insensitive PC-3 cell line (222) produces growth inhibition. Analysis by comparative genomic hybridization of specimens from local recurrences of prostate cancer in patients following endocrine treatment revealed that 30% had amplification at the chromosomal location (Xq11-q13) of the AR gene (223). Further studies hopefully will provide a clearer picture of possible subsets of prostate cancers that contain specific AR mutations and the possible significance of AR genomic amplification during prostate cancer progression.

Another possible aberration of steroid hormone responsiveness is the altered regulation of growth factor and growth factor receptor genes under androgen regulation. In the androgen-responsive human prostate cancer line LNCaP, the EGFR and its ligands amphiregulin, TGF-α, and EGF are induced by androgenic stimulation (201). The growth factor FGF-2 is also induced by androgens in this cell line (224), and it is possible that other growth factors and their receptors are regulated by androgenic steroids directly or indirectly.

The complex biologic endpoints for androgen action and the inadequate information about androgen target genes involved in carcinogenesis preclude a complete understanding of the role of steroid hormones in prostate cancer. Although additional studies may reveal a direct role for androgens in the transformation process, currently it appears that androgens modulate the activities of genes that are more directly involved in malignant progression.

19.5. A THREE-COMPARTMENT MODEL OF PROSTATE CANCER PROGRESSION

To integrate the specific components of the transformation pathways in prostate cancer within the context of clinical significance, we have organized these pathways into a three-compartment model (Fig. 19.3) for prostate cancer progression based on the metastatic cascade (90,225,226). In this model the first compartment represents the malignancy at the local site, where the tumor must acquire growth and angiogenic potential, as well as the ability to invade beyond the confines of the gland. Invasion is a complex process involving adhesion, proteolytic degradation, and motility. Upon gaining access (i.e., intravasation) to the second compartment,

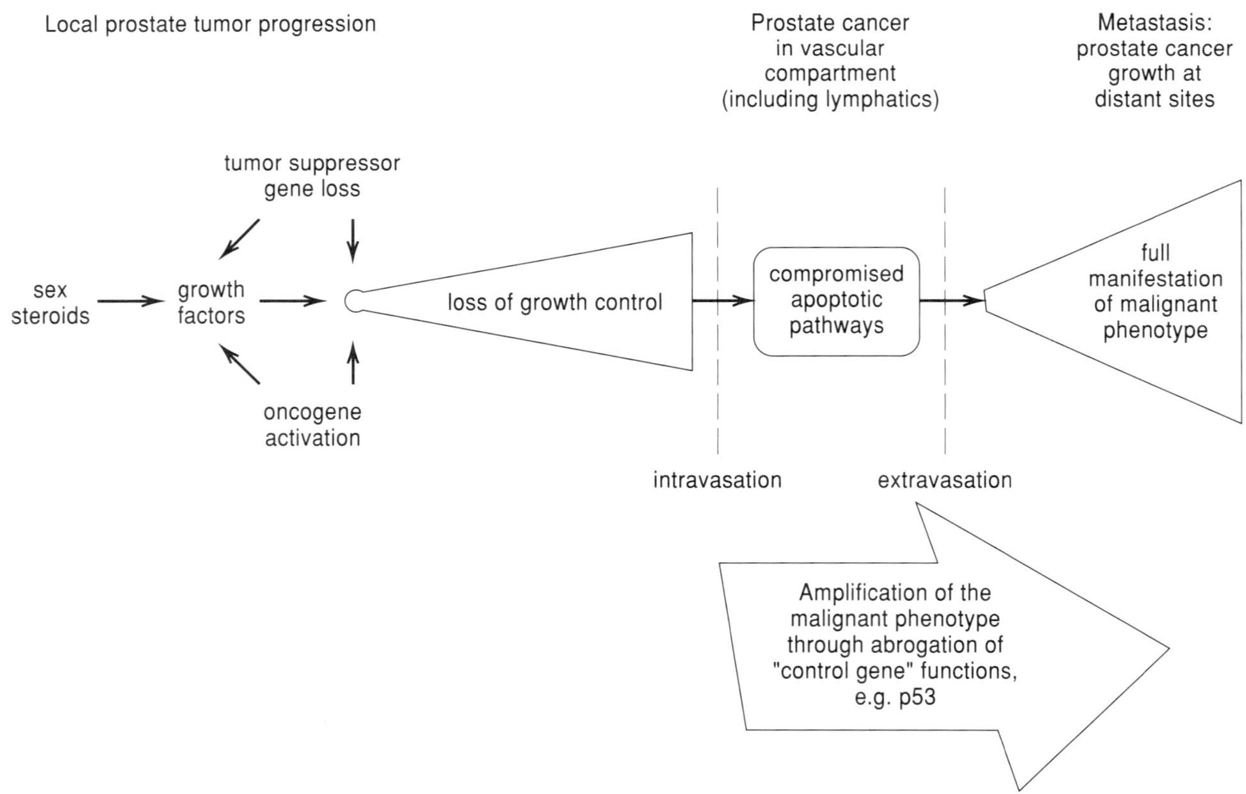

Fig. 19.3. A three-compartment model of prostate cancer metastasis. The initiation and progression of a local prostate tumor involve the interactions of altered tumor suppressor genes, activated oncogenes, growth factors, and steroid hormones, leading to a loss of growth control. Certain cells within the local compartment acquire the ability to invade the second compartment, the vascular system composed not only of the bloodstream but also the lymphatic system. In this compartment it is likely that cells that have acquired compromised apoptotic pathways, such as through mutation of the p53 gene, survive. A fraction of these surviving cells are capable of extravasation and growth at a distant metastatic site. The three distinct compartments impose unique selection pressures for progession that result in amplification of the malignant phenotype.

the host vasculature, the prostate cancer is now free in the blood and/or lymphatic systems to undergo another compartmentalized set of selection pressures. Exit from the host vasculature (i.e., extravasation) into the third compartment, the metastatic site, can lead to growth and clonal expansion of prostate cancer cells. In this third compartment, activities similar to those that occurred at the local site are involved, including angiogenesis and invasion. In addition, other poorly understood unique phenotypic properties are likely to be critical determinants of this final step in the selection process.

19.5.1. The Local Compartment: Progression to Intravasation

Within the prostate gland, premalignant and malignant cancer cells suffer genetic damage of diverse types, including oncogenic activation and mutations in tumor suppressor genes. During local prostate tumor progression, selection appears to be based mainly on changes in growth potential. Yet, specific genetic alterations at this stage do not appear to lead to significant clonal expansion. In most cases, the analysis of molecular alterations in specific genes has thus far failed to document extensive clonal growth. However, genetic alterations may involve activities that converge at the level of aberrant growth factor activities. Dysregulated growth factor expression and growth factor receptor-mediated signaling may, in turn, lead to a redistribution of specific cell types through the selection of cell populations that have lost growth controls and/or have acquired invasive properties (adhesion, motility, and proteolytic activities). Circulating androgenic hormones may affect both growth factors and oncogene products, although these activities are poorly understood. One clear example of the convergence of various oncogenic events is the overexpression and accumulation of TGF-β1. In some experimental systems, differential loss of growth inhibitory activities and/or acquisition of invasive activities in response to TGF-β1 (and possibly other growth factors) may be important in progression (227).

The proteolytic activities most often associated with tumor invasion involve the plasminogen activators and matrix metalloproteinases (MMPs) (228). These proteases target basement membrane constituents such as laminin, collagen, proteoglycan, and elastin. The type IV collagenase/gelatinases MMP-2 and MMP-9 have been associated with tumor metastasis (228,229). Activation of collagenase expression and activity by oncogenes (e.g., v-*src*, c-Ha-*ras*, or v-*mos* [230]) and by several polypeptide growth factors (TGF-β, EGF, TGF-α, TNF-α, interleukin [IL]-1β, and IL-1 [231,232]) suggests that multiple oncogenic pathways may promote matrix degradative activities important for metastasis.

The recognition and destruction of premalignant and malignant cells by the host immune system is a critical barrier that must be overcome during prostate cancer progression. Although understanding of the molecular basis or immunologic responses has dramatically increased in recent years, the specific mechanism(s) through which tumor cells escape immune surveillance has generally remained unclear. This is certainly true for prostate cancer, for which published papers are minimal (233–235). An immune response to prostate cancer may be muted in the local compartment through increased expression and accumulation of immunomodulatory cytokines, e.g., TGF-β1, by prostate cancer cells. We have demonstrated that TGF-β1 accumulates to very high levels in primary prostate cancer (133,236) and is associated with prostate cancer progression to metastasis (179). The immunosuppressive effects of TGF-β1 have been well documented (172,173). Immunosuppression results not only from lymphocyte growth inhibition but also from downregulation of MHC genes, as has been documented in TGF-β1 knockout mice (237) and several cancer cell lines (238). It is conceivable that control genes such as p53 regulate TGF-β1 (153) and may alter the immune response indirectly through TGF-β1-mediated immunosuppressive activities. Overexpression of c-*myc* may also impair an immune response through repression of MHC class I genes (66). Further studies regarding this potential mechanism as well as others require the establishment of relevant sophisticated animal models capable of addressing questions related to prostate cancer immunogenicity during progression.

Escape from the local compartment into the vasculature involves aberrant cell–cell and cell–matrix interactions. Alterations in the expression of various extracellular matrix attachment molecules have been reported to correlate with prostate cancer progression (239–241). Induction of apoptosis has been correlated with integrin signaling, and blocking attachment can induce apoptosis (242,243).

It appears that the capacity to induce apoptosis is still largely patent in the local compartment. This conclusion is based on the documented positive correlation of apoptotic activity (as well as mitotic activity) with increasing Gleason grade (244). We reported a statistically significant positive correlation between the frequency of apoptosis, apoptotic index (AI), and Gleason grade (244) and further demonstrated that AI may be a useful predictor of the prognosis in Gleason grade 3 prostate cancer (245). Whereas high levels of apoptosis are associated with more aggressive disease, overall, the relative balance between mitosis and apoptosis will dictate the extent of tumor growth in the local compartment.

The observation of increased apoptosis with increasing malignant potential in prostate cancer interestingly parallels a recent area of investigation that focuses on the roles of c-*myc and* p53 (see above). As c-*myc* can induce apoptosis through the p53 pathway and androgen-suppressed apoptosis in the prostate is modified in p53 knockout mice (246), it's conceivable that p53 mediated activities play a role in apoptosis in prostate cancer. Loss of normal p53 function alone or together with MAPK signaling activities (45) could compromise apoptosis in prostate cancer. However, experimental data indicate that castration can efficiently induce prostatic regression in the complete absence of p53, indicating that other effector proteins are operative (247). p53 can induce *bax*, which under some conditions can lead to apoptosis (248,249). Interestingly, expression of wild-type p53 can also lead to decreased expression of *bcl*-2 (248), which can block p53-dependent apoptosis in specific cell types (250). Overexpression of *bcl*-2 has been associated with metastatic compared to nonmetastatic variants of LNCaP cells derived from a nude mouse model (251). Thus the cells that escape the local compartment and survive in the vascular compartment likely acquire apoptosis resistance that may result from loss of function of p53 or other control genes. Further studies are necessary to extend these results to prostate cancer cells.

In the local compartment, diverse genetic alterations lead to loss of growth control, acquisition of invasive properties, and possibly resistance to immune surveillance. It is our opinion that loss of specific control gene functions occurs relatively late in the process of limited clonal expansion in the local tumors and may result in compromised apoptotic pathways in subsets of cells. The prototypic example of this type of alteration is mutation of the p53 gene, which has been demonstrated in localized prostate cancer but at a relatively lower frequency compared to cancer cells at metastatic sites (Fig. 19.2). In addition, the observation of reduced apoptotic potential in prostate cancer cells within the perineural regions (252) also supports loss-of-control gene function and further relates this phenomenon to apoptosis resistance. As the perineural region could be considered to be a "gateway" to disseminated disease (253–255), changes that occur in this region may be highly informative.

19.5.2. The Vascular Compartment: Potential Role of Apoptosis Resistance

During the second stage of progression, prostate cancer cells gain access to the vascular compartment, possibly through both active and passive mechanisms, and face different selection pressures to survive in this hostile environment. It has been speculated that millions of cancer cells enter the circulation daily from tumors that are larger than 1 cm^3 (256). Indeed, there have been numerous reports on the high prevalence of cancer cells in the circulation of prostate cancer patients based on detection of the mRNA for prostate specific antigen (257–260). However, in experimental models the majority (>99.9%) of circulating cancer cells die rather than produce metastases (226,261).

It has also been demonstrated that increased apoptosis resistance related to direct or possibly indirect inactivation of p53 may lead to increased metastatic potential in some experimental systems (145,250,251). Thus, although many prostate cancer cells gain access to the vascular compartment, only a small subset have the phenotypic properties that enable them to survive, extravasate, and grow at the metastatic site.

19.5.3. Metastasis to Distant Sites: Potential Role of P53 "Control Gene" Activities

In the final stages of progression, highly selected prostate cancer clones that have survived the vascular compartment extravasate and undergo expansion at a distant metastatic site such as bone or soft tissues. The site of metastasis is not random and the century-old (262) concept of *seed* and *soil* is still applicable, as the organ microenvironment has been shown to influence metastasis (263). Many of the invasive gene activities necessary for escape from the primary compartment may also enhance the ability of cancer cells to extravasate from the vasculature. However, our previous studies using genetically matched primary tumor-derived mouse prostate cancer cell lines and their metastasis-derived counterparts suggested that intravasation may involve phenotypic properties that are different from those required for extravasation (264). Extravasation and colonization at metastatic sites are likely to be as complex as earlier stages in progression and, at this point, poorly understood. One activity essential for colonization at a metastatic site is angiogenesis (90). Interestingly, this important activity may also be under p53 control.

As previously discussed, transcription factors such as p53 may control expression of specific growth factors, which, in turn, could provide the capacity for angiogenesis as well as cell survival. Since wild-type p53 has been shown to increase thrombospondin 1 gene activity (152) and suppress FGF-2 gene activity (154), loss of p53 regulation could have a stimulatory effect on angiogenesis. Other angiogenesis-enhancing effects could result from loss of p53 regulation of vascular endothelial growth factor (151,155). The possibility that p53 regulates TGF-β1 (153) is also of interest since TGF-β1 is overexpressed and accumulates to high levels in both experimental and human prostate cancer (see above), with the highest levels of accumulation in metastatic sites (179), and is a potent inducer of angiogenesis *in vivo* (265,266).

Therefore, p53 may not only suppress the transformed phenotype through the support of DNA repair and chromosome stability, but may also suppress prostate cancer progression, and possibly progression in other malignancies, through the regulation of specific growth factors that could play a role in acquisition of the metastatic phenotype. Abrogation of p53 function through direct mutation, or possibly even through aberrant transcriptional regulation (6), could generate relatively rapid changes in gene expression that are directly relevant to metastasis of prostate cancer. It is therefore conceivable that loss of p53 function could underlie the abrupt transition observed in prostate cancer in which only limited clonal expansion of primary tumor cells harboring specific mutations occurs locally, yet loss of p53 function leads to aberrant regulation of multiple gene activities that are able to enhance survival in the vasculature and/or promote clonal expansion at distant sites.

19.6. SUMMARY

In summary, amplification of the malignant phenotype in prostate cancer involves critical genetic alterations that lead to the selection of complementary transforming gene activities. In some cases, loss of function for critical control genes such as p53 may potentiate the progression to metastasis through loss of transcriptional control of multiple gene activities. Polypeptide growth factors may function as effectors of transformation by establishing autocrine and paracrine pathways and by activating additional genes that further drive malignant progression. Androgenic steroids also appear to modulate progression through the transcriptional regulation of downstream target genes. Further understanding of critical molecular pathways in prostate cancer will likely provide targets for combination therapies that impose barriers to genetic alteration-driven progression.

ACKNOWLEDGMENT

The authors thank Dr. Dov Kadmon for his critical review of the manuscript. The studies that led to this manuscript were supported by National Institutes of Health grants CA50588 and CA68814 and SPORE P50-58204.

REFERENCES

1. Parker SL, Tong T, Bolden S, Wingo PA: Cancer statistics, 1996. CA—Cancer J Clin 46:5–27, 1996.
2. Brawley OW, Kramer BS: Epidemiology of prostate cancer. In Vogelzang NJ, (eds): Comprehensive Textbook of Genitourinary Oncology. Baltimore: Williams & Wilkins, 1996, pp 565–572.
3. Sakr WA, Macoska JA, Benson P, Grignon DJ, Wolman SR, Pontes JE, Crissman JD: Allelic loss in locally metastatic, multisampled prostate cancer. Cancer Res 54: 3273–3277, 1994.
4. Qian J, Bostwick D, Takashi S, Borell T, Herath J, Lieber M, Jenkins R: Chromosomal anomalies in prostatic intraepithelial neoplasia and carcinoma detected by fluorescence in situ hybridization. Cancer Res 55:5408–5414, 1995.
5. Brothman A, Peehl D, Patel A, McNeal J: Frequency and pattern of karyotypic abnormalities in human prostate cancer. Cancer Res 50:3795–3803, 1990.
6. Thompson TC, Park SH, Timme TL, Ren C, Eastham JA, Donehower LA, Bradley A, Kadmon D, Yang G: Loss of p53 function leads to metastasis in *ras* + *myc*-initiated mouse prostate cancer. Oncogene 10:869–879, 1995.
7. Kinzler K, Vogelstein B: Lessons from hereditary colorectal cancer. Cell 87:159–170, 1996.
8. Fearon ER, Vogelstein B: A genetic model for colorectal tumorigenesis. Cell 61: 759–767, 1990.
9. Whitmore WF: Natural history of low stage prostatic cancer and the impact of early detection. Urol Clin North Am 17:689–700, 1990.
10. Whitmore WF: The natural history of prostate cancer. Cancer 32:1104–1112, 1973.
11. Thompson TC, Timme TL, Sehgal I: The role of p53 in prostate cancer progression. In Fortner JG, Sharp PA (eds): Accomplishments in Cancer Research 1996, Philadelphia: Lippincott-Raven, 280–288, 1997.
12. Parsons R, Li GM, Longley MJ, Fang WH, Papadopoulos N, Jen J, de la Chapelle A, Kinzler KW, Vogelstein B, Modrich P: Hypermutability and mismatch repair deficiency in RER+ tumor cells. Cell 75:1227–1236, 1993.
13. Fishel R, Lescoe M, Rao M, Copeland N, Jenkins N, Garber J, Kane M, Kolodner R: The human mutator gene homolog MSH2 and its association with hereditary nonpolyposis colon cancer. Cell 75:1027–1038, 1993.
14. Leach FS, Nicolaides NC, Papadopoulos N, Liu B, et al: Mutations of a mutS homolog in hereditary nonpolyposis colorectal cancer. Cell 75:1215–1225, 1993.
15. Bronner CE, Baker SM, Morrison PT, Waren G, et al: Mutation in the DNA mismatch repair gene homologue hMLH1 is associated with hereditary non-polypolsis colon cancer. Nature 368:258–261, 1994.
16. Papadopoulos N, Nicolaides NC, Wei YF, Ruben SM, et al: Mutation of a mutL homolog in hereditary colon cancer. Science 263:1625–1629, 1994.

17. Sager R, Sheng S, Anisowicz A, Sotiropoulou G, Zou Z, Stenman G, Swisshelm K, Chen Z, Hendrix MJ, Pemberton P, Rafidi P, Ryan K: RNA genetics of breast cancer: maspin as paradigm. Cold Spring Harbor Symp Quant Biol 59:537–546, 1994.
18. Tjian R, Maniatis T: Transcriptional activation: A complex puzzle with few easy pieces. Cell 77:5–8, 1994.
19. Latchman D: Transcription-factor mutations and disease. N Engl J Med 334:28–33, 1996.
20. Land H, Parada LF, Weinberg RA: Tumorigenic conversion of primary embryo fibroblasts requires at least two cooperating oncogenes. Nature 304:596–602, 1983.
21. Weinberg RA: Oncogenes, tumor suppressor genes, and cell transformation: Trying to put it all together. In Brugge J Curran T, Harlow E, McCormick, (eds): Origins of Human Cancer. Cold Spring Harbor Press, 1991, pp 1–16.
22. Robinson D, He F, Pretlow T, Kung HJ: A tyrosine kinase profile of prostate carcinoma. Proc Natl Acad Sci USA 93:5958–5962, 1996.
23. Mellon K, Thompson S, Charlton RG, Marsh C, Robinson M, Lane DP, Harris AL, Horne CH, Neal DE: p53, c-erbB2 and the epidermal growth factor receptor in the benign and malignant prostate. J Urol 147:496–499, 1992.
24. Zhau HE, Wan DS, Zhou J, Miller GJ, von Eschenbach AC: Expression of c-erb B-2/neu proto-oncogene in human prostatic cancer tissues and cell lines. Mol Carcinog 5:320–327, 1992.
25. Kuhn EJ, Kurnot RA, Sesterhenn IA, Chang EH, Moul JW: Expression of the c-erbB-2 (HER-2/neu) oncoprotein in human prostatic carcinoma. J Urol 150:1427–1433, 1993.
26. Ibrahim GK, Kerns B-JM, Macdonald JA, Kinney RB, Humphrey PA, Robertson CN: Differential immunoreactivity of epidermal growth factor receptor in benign, dysplastic and malignant prostatic tissues. J Urol 149:170–173, 1995.
27. Gallego C, Gupta SK, Heasley LE, Qian NX, Johnson GL: Mitogen-activated protein kinase activation resulting from selective oncogene expression in NIH 3T3 and rat 1a cells. Proc Natl Acad Sci USA 89:7355–7359, 1992.
28. Leevers SJ, Marshall CJ: Activation of extracellular signal-regulated kinase, ERK2, by $p21^{ras}$ oncoprotein. EMBO J 11:569–574, 1992.
29. Pomerance M, Schweighoffer F, Tocque B, Pierce M: Stimulation of mitogen-activated protein kinase by oncogenic Ras p21 in Xenopus oocytes. J Biol Chem 267:16155–16160, 1992.
30. Robbins DJ, Cheng M, Zhen E, Vanderbilt A, Feig LA, Cobb MH: Evidence for a Ras-dependent extracellular signal-regulated protein kinase (ERK) cascade. Proc Natl Acad Sci USA 89:6924–6928, 1992.
31. Shibuya EK, Polverino AJ, Chang E, Wigler M, Ruderman JV: Oncogenic ras triggers the activation of 42-kDa mitogen-activated protein kinase in extracts of quiescent *Xenopus* oocytes. Proc Natl Acad Sci USA 89:9831–9835, 1992.
32. Thomas SM, DeMarco M, D'Arcangelo G, Halegoula S, Brugge JS: Ras is essential for nerve growth factor- and phorbol ester-induced tyrosine phosphorylation of MAP kinases. Cell 68:1031–1040, 1992.
33. Wood KW, Sarnecki C, Roberts TM, Blenis J: ras mediates nerve growth factor receptor modulation of three signal-transducing protein kinases: MAP kinase, Raf-1, and RSK. Cell 68:1041–1050, 1992.
34. Dent P, Haser W, Haystead TA, Vincent LA, Roberts TM, Sturgill TW: Activation of mitogen-activated protein kinase by v-Raf in NIH/3T3 cells and *in vitro*. Science 257:1404–1407, 1992.
35. Howe LR, Leevers SJ, Gomez N, Nakielny S, Cohen P, Marshall CJ: Activation of MAP kinase pathway by the protein kinase raf. Cell 71:335–342, 1992.

36. Kyriakis JM, App H, Zhang X-F, Banerjee P, Brautigan DL, Rapp UR, Avruch J: Raf-1 activates MAP kinase-kinase. Nature 358:417–421, 1992.
37. Gupta SK, Gallego C, Johnson GL, Heasley LE: MAP kinase is constitutively activated in *gip2* and *src* transformed Rat 1a fibroblasts. J Biol Chem 267:7987–7990, 1992.
38. Wang H-C, Erickson RL: Activation of protein serine-threonine kinases p42, p63, and p87 in Rous sarcoma virus-transformed cells: Signal transduction/transformation dependent MAP kinases. Mol Cell Biol 3:1329–1337, 1992.
39. Nebreda AR, Hunt T: The c-*mos* proto-oncogene protein kinase turns on and maintains the activity of MAP kinase, but not MPF, in cell-free extracts of *Xenopus* oocyte and eggs. EMBO J 10:1979–1986, 1993.
40. Posada J, Yew N, Ahn NG, Vande Woude GF, Cooper JA: Mos stimulates MAP kinase in *Xenopus* oocytes and activates a MAP kinase in vitro. Mol Cell Biol 13:2546–2553, 1993.
41. Shibuya EK, Ruderman JU: Mos induces the in vitro activation of mitogen-activated protein kinases in frog oocytes and mammalian somatic cells. Mol Cell Biol 4:781–790, 1993.
42. Cowley S, Paterson H, Kemp P, Marshall CJ: Activation of MAP kinase kinase is necessary and sufficient for PC12 differentiation and transformation of NIH 3T3 cells. Cell 77:841–852, 1994.
43. Mansour SJ, Matten WT, Hermann AS, Candia JM, Rong S, Fukasawa K, VandeWoude GF, Ahn NG: Transformation of mammalian cells by constitutively active MAP kinase. Science 265:966–970, 1994.
44. Okazaki K, Sagata N: MAP kinase activation is essential for oncogenic transformation of NIH3T3 cells by Mos. Oncogene 10:1149–1157, 1995.
45. Fukasawa K, Vande Woude GF: Synergy between the mos/mitogen-activated protein kinase pathway and loss of p53 function in transformation and chromosome instability. Mol Cell Biol 17:506–518, 1997.
46. Carter BS, Epstein, JI, Isaacs WB: *ras* gene mutations in human prostate cancer. Cancer Res 50:6830–6832, 1990
47. Gumerlock PH, Poonamallee UR, Meyers FJ, deVere White RW: Activated *ras* alleles in human carcinoma of the prostate are rare. Cancer Res 51:1632–1637, 1991.
48. Anwar K, Nakakuki K, Shiraishi T, Naiki H, Yatani R, Inuzuka M: Presence of *ras* oncogene mutations and human papillomavirus DNA in human prostate carcinomas. Cancer Res 52:5991–5996, 1992.
49. Konishi N, Enomoto T, Buzard G, Oshima M, Ward JM, Rice JM: K-*ras* activation and *ras* p21 expression in latent prostatic carcinoma in Japanese men. Cancer 69:2293–2299, 1992.
50. Feig LA: The many roads that lead to *ras*. Science 260:767–768, 1994.
51. Gibbs JB, Oliff A, Kohl NE: Farnesyltransferase inhibitors: *ras* research yields a potential cancer therapeutic. Cell 77:175–178, 1994.
52. Kohl NE, Omer CA, Conner MW, Anthony NJ, Davide JP, deSolms SJ, Giuliani EA, Gomez RP, Graham SL, Hamilton K, Handt LK, Hartman GD, Koblan KS, Kral AM, Miller PJ, Mosser SD, O'Neill TJ, Rands E, Schaber MD, Gibbs JB, Oliff A: Inhibition of farnesyltransferase induces regression of mammary and salivary carcinomas in ras transgenic mice. Nature Med 1:792–797, 1996.
53. Kelley K, Cochran B, Stiles C, Leder P: Cell specific regulation of the c-myc gene by lymphocyte mitogens and platelet-derived growth factor. Cell 35:603–610, 1983.
54. Evan GI, Wyllie AH, Gilbert CS, Littlewood TD, Land H, Brooks M, Waters CM, Penn LZ, Hancock DC: Induction of apoptosis in fibroblasts by c-myc protein. Cell 69:119–128, 1992.

55. Luscher B, Eisenman RN: New light on Myc and Myb. Part I. Myc. Genes Dev 4:2025–2035, 1990.
56. Penn LJZ, Laufer EM, Land H: C-MYC: Evidence for multiple regulatory functions. Semin Cancer Biol 1:69–80, 1990.
57. Blackwood EM, Eisenman RN: Max: A helix-loop-helix zipper protein that forms a sequence-specific DNA binding complex with Myc. Science 251:1211–1217, 1991.
58. Prendergast GC, Ziff EB: Methylation-sensitive sequence-specific DNA binding by the c-Myc basic region. Science 251:186–189, 1991.
59. Kato GJ, Lee WMF, Chan L, Dang CV: Max: Functional domains and interaction with c-Myc. Genes Dev 6:81–92, 1992.
60. Amati B, Land H: Myc-Max-Mad: A transcription factor network controlling cell cycle progression, differentiation and death. Curr Opin Genet Dev 4:102–108, 1994.
61. Galaktionov K, Chen X, Beach D: Cdc25 cell cycle phosphatase as a target of c-myc. Nature 382:511–517, 1996.
62. Amati B, Littlewood TD, Evan GI, Land H: The c-myc protein induces cell cycle progression and apoptosis through dimerization with max. EMBO J 12:5083–5087, 1993.
63. Wyllie AH: The genetic regulation of apoptosis. Curr Opin Genet Dev 5:97–104, 1995.
64. Packham G, Porter CW, Cleveland JL: c-Myc induces apoptosis and cell cycle progression by separable, yet overlapping, pathways. Oncogene 13:461–469, 1996.
65. Facchini LM, Chen S, Marhin WW, Lear JN, Penn LZ: The Myc negative autoregulation mechanism requires Myc-Max association and involves the c-*myc* P2 minimal promoter. Mol Cell Biol 17:100–114, 1997.
66. Versteeg RI, Noordermeer A, Kruse-Wolters M, Ruiter DJ, Schrier PI: c-myc downregulates class I HLA expression in human melanomas. EMBO J 7:1023–1029, 1988.
67. Quarmby VE, Beckman WC, Wilson EM, French FS: Androgen regulation of c-myc messenger ribonucleic acid levels in rat ventral prostate. Mol Endocrinol 1:865–874, 1987.
68. Buttyan R, Zakeri Z, Lockshin R, Wolgemuth D: Cascade induction of c-fos, c-myc, and heat shock 70K transcripts during regression of the rat ventral prostate gland. Mol Endocrinol 2:650–657, 1988.
69. Fleming WH, Hamel A, MacDonald R, Ramsey E, Pettigrew NM, Johnston B, Dodd JG, Matusik RJ: Expression of the c-myc protooncogene in human prostatic carcinoma and benign prostatic hyperplasia. Cancer Res 46:1535–1538, 1986.
70. Buttyan R, Sawczuk IS, Benson MC, Siegal JD, Olsson CA: Enhanced expression of the c-myc protooncogene in high grade human prostate cancers. Prostate 11:327–337, 1987.
71. Davies P, Eaton CL, France TD, Phillips ME: Growth factor receptors and oncogene expression in prostate cells. Am J Clin Oncol 11S2:S1–S7, 1988.
72. Karp JE, Chiarodo A, Brawley O, Kelloff GJ: Prostate cancer prevention: Investigational approaches and opportunities. Cancer Res 56:5547–5556, 1996.
73. Eagle LR, Yin X, Brothman AR, Williams BJ, Atkin NB, Prochownik EV: Mutation of the MXI1 gene in prostate cancer. Nat Genet 9:249–255, 1995.
74. Sandberg AA: Chromosomal abnormalities and related events in prostate cancer. Hum Pathol 23:368–380, 1992.
75. Isaacs WB, Bova GS, Morton RA, Bussemakers MJ, Brooks JD, Ewing CM: Genetic alterations in prostate cancer. Cold Spring Harbor Symp Quant Biol 59:653–659, 1996.
76. Zenklusen JC, Thompson JC, Troncoso P, Kagan J, Conti CJ: Loss of heterozygosity in human primary prostate carcinomas: A possible tumor suppressor gene at 7q31.1. Cancer Res 54:6370–6373, 1994.
77. Kagan J, Stein J, Babaian RJ, Joe YS, Pisters LL, Glassman AB, von Eschenbach AC, Troncoso P: Homozygous deletions at 8p22 and 8p21 in prostate cancer implicate these

regions as the sites for candidate tumor suppressor genes. Oncogene 11: 2121–2126, 1995.
78. Macoska JA, Trybus TM, Benson PD, Sakr WA, Grignon DJ, Wojno KD, Pietruk T, Powell IJ: Evidence for three tumor suppressor gene loci on chromosome 8p in human prostate cancer. Cancer Res 55:5390–5395, 1995.
79. Suzuki H, Emi M, Komiya A, Fujiwara Y, Yatani R, Nakamura Y, Shimazaki J: Localization of a tumor suppressor gene associated with progression of human prostate cancer with a 1.2 Mb region of 8p22–p21.3. Genes Chromosomes Cancer 13:168–174, 1995.
80. Vocke CD, Pozzatti RO, Bostwick DG, Florence CD, Jennings SB, Strup SE, Duray PH, Liotta LA, Emmert-Buck MR, Lineham WM: Analysis of 99 microdissected prostate carcinomas reveals a high frequency of allelic loss on chromosome 8p12–21. Cancer Res 56:2411–2416, 1996.
81. Carter BS, Ewing CM, Ward WS, Treiger BF, Aalders TW, Schalken JA, Epstein JI, Isaacs WB: Allelic loss of chromosomes 16q and 10q in human prostate cancer. Proc Natl Acad Sci USA 87:8731–8755, 1991.
82. Phillips SM, Morton DG, Lee SJ, Wallace DM, Neoptolemos JP: Loss of heterozygosity of the retinoblastoma and adenomatous polyposis susceptibility gene loci and in chromosomes 10p, 10q and 16q in human prostate cancer. Br J Urol 73:390–395, 1994.
83. Gray IC, Phillips SM, Lee SJ, Neoptolemos JP, Weissenbach J, Spurr NK: Loss of the chromosomal region 10q23–25 in prostate cancer. Cancer Res 55:4800–4803, 1995.
84. Gao X, Zacharek A, Grignon DJ, Sakr W, Powell IJ, Porter AT, Honn KV: Localization of potential tumor suppressor loci to a <2 Mb region on chromosome 17q in human prostate cancer. Oncogene 11:1241–1247, 1995.
85. Bookstein R, Shew J-Y, Chen P-L, Scully P, Lee W-H: Suppression of tumorigenicity of human prostate carcinoma cells by replacing a mutated RB gene. Science 247:712–715, 1990.
86. Bookstein R, Rio P, Madreperia SA, Hong F, Allred C, Grizzle WE, Lee WH: Promoter deletion and loss of retinoblastoma gene expression in human prostate carcinoma. Proc Natl Acad Sci USA 87:7762–7766, 1990.
87. Phillips SM, Barton CM, Lee SJ, Morton DG, Wallace DM, Lemoine NR, Neoptolemos JP: Loss of the retinoblastoma susceptibility gene (RB1) is a frequent and early event in prostatic tumorigenesis. Br J Cancer 70:1252–1257, 1994.
88. Brooks JD, Bova GS, Isaacs WB: Allelic loss of the retinoblastoma gene in primary human prostatic adenocarcinomas. Prostate 26:35–39, 1995.
89. Ittmann MM, Wieczorek R: Alterations of the retinoblastoma gene in clinically localized, stage B prostate adenocarcinomas. Hum Pathol 27:28–34, 1996.
90. Weinstat-Saslow D, Steeg PS: Angiogenesis and colonization in the tumor metastatic process: Basic and applied advances. FASEB J 8:401–407, 1994.
91. Konishi N, Nakota S, Tsuzuki T, Matsumata K, Kitahori Y, Hiusa Y, Urana T, Shiku H: Expression of nm23-H1 and nm23-H2 proteins in prostate carcinoma. Jpn J Cancer Res 84:1050–1054, 1993.
92. Igawa M, Rukstalis DB, Tanabe T, Chodak GW: High levels of nm23 expression are related to cell proliferation in human prostate cancer. Cancer Res 54:1313–1318, 1994.
93. Dong J-T, Lamb PW, Rinker-Schaeffer CW, Vukanovic J, Ichikawa T, Isaacs JT, Barrett JC: KAI1, a metastatic suppressor gene for prostate cancer on human chromosome 11p11.2. Science 268:884–886, 1995.
94. Ueda T, Ichikawa T, Tamaru J-I, Mikata A, Akakura K, Akimoto S, Imai T, Yoshie O, Shiraishi T, Yatani R, Ito H, Shimazaki J: Expression of the KAI1 protein in benign prostatic hyperplasia and prostate cancer. Am J Pathol 149:1435–1440, 1996.

95. Dong J-T, Suzuki H, Pin SS, Bova GS, Schalken JA, Isaacs WB, Barrett JC, Isaacs JT: Down-regulation of the KAI1 metastasis suppressor gene during the progression of human prostatic cancer infrequently involves gene mutation or allelic loss. Cancer Res 56:4387–4390, 1996.
96. Vogelstein B, Kinzler KW: p53 function and dysfunction. Cell 70:523–526, 1992.
97. Hartwell L: Defects in a cell cycle checkpoint may be responsible for the genomic instability of cancer cells. Cell 71:543–546, 1992.
98. Nigro JM, Baker SJ, Presinger AC, Jessup JM, Hostetter R, Cleary K, Bigner SH, Davidson N, Baylin S, Deville P, Glover T, Collins FS, Weston A, Modali R, Haris CC, Vogelstein B: Mutations in the p53 gene occur in diverse human tumour types. Nature 342:705–708, 1989.
99. Levine AJ, Momand J, Finlay CA: The p53 tumor suppressor gene. Nature 351: 453–456, 1991.
100. Hollstein M, Sidransky D, Vogelstein B, Harris CC: p53 mutations in human cancers. Science 253:49–53, 1992.
101. Finlay CA, Hinds PW, Tan T-H, Eliyahu D, Oren EM, Levine AJ: Activating mutations for transformation by p53 produce a gene product that forms an hsc70-p53 complex with an altered half-life. Mol Cell Biol 8:531–535, 1988.
102. Iggo R, Gatter K, Bartek J, Lane DP, Harris AL: Increased expression of mutant forms of p53 oncogene in primary lung cancer. Lancet 335:675–679, 1990.
103. Visakorpi T, Kallioniemi O, Heikkinen A, Koivula T, Isola J: Small subgroup of aggressive, highly proliferative prostatic carcinomas defined by p53 accumulation. J Natl Cancer Inst 84:883–887, 1992.
104. Bookstein R, MacGrogan D, Hilsenbeck SG, Sharkey F, Allred DC: p53 is mutated in a subset of advanced-stage prostate cancers. Cancer Res 53:3369–3373, 1993.
105. Aprikian AG, Sarkis AS, Fair WR, Zhang Z-F, Fuks Z, Corden-Cardo C: Immunohistochemical determination of p53 protein nuclear accumulation in prostatic adenocarcinoma. J Urol 151:1276–1280, 1994.
106. Bauer J, Sesterhenn I, Mostofi F, McLeod D, Srivastava S, Moul J: Elevated levels of apoptosis regulator proteins p53 and bcl-2 are independent prognostic biomarkers in surgically treated clinically localized prostate cancer. J Urol 156:1511–1516, 1994.
107. Bauer JJ, Sesterhenn IA, Mostofi KF, McLeod DG, Srivastava S, Moul JW: p53 nuclear protein expression is an independent prognostic marker in clinically localized prostate cancer patients undergoing radical prostatectomy. Clin Cancer Res 1: 1295–1300, 1995.
108. Berner A, Geitvik G, Karlsen F, Fossa SD, Nesland JM, and Borresen A-L: TP53 mutations in prostatic cancer. Analysis of pre- and post-treatment archival formalin-fixed tumour tissue. J Pathol 176:299–308, 1995.
109. Heindenberg HB, Sesterhenn IA, Gaddipati JP, Weghorst CM, Buzard GS, Moul JW, Srivastava S: Alteration of the tumor suppressor gene p53 in a high fraction of hormone refractory prostate cancer. J Urol 154:414–421, 1995.
110. Hughes JH, Cohen MB, Robinson RA: p53 immunoreactivity in primary and metastatic prostatic adenocarcinoma. Mod Pathol 8:462–466, 1995.
111. Konishi N, Hiasa Y, Matsuda H, Tao M, Tsuzuki T, Hayashi I, Kitahori Y, Shiraishi T, Yatani R, Shimazaki J, Lin J-C: Intratumor cellular heterogeneity and alterations in *ras* oncogene and p53 tumor suppressor gene in human prostate carcinoma. Am J Pathol 147:1112–1122, 1995.
112. Moyret-Lalle C, Marcais C, Jacquemier J, Moles JP, Daver A, Soret JY, Jeanteur P, Ozturk M, Theillet C: ras, p53 and HPV status in benign and malignant prostate tumors. Int J Cancer 64:124–129, 1995.

113. Shurbaji MS, Kalbfleisch JH, Thurmond TS: Immunohistochemical detection of p53 protein as a prognostic indicator in prostate cancer. Hum Pathol 26:106–109, 1995.
114. Yang G, Stapelton AMF, Wheeler TM, Truong LD, Timme TL, Scardino PT, Thompson TC: Clustered p53 immunostaining: A novel pattern associated with prostate cancer progression. Clin Cancer Res 2:635–640, 1996.
115. Kumini K, Amano T, Uchibayashi T: Point mutation of the p53 gene is an infrequent event in untreated prostate cancer. Cancer Detect Prev 20:218–222, 1996.
116. Prendergast NJ, Atkins MR, Schatte EC, Paulson DF, Walther PJ: p53 immunohistochemical and genetic alterations are associated at high incidence with postirradiated locally persistent prostate carcinoma. J Urol 155:1685–1692, 1996.
117. Effert P, McCoy R, Walther P, Liu E: p53 gene alterations in human prostate carcinoma. J Urol 150:257–261, 1993.
118. Navone N, Troncoso P, Pisters L, Goodran T, Palmer J, Nichols W, von Eschenbach A, Conti C: p53 accumulation and gene mutation in the progression of human prostate cancer. J Natl Cancer Inst 85:1657–1669, 1993.
119. Chi SG, deVere White RW, Meyers FJ, Siders DB, Lee F, Gumerlock PH: p53 in prostate cancer: Frequent expressed transition mutations. J Natl Cancer Inst 86: 926–933, 1994.
120. Dinjens WNM, van der Weiden MM, Schroeder FH, Bosman FT, and Trapman J: Frequency and characterization of p53 mutations in primary and metastatic human prostate cancer. Br J Cancer 56:630–633, 1994.
121. Voeller HJ, Sugars LY, Pretlow T, Gelmann EP: p53 oncogene mutations in human prostate cancer specimens. J Urol 151:492–495, 1994.
122. Watanabe M, Ushijima T, Kakiuchi H, Shiraishi T, Yatani R, Kotake T, Sugimura T, Nagao M: p53 gene mutations in human prostate cancers in Japan: Different mutation spectra between Japan and Western countries. Jpn J Cancer Res 85:904–909, 1994.
123. Eastham J, Stapleton A, Gousse A, Timme T, Yang G, Slawin K, Wheeler T, Scardino P, Thompson: Association of p53 mutations with metastatic prostate cancers. Clin Cancer Res 1:1111–1118, 1995.
124. Mirchandani D, Zheng J, Miller G, Ghosh A, Shibata DK, Cote RJ, Roy-Burman R: Heterogeneity in intratumor distribution of p53 mutations in human prostate cancer. Am J Pathol 147:92–101, 1995.
125. Brooks JD, Bova GS, Ewing CM, Piantadosi S, Carter BS, Robinson JC, Epstein JI, Isaacs WB: An uncertain role for p53 gene alterations in human prostate cancers. Cancer Res 56:3814–3822, 1996.
126. Dahiya R, Deng G, Chen KM-K, Chui RM, Haughney PC, Narayan P: p53 tumor-suppressor gene mutations are mainly localised on exon 7 in human primary and metastatic prostate cancer. Br J Cancer 74:264–268, 1996.
127. Suzuki H, Komiya A, Aida S, Ito H, Yatani R, Shimazaki J: Detection of human papillomavirus DNA and p53 gene mutations in human prostate cancer. Prostate 28: 318–324, 1996.
128. Stapleton AMF, Timme TL, Gousse AE, Li Q-F, Tobon AT, Slawin KM, Thompson TC: p53 mutants in primary human prostate cancer are clonally expanded in metastases. Clin Cancer Res 3:1389–1397, 1997.
129. Thompson T, Southgate J, Kitchener G, Land H: Multi-stage carcinogenesis induced by *ras* and *myc* oncogenes in a reconstituted organ. Cell 56:917–930, 1989.
130. Merz VW, Miller GJ, Krebs T, Timme TL, Kadmon D, Park SH, Egawa S, Scardino PT, Thompson TC: Elevated TGF-β1 and β3 mRNA levels are associated with *ras* and *myc*-induced carcinomas in reconstituted mouse prostate: Evidence for a paracrine role during progression. Mol Endocrinol 5:503–513, 1991.

131. Thompson TC, Timme TL, Kadmon D, Park SH, Egawa S, Yoshida K: Genetic predisposition and mesenchymal–epithelial interactions in *ras* + *myc*-induced carcinogenesis in reconstituted mouse prostate. Mol Carcinog 7:165–179, 1993.
132. Thompson TC, Truong LD, Timme TL, Kadmon D, McCune BK, Flanders KC, Scardino PT, Park SH: Transforming growth factor-β1 as a biomarker for prostate cancer. J Cell Biochem 16(S):54–61, 1992.
133. Truong LD, Kadmon D, McCune BK, Flanders KC, Scardino PT, Thompson TC: Association of transforming growth factor-beta with prostate cancer: An immunohistochemical study. Hum Pathol 24:4–9, 1993.
134. Donehower LA, Harvey M, Slagle BL, McArthur MJ, Montgomery CAJ, Butel JS, Bradley A: Mice deficient for p53 are developmentally normal but susceptible to spontaneous tumors. Nature 356:215–221, 1992.
135. Deffie A, Wu H, Reinke V, Lozano G: The tumor suppressor p53 regulates its own transcription. Mol Cell Biol 13:3415–3423, 1993.
136. Momand J, Zambetti GP, Olson DC, George D, Levine AJ: The mdm-2 oncogene product forms a complex with the p53 protein and inhibits p53-mediated transactivation. Cell 69:1237–1245, 1992.
137. Wu X, Bayle J, Olson D, Levine A: The p53-mdm-2 autoregulatory feedback loop. Genes Dev 7:1126–1132, 1993.
138. El-Diery WS, Tokino T, Velculescu V, Levy D, Parsons R, Trent J, Lin D, Mercer W, Kinzler K, Vogelstein B: WAF1, a potent mediator of p53 tumor suppression. Cell 75:817–825, 1993.
139. Harper J, Adami G, Wei N, Keyomarski K, SJ. E: The p21 Cdk-interacting protein Cip1 is a potent inhibitor of G1 cyclin-dependent kinases. Cell 75:805–816, 1993.
140. Xiong Y, Hannon G, Zhang D, Casso R, Kobayashi R, Beach D: p21 is a universal inhibitor of cyclin kinases. Nature 366:701–704, 1993.
141. Okamoto K, Kamibayashi C, Serrano M, Prives C, M.C. M, Beach D: p53-dependent association between cyclin G and the B′ subunit of protein phosphatase 2A. Mol Cell Biol 16:6593–6602, 1996.
142. Ko L, Prives C: p53: Puzzle and paradigm. Genes Dev 10:1054–1072, 1996.
143. Rotter V, Aloni-Grinstein R, Schwartz D, Elkind N, Simon A, Wolkowicz R, Lavigne M, Beserman P, Kapon A, Goldfinger N: Does wild-type p53 play a role in normal cell differentiation? Semin Cancer Biol 5:229–236, 1994.
144. Oren M: Relationship of p53 to the control of apoptotic cell death. Semin Cancer Biol 5:221–227, 1994.
145. Nikiforov M, Hagen K, Ossovskaya V, Connor T, Lowe S, Deichman G, Gudkov A: p53 modulation of anchorage independent growth and experimental metastasis. Oncogene 13:1709–1719, 1996.
146. Friedlander P, Haupt Y, Prives C, Oren M: A mutant p53 that discriminates between p53-responsive genes cannot induce apoptosis. Mol Cell Biol 16:4961–4971, 1996.
147. Amson R, Nemani M, Roperch J, Israeli D, Bougueleret L, Le Gall I, Medhioub M, Linares-Cruz G, Lethrosne F, Pasturaud P, Piouffre L, Prieur S, Susini L, Alvaro V, Millasseau P, Guidicelli C, Bui H, Massart C, Cazes L, Dufour F, Bruzzoni-Giovanelli H, Owadi H, Hennion C, Charpak G, Dausset J, Calvo F, Oren M, Cohen D, Telerman A: Isolation of 10 differentially expressed cDNAs in p53-induced apoptosis: Activation of the vertebrate homologue of the drosophila seven in absentia gene. Proc Natl Acad Sci USA 93:3953–3957, 1996.
148. Roemer K, Mueller-Lantzsch N: p53 transactivation domain mutant Q22, S23 is impaired for repression of promoters and mediation of apoptosis. Oncogene 12:2069–2079, 1996.

149. Yonish-Rouach E, Deguin V, Zaitchouk T, Breugnot C, Mishal Z, Jenkins JR, May E: Transcriptional activation plays a role in the induction of apoptosis by transiently transfected wild-type p53. Oncogene 12:2197–2205, 1996.
150. Santhanam U, Ray A, Sehgal PB: Repression of the interleukin 6 gene promoter by p53 and the retinoblastoma susceptibility gene product. Proc Natl Acad Sci USA 88:7605–7609, 1991.
151. Kieser A, Weich HA, Brandner G, Marme D, Kolch W: Mutant p53 potentiates protein kinase C induction of vascular endothelial growth factor expression. Oncogene 9:963–969, 1994.
152. Dameron K, Volpert O, Tainsky M, Bouck N: Control of angiogenesis in fibroblasts by p53 regulation of thrombospondin-1. Science 265:1582–1584, 1994.
153. Fujiwara T, Mukhopadhyay T, Cal DW, Morris DK, Roth JA, G, Grimm EA: Retroviral-mediated transduction of p53 gene increases TGF-β expression in a human glioblastoma cell line. Int J Cancer 56:834–839, 1994.
154. Ueba T, Nosaka T, Takahashi J, Shibata F, Florkiewicz R, Vogelstein B, Oda Y, Kikuchi H, Hatanaka M: Transcriptional regulation of basic fibroblast growth factor gene by p53 in human glioblastoma and hepatocellular carcinoma cells. Proc Natl Acad Sci USA 91:9009–9013, 1994.
155. Mukhopadhyay D, Tsiokas L, Sukhatme VP: Wild-type p53 and v-src exert opposing influences on human vascular endothelial growth factor gene expression. Cancer Res 55:6161–6165, 1995.
156. Ludes-Meyers JH, Subler MA, Shivakumar CV, Munoz RM, Jiang P, Bigger JE, Brown DR, Deb SP, Deb S: Transcriptional activation of the human epidermal growth factor receptor promoter by human p53. Mol Cell Biol 16:6009–6019, 1996.
157. Metz T, Harris AW, Adams JM: Absence of p53 allows direct immortalization of hematopoietic cells by the myc and raf oncogenes. Cell 82:29–36, 1995.
158. Gao X, Wu N, Grignon D, Zacharek A, Liu H, Salkowski A, Li G, Sakr W, Sarkar F, Porter AT, Chen YQ, Honn KV: High frequency of mutator phenotype in human prostatic adenocarcinoma. Oncogene 9:2999–3003, 1994.
159. Egawa S, Uchida T, Suyama K, Wang C, Ohori M, Irie S, Iwamura M, Koshiba K: Genomic instability of microsatellite repeats in prostate cancer: Relationship to clinicopathological variables. Cancer Res 55:2418–2421, 1995.
160. Uchida T, Wada C, Wang C, Ishida H, Egawa S, Yokoyama E, Ohtani H, Koshiba K: Microsatellite instability in prostate cancer. Oncogene 10:1019–1022, 1995.
161. DeLarco JE, Todaro GJ: Growth factors from murine sarcoma virus-transformed cells. Proc Natl Acad Sci USA 75:4001–4005, 1978.
162. Anzano MA, Roberts AB, Smith JM, Sporn MB, DeLarco JE: Sarcoma growth factor from conditioned medium of virally transformed cells is composed of both type α and type β transforming growth factor. Proc Natl Acad Sci USA 80:6264–6268, 1980.
163. Moses HL, Branum EB, Proper JA, Robinson RA: Transforming growth factor production by chemically transformed cells. Cancer Res 41:2842–2848, 1981.
164. Geiser AG, Kim S-J, Roberts AB, Sporn MB: Characterization of the mouse transforming growth factor $\beta 1$ promoter and activation by the Ha-*ras* oncogene. Mol Cell Biol 11:84–92, 1991.
165. Beckmann MP, Betsholtz C, Heldin CH, Westermark B, DiMarco E, DiFiore PP, Robbins KC, Aaronson SA: Comparison of biological properties and transforming potential of human PDGF-A and PDGF-B chains. Science 241:1346–1349, 1988.
166. Tsao MS, Zhang XY: The effects of continuous exposure to epidermal growth factor on the spontaneous transformation of cultured rat liver epithelial cells. Am J Pathol 140:85–94, 1992.

167. Yang D, Kohler SK, Maher VM, McCormick JJ: v-sis oncogene-induced transformation of human fibroblasts into cells capable of forming benign tumors. Carcinogenesis 15:2167–2175, 1994.
168. Zhang X, Wang T, Batist G, Tsao MS: Transforming growth factor beta 1 promotes spontaneous transformation of cultured rat liver epithelial cells. Cancer Res 54:6122–6128, 1994.
169. Haley JD, Hsuan JJ, Waterfield MD: Analysis of mammalian fibroblast transformation by normal and mutated human EGF receptors. Oncogene 4:273–278, 1989.
170. Seger R, Krebs EG: The MAPK signaling cascade. FASEB J 9:726–735, 1995.
171. Marshall CJ: Specificity of receptor tyrosine kinase signaling: Transient versus sustained extracellular signal-regulated kinase activation. Cell 80:179–185, 1995.
172. Massague J: The transforming growth factor-β family. Annu Rev Cell Biol 6:597–641, 1990.
173. Roberts AB, Sporn MB: Transforming growth factor-β. Adv Cancer Res 51:107–145, 1988.
174. Moses HL, Tucker RF, Leof EB, Coffey RJJ, Halper J, Shipley GD: Type-β transforming growth factor is a growth stimulator and growth inhibitor. Cancer Cells 3:65–71, 1985.
175. Salo T, Lyons JG, Rahemtulla F, Birkendal-Hansen H, Larjava H: Transforming growth factor-β1 upregulates type IV collagenase expression in cultured human keratinocytes. J Biol Chem 266:11436–11441, 1991.
176. Sieweke MH, Thompson HL, Sporn MB, Bissell MJ: Mediation of wound-related Rous sarcoma virus tumorigenesis by TGF-β. Science 248:1656–1660, 1990.
177. Welch DR, Fabra A, Nakajima M: Transforming growth factor-β stimulates mammary adenocarcinoma cell invasion and metastatic potential. Proc Natl Acad Sci USA 87:7678–7682, 1990.
178. Ikeda T, Lioubin MN, Marquardt H: Human transforming growth factor type-beta 2: Production by a prostatic adenocarcinoma cell line, purification, and initial characterization. Biochemistry 26:2406–2410, 1987.
179. Eastham JA, Troung LD, Rogers E, Kattan M, Flanders KC, Scardino PT, Thompson TC: Transforming growth factor-β1: Comparative immunohistochemical localization in human primary and metastatic prostate cancer. Lab Invest 73:628–635, 1995.
180. Ivanovic V, Melman A, Davis-Joseph B, Valcic, M, Geliester J: Elevated plasma levels of TGF-β1 in patients with invasive prostate cancer. Nature Med 1:282–284, 1995.
181. Steiner MS, Barrack ER: Transforming growth factor-β1 overproduction in prostate cancer; effects in growth in vivo and in vitro. Mol Endocrinol 6:15–25, 1992.
182. Kyprianou N, Isaacs JT: Expression of transforming growth factors-β1 in the rat ventral prostate during castration-induced programmed cell death. Mol Endocrinol 3:1515–1522, 1989.
183. Egawa S, Kadmon D, Miller GJ, Scardino PT, Thompson TC: Alterations in mRNA levels for growth-related genes after transplantation into castrated hosts in oncogene-induced clonal mouse prostate carcinoma. Mol Carcinog 5:52–61, 1991.
184. Kyprianou N, English HF, Isaacs JT: Programmed cell death during regression of PC-82 human prostate cancer following androgen ablation. Cancer Res 50:3748–3758, 1990.
185. Williams RH, Stapleton AMF, Yang G, Truong LD, Rogers E, Timme TL, Wheeler TM, Scardino PT, Thompson TC: Reduced levels of transforming growth factor β receptor type II in human prostate cancer: An immunohistochemical analysis. Clin Cancer Res 2:635–640, 1996.

186. Park K, Kim S, Bang Y, Park J, Kim NK, Roberts AB, Sporn MB: Genetic changes in the transforming growth factor β (TGF-β) type II receptor gene in human gastric cancer cells: Correlation with sensitivity to growth inhibition by TGF-β. Proc Natl Acad Sci USA 91:8772–8776, 1994.
187. Markowitz S, Wang J, Myeroff L, Parsons R, Scen L, Lutterbaugh J, Fon RS, Zborowska E, Kinzler KW, Vogelstein B, Brattain M, Wilson KV: Inactivation of the type II TGF-β receptor in colon cancer cells with microsatellite instability. Science 268:1336–1338, 1995.
188. Kim IY, Ahn H-J, Zelmner DJ, Shaw JW, Sensibar JA, Kim J-H, Kato M, Lee C: Genetic change in transforming growth factor β (TGF-β) receptor type I gene correlates with insensitivity to TGF-β1 in human prostate cancer cells. Cancer Res 56:44–48, 1996.
189. Mydlo JH, Bulbul MA, Richon VM, Heston WD, and Fair WR: Heparin-binding growth factor isolated from human prostatic extracts. Prostate 12:343–355, 1988.
190. Mori H, Maki M, Oishi K, Jaye M, Igarashi K, Yoshida O, Hatanaka M: Increased expression of genes for basic fibroblast growth factor and transforming growth factor type beta in human benign prostatic hyperplasia. Prostate 16:71–80, 1990.
191. Weldner N, Semple JP, Welch WR, Folkman J: Tumor angiogenesis and metastasis—correlation in invasive breast carcinoma. N Engl J Med 324:1–8, 1991.
192. Brawer MK, Deering RE, Brown M, Prestron SD, Bigler SA: Predictors of pathologic stage in prostatic carcinoma. The role of neovascularity. Cancer 73:678–687, 1994.
193. Nakamoto T, Chang CS, Ki AK, Chodak GW: Basic fibroblast growth factor in human prostate cancer cells. Cancer Res 52:571–577, 1992.
194. Muller WJ, Lee FS, Dickson C, Peters G, Pattengale P, Leder P: The int-2 gene product acts as an epithelial growth factor in transgenic mice. EMBO J 9:907–913, 1990.
195. Yan G, Fukabori Y, McBride G, Nikolaropolous S, McKeehan W: Exon switching and activation of stromal and embryonic fibroblast growth factor (FGF)-FGF receptor genes in prostate epithelial cells accompany stromal independence and malignancy. Mol Cell Biol 13:4513–4522, 1993.
196. Yan G, Fukabori Y, Nikolaropoulos S, McKeehan WL: Heparin-binding keratinocyte growth factor is a candidate stromal-to-epithelial cell andromedin. Mol Endocrinol 6:2123–2128, 1992.
197. Leung HY, Dickson C, Robson CN, Neal DE: Over-expression of fibroblast growth factor-8 in human prostate cancer. Oncogene 12:1833–1835, 1996.
198. Thompson TC: Growth factors and oncogenes in prostate cancer. Cancer Cells 2:345–354, 1990.
199. Ware JL: Growth factors and their receptors as determinants in the proliferation and metastasis of human prostate cancer. Cancer Metastasis Rev 12:287–301, 1993.
200. Sandgren EP, Luetteke NC, Palmiter RD, Brinster RL, Lee DC: Overexpression of TGF alpha in transgenic mice: Induction of epithelial hyperplasia, pancreatic metaplasia and carcinoma of the breast. Cell 61:1121–1135, 1990.
201. Sehgal I, Bailey J, Hitzemann K, Pittelkow MR, Maihle NJ: Epidermal growth factor receptor-dependent stimulation of amphiregulin expression in androgen-stimulated human prostate cancer cells. Mol Cell Biol 5:339–347, 1994.
202. Scher H, Sarkis A, Reuter V, Cohen D, Netto G, Petrylak D, Lianes P, Fuks Z, Mendesohn J, Cordon-Cardo C: Changing pattern of expression of the epidermal growth factor receptor and transforming growth factor-α in the progression of prostatic neoplasms. Clin Cancer Res 1:545–550, 1995.

203. Deb SP, Munoz RM, Brown DR, Subler MA, Deb S: Wild-type p53 activates the human epidermal growth factor receptor promoter. Oncogene 9:1341–1349, 1994.
204. Peehl DM, Cohen P, Rosenfeld RG: The insulin-like growth factor system in the prostate. World J Urol 13:306–311, 1995.
205. Cohen P, Peehl DM, Lamson G, Rosenfeld RG: Insulin-like growth factors (IGFs), IGF receptors, and IGF binding proteins in primary cultures of prostate epithelial cells. J Clin Endocrinol Metab 73:401–407, 1991.
206. Cohen P, Graves HC, Peehl DM, Kamaei M, Giudice LM, Rosenfield RG: Prostate-specific antigen (PSA) is an IGF-binding protein-3 (IGFBP-3) protease found in seminal plasma. J Clin Endocrinol Metab 73:491–497, 1992.
207. di Sant'Agnese PA: Neuroendocrine differentiation in human prostatic carcinoma. Hum Pathol 23:287–296, 1992.
208. Abrahamsson P-A, Wadstrom LB, Alumets J, Falkemer S, Grimelius L: Peptide-hormone- and serotinin-immunoreactive tumor cells in carcinoma of the prostate. Pathol Res Pract 182:298–307, 1987.
209. Kapuscinski M, Shulkes A, Read D, Hardy, KJ. Expression of neurotensin in endocrine tumors. J Clin Endocrinol Metab 70:100–106, 1990.
210. Noordzij MA, van Steenbrugge GJ, van der Kwast TH, Schröder FH: Neuroendocrine cells in the normal, hyperplastic and neoplastic prostate. Urol Res 22:333–341, 1995.
211. Bang Y-J, Pirnia F, Fang W-G, Kang WK, Sartor O, Whitesell L, Ha MJ, Tsokos M, S, Sheahan MD, Nguyen P, Niklinski WT, Myers CE, Trepel JB: Terminal neuroendocrine differentiation of human prostate carcinoma cells in response to increased intracellular cyclic AMP. Proc Natl Acad Sci USA 91:5330–5334, 1994.
212. Sehgal I, Powers S, Huntley B, Powis G, Pittelkow M, Maihle NJ: Neurotensin is an autocrine trophic factor stimulated by androgen withdrawal in human prostate cancer. Proc Natl Acad Sci USA 91:4673–4677, 1994.
213. Djakiew D, Delsite R, Pflug B, Wrathall J, Lynch JH, Onoda M: Regulation of growth by a nerve growth factorlike protein which modulates paracrine interactions between a neoplastic epithelial cell line and stromal cells of the human prostate. Cancer Res 52:3304–3308, 1991.
214. Djakiew D, Pflug BR, Delsite R, Onoda M, Lynch JH, Arand G, Thompson EW: Chemotaxis and chemokinesis of human prostate tumor cell lines in response to human prostate stromal cell secretory proteins containing a nerve growth factorlike protein. Cancer Res 53:1416–1420, 1993.
215. Huggins C, Hodges CV: Studies on prostate cancer: Effect of castration, of estrogen, and of androgen injection on serum phosphatase in metastatic carcinoma of the prostate. Cancer Res 1:293–297, 1941.
216. Noble RL: The development of prostatic adenocarcinoma in Nb rats following prolonged sex hormone administration. Cancer Res 37:1929–1933, 1977.
217. Hsing AW: Hormones and prostate cancer: Where do we go from here. J Natl Cancer Inst 88:1093–1094, 1996.
218. Newmark JR, Hardy DO, Tanb DC, Carter BS, Epstein JI, Isaacs WB, Braun TR, Barrack ER: Androgen receptor gene mutations in human prostate cancer. Proc Natl Acad Sci USA 89:6319–6323, 1992.
219. Suzuki H, Sato N, Watabe Y, Masai M, Seino S, Shimazaki J: Androgen receptor gene mutations in human prostate cancer. J Steroid Biochem Mol Biol 46:759–765, 1993.
220. Culig Z, Hoblsch A, Cronauer MV, Cato ACB, Hittmair A, Radmayr C, Eberle J, Bartsch G, Klocker H: Mutant androgen receptor detected in an advanced-stage prosta-

tic carcinoma is activated by adrenal androgens and progesterone. Mol Endocrinol 7:1541–1550, 1993.

221. Lim-Tio SS, Horsfall DJ, Aspinall JO, Marshall VR, Skinner JM: Detection of discrete androgen receptor epitopes in prostate cancer by immunostaining: Measurement by color video image analysis. Cancer Res 54:4096–4102, 1994.

222. Yuan S, Trachtenberg J, Mills GB, Brawn TJ, Xu F, Keating A: Androgen-induced inhibition of cell proliferation in an androgen-insensitive prostate cancer cell line (PC-3) transfected with a human androgen receptor complementary DNA. Cancer Res 53:1304–1311, 1993.

223. Visakorpi T, Hyytinen E, Koivisto P, Tanner M, Keinanen R, Palmberg C, Palotie A, Tammela T, Isola J, Kallioniemi O-P: *In vivo* amplification of the androgen receptor gene and progression of human prostate cancer. Nature Genet 9:401–406, 1995.

224. Zuck B, Goepfert C, Nedlin-Chittka A, Sohrt K, Voigt KD, Knabbe C: Regulation of fibroblast growth factor-like protein(s) in the androgen-responsive human prostate carcinoma cell line LNCaP. J Steroid Biochem Mol Biol 41:659–663, 1992.

225. Liotta LA: Tumor invasion and metastasis: Role of the extracellular matrix. Cancer Res 46:1–7, 1986.

226. Fidler IJ: Critical factors in the biology of human cancer metastasis: Twenty-eighth G.H.A. Clowes memorial award lecture. Cancer Res 50:6130–6138, 1990.

227. Sehgal I, Baley PA, Thompson TC: TGF-β1 stimulates contrasting responses in metastatic versus primary mouse prostate cancer derived cell lines in vivo. Cancer Res 56:3359–3365, 1996.

228. Liotta LA, Steeg PA, Stetler-Stevenson WG: Cancer metastasis and angiogenesis: An imbalance of positive and negative regulation. Cell 64:327–336, 1991.

229. Bernhard EJ, Gruber SB, Muschel RJ: Direct evidence linking expression of matrix metalloproteinase 9 (92-kDa gelatinase/collagenase) to the metastatic phenotype in transformed rat embryo cells. Proc Natl Acad Sci USA 91:4293–4297, 1994.

230. Schonthal A, Herrlich P, Rahmsdorf AJ, Ponta H: Requirement for *fos* gene expression in the transcriptional activation of collagenase by other oncogenes and phorbol esters. Cell 54:325–334, 1988.

231. Lyons JG, Birkedal-Hansen B, Pierson MC, Whitelock JM, Birkedal-Hansen H: Interleukin-1 beta and transforming growth factor-α/epidermal growth factor induce expression of (M)r 95,000 Type IV collagenase/gelatinase and interstitial fibroblast-type collagenase by rat mucosal keratinocytes. J Biol Chem 268:19143–19151, 1993.

232. Stetler-Stevenson WG: Type IV collagenases in tumor invasion and metastasis. Cancer Metastasis Rev 9:289–303, 1990.

233. Blades RA, Keating PJ, McWilliam LJ, George NJ, Stern PL: Loss of HLA class I expression in prostate cancer: Implications for immunotherapy. Urology 46:681–686, 1995.

234. Sanda MG, Restifo NP, Walsh JC, Kawakami Y, Nelson WG, Pardoll DM, Simons JW: Molecular characterization of defective antigen processing in human prostate cancer. J Natl Cancer Inst 87:280–285, 1995.

235. Twillie DA, Eisenberger MA, Carducci MA, Hseih WS, Kim WY, Simons JW: Interleukin-6: A candidate mediator of human prostate cancer morbidity. Urology 45:542–549, 1995.

236. Merz VW, Arnold AM, Studier UE: Differential expression of transforming growth factor-β1 and -β3 as well as c-*fos* mRNA in normal human prostate. Benign prostatic hyperplasia and prostatic cancer. World J Urol 12:96–98, 1994.

237. Geiser AG, Letterio JJ, Kulkarni AB, Karlsson S, Roberts AB, Sporn MB: Transform-

ing growth factor beta 1 (TGF-beta 1) controls expression of major histocompatibility genes in the postnatal mouse: Aberrant histocompatibility antigen expression in the pathogenesis of the TGF-beta 1 null mouse phenotype. Proc Natl Acad Sci USA 90:9944–9948, 1993.

238. Darley R, Morris A, Passas J, Bateman W: Interactions between interferon gamma and retinoic acid with transforming growth factor beta in the induction of immune recognition molecules. Cancer Immunol Immunother 37:112–118, 1993.
239. Umbas R, Schalken JA, Aalders TW, Carter BS, Karthaus HF, Schaafsma HE, Debruyne FM, Isaacs WB: Expression of the cellular adhesion molecule E-cadherin is reduced or absent in high-grade prostate cancer. Cancer Res 52:5104–5109, 1992.
240. Cress AE, Rabinovitz I, Zhu W, Nagle RB: The alpha 6 beta 1 and alpha 6 beta 4 integrins in human prostate cancer progression. Cancer Metastasis Rev 14:219–228, 1995.
241. Kleinerman DI, Troncoso P, Lin SH, Pisters LL, Sherwood ER, Brooks T, von Eschenbach AC, Hsieh JT: Consistent expression of an epithelial cell adhesion molecule (C-CAM) during human prostate development and loss of expression in prostate cancer: Implication as a tumor suppressor. Cancer Res 55:1215–1220, 1995.
242. Bates RC, Lincz LF, Burns GF: Involvement of integrins in cell survival. Cancer Metastasis Rev 14:191–203, 1995.
243. Boudreau N, Sympson CJ, Werb Z, Bissell MJ: Suppression of ICE and apoptosis in mammary epithelial cells by extracellular matrix. Science 267:891–893,1995.
244. Aihara M, Truong LD, Dunn JK, Wheeler TM, Scardino PT, Thompson TC: Frequency of apoptotic bodies positively correlates with Gleason grade in prostate carcinoma. Hum Pathol 25:797–801, 1994.
245. Aihara M, Scardino PT, Truong LD, Wheeler TM, Goad JR, Yang G, and Thompson TC: The frequency of apoptosis correlates with the prognosis of Gleason Grade 3 adenocarcinoma of the prostate. Cancer 75:522–529, 1995.
246. Colombel M, Radvanyi F, Blanche M, Abbou C, Buttyan R, Donehower LA, Chopin D, Thiery JP: Androgen suppressed apoptosis is modified in p53 deficient mice. Oncogene 10:1269–1274, 1995.
247. Berges RR, Furuya Y, Remington L, English HF, Jacks T, Isaacs JT: Cell proliferation, DNA repair, and p53 function are not required for programmed death of prostatic glandular cells induced by androgen ablation. Proc Natl Acad Sci USA 90:910–914, 1993.
248. Miyashita T, Krajewski S, Krajewski M, Wang HG, Lin HK, Liebermann DA, Hoffman B, Reed JC: Tumor suppressor p53 is a regulator of *bcl-2* and *bax* gene expression in vitro and in vivo. Oncogene 9:1799–1805, 1994.
249. Miyashita T, Reed JC: Tumor suppressor p53 is a direct transcriptional activator of the human *bax* gene. Cell 80:293–299, 1995.
250. Chiou S-K, Rao L, White E: Bcl-2 blocks p53-dependent apoptosis. Mol. Cell. Biol. 14:2556–2563, 1994.
251. McConkey DJ, Greene G, Pettaway CA: Apoptosis resistance increases with metastatic potential in cells of the human LNCaP prostatic carcinoma cell line. Cancer Res 56:5594–5599, 1996.
252. Yang G, Wheeler TM, Kattan MW, Scardino PT, Thompson TC: Perineural invasion of prostate carcinoma cells is associated with reduced apoptotic index. Cancer 78:1267–1271, 1996.
253. Rodin AE,Larson DL, Roberts DK: Nature of the perineural space invaded by prostatic carcinoma. Cancer 20:1772–1779, 1967.
254. Hassan MO, Maksem J: The prostatic perineural space and its relation to tumor spread. An ultrastructural study. Am J Surg Pathol 4:143–148, 1980.

255. Villers A, McNeil JE, Redwine EA, Freiha FS, Stamey TA: The role of perineural space invasion in the local spread of prostatic carcinoma. J Urol 142:763–768, 1989.
256. Fidler IJ: The evolution of biological heterogeneity in metastatic neoplasm. In Nicolson GL, Milas L (eds): Cancer Invasion and Metastasis: Biologic and Therapeutic Aspects. New York: Raven Press, 1984, p 5.
257. Moreno J, Croce CM, Fischer R, Monne M, Vihko P, Mullholland SG, Gomella LG: Detection of hematogenous micrometastasis in patients with prostate cancer. Cancer Res 52:6110–6112, 1992.
258. Katz AE, Seaman E, Olsson CA, O'Toole KM, Raffo AJ, McMahon D, Cama C, Benson MC, Perlman H, Buttyan R: Molecular staging of prostate cancer with the use of an enhanced reverse transcription-PCR assay. Urology 43:765–775, 1994.
259. Seiden MV, Kantoff PW, Krithivas K, Propert K, Bryant M, Haltom E, Gaynes L, Kaplan I, Bubley G, DeWolf W, Sklar J: Detection of circulating tumor cells in men with localized prostate cancer. J Clin Oncol 12:2634–2639, 1994.
260. Israeli RS, Miller WH, Su SL, Samadi DS, Powell CT, and Heston WDW: Sensitive detection of prostatic hematogenous tumor cell dissemination using prostate-specific antigen and prostate-specific membrane-derived primers in the polymerase chain reaction. J Urol 153:573–577, 1995.
261. Fidler IJ: Metastasis: Quantitative analysis of distribution and fate of tumor emboli labled with ^{125}I-5-iodo-2′-deoxyuridine. J Natl Cancer Inst 45:775, 1970.
262. Paget S: The distribution of secondary growths in cancer of the breast. Lancet 1:571–573, 1889.
263. Fidler IJ: Modulation of the organ microenvironment for treatment of cancer metastasis. J Natl Cancer Inst 87:1588–1592, 1995.
264. Hall SJ, Thompson TC: Spontaneous metastatic activities but not experimental metastatic activities differentiate primary tumor-derived versus metastasis-derived mouse prostate cancer cell lines. Clin Exp Metastasis 15:484–492, 1997.
265. Roberts AB, Sporn MB, Assoian RK, Smith JM, Roche NS, Wakefield LM, Heive UI, Liotta LA, Falanga V, Kehrl JH, Fauci AS: Transforming growth factor type β: Rapid induction of fibrosis and angiogenesis *in vivo* and stimulation of collagen formation *in vitro*. Proc Natl Acad Sci USA 83:4167–4171, 1986.
266. Yang EY, Moses HL: Transforming growth factor-β1 induces changes in cell migration, proliferation and angiogenesis in the chicken chorioallantoic membrane. J Cell Biol 111:731–741, 1990.

20

WNT AND FIBROBLAST GROWTH FACTOR GENE EXPRESSION DURING DEVELOPMENT OF THE MAMMARY GLAND AND ROLE OF WNTs IN HUMAN CANCER

MARZIEH JÖNSSON

Department of Oncology, University of Lund, SE-221 85 Lund, Sweden

KENNETH SMITH AND ADRIAN L. HARRIS

Growth Factors Group, Imperial Cancer Research Fund, University of Oxford, Institute of Molecular Medicine, John Radcliffe Hospital, Headington, Oxford OX3 9DU, United Kingdom

20.1. THE WNT FAMILY

20.1.1. Background

Fourteen years ago, the discovery of a locus that was activated in response to proviral insertion of mouse mammary tumor virus (MMTV) led to identification of a mouse proto-oncogene called *int-1* (for MMTV integration site) (1). The *int-1* gene was later shown to be homologous to the *Drosophila wingless (wg)* gene, which has a role in embryonic segmentation, and the combination of these two terms gave rise to the name WNT (2). With the cloning of mouse *wnt-1* (*int-1*) and *wnt-3*, proto-oncogenes with conserved sequences were identified, and use of the polymerase chain reaction produced several WNT family members. Searches initiated in the mouse and extended to many other species have led to cloning of several WNT coding sequences listed in Table 20.1. *Wnt* genes have attracted much attention by virtue of their high degree of conservation during evolution, their involvement in embryonic development (e.g., in normal development of mammalian brains, normal

Hormones and Growth Factors in Development and Neoplasia, Edited by Robert B. Dickson and David S. Salomon
ISBN 0-471-16899-8

TABLE 20.1. Known *wnt* Genes[a] in Different Species

Human	Mouse	*Xenopus*	*Drosophila*
wnt-1	*wnt-1*	*wnt-1*	*wnt-1 (wg)*
wnt-2	*wnt-2*	*wnt-2*	*wnt-2*
wnt-3	*wnt-3*		*wnt-3*
wnt-3a	*wnt-3a*	*wnt-3a*	
wnt-4	*wnt-4*	*wnt-4*	*wnt-4*
wnt-5a	*wnt-5a*	*wnt-5a*	
wnt-5b	*wnt-5b*		
	wnt-6	*wnt-6*	
wnt-7a	*wnt-7a*	*wnt-7a*	
wnt-7b	*wnt-7b*	*wnt-7b*	
		wnt-8	
wnt-8b		*wnt-8b*	
wnt-10		*wnt-10*	
wnt-10b	*wnt-10b*		
wnt-11	*wnt-11*	*wnt-11*	
	wnt-12		
wnt-13			

[a]In addition to the *wnt* genes listed in Table 20.1, *wnt 8c* in the chick, *wnt-10a* in the zebrafish, and *wnt-9* have been identified.

segmentation of the insect embryo, and induction of axis duplication in frog embryos) and their implication in several types of cancer (reviewed in ref. 3).

Although WNT proteins contain a secretory signal sequence, they associate with the cell surface and extracellular matrix and are, therefore, not freely diffusible but exert their influence within a few cell diameters. Autocrine and paracrine effects have been proposed to explain the action of WNT proteins. Addition of heparin or suramin to cell cultures increases the level of soluble WNT proteins and may play a role in the interaction and biologic activity of WNTs. The *wnt* open reading frame encodes a protein containing a secretory leader sequence and up to 24 highly conserved cysteines, suggesting that these cysteines are indispensable. One or more N-linked glycosylation sites are present. Most of the deduced proteins contain 350–380 amino acids (aa), with over 100 conserved residues evenly distributed throughout the entire sequence. Several WNT proteins have additional internal amino-terminal or carboxy-terminal domains; for example the WG protein has 85 aa inserted before the region encoded by its final exon, the *Drosophila* WNT-3 protein has a 155-aa insertion and a long amino-terminal extension, and the mouse WNT-2 and *Xenopus* WNT-8 proteins have short carboxy-terminal extensions (3). The aa identity between WNT proteins is approximately 50–60%. However, the homology between five pairs of *wnt* genes—*wnt-3* and *wnt-3a*, *wnt-5a* and *wnt-5b*, *wnt-7a* and *wnt-7b*, *wnt-2* and *wnt-2a* (also called *wnt-13*), and *wnt-10* and *wnt-10b*—is 80–90% (4), indicating that these paired genes have been duplicated during evolution (5). WNT proteins enter the endoplasmic reticulum (ER) and are glycosylated. Most of the WNT proteins retained in the ER associate with BiP, a 78-kDa

ER-based chaperonin that binds to improperly folded proteins and prevents their secretion and/or promotes proper folding (6). Despite sequence similarity, the individual WNT proteins have distinct functions in normal embryo development and differentiation as well as in tumorigenesis.

20.2 CLASSIFICATION OF *wnt* GENES

20.2.1. Classification into Various Subfamilies

Classification has been achieved on the basis of sequence homology and, to some extent, expression pattern. An example is the WNT-8 subfamily, which comprises the *wnt-8, 8b,* and *8c* sequence of the zebrafish *Danio rerio* (*Zwnt-8, 8b*), the frog *Xenopus laevis* (*Xwnt-8, 8b*) and the chicken *Gallus gallus* (*Cwnt-8b, 8c*). Within this subfamily, the three *wnt-8b* genes have a similar pronounced expression pattern in the developing forebrain (7).

20.2.2. Classification into Transforming and Nontransforming Subgroups

The potential of each member of the WNT family to induce transformation of cultured mouse mammary epithelial C57MG cells varies. Transformation of C57MG cells was defined as conversion of cuboidal mammary epithelial cells to elongated, highly refractile cells that continue to grow to a dense culture. The results demonstrated that different *wnt* genes transformed this cell line differently. On the basis of this observation, the WNT family was divided into three groups. The transforming group including *wnt-1, wnt-3a,* and *wnt-7a* induced colonies that formed and shed balls of cells. *Wnt-2, wnt-5b* and *wnt-7b* induced partial transformation with an apparent increase in saturation density. *Wnt-6, wnt-4,* and *wnt-5a,* which are normally expressed in the mammary gland and in C57MG cells, failed to induce transformation (8). Another study by Dale et al. showed that fibroblast C3H 10T1/2 cells expressing *wnt-1, wnt-6,* and *wnt-7b* in 5% serum grew to higher density than control or *wnt-5a* expressing cells (9). Moreover, *wnt-11* has also been shown to be transforming. These data together clearly demonstrate the distinct effects of individual *wnt* genes on cell growth and proliferation, depending on the cell line.

20.2.3. Classification Based on Function

Overexpression of WNT family members in *Xenopus* embryos demonstrated that *Xwnt-1, wnt-3a, wnt-8,* and *wnt-8b* promote duplication of the embryonic axis whereas *Xwnt-5a, wnt-4,* and *wnt-11* block morphogenetic movements in animal caps in response to activin (10). This was the basis of dividing WNTs into two functional classes: XWNT-1 and XWNT-5a. Since some embryonic cells may be exposed to both signals during development, this raises the question as to how the signaling pathways of these classes may interact. It has been shown that members of the WNT-5a class antagonize the WNT-1 class signal, thus decreasing cell adhesion. The cellular mechanism of this antagonism is not known (11). Domain swap experiments showed that the carboxy-terminal halves are functional domains for specifying the responses to XWNT signals (12).

20.3. THE *Drosophila WNT* GENE FAMILY

Screening for zygotic lethal mutations affecting the cuticular pattern of the *Drosophila melanogaster* embryo resulted in the identification of more than 130 genes, most involved in embryonic development. These genes provide an indication of the potential of the *Drosophila* system, which offers a combination of molecular tools, *in vivo* cell biology, and genetic analysis. Segmentation is one of the processes involved in pattern formation in *Drosophila* where coordination between a cascade of signals divides the embryos into 15 segments (13). Phenotypically, three classes of segmentation genes can be defined: the gap, pair rule, and segment polarity genes. The best-known example of the segment polarity gene is *wg*, the *Drosophila* ortholog of *wnt*-1 (reviewed in ref. 14). WG protein is expressed only in a subset of cells within each segment and secreted early in embryogenesis. The protein assumes a graded distribution across the entire segment, with the highest level present in those cells that secrete it, and successively lower levels are found in cells distant from this position (Fig. 20.1). By affecting the behavior of cells at a distance, WG establishes an anterior-posterior polarity in the embryonic segment in cooperation with the other segment polarity genes that receive and interpret the WG signal. Some of the segment polarity gene products, such as Porcupine (PORC) and Engrailed (EN), are required for production of WG and for maintaining and accurately positioning the signaling centers, whereas others, such as Disheveled (DSH) and Armadillo (ARM), are involved in the WG pathway as components of signal transduction (reviewed in ref. 15).

Fig. 20.1. (A) Wingless RNA is expressed by a subset of cells in each segment; there is thus one stripe of *wg* expression per segment. At the onset of gastrulation there are four rows of cells in each segment. The third row of cells in each segment expresses *wg*. (B) By the end of embryogenesis there is a row of 12 cells. (C) The proposed model for the autocrine and paracrine WG signal transduction pathways. Anterior is left. Secretion of active WG produced in the cells at the posterior edge of each parasegment requires PORC in both autocrine (left cell) and paracrine (right cell) roles. WG interacts with FZ2 to activate DSH, which in turn inactivates ZW3 kinase, leading to ARM dephosphorylation and accumulation. Thus, *dsh* and *arm* are required for induction of EN, wherease *hh* and *porc* are not. PTC represses *smo* expression and HH release this repression by binding either to SMO or PTC (or indeed to a complex of these two proteins). HH signal increase the CI level, which in turn activates transcription of *wg*.

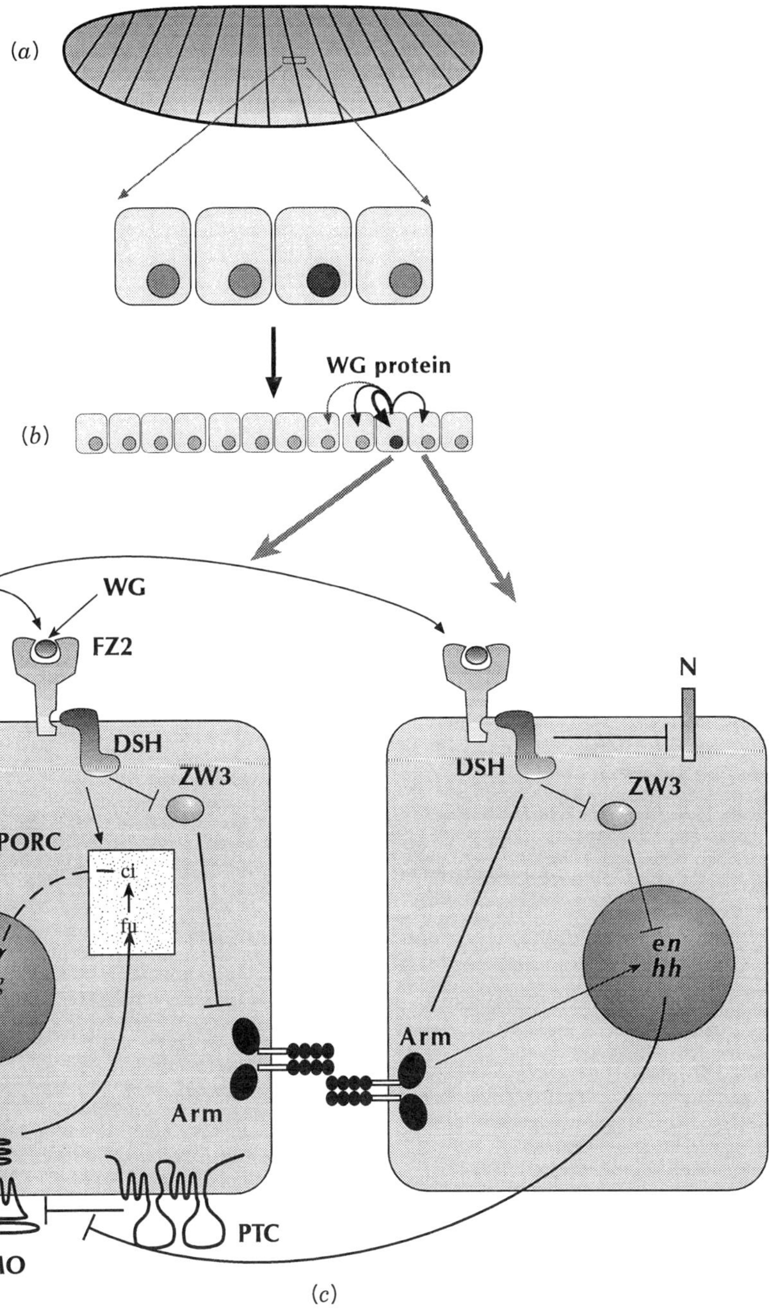
(a)
WG protein
(b)
WG
FZ2
DSH
ZW3
PORC
ci
fu
wg
Arm
SMO
PTC
N
DSH
ZW3
en
hh
Arm
(c)

20.4. COMPONENTS OF WG/WNT-1 SIGNALING IN *Drosophila* AND MAMMALIAN CELLS

The analysis of double mutations using a WG transgene in *Drosophila* made it possible to analyze epistasis between a dominant gain-of-function mutant for one gene in combination with a loss-of-function for the other. Comparing the phenotype with the WG-like phenotype clarifies the order of action of the different genes in the WG signaling cascade (Fig. 20.1). The components of WNT-1/WG signaling are described below (15,16).

20.4.1. WNT Receptors, Frizzled

In *Drosophila,* the *frizzled* (*fz*) genes play an important role in the development of tissue polarity. They are a family of putative receptors approximately 587 aa in size, containing a leader signal peptide and seven transmembrane segments. Each FZ protein has a highly conserved extracellular cysteine-rich domain (CRD) joined to the transmembrane domain by a variable linker. The N-terminal end faces extracellularly, and the C-terminal end faces the cytoplasmic side. Sequence analysis suggests a working model in which a CRD extracellular domain is tethered by a variable linker region to seven membrane-spanning α-helices. This structure suggests a dynamic model of transmembrane signal transduction analogous to that proposed for the G-protein-coupled receptors with N-terminal ligand binding domain (17). How the mechanisms of the FZ and WNT proteins interact to initiate signal transduction is unknown. It has been shown that the receptor ligand binding domain in FZ is in the CRD segment and part of the linker. Based on this finding, a signal transduction model has been proposed as follows: Ligand binding changes the interaction of the CRD with the transmembrane domain, resulting in rearrangement of the transmembrane α-helices and altering the cytosolic face of the protein. In this way, the WNT signal is transduced to a second messenger component of the pathway (18).

In the mouse, eight members of the FZ family are expressed in distinct tissues such as brain, eye, kidney, and lung (17). The existence of multiple FZ proteins implies the multifunctionality of this family and the existence of either a large number of effector pathways or convergence on a smaller number of pathways, as seen in G-protein-linked signal transduction. The expression of mouse *fz* genes in adult tissues and during embryonic development, as well as the primary sequence homology, indicates that this family plays the same role in mammals as in *Drosophila,* i.e., as a component in the WNT signaling pathway. The mechanism whereby the FZ protein receives and transmits WNT signals needs to be clarified.

20.4.2. Disheveled

In *Drosophila* a single *disheveled* (*dsh*) segment polarity gene has been isolated (19). Due to the high degree of conservation, it has been possible to isolate several *other dsh* genes as well, e.g. one in mouse (*dvl-1*), one in *Xenopus (Xdsh)* and three in human: *DVL-1*, *DVL-2* (20), and *DVL-3;* the last one has recently been identified by our laboratory (21). *Dsh* genes encode for cytoplasmic phosphoproteins resembling the *Drosophila* tumor suppressor gene *discs-large*. During WNT signaling DSH becomes phosphorylated, and the phosphorylated form associates with the cell membrane. Through an unknown mechanism, DSH antagonizes the function of ZW3/GSK3, as well as of Notch (N), to facilitate transduction of the WNT-1/WG-mediated signal (22,23). It is not yet clear whether DSH interacts directly with WNT receptors or is phosphorylated by another upstream component. DSH protein contains disc homologous repeats (DHR) that have been shown to interact with a S/T-X-V motif found at the C-terminal end of many proteins including FZ proteins, raising the possibility that DSH may interact directly with FZ (18). However, the presence of a gene coding for a protein kinase, NEMO, downstream of FZ in *Drosophila* has been shown to mediate the effects of FZ on ommatidial rotation. Thus, the possible existence of a protein kinase downstream FZ that might function as a link between FZ and DSH in the WNT signaling pathway cannot be excluded.

20.4.3. Zeste White 3 Kinase/Glycogen Synthase Kinase 3

The *Drosophila shaggy/zeste-white 3* (*zw3*) gene encodes a serine/threonine protein kinase involved in WG signaling (24). Mutation analysis in *Drosophila* embryos shows that loss of ZW3 kinase activity produces an embryonic phenotype opposite to that of WG, indicating that WG activity causes inactivation of ZW3 kinase. It has been shown that the WG signal is transduced to ZW3 kinase through DSH, leading to inactivation of ZW3 kinase and accumulation of ARM. During pattern formation, ZW3 also represses the *en* autoregulation phase. WG signaling stabilizes *en* expression in a paracrine manner by inactivation of this repression through an unknown mechanism (23). The mammalian ortholog of the *zw3* gene is referred to as *glycogen synthase kinase 3* (*gsk3*), whose product exists in two distinct isoforms; GSK3a and GSK3b, with 98% homology in their kinase domain sequence GSK3 is a multisubstrate enzyme participating in many different signaling pathways regulated by hormones and other growth factors such as insulin and epithelial growth factor (EGF) (25) and which play an important role in proliferation, differentiation, and development. GSK3 is also involved in the regulation of glycogen synthase, as well as in the regulation of transcription factors c-*jun* and L-*myc*. The molecular mechanism by which WNT signals inactivate ZW3 kinase is not clear. However, phosphorylation of the regulatory subunit on GSK-3b *in vitro* by protein kinase C (PKC) decreases its activity by 50%. The ribosomal protein S6 kinase, $RSK^{mo\text{-}2}$, a major substrate of mitogen-activated protein kinase (MAPK), was also shown to directly phosphorylate and inactivate both GSK-3 isoforms *in vitro* and thereby regulate its kinase activity (26).

20.4.4. Armadillo/Catenins

β-Catenin and plakoglobin are the mammalian orthologs of ARM, one of the most important components in the WNT signaling pathway (Fig. 20.2). In addition to

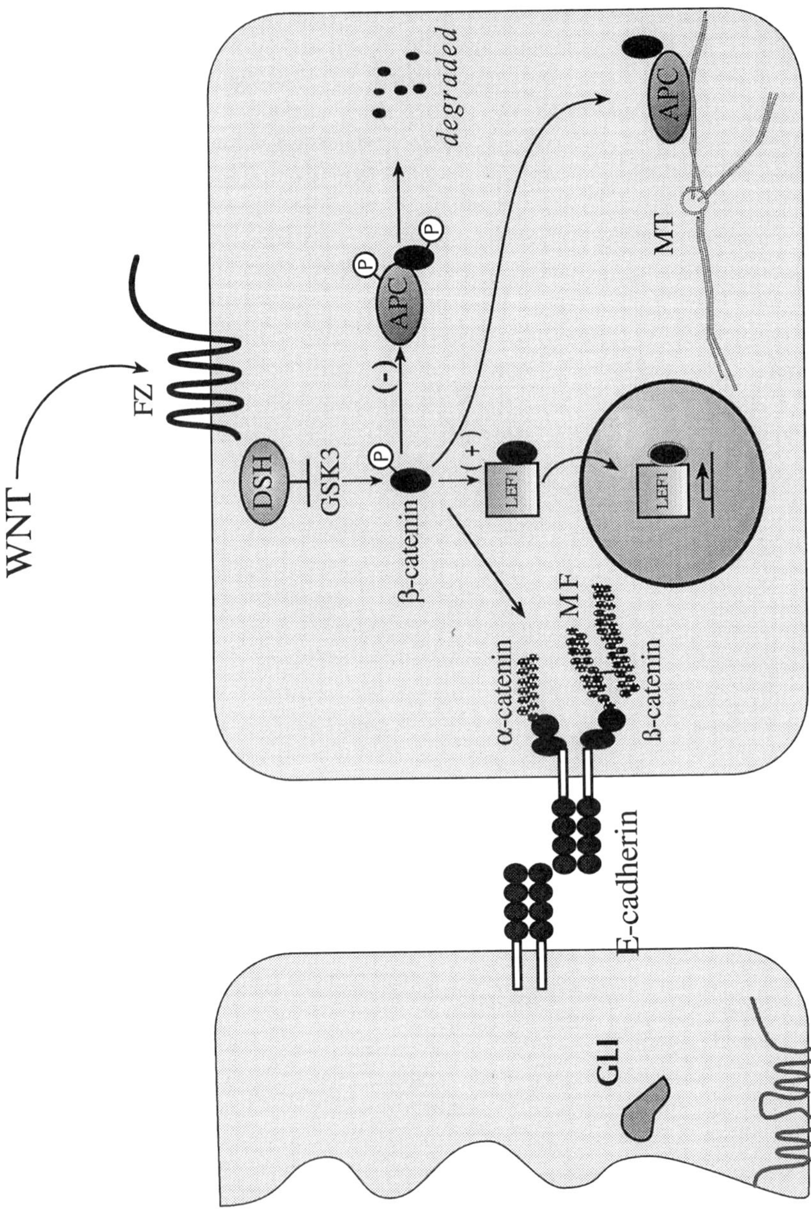
WNT
FZ
DSH
GSK3
P
β-catenin
(-)
APC
P
P
degraded
(+)
LEF1
LEF1
MT
APC
α-catenin
M F
β-catenin
E-cadherin
GLI
PTC

their role in WNT signaling, they are components of adherens junctions that play a role in cell–cell adhesion. ARM is a phosphoprotein whose phosphorylation status varies both during embryonic development and from tissue to tissue. WG signaling negatively regulates ARM phosphorylation by ZW3 kinase; that is, in the presence of active WG signaling, ZW3 kinase is inactivated and ARM is dephosphorylated and consequently accumulates in the cytoplasm and cell–cell contact sites (27). In mammalian cells, the first evidence implicating WNT-1 in the regulation of catenin (α, β, and plakoglobin) function derived from the effect of WG on ARM. For example, in the mouse, it was shown that WNT-1 expression in AT-20 (neuroendocrine) cells or in C57MG (mammary epithelial) cells resulted in an accumulation of plakoglobin, as well as an increase in the β-catenin cytoplasmic pool size (28). Later, it was shown that WNT-1 signaling increases the stability and steady-state levels of β-catenin at the posttranscriptional level (29).

The *adenomatous polyposis coli* (*apc*) gene, which is a tumor suppressor gene and is mutated in colon cancer, encodes for a protein that has also been shown to be associated with the β-catenin, γ-cat/plakoglobin, and microtubules (30). The interaction of catenins with APC may also modulate their stability and reduce the availability of catenins for interaction with E-cadherin, thereby affecting cell–cell adhesion and migration. As APC is a substrate of GSK3 kinase, the phosphorylation of the β-catenin binding site of the APC protein by this kinase is thought to cause downregulation of cellular β-catenin levels (31). The functional implication for the colocalization of APC protein with cytoskeletal microtubules is unknown. One possibility is that the APC/β-catenin complex could relay signals from catenins to the cytoskeleton that, in turn, interface with extracellular signals communicated through E-cadherin, which also binds to catenins. The WNT-1 signal transduction pathway leads to interaction of cytoplasmic β-catenin with APC to regulate its stability, as well as with β-catenin entering the nucleus and participating in regulating gene expression and cell fate. The direct binding of the transcription factor LEF-1, a mammalian HMG box factor, to β-catenin and translocation of β-catenin to the nucleus has provided evidence linking WNT signals to change in gene expression (32) (Fig. 20.2).

20.4.5. Other Components of the WG/WNT Signaling Pathway

Intrasegmental patterning in the *Drosophila* embryo is regulated by cell–cell communication. The anterior and posterior compartments of *Drosophila* wing and leg

Fig. 20.2. The WNT signal activates DSH through the FZ receptor, which then antagonizes GSK3, leading to an increased steady-state level of the cytoplasmic pool of β-catenin. In the presence of WG signal (+), cytoplasmic underphosphorylated free β-catenin can form a complex with E-cadherins and α-catenin that connects the complex to actin microfilaments (MF). β-Catenin links WNT signaling to the nucleus by binding to transcription factors such as LEF-1 and translocating into the nucleus, modulating expression of other effector molecules. In the absence of WG signal (−), APC binds to β-catenin and regulates its stability, thereby reducing the availability of β-catenin to E-cadherin. The function of binding of APC–β-catenin complexes to microtubules (MT) is not known.

imaginal discs are derived from founder cells located on either side of the embryonic parasegment border. Transcription factors encoded by gap and pair rule genes activate the segment polarity genes *wg* and *en* at the blastoderm stage. En expression is initiated by the pair rule genes, but its maintenance is dependent on WG signaling from adjacent cells. Two homologs of the *Drosophila en* gene have been reported in the mouse, *EN-1* and *EN-2*. *EN-1* is expressed in the developing midbrain in the mouse, suggesting that the Wnt-1/EN regulatory pathway has been conserved from fly to mouse. The WG-dependent phase of *en* expression is transient, and its autoregulation gradually becomes independent of WG signaling. It has been suggested that EN (a nuclear DNA-binding protein) acts as a determinant in all cells of the posterior compartment and regulates the expression of the *hedgehog* (*hh*) gene, which, in turn, is required for WG signaling.

The localized expression of secreted HH protein by cells in the posterior compartment induces expression of secreted signaling factors, such as WG or Decapentaplegic (a member of the transforming growth factor β family, DPP), in nearby anterior cells, which, in turn, organize the spatial pattern in the wing and leg imaginal discs (reviewed in ref. 33). cAMP-dependent protein kinase (PKA) was also identified as a key component of the signal transduction pathway through which HH and Patched (PTC) directly localized expression of *dpp* or *wg* to establish the anterior-posterior boundary organizer. PKA is thought to oppose the effect of HH on Cubitus Interruptus (*Drosophila* GLI-related protein, a transcription factor, CI) and to keep the intracellular level of CI low. Two transmembrane receptors in the WG pathway are PTC, which acts as an anterior determinant, and Smooth (SMO), which is located on wg-expressing cells and responds to HH signals transduced from adjacent cells.

PORC activity is also required upstream of ZW3 kinase for the distribution of WG to enable even low levels of WG protein to reach the cell surface, where it signals to adjacent cells and consequently maintains *en* expression. In the late stage of pattern formation, *wg* expression also autoregulates and becomes independent of *hh* and *en* expression by activating the *gooseberry* (*gsb*) gene, which encodes a homeodomain-containing protein (Pax-1 in vertebrates) and is expressed in *wg*-expressing cells. *GSB* transcription factor begins to activate *wg* independently of EN and HH, presumably by binding directly to cis-regulatory elements of *wg* and thus generating an autoregulatory loop between *wg* and *gsb* (34). The other putative components in the WG and HH signaling pathway are Fused (FU), a serine/threonine kinase (*cdc2* in vertebrate) that is phosphorylated in response to HH, and a less well characterized component, Costal-2 (COS^2).

Furthermore, several observations have suggested a role for the members of other signaling pathways, such as N, in WG-mediated signaling events. The *n* gene encodes a receptor with a large extracellular domain containing 36 tandem epidermal growth factor (EGF)-like repeats that, through local cell interactions, control the fate of a broad spectrum of cells. Several genes, such as *delta* (*dl*) and *serrate* (*ser*), encode elements of the N signaling pathway. Genetic analysis has implicated both *n* and *wg* in establishment of the dorsoventral pattern in the wing, as well as in induction of bristles at the wing margin. DSH is a key component of both the WG and N signaling pathways. WG controls expression of the Achaetescute Complex (AS-C) proneural genes at the wing margin, whose activation is required to achieve the neural fate and Sensory Mother Cell (SMC) development. N acts as a suppressor of AS-C and SMC development by an unknown mechanism. DSH inhibits the effect of N through direct binding to the C-terminal domain of N and facilitates mediation of WG signal to SMC development, thus providing a molecular mechanism for the inhibitory cross-talk between the two pathways (22).

20.5. DISTINCT PATHWAYS FOR AUTOCRINE AND PARACRINE WG SIGNALING IN *Drosophila* EMBRYOS

Early in segmentation, a cascade of regulatory genes generate stripes of localized transcription factors that define repeating units called *parasegments* along the body axis. Within each segment, the transcription factors initiate the expression of WG in one row of cells and the expression of *hh* and the transcription factor *en* in the adjacent row of cells. WG exerts its biologic activity in a para- and autocrine route (Fig. 20.1). Reliable *wg* transcription requires simultaneous activation of both pathways. The initiation of *wg* and *en* expression is regulated by pair rule genes. Later, the expression of *en* is maintained by WG signaling and the expression of *wg* becomes dependent on EN, thus establishing a WG-EN loop in a paracrine manner. EN may achieve this maintenance of *wg* via HH, which is thought to antagonize the inhibitory effect of PTC on SMO by binding either to SMO or to PTC or to the complex of these proteins.

After overcoming the inhibition of PTC, HH increases the intracellular level of CI, which, in turn, increases the expression of HH target genes including *wg* and *dpp*. Expression of *hh* in the paracrine signaling loop is maintained by WG activity, and *wg* expression is maintained by HH. That is, these mutual regulatory interactions stabilize the expression domains of the two genes in an intimate regulatory loop that acts to ensure the coordinated expression of inductive signals during the patterning of a field of cells (35).

The intracellular changes in the paracrine signaling pathway generated by secreted WG start by binding to FZ2 and activation of the intracellular DSH that is required for transduction of the WG signal between cells. Activation of DSH leads, in turn, to inactivation of ZW3 kinase. Consequently, the inactivation of ZW3 kinase results in dephosphorylation and accumulation of ARM. In the absence of WG, most ARM is phosphorylated and assembled into adherens junctions. Phosphorylated ARM is unstable in the cytoplasm. The WG signal suppresses phosphorylation of ARM, thereby stabilizing cytoplasmic ARM, which results in the maintenance of *en* expression either directly or by the formation of a second signaling complex. The involvement of ARM in regulation of *en* expression was shown in arm^- embryos in which the expression of *en* decayed. The presence of ARM at adhesive junctions promotes intimate cell–cell contact,which probably is a prerequisite to WG signaling between cells.

Autocrine WG regulates its own expression through DSH to activate *fu* and *ci*. The mechanism of activation of *fu* and *ci* is not known. In parallel, HH counteracts the PTC-mediated repression on *fu* and *ci*, and thereby CI increases the expression of *wg*. In the paracrine pathway, *fu* and *ci* are not involved (36).

20.6. *Wnt* GENE FAMILY AND NORMAL VERTEBRATE DEVELOPMENT

20.6.1. *Wnt* Genes and Central Nervous System Development in Mammals

The site of expression of most members of the WNT family is still incompletely cataloged. Most of the studies of central nervous system (CNS) development have been carried out in mouse embryos (reviewed in ref. 43). Transcripts of all mouse *wnt* genes have been found in a spatially and temporally restricted pattern during gastrulation, neurulation, or early organogenesis, indicating that WNT signaling plays an important role throughout early CNS development. Eight WNT family members are expressed with a distinct anterior-posterior or dorsoventral restricted pattern in the neural plate and neural tube. Among these, *wnt-5a* and *wnt-7a* are expressed in overlapping domains in the ventral and lateral diencephalon, whereas *wnt-1, wnt-3, wnt-3a,* and *wnt-4* mRNA are present in the dorsal diencephalon (37). *Wnt-7b* is expressed throughout much of the diencephalon, the optic stalks, and the dorsal optic vesicle. It has been shown that *wnt-7a* participates with other regulatory genes—*en-2, en-1, pax-2 and gli*—to divide the embryonic cerebellum into 11 sagittal domains and at least 2 anterior/posterior regions. These domains of gene expression form part of the basic system of positional information similar to that of the *Drosophila* embryo, regulating the patterning of cerebellar foliation.

20.6.2 Role of *wnt* Genes in Axis Formation and Mesoderm Induction in *Xenopus*

Owing to their size, abundance, and ease of microinjection, embryos of *Xenopus laevis* are a suitable model for studying WNT signaling pathways. The first indication that WNT may be involved in mesodermal patterning came from McMahon (4), who showed that injection of *int-1* into *Xenopus* embryos causes axis duplication. It became clear that WNT proteins have a powerful inductive action and that this action is involved in converting ventral into dorsal tissue. When *wnt-1, wg,* or *Xwnt-8* mRNA was injected into the ventral side, axis duplication was prompted, whereas injection into the dorsal side had little effect (4). This experiment strongly supported the hypothesis that multiple WNT signaling pathways exist and that *wnt* genes are functional during early development. Another experiment that confirmed this important observation was that the centralization of embryos by ultraviolet light radiation could be rescued by overexpression of *Xwnt-8, Xwnt-1,* and *Xwnt-3a*. These genes rescued the dorsal axis formation by restoring a Nieuwkoop center in embryos. What makes these results so important is the ability of Wnt proteins to induce complete dorsal axes, and even head structures, whereas attempts to rescue centralized embryos or to induce additional axes by injection of other mesoderm-inducing growth factors such as activin resulted in trunk structures (reviewed in ref. 38).

20.6.3. *Wnt* Genes and Mammary Gland Development in Mammals

There are three distinct periods of mouse postnatal mammary gland development. Ductal growth is dependent on the presence of estrogen and occurs from 6 to 8 weeks of age to create a bush-like network of ducts. The onset of pregnancy initiates a second phase of development in which lobular-alveolar structures develop from the existing ductal system in a process driven by pregnancy-induced hormones including estrogen. The third phase is the growth and differentiation of the lobular system to form milk-producing alveoli during lactation. Each phase of mammary gland development requires a specific combination of hormones and locally acting signaling molecules. The signaling molecules in the local action include WNTs, epidermal growth factor (EGF), fibroblast growth factor (FGF), and transforming growth factor-β (TGF-β).

The regulation of *wnt* genes and their role in normal mammary gland or mammary cancer development has been well studied (reviewed in refs. 4 and 39). The generation of mammary adenocarcinomas by *wnt-1* and *wnt-3* is evidence that mammary cells respond to Wnt signals. Later, it was shown that six mouse *wnt* genes are differentially expressed during mammary gland development, pregnancy, and lactation (Table 20.2). Of these, *wnt-4* and *wnt-5a* are expressed during lobular development and are hormonally regulated, while *wnt-5b* and *wnt-6* are expressed at the onset of pregnancy and subsequently decline during lactation (40). A study by Gavin and MacMahon (41) showed that *wnt-4* and *wnt-5b* are induced during pregnancy. The variation between the results obtained from different studies may be due to the differences between the ages and strains of animals used for experiments or the sensitivity of the method used for the detection of expression. *Wnt-7b* was expressed in the epithelium of mature virgin animals in which ductal growth had ceased. Thus, the temporal change of the expression pattern of *wnt* genes during pregnancy and lactation and the expression of *wnt-7b* in nonproliferating cells indicate that *wnt* genes are involved in the control of lobuloalveolar development and that their expression does not always correlate with cell proliferation. During mammary gland development, several *wnt* genes including *wnt-2, wnt-5a,* and *wnt-6* are expressed in the stroma prior to ductal outgrowth and may thus be involved in controlling epithelial proliferation by the stroma. In support of this conclusion, *wnt-2* is expressed in human lung and breast fibroblasts (42), suggesting that it is involved in

TABLE 20.2. *Wnt* Expression in Different Stages of Mouse Mammary Gland Development

	Mammary Tissue Stage					
		Pregnancy				
wnt Genes	Virgin	Early	Middle	Late	Lactation	Involution
wnt-2	++	+++	+	−	−	++
wnt-4	[1]	+++	+	−	−	+/−
wnt-5a	−	+++	+	−	−	+++
wnt-5b	+	+	+++	++	−	+++
wnt-6	+	++	+++	+	+	+
wnt-7b	++	++	−	−	−	++

[1]Expression of *wnt-4* in virgin mice is differentially regulated, depending on the age and strain.

Source: Refs. 4 and 40.

lung and breast development by mediating intercellular interaction(s) between mesenchyme and epithelium.

In humans, several *wnt* genes have been reported to be expressed in breast tissue, suggesting that these genes may play the same role in mammary gland development in humans as they do in the mouse. The fact that not all *wnt* genes are expressed suggests some interspecies differences. The duplicated pairs with very high homology, such as *wnt-3* and *wnt-3a* or *wnt-7a* and *wnt-7b,* show different expression profiles; *wnt-3* and *wnt-7b* are expressed, whereas *wnt-3a* and *wnt-7a* are absent, indicating a lack of redundancy in WNT function. The expression profile of *wnt* genes in human breast tissue resembles that in the mouse; *wnt-2, wnt-4,* and *wnt-7b* are expressed, while *wnt-1, wnt-3a,* and *wnt-7a* are not. *Wnt-3* is expressed in human but not in mouse mammary tissue (Table 20.3). In humans, the profile of *wnt* expression is similar in pre- and postmenopausal females, in contrast to the mouse, where the expression pattern varies depending on the stage of mammary gland development. The difficulty in obtaining samples and the low expression have been responsible for the lack of data on *wnt* expression during different hormonal periods of human breast development (42).

However, studies of *wnt* expression in normal human breast cell lines have shown an expression pattern similar to that of breast tissue with the exception of *wnt-2,* which is absent in cell lines but present in tissues. This may indicate that *wnt-2* is expressed by stromal cells, as mentioned above. The exact role of *wnt* in mammary gland development is not clear. The *wnt-1* and *wnt-3* genes, which were originally isolated as mammary oncogenes, are not expressed in the mammary gland (4). However, activation of transcription of the normally silent *wnt-1* genes by proviral integration at the *wnt-1* locus induces the activity of *wnt-1* in mammary glands, suggesting that *wnt-1* and *wnt-3* may function as oncogenes by perturbing the regulatory pathways of other mammary gland-expressed *wnt* genes. The fact that WNT-1 hyperplasia can occur in ovarietomized females suggests that WNT-1 is capable of bypassing hormonal regulation and activating local developmental control pathways.

TABLE 20.3. *Wnt* Gene Expression in the Mouse Mammary Gland and in Human Breast Cancer Cell Lines and Breast Tissues

	wnt Genes						
	2	3	3*a*	4	7*a*	7*b*	5*a*
Mouse breast	+	−	−	+	−	+	+/−[1]
Human cell lines	−	+	−	+	−	+	−
Human tissue							
Normal	+	+	−	+	−	+	+
Benign	+	+	−	+	−	+	+
Malignant	+	+	−	+	−	+	+

[1]*Wnt-5a* is expressed differentially in the mouse breast, depending on mammary gland development (see Table 20.2).

Source: Refs. 42 and 54.

Recently, it has been reported that the oncogene *int-3*, which causes mouse mammary tumors when activated by MMTV, encodes the intracellular domain of a novel mouse gene referred to as *n 4*, which is a member of the N family. Structural comparison of INT-3 and N 4 showed that the extracellular domain of N 4 was lost in the INT-3 oncogene-activated form, leading to constitutive activation of the N 4 protein. Although the WG and N signaling pathways interact during pattern formation in *Drosophila,* the mechanism of this interaction in tumorigenesis is not known (43).

20.7. THE FIBROBLAST GROWTH FACTOR FAMILY

The FGF family currently comprises nine members, FGFs 1–9, which are closely related heparin binding polypeptides (reviewed in refs. 44 and 45). FGF members interact with five different tyrosine kinase receptors (FGFRs) to stimulate mitogenic, differentiation, migrational, and survival responses, depending on the target cells. The genes encoding FGFs and FGFRs are conserved in both vertebrates and invertebrates. FGF-1 and FGF-2 (acidic and basic FGF) were originally discovered as polypeptides in pituitary and brain extracts and could induce proliferation of fibroblasts in cell cultures. Further members of the FGF family were identified on the basis of sequence homology. *Fgf-3* (*int-2*) was identified by gene activation following insertion of MMTV. *Fgf3* is located tandemly linked to *hst-1* (*kfgf* or *fgf-4*). Not only does it have the capacity to transform 3T3 fibroblasts, but it is also found to be frequently amplified in human stomach tumors. FGF-8 is an androgen-induced growth factor. Similarly to FGF-3, it was identified as a gene activated in MMTV-integrated mammary tumors. FGF-9 is a recently discovered member that was purified from a human glioma cell line. FGF and WNT proteins have several common features, such as the ability to bind to the extracellular matrix and to heparin. They are also involved in proliferation, and changes in their normal signaling pathways lead to oncogenic behavior. Most members of the FGF family are involved in mesoderm induction during embryogenesis. Introduction of dominant-negative forms of FGF receptor into developing embryos perturbs mesoderm formation, providing direct evidence that FGF molecules are required for mesoderm induction. The interaction of the WNT and FGF families during mesoderm induction will be discussed later.

20.8. THE *FGF* GENE FAMILY AND MAMMARY GLAND DEVELOPMENT

During normal mouse mammary development and differentiation, high levels of *bfgf* mRNA are present in virgins but drop markedly during pregnancy. The fact that bFGF is not detectable in cells from the lactating female suggests that it is the predominant FGF involved in ductal development of the virgin mammary gland. In contrast to *bfgf, afgf* is expressed in both virgin and pregnant mammary glands, but very low levels of expression are detectable during lactation. Moreover, *bfgf* is expressed predominantly in the stroma of virgin glands, indicating that this expression may require the presence of an inductive effect of the factors expressed by the epithelial compartment (46). Thus, it is attractive to speculate that stromal–epithelial interaction in developing glands may result in a paracrine loop in which members of both the WNT and FGF families participate.

20.9. FGF/WNT INTERACTION IN DEVELOPMENT

The expression of both WNT and FGF factors during normal mammary gland development and tumorigenesis suggests that the respective signaling pathways may interact. Evidence indicating cooperation between the FGF and WNT families in development comes from studies of mesoderm induction in *Xenopus* embryos. Embryos treated with bFGF develop a ventral character, whereas embryos injected with *Xwnt-8* mRNA result in mesoderm formation with a distinctly more dorsal nature. The fact that bFGF is present in embryos at the right time to act as a mesoderm inducer but cannot induce dorsal mesoderm by itself suggests that WNT signals cooperate with FGF signaling to complete this process. The interaction between *wnt* genes and other mesoderm-inducing growth factors is important but not yet clear. However, the fact that *Xwnt-8* can be induced by bFGF suggests that WNT activity is downstream of FGF family members, and it is possible that the signal induced by dorsal vegetal cells is transduced by an FGF mesoderm-inducing factor in cooperation with a member of the WNT family (10,46).

20.10. EXPRESSION OF *wnt/fgf* GENES IN MAMMARY CARCINOMAS

Tumorigenesis by MMTV insertion has been reported for a number of *int* genes including *wnt-1* (*int-1*), *wnt-3*, *fgf3* (*int-2*), *fgf-4*, and *n 4* (*int-3*) (43,47). The MMTV enhancer acting on the normal WNT promoter is presumed to result in activation of transcription of these genes, thereby conferring a growth advantage to individual cells, accounting for the clonal character of the tumors. In the case of *wnt-1*, the insertion does not interrupt the *wnt-1* open reading frame and overexpression of the protein has been postulated to play a causative role in tumorigenesis.

In vitro transformation of C57MG cells, i.e., conversion of cuboidal mammary epithelial cells to elongated, highly refractile cells that continue to grow to a dense culture provides a useful assay to monitor the involvement of WNTs in transformation. Although the *Drosophila* WG and mouse WNT-1 are only 54% identical at the aa level, the overexpression of WG in mouse mammary RAC311c cells showed that WG was able to transform this cell line to the same extent as WNT-1. Cells expressing *wg* were also tumorigenic, indicating that *Drosophila* WG is capable of inducing full neoplastic transformation (48). These results imply a striking conservation in the properties of the WNT proteins and in their mechanism of action.

Transgenic mice have also provided conclusive evidence for the involvement of *wnt* genes in cancer. For example, mice carrying a *wnt-1* transgene show hormone-independent hyperplasia of mammary epithelium together with an increased incidence of breast tumors. However, the long latency period required for the generation of tumors in these mice suggests that overexpression of *wnt-1* can contribute to the generation of mammary tumors but is not by itself sufficient for complete transformation. WNTs and other growth factors, such as FGF in synergy with ovarian hormones, seem to be powerful collaborators in mammary carcinomas. An example of this process is the coactivation of WNT-1 and FGF3 in many spontaneous mouse mammary tumors by integration of MMTV. Mice expressing both *wnt-1* and *fgf-3* transgenes developed mammary carcinogenesis earlier and more frequently than mice bearing either transgene alone. MMTV infection of mice carrying a *wnt-1* transgene results in tumors with a high frequency of insertional activation of *fgf-3*. Since *wnt-1* and *fgf-3* (3) are not expressed in mammary glands during normal development, the activation of these factors during tumorigenesis may represent their improper temporal expression, resulting in the disruption of other signaling pathways regulated by other members of the WNT and FGF families. The reason *wnt-1* and *fgf3* are more frequent targets for activation by MMTV insertion than the other members of these families is not known. One possible explanation is that, e.g., *fgf3* expression per se can only give rise to hormone-dependent hyperplasia often in pregnancy-dependent mammary tumors and that further events, including stimulation by *wnt-1* or insertional mutagenesis by MMTV, may be required to achieve hormone independence. The amplification of the *fgf3* cluster, including the linked *hst* and *bcl-1* oncogenes, has been observed in up to 19% of breast tumors, but the degree of amplification is low and is not paralleled by increased gene expression (49).

Fgf8 was identified as a cooperating gene in MMTV-induced mammary tumors in WNT-1 transgenic mice. Activation of *fgf8* is significantly more frequent in tumors from WNT-1 transgenic mice than in tumors from nontransgenic controls, indicating an apparent selection for *fgf8* activation in WNT-1 transgenic mice. The exact mechanism of this cooperation is not clear. However, it has been shown that cells expressing *wnt-1* are sensitized to stimulation by FGF8 compared with other growth factors (50).

Mouse *wnt-10b* has also been identified immediately adjacent to MMTV, suggesting that MMTV integration contributes to the oncogenic process in induced mouse mammary tumors (51). The recently identified human homolog of mouse *wnt-10b* by our laboratory and its upregulation in a subset of mammary tumors supports a role for the *wnt-10b* gene in the genesis of mammary tumors in both mice and humans (52).

Several of the *wnt* genes, including *wnt-2, wnt-4, wnt-5a,* and *wnt-7b,* show elevated expression in disease states in the human mammary gland. The expression of *wnt-2* and *wnt-4* is 10- to 20-fold higher in benign fibroadenomas than in normal or malignant breast tissues. As benign tumors contain a greater proportion of stromal cells and as *wnt-2* is expressed by stroma, this may explain the higher level of *wnt-2* expression in stromal hyperplasia. The generation of mouse mammary hyperplasias by *wnt-1,* together with the high *wnt-4* and *wnt-2* expression in fibroadenomas that are hyperplastic rather than neoplastic, supports a causal role for *wnt* genes in the pathology of these lesions. *Wnt-7b* is expressed in benign and malignant tumors at a comparable level with normal tissue except for a subset of malignant tumors showing 30-fold higher expression of *wnt-7b* (42).

The expression of *wnt-7b* has been shown to be upregulated fourfold in superficial bladder tumors compared to normal bladder tissue. The level of *wnt-7b* message in superficial tumors is fivefold higher than in invasive bladder tumors, suggesting that alteration in WNT-7b signaling is an early event in the development of bladder cancer (53).

Wnt-5a is one of the *wnt* genes highly expressed in several types of cancer including breast cancer. Elevation of the *wnt-5a* message has been reported to be 10- and 4-fold in benign and malignant breast tissue, respectively (54). The level of *wnt-5a* expression in breast cancer cell lines is undetectable, in contrast to its upregulation in tumors. A secondary inductive pathway or interaction between stroma and epithelium might be involved in upregulation of this gene in cancer. Aberrant expression of *wnt-5a* has also been reported in several other cancer types including in lung cancers, prostate carcinomas, and melanomas (55). The *wnt-5a* upregulation was not due to either gene rearrangement or amplification. Thus these data together indicate a role for WNT-5a in human cancer.

Upregulation of *wnt-2* has also been reported in colon cancer at different stages, as well as in other digestive neoplasms such as gastric and esophageal carcinomas. In normal human breast, *Wnt-2* was found to be expressed at a low level in normal breast fibroblasts but not in epithelial cells, whereas in breast cancer a subset of tumors was found to express high levels of *wnt-2* in both fibroblasts and epithelial cells. The increase in epithelial expression of *wnt-2* in tumors suggests a role for *wnt-2,* which may act through the creation of an autocrine WNT signaling loop in breast cancer (56).

20.11. ALTERATIONS OF DOWNSTREAM COMPONENTS OF THE WNT PATHWAYS IN HUMAN CANCER

The signaling pathway of WNTs in cancer is poorly understood. The downstream elements of WNT signaling are commonly mutated and altered in many cancers, but whether WNT signals interact and are altered is unknown. However, recent work has linked the WNT pathway to β-catenin and APC, both of which are altered in several types of cancer such as breast and colon cancer. In normal cells, β-catenin/ARM has been identified as a key component in the WNT/WG signaling pathway (Fig. 20.2). Disruption of the *β-catenin* gene in mouse embryos results in detachment of ectodermal cells in the embryo, consistent with reduced cell adhesion. Moreover, expression of a truncated *β-catenin* leads to loss of cell adhesion in human cancer cell lines (57).

In the cytoplasm, APC functions to maintain low levels of cytoplasmic β-catenin in the cell, thereby suppressing WNT signaling. Mutations in APC, best studied in colon cancer, which disrupt the balance between the cytoplasmic β-catenin–APC complex and the free pool of β-catenin, may lead to decreased E-cadherin activity in adherens junctions, as well as modulation of desmosomal cadherin activity in cancer. It has been shown that the wild-type APC is associated with the microtubule cytoskeleton, whereas mutants in tumors have lost this activity (58).

The overexpression of *wnt-1* in mouse mammary cancer (or occurrence of the APC mutations) in colon cancer results in an increased pool of β-catenin, which may interact and become tyrosine phosphorylated in response to secreted signaling molecules including hepatocyte growth factor (HGF) and EGF. The phosphorylation of free cytoplasmic β-catenin in response to EGF appears to involve the association of the β-catenin–cadherin complex with the EGF receptor and direct phosphorylation of β-catenin by the EGF receptor. The impact of β-catenin phosphorylation on its ability to bind to E-cadherin and cell adhesion is not clear (reviewed in refs. 59 and 60). However, transformation of cells by activated Src and Ras results in tyrosine phosphorylation of β-catenin and a coincidental decrease in cell adhesion, disassembly of adherens junctions, and conversion of epithelial cells to a mesenchymal phenotype. Moreover, catenin binding is essential for E-cadherin function since deletion of the intracellular catenin-binding domain of E-cadherin results in a loss of activity even though the extracellular binding domain is intact. In breast cancer, reduced E-cadherin expression correlates with tumor size and is more frequent in invasive ductal carcinomas and poorly differentiated carcinomas.

Analysis of mutations in the PTC gene (the human homolog of *Drosophila ptc*) in the human nevoid basal cell carcinoma syndrome (NBCCS) revealed alterations in PTC in NBCCS patients and in related tumors. The reduction in expression of *ptc* may lead to the developmental abnormalities observed in the syndrome, and complete loss of PTC function may contribute to transformation of certain cell types (61).

A gene related to *Drosophila ci* downstream of WG signaling referred to as *gli* has been isolated in humans by virtue of its amplification in a human glioblastoma. It has been shown that *gli* is expressed in embryonal carcinoma cells but not in most adult tissues. This reveals an interesting similarity between *gli* gene expression in developing embryonic tissues and in neoplastic tissues (62). Alteration of the other WNT signaling components in human cancer is not yet known.

Finally, most of the *wnt* genes have been mapped and sequence analyzed, and many of the signaling components in the WNT signaling pathway have been characterized. Further research is needed to clarify how *wnt* genes fit into the complex picture of tumorigenesis and whether, because of their extracellular functions and heparin binding properties, they may be therapeutic targets.

REFERENCES

1. Nusse R, Varmus HE: Many tumours induced by mouse mammary tumor virus contain a provirus integrated in the same region of the host chromsome. Cell 31:99–109, 1982.
2. Nusse R, Brown A, Papkoff J, Scrambler P, Shackleford G, McMahon A, Moon R, Varmus HA: A new nomenclature for *int*-1 and related genes: The *Wnt* gene family. Cell 64:231–232, 1991.
3. Nusse R, Varmus HE: *Wnt* genes. Cell 69:1073–1087, 1992.
4. McMahon AP: The *Wnt* family of development regulators. Trends Genet 8:236–242, 1992.
5. Sidow A: Diversification of the *Wnt* gene family on the ancestral lineage of vertebrates. Proc Natl Acad Sci USA 89:5098–5102, 1992.
6. Burrus LW, McMahon AP: Biochemical analysis of murine Wnt proteins reveals both shared and distinct properties. Exp Cell Res 220:363–373, 1995.
7. Lako M, Strachan T, Curtis ARJ, Lindsay S: Isolation and characterization of WNT 8B, a novel *Wnt* gene that maps to 10q24. Genomics 35:386–388, 1996.
8. Wong GT, Gavin BJ, McMahon AP: Differential transformation of mammary epithelial cells by *Wnt* genes. Mol Cell Biol 14:6278–6286, 1994.

9. Bradbury JM, Niemeyer CC, Dale TC, Edwards PA: Alterations of the growth characteristics of the fibroblast cell line C3H 10T1/2 by members of the *Wnt* gene family. Oncogene 9:2597–2603, 1994.
10. Moon RT, DeMarais A, Olson DJ: Responses to *Wnt* signals in vertebrate embryos may involve changes in cell adhesion and cell movement. J Cell Sci 17:183–188, 1993.
11. Torres MA, Yang-Snyder JA, Purcell SM, DeMarais AA, McGrew LL, Moon RT: Activities of the *Wnt*-1 class of secreted signaling factors are antagonized by the *Wnt*-5A class and by a dominant negative cadherin in early *Xenopus* development. J Cell Biol 133:1123–1137, 1996.
12. Du SJ, Purcell SM, Christian JL, McGrew LL, Moon RT: Identification of distinct classes and functional domains of *Wnts* through expression of wild-type and chimeric proteins in *Xenopus* embryos. Mol Cell Biol 15:2625–2634, 1995.
13. Martinez-Arias A, Lawrence PA: Parasegments and compartments in the *Drosophilia* embryo. Nature 313:639–642, 1985.
14. Peifer M, Bejsovec A: Knowing your neighbors: Cell interactions determine intrasegmental patterning in *Drosophila*. Trends Genet 8:243–249, 1992.
15. Klingensmith J, Nusse R: Signaling by *wingless* in *Drosophila*. Dev Biol 166:396–414, 1994.
16. Siegfried E, Wilder EL, Perrimon N: Components of *wingless* signalling in *Drosophila*. Nature 367:76–79, 1994.
17. Wang Y, Macke JP, Abella BS, Andreasson K, Worley P, Gilbert DJ, Copeland NG, Jenkins NA, Nathans J: A large family of putative transmembrane receptors homologous to the product of the *Drosophila* tissue polarity gene *frizzled*. J Biol Chem 271:4468–4476, 1996.
18. Bhanot P, Brink M, Samos CH, Hsieh J-C, Wang Y, Macke JP, Andrew D, Nathans J, Nusse R: A new member of the *frizzled* family from *Drosophila* functions as a Wingless receptor. Nature 382:225–230, 1996.
19. Klingensmith J, Nusse R, Perrimon N: The *Drosophila* segment polarity gene *dishevelled* encodes a novel protein required for response to the *wingless* signal. Genes Dev 8:118–130, 1994.
20. Pizzuti A, Amati F, Calabrese G, Mari A, Colosimo A, Silani V, Giardino L, Ratti A, Penso D, Calza L, Palka G, Scarlato G, Novelli G, Dallapiccola B: cDNA characterization and chromosomal mapping of two human homologues of the *Drosophila dishevelled* polarity gene. Hum Mol Genet 5:953–958, 1996.
21. Bui TD, Beier DR, Jonsson M, Smith K, Dorrington S, Kaklamanis L, Kearney L, Regan R, Sussman DJ, Harris AL: cDNA cloning of a human *dishevelled DVL-3* gene, mapping to 3q27, and expression in human breast and colon carcinomas. Accepted by B.B. Research communications.
22. Axelrod JD, Matsuno K, Artavanis-Tsakonas S, Perrimon N: Interaction between wingless and notch signaling pathways mediated by dishevelled. Science 271:1826–1832, 1996.
23. Siegfried E, Chou T-B, Perrimon N: *wingless* signaling acts through *zeste-white 3*, the Drosophila homolog of *glycogen synthase kinase-3*, to regulate *engrailed* and establish cell fate. Cell 71:1167–1179, 1992.
24. Perrimon N: The genetic basis of patterned baldness in Drosophila. Cell 76:781–784, 1994.
25. Saito Y, Vandenheede JR, Cohen P: The mechanism by which epidermal growth factor inhibits glycogen synthase kinase 3 in A431 cells. Biochem J 303:27–31, 1994.
26. Eldar-Finkelman H, Seger R, Vandenheede JR, Krebs EG: Inactivation of glycogen synthase kinase-3 by epidermal growth factor is mediated by mitogen-activated protein kinase/p90 ribosomal protein S6 kinase signaling pathway in NIH/3T3 cells. J Biol Chem 270:987–990, 1995.

27. Peifer M, Pai L-M, Casey M: Phosphorylation of the *Drosophila* adherens junction protein armadillo: Roles for wingless signal and zeste-white 3 kinase. Dev Biol 166: 543–556, 1994.
28. Hinck L, Nelson WJ, Papkoff J: Wnt-1 modulates cell–cell adhesion in mammalian cells by stabilizing b-catenin binding to the cell adhesion protein cadherin. J Cell Biol 124:729–741, 1994.
29. Papkoff J, Rubinfeld B, Schryver B, Polakis P: Wnt-1 regulates free pools of catenins and stabilizes APC-catenin complexes. Mol Cell Biol 16:2128–2134, 1996.
30. Su L-K, Vogelstein B, Kinzler KW: Association of the APC tumor suppressor protein with catenins. Science 262:1734–1736, 1993.
31. Rubinfeld B, Albert I, Porfiri E, Fiol C, Munemitsu S, Polakis P: Binding of GSK3b to the APC-β-catenin complex and regulation of complex assembly. Science 272: 1023–1026, 1996.
32. Behrens J, von Kries JP, Kühl M, Bruhn L, Wedlich D, Grosschedl R, Birchmeier W: Functional interaction of b-catenin with the transcription factor LEF-1. Nature 382: 638–642, 1996.
33. Siegfried E, Wilder EL, Perrimon N: Components of *wingless* signalling in *Drosophila*. Nature 367:76–80, 1994.
34. Li X, Noll M: Role of the *gooseberry* gene in *Drosophila* embryos: Maintenance of *wingless* expression by a *wingless-gooseberry* autoregulatory loop. EMBO J 12: 4499–4509, 1993.
35. van den Heuvel M, Ingham PW: 'Smoothening' the path for hedgehogs. Trends Cell Biol 6:451–452, 1996.
36. Hooper JE: Distinct pathways for autocrine and paracrine Wingless signalling in *Drosophila* embryos. Nature 372:461–463, 1994.
37. Dickinson ME, McMahon AP: The role of *Wnt* genes in vertebrate development. Curr Opin Genet Dev 2:562–566, 1992.
38. Smith JC: A *wnt*er's tale. Curr Biol 2:17–21, 1992.
39. Rosen JM, Humphreys R, Krnacik S, Juo P, Raught B: The regulation of mammary gland development by hormones, growth factors, and oncogenes. Prog Clin Biol Res 387:95–111, 1994.
40. Weber-Hall SJ, Phippard DJ, Niemeyer CC, Dale TC: Developmental and hormonal regulation of *Wnt* gene expression in the mouse mammary gland. Differentiation 57: 205–214, 1994.
41. Gavin BJ, McMahon AP: Differential regulation of the *Wnt* gene family during pregnancy and lactation suggests a role in postnatal development of the mammary gland. Mol Cell Biol 12:2418–2423, 1992.
42. Huguet EL, McMahon JA, McMahon AP, Bicknell R, Harris AL: Differential expression of human *Wnt* genes 2, 3, 4, and 7B in human breast cell lines and normal and disease states of human breast tissue. Cancer Res 54:2615–2621, 1994.
43. Uyttendaele H, Marazzi G, Wu G, Yan Q, Sassoon D, Kitajewski J: *Notch4/int-3*, a mammary proto-oncogene, is an endothelial cell-specific mammalian *Notch* gene. Development 122:2251–2259, 1996.
44. Yamaguchi TP, Rossant J: Fibroblast growth factors in mammalian development. Curr Opin Genet Dev 1995, 5:485–491, 1995.
45. Burgess WH, Maciag T: The heparin-binding (fibroblast) growth factor family of proteins. Annu Rev Biochem 58:575–606, 1989.
46. Christian JL, McMahon JA, McMahon AP, Moon RT: *Xwnt-8*, a *Xenopus Wnt-1/int-1-*

related gene responsive to mesoderm inducing growth factors, may play a role in ventral mesodermal patterning during embryogenesis. Development 111:1045–1055, 1991.

47. Shackleford GM, MacArthur CA, Kwan HC, Varmus HE: Mouse mammary tumor virus infection accelerates mammary carcinogenesis in *Wnt-1* transgenic mice by insertional activation of *int-2/Fgf-3* and *hst/Fgf-4*. Proc Natl Acad Sci USA 90:740–744, 1993.
48. Ramakrishna NR, Brown MC: *Wingless,* the Drosophila homolog of the proto-oncogene *Wnt-1,* can transform mouse mammary epithelial cells. Dev Suppl 95–103, 1993.
49. Van de Vijver MJ, Nusse R: The molecular biology of breast cancer. Biochim Biophys Acta 1072:33–50, 1991.
50. MacArthur CA, Shankar DB, Shackleford GM: Fgf-8, activated by proviral insertion, cooperates with the Wnt-1 transgene in murine mammary tumorigenesis. J Virol 69:2501–2507, 1995.
51. Hardiman G, Albright S, Tsunoda J-I, McClanahan T, Lee F: The mouse *Wnt-10b* gene isolated from helper T cells is widely expressed and a possible oncogene in BR6 mouse mammary tumorigenesis. Gene 172:199–205, 1996.
52. Bui TD, Rankin J, Smith K, Huguet EL, Rubben S, Strachan T, Harris AL, Lindsay S: A novel Wnt gene, Wnt-10b, map to 12q13 is expressed in human breast cancer. Oncogene 14:1249–1253, 1997.
53. Bui TD, O'Brien T, Crew J, Cranston D, Harris AL: High expression of Wnt7b in human superficial bladder cancer versus invasive bladder cancer. Submitted.
54. Lejeune S, Huguet EL, Hamby A, Poulsom R, Harris AL: *Wnt5a* cloning, expression, and up-regulation in human primary breast cancers. Clin Cancer Res 1:215–222, 1995.
55. Iozzo RV, Eichstetter I, Danielson KG: Aberrant expression of the growth factor *Wnt-5a* in human malignancy. Cancer Res 55:3495–3499, 1995.
56. Dale TC, Weber-Hall SJ, Smith K, Huguet EL, Jayatilake H, Gusterson BA, Shuttleworth G, O'Hare M, Harris AL: Compartment switching of *WNT-2* expression in human breast tumors. Cancer Res 56:4320–4323, 1996.
57. Oyama T, Kanai Y, Ochiai A, Akimoto S, Oda T, Yanagihara K, Nagafuchi A, Tsukita S, Shibamoto S, Ito F, Takeichi M, Matsuda H, Hirohashi S: A truncated b-catenin disrupts the interaction between E-cadherin and a-catenin: A cause of loss of intercellular adhesiveness in human cancer cell lines. Cancer Res 54:6282–6287, 1994.
58. Smith KJ, Levy DB, Maupin P, Pollard TD, Vogelstein B, Kinzler KW: Wild type but not mutant APC associates with the microtubule cytoskeleton. Cancer Res 54: 3672–3675, 1994.
59. Miller JR, Moon RT: Signal transduction through b-catenin and specification of cell fate during embryogenesis. Genes Dev 10:2527–2539, 1996.
60. Hinck L, Näthke IS, Papkoff J, Nelson WJ: b-Catenin: A common target for the regulation of cell adhesion by Wnt-1 and Src signaling pathways. Trends Biochem Sci 19:538–542, 1994.
61. Hahn H, Wicking C, Zaphiropoulos PG, Gailani MR, Shanley S, Chidambaram A, Vorechovsky I, Holmberg E, Unden AB, Gillies S, Negus K, Smyth I, Pressman C, Leffell DJ, Gerrard B, Goldstein AM, Dean M, Toftgard R, Chenevix Trench G, Wainwright B, Bale AE: Mutations of the human homolog of Drosophila *patched* in the nevoid basal cell carcinoma syndrome. Cell 85:841–851, 1996.
62. Kinzler KW, Ruppert JM, Bigner SH, Vogelstein B: The *GLI* gene is a member of the *Kruppel* family of zinc finger proteins. Nature 332:371–374, 1988.

21

INT3, A NOVEL-NOTCH-RELATED GENE IN MAMMARY GLAND DEVELOPMENT AND NEOPLASIA

EDITH C. KORDON AND GILBERT H. SMITH

Laboratory of Tumor Immunology and Biology, National Cancer Institute, National Institutes of Health, Bethesda, MD 20892

21.1. INTRODUCTION

The mouse mammary gland has proven to be a useful model to study both normal development and neoplastic transformation. The role of growth factors and hormones in these processes has been well characterized (1). The proviral DNA of the mouse mammary tumor virus (MMTV) induces neoplastic transformation in acutely infected mouse mammary glands by acting as an insertional mutagen (2). Premalignant lesions and mammary tumors develop as a result via the clonal expansion of mutated mammary epithelial cells that contain one or more MMTV DNA insertions. Genetic loci that are frequently affected in mouse mammary tumors by insertion of MMTV DNA and that are commonly found in the clonally dominant mammary tumor cell populations are called INT genes. Several INT genes have been cloned and characterized at the molecular and biological level (2–4). Previously, our laboratory described and characterized a common MMTV insertion locus designated *INT3* (5,6).

Hormones and Growth Factors in Development and Neoplasia, Edited by Robert B. Dickson and David S. Salomon
ISBN 0-471-16899-8

21.2. DISCOVERY OF *INT3*

The *INT3* gene was originally identified as a common insertion site for infectious MMTV provirus in five independent spontaneous mammary tumors arising in Czech II mice. Restriction enzyme mapping of the MMTV integration sites within *INT3* in these tumors demonstrated that they all were within a 500-bp region of the cellular genome and in the same transcriptional orientation. One consequence of this integration event is the transcription of a 2.4-kb species RNA, corresponding to genomic sequences 3′ of the viral integration site (5) (Fig. 21.1). Subsequently, a flanking genomic DNA fragment was used to probe an *INT3* rearranged mammary tumor cDNA library. The sequence analysis of the cDNA clones obtained by this method revealed that the *INT3*-activated message is initiated within the 3′ MMTV long terminal repeat (LTR) sequence. The 5′ of the cDNA corresponded to the U5 region of the viral LTR sequence, while the remainder of the cDNA represented eight exons located within 4.8 kbp of cellular DNA sequences 3′ of the viral insertion site (6) (Fig. 21.1). This finding demonstrated that while the primary mechanisms of activation of the other INT genes by MMTV is enhancer insertion, *INT3* expression is activated by promoter insertion. The nucleotide sequence analysis demonstrated that the activated *INT3* cDNA sequence was most similar to the region of the *NOTCH* gene from *Drosophila melanogaster, Xenopus laevis,* rats, and humans (57–63% identity), which encodes the intracytoplasmic domain of the protein (7–10). This novel *INT3* transcript encodes the intracellular domain of the protein characterized by six tandem repeats of a 33-amino acid repeat sequence. This repeat motif was first recognized in yeast *cdc-10/SW16* cell cycle transcriptional regulatory proteins (38% and 36% identical, respectively) (11–13).

21.3. *INT3* AS A MEMBER OF THE *NOTCH* GENE FAMILY

Using the activated *INT3* cDNA as a probe in Northern blot analysis, the expression of the complete (not rearranged) *INT3* gene was found in mouse lung, heart, skeletal muscle, testis, and brain. The detected band corresponded to a 6.5-kb RNA species. Significantly lower steady-state levels were detected in the liver and in normal mammary gland tissue from virgin, pregnant, and lactating mice by a reverse transcription-polymerase chain reaction (RT-PCR) assay (14).

Recently, (14,15) the nucleotide sequence of the 6.5 kb-*INT3* RNA transcript has been completed. A comparison of the entire amino acid sequence of *INT3* with the sequence of *Drosophila NOTCH* (N) another rodent N protein, showed that they are overall 58% and 60% similar respectively, whereas the other rodent N gene products are more similar to one another (69–70%) and to *Drosophila* N (62–63%) (14). This is consistent with *INT3* being a distantly related member of the N gene family, and therefore, it was renamed *NOTCH-4/INT3*. The *NOTCH-4/INT3* gene spans 24 kb of genomic DNA and is composed of 27 exons/introns (14). It is located on chromosome 17 (16) in the mouse genome and on chromosome 6p21.3 at the junction of class II and class III genes of the major histocompatibility complex (MHC) locus in the human genome (17).

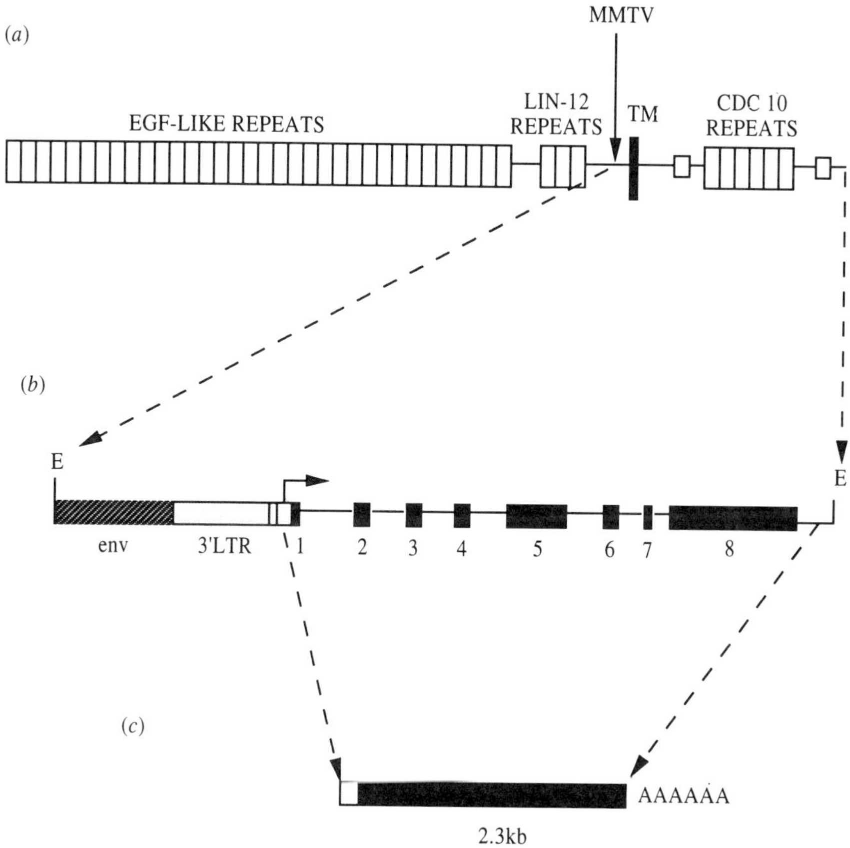

Fig. 21.1. (a) Organization of the *INT3/NOTCH-4* genomic sequences. The open rectangles correspond to the EGFL repeats, LIN-12 repeats, and CDC10 repeats. The solid rectangle corresponds to the transmembrane domain. (b) Representation of the 7.5-kb *Eco*RI-to *Eco*RI fragment cloned from an *INT3* rearranged tumor (5). This clone contains partial MMTV sequences from the 3′ side of the integrated virus, along with the flanking cellular sequences that encode the *INT3* tumor transcript. The black boxes represents the eight exons inferred from the cDNA and genomic DNA sequences. The black arrow indicates the start of transcription. (c) The bottom line shows the 2.3-kb *INT3* truncated message found in the *INT3* rearranged tumors. The stippled box at the 5′ end of the activated *INT3* RNA represents the portion of the U5 region of the MMTV LTR that is also transcribed.

Members of the N gene family are known to participate in cell fate determination, cellular differentiation, and transcriptional regulation. The *Drosophila* N gene, its vertebrate homologs, and the related LINL2 and GLP-1 genes of *Caenorhabditis elegans* encode a transmembrane receptor. These proteins are the starting point of a signaling pathway that, through local cell interaction, plays a pivotal role in cell fate determination during development (reviewed in ref. 18). N is widely expressed in the *Drosophila* embryo and continues to be expressed in uncommitted and proliferating cell populations at later stages of development. *Drosophila* N mutants showed defective formation of the nervous system, mesoderm, germ line, ovarian follicle cells, adult sensory bristles, and eye structure, demonstrating the relevance of the protein during development. It is postulated that N may affect the developmental programs in these complex tissues through either a process termed *lateral inhibition or specification* between developmentally equivalent cells or *inductive signaling* between nonequivalent cells (18,19). In the former case, stochastic fluctuations in components of a signaling pathway within a cell or group of cells induce differences to arise between adjacent, otherwise equivalent cells. These differences are subsequently enhanced by local inhibitor signals and result in a spaced pattern of two different cell types. In the latter case, the pattern of cell fate determination is determined by inductive signaling between nonequivalent cells. This process depends on timing, spatial arrangement, and specificity of the signals between the two cell types.

The N protein is composed of an extracellular domain containing 36 tandem epidermal growth factor (EGF)-like repeats and 3 cysteine-rich LIN12 repeats; a transmembrane domain; and a intracellular domain containing 6 tandem CDC10/ankyrin repeats, an Opa sequence, and a PEST sequence. Similar motifs were found in *INT3/NOTCH-4* (Fig. 21.1), although it has several novel characteristiques that distinguish it from other members of this gene family. The extracellular domain of *INT3/NOTCH-4* contains 29 instead of the 36 EGF-like repeats found in *NOTCH-!*. Five novel EGF-like repeats have been created as consequence of apparent small deletions that have occurred within the coding region for the extracellular domain of the original *NOTCH* gene during evolution (14).

The N signaling pathway seems to be quite unlike the classical ones, which involve cascades of protein kinases (18,20–22). The molecular models of the N signaling pathway suggest that in the absence of a ligand–receptor interaction, the N transmembrane receptor exists as monomer on the cell surface. In this condition, the Suppressor of Hairless (Su(H)) protein is bound to the CDC10/ankyrin repeats of the N intracellular domain. The Su(H) protein is closely related to the KBF-2/RBP-Jk family of mammalian DNA binding transcription factors (23,24). Activation of N function is initiated with the physical interaction of its extracellular domain with either of its two known ligands, DELTA (D1) and SERRATE (Ser). In *Drosophila* these ligands play equivalent but complementary roles in compartment-specific signaling for wing margin development (25). The presence of EGF repeats 10 and 11 in the N extracellular domain is both necessary and sufficient for ligand binding to occur. It is postulated that this interaction results in receptor multimerization, a process that is necessary for N activation and the translocation of Su(H) protein to the nucleus. Based on studies of the role murine N plays in myogenesis, Jarriault et al. (22) and Kopan et al. (21) postulate that for nuclear translocation to occur, the activated N intracellular domain–Su(H) complex must be proteolytically processed to release a protein complex capable of nuclear localization. Genetic studies in *Drosophila* have showed that the Enhancer of Split complex [E(spl)–C], a genetic locus composed of eight genes, is the target of N/Su(H) action. Bailey and

Posakony (20) have established that *Drosophila* Su(H) activates transcription of genes within E(spl)–C. in response to N activation and that three of these genes contain multiple specific binding sites for Su(H). E(spl)–C encodes seven functionally redundant basic helix-loop-helix (bHLH) proteins that are transcriptional repressors and another nuclear protein having WD-40 repeats. Upregulation of E(spl)–C suppresses the expression of the achaete–scute gene complex (AS–C), which encodes tissue-specific transcription factors involved in neuroblastic differentiation, indicating once again that the role of N in the developmental process is to inhibit differentiation. In addition, Heitzler et al. (26) have provided evidence for a regulatory loop between N and D1 that is under the transcriptional control of E(spl)–C and AS–C, such that within a cell or group of cells, activation of N leads to suppression of D1 expression.

21.4. *INT3/NOTCH-4* INVOLVEMENT IN TUMORIGENESIS

The *INT3/NOTCH-4* rearranged mouse mammary tumors discovered in our lab offered an excellent opportunity to study the possible role of the N genes in tumorigenesis and differentiation. To ascertain whether the position of the target region within the *INT3/NOTCH-4* gene could provide insights into the apparently strong selection for viral integration and tumorigenesis, the nucleotide sequence of the host-virus junction was determined in nine independent *INT3* rearranged MMTV-induced mammary tumors. In eight of these tumors viral integration occurred within the 177 bp of exon 22. Furthermore, in six of the tumors, the viral integration site corresponded to either bp 4297 or 4318 of *INT3-NOTCH-4*. In the ninth MMTV-induced tumor, viral insertion occurred at bp 4186 in exon 21 (14). We have previously reported a mouse mammary tumor in which NOTCH-4/*INT3* is activated by the integration of an intracisternal type A particle (IAP) genome (27). Nucleotide sequence analysis of the IAP-host junction showed that the insertion occurred at bp 4418 in exon 23. The region of *INT3/NOTCH-4* RNA from 4186 to 4418 encodes the portion of the extracellular domain of the protein that lies between the LIN12 repeats and the transmembrane domain. Thus, analysis of the insertion region within *INT3/NOTCH-4* in all 10 of the spontaneous mammary tumors studied reveals that the gene product present in the transformed cells would be composed of a few residues of the extracellular domain, the transmembrane domain, and the intracellular domain but would not be inserted in the membrane for lack of signal peptide sequence (14).

Therefore, the only tumorigenic *INT3/NOTCH-4* mutations are those that result in the expression of the intracellular domain under the control of the proviral promoter. This strongly suggests that MMTV-induced activation of *INT3/NOTCH-4* necessitates the absence of the regulatory action of the extracellular domain, including the LIN12 repeat sequences, leaving the expressed intracellular domain free to function in its role in mammary tumorigenesis. Consistent with this conclusion, expression of the tumor-specific *INT3* genomic DNA fragment encoding the 2.3-kb RNA species in HC11 mouse mammary epithelial cells *in vitro* induces anchorage-independent growth in soft agar (28).

21.5. *MMTV-LTR/INT3* TRANSGENIC MICE

To determine the *in vivo* consequences of activated (truncated) *INT3* expression, transgenic mice were generated harboring a mammary tumor derived-genomic DNA fragment composed of the 3′ MMTV-LTR and flanking downstream *INT3* cellular sequences. All six *MMTV-LTR/INT3* founders and the progeny of one established line of transgenic mice (called FVB-3) bearing this genomic tumor DNA fragment developed similar phenotypic abnormalities in tissues expressing the transgene. The affected tissues included the mammary and salivary glands, the epithelial glands of the nasal mucosa and salivary sinus, the extraorbital lacrimal glands, and the Harderian glands (29). All of these tissues contained proliferating immature ductule epithelia and were incompletely differentiated. All male *MMTV-LTR/INT3* transgenic mice were sterile, apparently due to severe hyperplasia of the epididymal epithelium. In addition, the females were lactation deficient and were unable to suckle their litters successfully.

The *in vivo* expression of the truncated *INT3* induced multiple dramatic hyperproliferative abnormalities; the transgenic females frequently exhibited intraductal hyperplasia of mammary epithelium and development of poorly differentiated adenocarcinomas of the mammary gland. Focal and often multiple, poorly differentiated mammary and salivary adenocarcinomas appeared in the majority of transgenic mice between 2 and 7 months of age. Finally, all female mice containing this transgene developed mammary tumors (29,30). These findings strongly suggest that the virus-induced expression of the intracellular domain of *INT3/NOTCH-4* is a seminal contributory factor in the development of those Czech II mouse mammary tumors in which the rearrangement of the *INT3* locus was originally observed (5).

Besides confirming the involvement of truncated *INT3/NOTCH-4* expression in mammary tumorigenesis, these transgenic mice showed that this gene might convey developmental information. This was shown by the fact that both simple and compound exocrine glands expressing the activated *INT3* transgene appear to be incompletely differentiated and possess proliferative, immature ductal epithelial cells (29). This effect is especially noticeable in the mammary glands of *INT3* transgenic mice.

21.6. EFFECT OF THE EXPRESSION OF THE TRUNCATED *INT3/NOTCH-4* DURING THE DEVELOPMENT OF THE MOUSE MAMMARY GLAND

Growth and development of the mammary gland in the mouse occurs primarily after birth and progresses through defined stages with the attainment of puberty and the onset of pregnancy, leading ultimately to the functional purpose of the gland: lactation (31). At birth, female mice possess rudimentary mammary glands composed of a simple system of branching tubes opening at the nipple. At approximately 3 weeks of age, correlating with the onset of puberty, the immature gland begins to grow rapidly due to intense mitotic activity that occurs within the end buds. Within the end buds are undifferentiated cap or stem cells that give rise to myoepithelial and parenchymal cells. By 12 weeks of age in virgin females, the fat pad is filled with a tree-like duct structure and the end buds disappear. Despite the absence of end buds and cap cells, any portion of the mammary gland is capable of regenerating a complete, functional epithelial tree when transplanted, indicating the presence of functional stem cells throughout the gland (31,32). During pregnancy, the secretory alveolar lobules develop upon the structure of branching ducts until they completely fill the fatty stroma. This lobuloalveolar architecture persists until the end of lactation, at which time there is massive apoptosis leading to mammary gland involution.

In the transgenic *MMTV-LTR-INT3* (FVB-3) mice, the mammary ducts do not grow normally and fail to fill the mammary fat pads completely in virgin transgenic females. FVB-3 virgin glands were unable to form and maintain active terminal end buds, although there was considerable evidence of mitotic activity in the terminal regions of the ducts. At least one possible reason for the disability may reside in the absence of intercellular junctional complexes, as revealed by electron microscopy (30). Indeed, the "absence of specialized cell junctions or adhesions and intercellular spaces frequently observed at the cell's lateral borders" are features that distinguish "cap cells" from the luminal epithelial cells of the subtending ducts and the end bud body cells in the normal immature gland (33). The majority of the epithelial cells observed within the FVB-3 transgenic mammary ducts possess ultrastructural features characteristic of cap cells. This suggests that FVB-3 mammary ductal cells can assume some of the characteristics of cap cells but are, for some reason, unable to complete the developmental transition necessary to form and keep functional end buds in the virgin female. Curiously, ductal growth and penetration of all FVB-3 mammary fat pads occurred during the first pregnancy, suggesting that hormonally regulated mechanisms might account for the inability of ductal extension in the virgin. Nevertheless, despite the ductal proliferation observed in the pregnant gland, functional lobular development was absent and the FVB-3 female mice were unable to lactate (29,30).

21.7. *WAP-INT3* TRANSGENIC MICE

To determine whether the lobular developmental defect was associated with early events correlating with the early expression of the transgene or was a consequence of transgene expression during lobular development, we developed transgenic mice expressing the intracellular region of *INT3* under the transcriptional control of the whey acid protein *(WAP)* gene promoter, which is specifically active in differentiating secretory mammary epithelium.

Expression from the WAP promoter is tightly restricted to the mammary epithelium (28,34,35) and, more specifically, to the differentiating secretory epithelial cells in the developing gland of pregnant females. In virgin female mammary glands, scattered individual ductal luminal epithelial cells transiently show WAP expression at estrus (6,36). As a result of WAP promoter specificity, the only organ system affected by transgene expression is the mammary gland. Therefore, *WAP/INT3* males are fertile and tumor free, and females show transgene-induced effects only in their mammary epithelium, in contrast to the multiple tissue consequences seen in the *MMTV-LTR/INT3* mice.

In addition to the absence of multiple organ defects, there were several striking differences between the *WAP/INT3* and *MMTV-LTR/INT3* transgenic model systems with respect to mammary gland epithelial growth and development. A full, completely branched mammary ductal system developed in virgin *WAP/INT3* females, and the *WAP/INT3* ductal epithelium formed the appropriate intercellular connections necessary for a proper ductal morphology. Neither of these events occurred in the virgin *MMTV-LTR/INT3* transgenic mouse mammary glands (30). However, in both models, expression of the intracellular portion of *INT3* severely depressed secretory lobular development during pregnancy, preventing the attainment of a lactational phenotype. The normal development of ducts in *WAP/INT3* mice suggests that the transgene is not expressed in the nonsecretory ductal epithelial progenitor cells. In agreement with the restriction of WAP expression to mammary secretory epithelial progenitors, the mammary ducts of *WAP/INT3* males remain tumor free, in contrast to those in *MMTV-LTR/INT3* males. *WAP/INT3* and *MMTV-LTR/INT3* transgenic mouse lines are therefore appropriate for distinguishing the effects of *INT3* on different epithelial cell subsets in the differentiating mammary gland.

Coincident with the limited lobular development of the *WAP/INT3* mammary gland during pregnancy, dysplastic lesions appear throughout the gland. These lesions do not regress after weaning but progress to frank carcinoma. Although mammary tumors also appear in 100% of nonbreeding virgin *WAP/INT3* females, the latent period for reaching 50% tumor incidence is more than double that required for the same incidence in breeders. In contrast, there is no significant difference in age of tumor onset between virgin and parous *MMTV-LTR/INT3* females (Smith G.H., unpublished data). This is partly due to the difference in the temporal pattern of transgene expression between the two lines. In addition, variations in the morphology of the tumors and their gene expression suggest that *WAP/INT3* and *MMTV-LTR/INT3* neoplasias arise from distinctly different mammary epithelial cell populations (Smith G.H., unpublished data).

21.8. ROLE OF *INT3/NOTCH-4* IN NORMAL MAMMARY GLAND DEVELOPMENT

As previously mentioned, there are two distinct patterns of growth and development in the mouse mammary gland. The first involves the penetration of the mammary fat pad by branching ductal morphogenesis and is under the control of ovarian hormones and local growth factors (37). The second involves the growth and expansion of the secretory lobules at the onset of pregnancy. This is dependent on the hormonal status induced by pregnancy and continues until parturition. Recently, we designed a model for the existence and behavior of mammary epithelial stem cells and their developmentally committed proliferative daughters in the mouse mammary gland. Beginning in the embryo and continuing throughout the life span of the adult, mammary epithelium-committed stem cells are present within the glandular population. These cells divide and give rise to daughters with specific developmental potentials and a limited proliferative life span. The stem cell divisions are unequal in the sense that one of the two cells produced by the mitotic event remains a stem cell, whereas the other becomes a progenitor cell with a specific differentiative program and a transient proliferative capacity. During the early development of the prepubertal gland the stem cell divides, producing a ductal progenitor(s) programmed for terminal end bud formation, ductal extension, and myoepithelial cell development (31). At the onset of puberty, the ovarian hormones activate local growth promoters in the fatty stroma and mammary epithelial-associated stroma (38,39), leading to increased activity of ductal extension. During this period, some of the stem cell progeny receive signals for lobular, i.e., secretory, differentiative programs. This activity occurs primarily in the epithelium of the ducts subtending the terminal end buds during the peak of estrus. We have reported evidence of transient milk protein production and secretory-specific gene promoter activity in the individual ductal epithelial cells during this point in the ovarian cycle (36) and during pregnancy. Therefore, each of these periods of expansion (postpuberal and pregnancy) involves the production of epithelial progenitors committed to a specific developmental lineage (i.e., ducts and lobules) that are responsive to specific inductive and growth signals (40).

In the *MMTV-LTR/INT3* transgenic line, the transgene is expressed very early during development. the fact that even male FVB-3s had mammary adenocarcinomas indicated that the transgene was expressed in the male embryo before the involution of the mammary gland, which occurs at day 14 of gestation. Therefore, we hypothesize that during the development of the mammary epithelium, both types of progenitors and their progeny express the *MMTV-LTR/INT3* and limit the capacity of each to fulfill its differentiative program. In contrast, in the *WAP/INT3* line, the expression of the transgene is restricted to the pregnancy-induced secretory progenitor, which is responsible for the fabrication of the alveolar structures. Presumably, the expression of the truncated *INT3/NOTCH-4* leads to a gain in function that results in the attenuation of functional differentiation in the lobular progenitor cells and their daughters. Consistent with this conclusion, full-term pregnant *WAP/INT3* glands possess lobular epithelium that is strongly positive for casein gene expression, an early differentiation marker in mammary secretory differentiation (28), but are negative for WAP, a protein expressed during terminal differentiation of the lactogenic phenotype. Similarly, *WAP/INT3* glands with mammary dysplasia are positive for casein expression and negative for WAP in postpartum females. In contrast, *MMTV-LTR/INT3* mammary tumors are negative for the expression of either of these markers, which are specific for the lactational phenotype. This suggests that the cells giving rise to *WAP/INT3* mammary tumors originate from the secretory epithelial cell progenitors, whereas the *MMTV-LTR/INT3* tumors appear early and arise from the ductal progenitor cells or from the individual mammary epithelial stem cells. Evidence for the latter conclusion comes from the observation that

MMTV-LTR/INT3 glands are not able to grow and fill epithelium-divested mammary fat pads when transplanted to normal virgin hosts, suggesting that the clonogenic mammary stem cells are nonfunctional in *MMTV-LTR/INT3* mammary transplants (30). In contrast, *WAP/INT3* mammary transplants are successful in reconstituting a full mammary ductal system in more than 88% of the normal gland-free mammary fat pads implanted.

Antibodies against a portion of the CDC10 region and the pest region of the intracellular domain were used to localize the *INT3/NOTCH-4* protein in the differen-

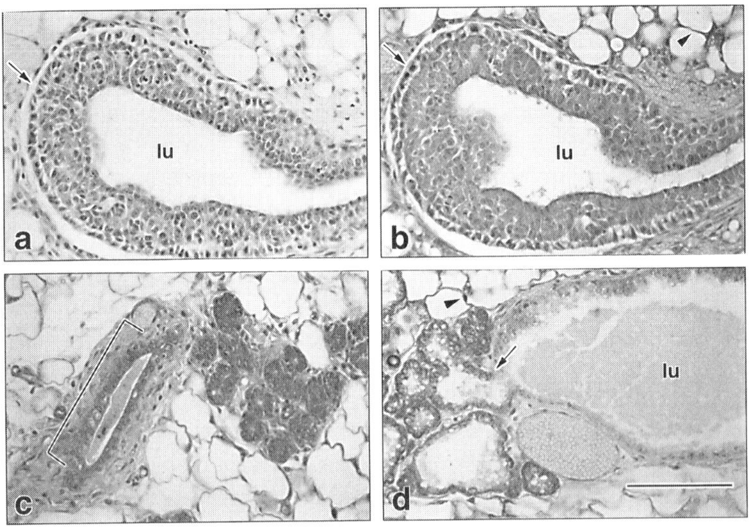

Fig. 21.2. Immunoperoxidase staining (dark brown) of growing and differentiating normal mouse mammary gland with anti-*INT3/Notch-4* shows that the protein is present in many cell types in mammary tissue. (a) A growing terminal end bud in a 4-week-old virgin female stained with anti-*INT3/Notch-4* after incubation of the antibody with an excess of the immunogenic *INT3*-specific synthetic peptide showing that *INT3/Notch-4* staining is completely suppressed by preincubation with the immunizing epitope. All sections were counterstained with hematoxylin (light blue). (b) A consecutive section containing the same terminal end bud shown in (a) stained with anti-*INT3/Notch-4* after incubation with an irrelevant *INT3/Notch-4* peptide. Arrows in (a) and (b) indicate the thin layer of "cap cells" typical of the rapidly growing terminal bud. These are stained strongly for *INT3*, both in the cytoplasm and sometimes in the nucleus as well. In addition, the cytoplasm of the adipocytes of the surrounding fat pad (arrowhead) and the luminal surface of the epithelial cell lining the ductal lumen (lu) are strongly stained. Moderate staining was also observed in the cytoplasm of the epithelial cells in the body of the growing end bud. (c) In early pregnancy, many of the epithelial cells of the duct (bracket), particularly those in the developing secretory acini (right center), show strong cytoplasmic staining. (d) At lactation, the ductal epithelium becomes negative for *INT3/Notch-4*-staining; only the cells of the secretory lobules are positive (the arrow indicates the transitional region between the duct and a secretory acinus). Staining in the secretory lobules is particularly strong in the basolateral portions of the epithelial cytoplasm; no nuclear staining is evident in the epithelium, although it is present in adipocyte nuclei (arrowhead). The bar equals 100 μm. All figures are at the same magnification.

tiating normal mammary epithelium by immunohistochemistry. These studies demonstrated that *INT3/NOTCH-4* expression is strongly associated with the cap cells at the tip of the proliferating end buds of developing ducts (30) (Fig. 21.2a,b). During early pregnancy, the immunoreactivity shifts to the apical cytoplasm of the mammary epithelium of the proliferating alveolar buds, suggesting an intimate association of *INT3/NOTCH-4* with secretory mammary epithelial growth and development (Fig. 21.2c). This conclusion is strengthened in the fully lactating gland, where immunoreactive *INT3/NOTCH-4* is relocated to a basal position in the secretory epithelium of functional lobules and disappears entirely from the mammary ductal epithelium (Fig. 21.2d). The shift in immunoreactive *INT3/NOTCH-4* from a luminal to a basal position in secretory mammary epithelium at lactational maturity probably reflects a dynamic change in the functional activity of this protein, induced either intrinsically or by an accompanying alteration in the three-dimensional pattern of *INT3/NOTCH-4*-specific ligand localization.

In our transgenic models, a gain-of-function mutation of *INT3* has a pronounced effect on the capacity for functional differentiation of secretory mammary epithelium. The result of a gain of function of this effect in our transgenic models is mammary epithelial dysplasia, absence of functional differentiation, and tumorigenesis. This suggests that *INT3* normally functions in the presence of its ligand to maintain mammary epithelial cells in a proliferatively responsive state by limiting their response to differentiative environmental signals. The corollary, i.e., the absence of *INT3/NOTCH-4* activity, is currently under investigation in a null *INT3/NOTCH-4* mouse knockout model. It is expected that this model will provide further clues to the role of this protein in mammary gland development and the subsequent acquisition of functional secretory cell differentiation.

REFERENCES

1. Callahan R, Salomon DS: Oncogenes, tumor suppresser genes and growth factors in breast cancer. Novel targets for diagnosis, prognosis, and therapy. Cancer Surv 18:35–56, 1993.
2. Varmus HE: Recent evidence of oncogenesis by insertional mutagenesis. Cancer Surv 1:309–320, 1982.
3. Dickson C, Smith R, Brookes S, Peters G: Tumorigenesis by mouse mammary tumor virus: Proviral activation of a cellular gene in the common integration region int-2. Cell 37:529–536, 1984.
4. Nusse R: The int genes in mammary tumorigenesis and in normal development. Trends Genet 4:291–295, 1988.
5. Gallahan D, Callahan R: Mammary tumorigenesis in feral mice: Identification of a new int locus in mouse mammary tumor virus (CzechII)-induced mammary tumors. J Virol 61:66–74, 1987.
6. Robbins J, Blondel BJ, Gallahan D, Callahan R: Mouse mammary tumor gene Int-3: A member of the Notch gene family transforms mammary epithelial cells. J Virol 66: 2594–2599, 1992.

7. Coffman C, Harris W, Kintner C: Xotch, the Xenopus homolog of Drosophila Notch. Science 249:1438–1441, 1990.
8. Ellison LW, Bird J, West DC, Soreng AL, Reynolds TC, Smith SD, Sklar J: TAN-1, the human homolog of the Drosophila Notch gene, is broken by chromosomal translocations in T lymphoblastic neoplasms. Cell 66:649–661, 1991.
9. Weinmaster G, Roberts VJ, Lemke G: A homolog of Drosophila Notch expressed during mammalian development. Development 113:199–205, 1991.
10. Wharton KA, Johansen KM, Xu T, Artavanis-Tsakonas S: Nucleotide sequence from the neurogenic locus Notch implies a gene product that shares homology with proteins containing EGF-like repeats. Cell 43:567–581, 1985.
11. Aves SJ, Durckacz BW, Carr A, Nurse P: Cloning, sequencing and transcriptional control of the Schizosaccharomyces pombe cdc10 'Start' gene. EMBO J 4:457–463, 1985.
12. Breeden L, Nasmyth K: Similarity between cell-cycle genes of budding yeast and fission yeast and the Notch gene of Drosophila. Nature 329:651–654, 1987.
13. Breeden L, Nasmyth K: Cell cycle control of the yeast HO gene: Cis and trans-acting regulators. Cell 48:389–397, 1987.
14. Galalhan D, Callahan R: The mouse mammary tumor gene Int3 is a truncated member of the Notch gene family (Notch-4). Oncogene 14:1883–1890, 1997.
15. Uyttendaele H, Marazzi G, Wu G, Yan Q, Sassoon D, Kitajewski J: Notch4/Int3, a mammary proto-oncogene, is an endothelial cell specific mammalian Notch gene. Development 122:2251–2259, 1996.
16. Gallahan D, Kozak C, Callahan R: A new common integration region (int-3) for mouse mammary tumor virus on mouse chromosome 17. J Virol 61:218–220, 1987.
17. Sugaya K, Fukagawa T, Matsumoto K, Mita K, Takahashi E, Ando A, Inoko H, Ikemura T: Three genes in the MHC class III region near the junction with the class II: Gene for receptor of advanced glycosylation end products, PBX2 homeobox gene and a NOTCH homolog, human counterpart of mouse mammary tumor gene int-3. Genomics 23:408–419, 1994.
18. Artavanis-Taskonas S, Matsuno K, Fortini ME: NOTCH signaling. Science 268: 225–232, 1995.
19. Wilkinson HA, Fitzgerald K, Greenwald I: Reciprocal changes in expression of lin-2 and its ligand lag-2 prior to commitment in a C. elegans cell fate decision. Cell 79: 1187–1198, 1994.
20. Bailey AM, Posakony JW: Suppressor of Hairless directly activates transcription of Enhancer of Slit Complex genes in response to NOTCH receptor activity. Genes Dev 9: 2609–2622, 1995.
21. Kopan R, Schroeter EH, Weintraub H, Nye JS: Signal transduction by activated MNOTCH: Importance of proteolytic processing and its regulation by the extracellular domain. Proc Natl Acad Sci USA 93:1683–1688, 1996.
22. Jarriault S, Brou C, Logeat F, Schroeter EH, Kopan R, Israel A: Signaling downstream of activated mammalian NOTCH. Nature 377:355–358, 1995.
23. Schweisguth F, Posakony JW: Suppressor of hairless, the Drosophila homolog of the mouse recombination signal-binding protein gene, controls sensory organ cell fate. Cell 69:1199–1212, 1992.
24. Furukawa T, Maruyama S, Kawaichi M: The Drosophila homolog of the immunoglobulin recombination signal-binding protein regulates peripheral nervous system development. Cell 69:1191–1197, 1992.

25. Doherty D, Feger G, Younger-Shepherd S, Jan LY, Jan YN: Delta is a ventral to dorsal signal complementary to Serrate, another NOTCH ligand, in Drosophila wing formation. Genes Dev 10:421–434, 1996.
26. Heitzler P, Bourouis M, Ruel L, Carteret C, Simpson P: Genes of the Enhancer of Split and achaete-scute complexes are required for a regulatory loop between NOTCH and DELTA during lateral signaling in Drosophila. Development 122:161–171, 1996.
27. Kordon E, Smith GH, Callahan R, Gallahan D: A novel non-MMTV activation of the Int-3 gene in a spontaneous mouse mammary tumor. J Virol 69:8066–8069, 1995.
28. Robinson GW, McKnight RA, Smith GH, Hennighausen L: Mammary epithelial cells undergo secretory differentiation in cycling virgins but require pregnancy for the establishment of terminal differentiation. Development 121:2079–2090, 1995.
29. Jhappan C, Gallahan D, Stahle C, Chu E, Smith GH, Merlino GT, Callahan R: Expression of an activated Notch-related int-3 transgene interferes with cell differentiation and induces neoplastic transformation in mammary and salivary glands. Genes Dev 6:345–355, 1992.
30. Smith G, Gallahan D, Diella F, Jhappan C, Merlino G, Callahan R: Constitutive expression of truncated Int3 gene in mouse mammary epithelium impairs differentiation and functional development. Cell Growth Differ 6:563–577, 1995.
31. Medina D, Smith GH: In search of mammary gland stem cells. Protoplasms 159:77–84, 1990.
32. Smith G, Medina D: A morphologically distinct candidate for an epithelial stem cell in mouse mammary gland. J Cell Sci 89:173–183, 1988.
33. Williams JM, Daniel CW: Mammary ductal elongation: Differentiation of myoepithelium and basal lamina during branching morphogenesis. Dev Biol 97:274–290, 1983.
34. Pittius C, Hennighausen L, Lee E, Westphal H, Nicols E, Vitale J, Gordon K: A milk protein gene promoter directs the expression of human tissue plasminogen activator cDNA to the mammary gland in transgenic mice. Proc Natl Acad Sci USA 85:5874–5878, 1988.
35. Hennighausen L, McKnight R, Burdon T, Baik M, Wall RJ, Smith GH: Whey acid protein extrinsically expressed from the mouse mammary tumor virus long terminal repeat results in hyperplasia of the coagulation gland epithelium and impaired mammary development. Cell Growth Differ 5:607–613, 1994.
36. Kordon EC, McKnight R, Jhappan C, Hennighausen L, Merlino G, Smith GH: Ectopic TGFβ1 expression in the secretory mammary epithelium induces early senescence of the epithelial stem cell population. Dev Biol 168:47–61, 1995.
37. Daniel CW, Silberstein GB: Postnatal development of the rodent mammary gland. In Neville MC, Daniel CW (eds): The Mammary Gland: Development, Regulation and Function. New York: Plenum Press, 1987, pp 3–36.
38. Sakakura T, Nishizuka Y, Dawe CJ: Mesenchyme-dependent morphogenesis and epithelium-specific cytodifferentiation in mouse mammary gland. Science 194:1439–1441, 1976.
39. Sakakura T, Kusano I, Kusakabe M, Inaguma Y, Nishizuka Y: Biology of mammary fat pad in fetal mouse: Capacity to support development of various fetal epithelia in vivo. Development 100:421–430, 1987.
40. Smith GH: Experimental mammary epithelial morphogenesis in an in vivo model: Evidence for distinct cellular progenitors of the ductal and lobular phenotype. Breast Cancer Res Treat 39:21–31, 1996.

22

TYROSINE KINASES AND SIGNAL TRANSDUCTION IN MOUSE MAMMARY TUMORIGENESIS

PETER M. SIEGEL AND WILLIAM J. MULLER

Departments of Biology (P.M.S.) and Pathology (W.J.M.), Institute for Molecular Biology and Biotechnology, McMaster University, Hamilton, Ontario L8S 4K1, Canada

22.1. INTRODUCTION

The view that cancer is the product of a series of mutations occurring within a single progenitor cell is now well established. The accumulation of genetic damage serves to dismantle or override control mechanisms that govern normal cellular proliferation. Breast cancer is one disease in which significant efforts have been made to reveal the differences between normal and neoplastic cells. Understanding the molecular requirements for transformation is crucial if attempts are to be made to impede or prevent this process. Although clinical studies have indicated that overexpression of certain oncogenes may be involved in tumor formation, a causal relationship cannot be established from such data. Direct evidence for the involvement of oncogenes in mammary tumorigenesis has been obtained through the study of transgenic mouse models. In particular, the generation of transgenic mice that express activated tyrosine kinases has provided important insights into the signal transduction pathways that are activated during the transformation process. The purpose of this chapter is to review recent work detailing the involvement of protein tyrosine kinases in mouse mammary tumorigenesis and the signal transduction pathways in which they operate. The primary focus will be on transgenic models

Hormones and Growth Factors in Development and Neoplasia, Edited by Robert B. Dickson and David S. Salomon
ISBN 0-471-16899-8

involving two tyrosine kinases that are frequently associated with the induction of breast cancer in humans: The Neu/ErbB-2 receptor tyrosine kinase (RTK) and the c-Src tyrosine kinase.

22.2. THE EGFR FAMILY OF GROWTH FACTOR RECEPTORS IN HUMAN BREAST CANCER

The RTKs most closely associated with cancers affecting the mammary gland are those of the epidermal growth factor receptor (EGFR) family. At present, this family consists of EGFR (ErbB-1, HER1) (1), Neu (ErbB-2, HER2) (2–4), ErbB-3 (HER3) (5,6), and ErbB-4 (HER4) (7). These proteins have a similar structural organization, all of them containing a cysteine rich extracellular domain, a single transmembrane domain, a tyrosine kinase domain, and a carboxyl tail. The characteristic features of this family have led to their classification as type I receptors (8). The primary role of these receptors is to convert signals present in the extracellular environment into appropriate cellular responses. Through the binding of a variety of growth factors, these receptors undergo dimerization, which, in turn, activates the intrinsic tyrosine kinase activity of the component monomers. As a consequence, the active receptor complex can recruit specific cytoplasmic molecules, which then initiate appropriate downstream signaling cascades.

The relationship between the expression of type I receptors and the induction of mammary carcinoma has been explored in a variety of clinical studies. Early investigations revealed a strong correlation between EGFR overexpression and a poor clinical outcome for the patient (9,10). However, more recent studies suggest that EGFR overexpression may have significant value only for predicting poor relapse-free and overall survival after short-term follow-up and is only weakly associated with a poor prognosis when examined over a longer follow-up interval (reviewed in ref. 11). Still others contend that expression of the EGFR is actually lower in the majority of tumors compared to nonmalignant mammary tissue (12,13). It remains possible that the low levels of EGFR detected in these latter reports are still functionally relevant to the disease process. In addition to studies of the EGFR, there has been a limited number of clinical studies investigating a role for ErbB-3 in the genesis of human breast cancers. When *erb*B-3 was initially cloned, the gene was found to be overexpressed in a number of mammary carcinoma cell lines (6). Subsequent studies have demonstrated that overexpression of *erb*B-3 can be detected in approximately 13–29% of human breast cancers (14–16), but this elevated expression does not appear to result from gene amplification (6,14). In addition, no significant correlation was established between *erb*B-3 overexpression and clinical outcome in node-negative individuals (15,16). The importance of the most recently described EGFR family member, *erb*B-4, to mammary tumorigenesis has also been inferred by its expression in a variety of mammary carcinoma cell lines (7).

In studies of the four members of the EGFR family, by far the most convincing evidence supports a role for ErbB-2 in human breast cancer. Various studies have estimated that the *erb*B-2 gene is overexpressed in 20–30% of human breast cancers

(17–19), leading to a poor clinical outcome for the patient (17–21). Although some controversy exists regarding the association of *erb*B-2 amplification and clinical outcome (22–25), it appears that a strong correlation does exist between *erb*B-2 expression and a poor prognosis in lymph node-positive patients (26). Two different mechanisms appear to facilitate the overexpression of *erb*B-2: The first relies on gene amplification, while the second achieves overexpression in the absence of increased *erb*B-2 copy number (13,18,27). A wealth of information now exists establishing a link between several members of the EGFR family and human breast cancer. However, determining whether a causal relationship exists between expression of these genes and tumor formation has relied on the generation of transgenic mouse models of the disease.

22.3. EXPRESSION OF THE Neu/ErbB-2 RTK IN THE MAMMARY EPITHELIUM OF TRANSGENIC MICE

The *neu* gene, the rat homolog of *erb*B-2, was originally cloned as an oncogene (2). Compared to the wild-type form of the gene, the activated *neu* cDNA was found to possess a single amino acid substitution in the transmembrane domain that replaces a valine residue with glutamic acid (28). This mutation renders the Neu receptor constitutively kinase active (29–31) due to its ability to undergo ligand-independent aggregation (32).

In an effort to define the importance of *erb*B-2/*neu* expression in mammary tumorigenesis more rigorously, transgenic mice have been generated that express the activated *neu* cDNA under the control of the mouse mammary tumor virus-long terminal repeat (MMTV-LTR) (33–35). In the first report, all the female carriers from one transgenic line (TG.NF) demonstrated global mammary transformation involving all glands after 95 days (33). The appearance of mammary gland abnormalities coincided with regions of transgene expression, as determined by *in situ* hybridization (33). An independent transgenic line (TG.NK), characterized in the same study, demonstrated a nonuniform pattern of transgene expression that translated into a stochastic tumor phenotype in which mammary tumors arose next to morphologically normal epithelium (33). The phenotypes displayed by these two lines of transgenic mice suggested that expression of activated *neu* requires few, if any, additional events to transform the primary epithelial cell. In contrast, other transgenic strains carrying an MMTV/activated *neu* transgene developed focal mammary tumors that arose asynchronously in animals between 5 and 10 months of age (34). More recent observations obtained from the characterization of additional strains of MMTV/activated *neu* mice (35) support the single step model of mammary gland transformation proposed by Muller et al. (33). The discrepancy in the phenotypes displayed by these strains may reflect differences in the levels of activated Neu protein expressed in the mammary epithelium (of these) animals. For example, transformation of the primary epithelial cell may require a critical threshold of activated Neu expression that is not achieved in those animals exhibiting the stochastic tumor phenotype. Regardless of the explanation, these studies argue that activated *Neu* is a potent oncogene in the mammary epithelium of transgenic mice.

The strong transforming ability demonstrated by the *neu* oncogene in the mammary epithelium has prompted several groups to examine whether a similar point mutation within *erb*B-2 might occur in human breast tumors. Screens based on direct sequencing (18), sequence-specific oligonucleotide hybridization (36), or differential endonuclease digestion (37) have failed to detect the presence of the analogous mutation within the transmembrane domain of ErbB-2 in human breast cancers. The lack of activating mutations in the transmembrane domain, coupled with clinical observations suggesting that *erb*B-2 is often amplified in human tumors, suggests that the primary mechanism operating in human breast cancer is overexpression of the normal, unaltered receptor. To directly test the oncogenic potential of wild-type *neu* in the mammary epithelium, several strains of transgenic mice expressing the wild-type *neu* cDNA under the control of the MMTV-LTR have been established (38). These animals develop focal mammary tumors with an average latency period of approximately 7 months (38). Expression analysis of these mice revealed that in many cases the level of transgene RNA was significantly higher in the tumor than in the surrounding epithelium, although in certain instances both the tumor and adjacent tissue expressed similar levels of the *neu* transgene. In animals belonging to the latter group, Neu protein was also detected in both tumor and adjacent mammary epithelium. However, analysis of the receptor's catalytic activity within these tissues revealed that the tumors alone exhibited elevated levels of Neu tyrosine kinase activity (38).

One possible explanation for the differential activation of Neu observed in mammary tumors compared to the surrounding epithelium is the occurrence of somatic activating mutations within the transgene. Given the observation that Neu can be activated by the occurrence of a single point mutation in the transmembrane domain, a reverse transcription-polymerase chain reaction (RT-PCR) was conducted on matched RNA samples from both tumor and adjacent tissue followed by sequence-specific oligonucleotide hybridization in an effort to detect this well-characterized mutation (V664E). Although no evidence was obtained for the presence of this mutation in any of the tumors tested, small in-frame deletions were discovered in *neu* transcripts originating from tumor RNA samples (39). These deletions were not observed in any of the adjacent mammary tissue samples examined. Subsequent RNase protection analysis revealed that these alterations occurred with a frequency of about 65% and that individual tumors within the same animal could possess distinct deletions (39). Furthermore, when representative deletions were introduced into an otherwise wild-type *neu* cDNA and tested in focus-forming assays, the altered Neu receptors were able to transform Rat-1 cells in culture. The altered Neu receptors were also constitutively phosphorylated, suggesting that they possessed elevated kinase activities (39). These results argue that somatic mutations capable of activating the Neu tyrosine kinase are involved in tumor formation observed in the MMTV/wild-type *neu* mice.

A recent study has proposed a mechanism by which these in-frame mutations lead to the activation of the Neu tyrosine kinase (40). Examination of the various deletions revealed that most of the alterations affect a stretch of conserved cysteine residues located within the extracellular domain close to the transmembrane domain. In fact, almost all of the deletions remove one of three conserved cysteine residues in this region. In addition, one alteration was discovered that resulted in a three-amino acid insertion within this same region. Significantly, one of the three inserted amino acids was a cysteine residue. These observations indicate that the balance of cysteine residues within this region is a key determinant in the regulation of the oncogenic potential of Neu. It is possible that the removal or mutation of a cysteine residue would disrupt an intramolecular disulfide bond and allow the formation of an intermolecular disulfide bond between two altered receptors, thereby promoting dimerization. Indeed, this study demonstrated that both deletion mutants and engineered point mutants, which targeted cysteine residues in the juxtatransmembrane region, underwent receptor dimerization in a manner that correlated with their ability to transform cells (40). Further evidence reinforcing the importance of disulfide bond formation in stabilizing the dimer was provided by the observation that inclusion of a reducing agent in the tissue culture medium severely impaired transformation mediated by these altered receptors (40). Finally, the observation that transgenic mice expressing selected Neu deletion mutants under the control of the MMTV-LTR develop mammary tumors with faster kinetics compared to the original MMTV/wild-type *neu* mice establishes the importance of these activating mutations in the induction of mammary cancers (P.M. Siegel and W.J. Muller, unpublished observations). It is apparent from these studies that activation of the Neu receptor's tyrosine kinase activity is a critical step during mammary gland transformation.

22.4. EGFR FAMILY LIGANDS IN MOUSE MAMMARY TUMORIGENESIS

Additional evidence implicating the EGFR family in mammary tumorigenesis has been provided by observations made from transgenic mice expressing different EGFR family ligands. Two groups of ligands are responsible for modulating the activity of the four members of the EGFR family, which include the epidermal growth factor (EGF) family of peptide growth factors (41) and the Neu differentiation factors (NDFs) (42–44). Although originally cloned as ligands for Neu (42,43), the NDFs have subsequently been shown to be ligands specific for ErbB-3 and ErbB-4 (45–47). At the present time, a cloned and completely characterized ligand that binds directly to Neu has not been described; however, a likely candidate exists in the form of Neu-activating factor (NAF) (48,49). This ligand has been shown to interact directly with the Neu receptor produced in insect cells, to induce tyrosine phosphorylation of Neu expressed in mammalian cells, and to stimulate Neu dimerization (49).

The importance of these ligands, and hence of the receptors to which they bind, in mammary tumorigenesis has been examined using transgenic mice. Several groups have established transgenic mice that overexpress transforming growth factor-α (TGFα), an EGFR-specific ligand, under either the MMTV-LTR or the metallothionein promoter (50–52). In two instances (50,52) TGFα expression resulted in epithelial hyperplasia and dysplasia, with only a small number of multiparous animals developing mammary adenocarcinomas. The remaining study reported reduced penetration of the mammary ducts into the fat pad and an increase in the number of actively dividing epithelial cells that were not restricted to the terminal end buds (51). An interesting observation from one report suggested that in mammary glands expressing high levels of TGFα, a concomitant increase in endogenous EGFR mRNA was observed (52). More recently, a member of the NDF family of ligands (NDFβ2) has also been expressed in the mammary glands of transgenic mice (53). Multiparous animals develop palpable mammary tumors at a median age of 357 days. When mammary tumor cell lines were established and examined for receptor expression, ErbB-3 and ErbB-4 were found to be expressed at high levels, while ErbB-2 was present in relatively low amounts (53). However, only the ErbB-3 receptor appeared to be tyrosine phosphorylated in these tumor lines. The above studies suggest that, while required, the ligands are not sufficient for mammary gland transformation in these animals.

22.5. INTERRECEPTOR INTERACTIONS AMONG THE EGFR FAMILY MEMBERS

As mentioned earlier, one of the first events believed to occur during receptor activation, whether by mutation or ligand stimulation, is dimerization (54,55). Within a family composed of structurally similar members, such as the EGFR family, the consequences of receptor dimerization are likely to be complex. In addition to the formation of homodimers, it is becoming increasingly clear that members of the EGFR family can also heterodimerize with one another (56,57). The possibility of heterodimerization within this family was first raised by the observation that EGF stimulation of Rat-1 fibroblasts (58) or mammary tumor cell lines (59,60) led to the phosphorylation of ErbB-2. The inability of EGF to bind directly to ErbB-2 suggested that this ligand must exert its effects through a mechanism of receptor transphosphorylation or by the formation of EGFR/ErbB-2 heterodimers. Subsequent studies have revealed the presence of physical complexes between the EGFR and ErbB-2 in both NIH 3T3 (61) and human breast carcinoma cells (62), providing direct evidence for heterodimerization. The consequence of such an interaction between the EGFR and ErbB-2 had previously been demonstrated by the ability of these two receptors to induce transformation when coexpressed in mouse fibroblast cells (63). The remaining two receptors of the EGFR family also interact with the EGFR. Ligand activation of the EGFR results in the phosphorylation of ErbB-3 (64,65) and ErbB-4 (66). Conversely, neuregulin (NRG) stimulation of either ErbB-3 (67) or ErbB-4 (67,68) promotes the phosphorylation of the EGFR. However, the ability to detect EGFR–ErbB-3 and EGFR/ErbB-4 complexes by ligand affinity labeling has proven to be difficult (66,69,70).

The ErbB-2 receptor can also interact with ErbB-3 and ErbB-4. ErbB-2 can be phosphorylated by both receptors in response to NRG stimulation and ErbB-2/ErbB-3 and ErbB-2/ErbB-4 heterodimers have been detected by ligand affinity labeling (66,67,69,70). The importance of ErbB-2/ErbB-3 associations was demonstrated by the observation that coexpression of these two receptors resulted in transformation of NIH3T3 cells (71). Similarly, NIH3T3 cells coinfected with ErbB-2 and ErbB-3 were able to form colonies in soft agar (72). The requirement of ErbB-2-mediated signaling from both EGF and NDF has been examined in T47D cells (73), as well as NDF signaling in MCF7 cells (74), both of which are mammary carcinoma cell lines that express all four members of the EGFR family. Using a single-chain antibody that specifically retains ErbB-2 in the endoplasmic reticulum, it was demonstrated that lack of ErbB-2 surface expression impaired the ability of the EGFR and ErbB-3 to undergo phosphorylation when stimulated with EGF or NDF, respectively (73,74). This resulted in decreased activation of the Ras pathway, as measured by reduced ERK-1 and ERK-2 phosphorylation. Collectively, these observations have led some investigators to speculate that ErbB-2 is a common and essential subunit for the remaining three members of the EGFR family (70). These findings reveal a complex network of interreceptor interactions that presumably allows for enormous signal diversification.

Recently, transgenic mice have been utilized to examine the *in vivo* interaction of certain members of the EGFR family. Bigenic animals have been generated that overexpress both TGFα and the wild-type Neu receptor in the mammary epithelium of these transgenic mice (75). These animals develop mammary tumors with greatly accelerated kinetics compared to transgenic mice expressing either TGFα and *neu* alone. Notably, the tumors that arise in the bigenic animals are multifocal and involve the entire mammary gland, unlike the focal tumors observed in both of the monogenic strains (75). In contrast to the MMTV/wild-type *neu* transgenic mice, which frequently display evidence of somatic activating mutations (39), no comparable mutations were observed in mammary tumors coexpressing wild-type *neu* and TGFα (75). Nonetheless, the tumors coexpressing these two transgenes exhibited elevated levels of tyrosine phosphorylated Neu. Thus, it would appear that activation of Neu in mammary tumorigenesis can occur through multiple molecular mechanisms that include somatic mutation of Neu or transphosphorylation of Neu through overexpression of an EGFR ligand. Although physical complexes between the EGFR and Neu could not be demonstrated by chemical cross-linking approaches or by reciprocal immunoprecipitation/immunoblot analysis, the presence of elevated levels of tyrosine phosphorylated Neu in the absence of somatic activating mutations suggests that induction of mammary tumors in these strains is occurring through a mechanism of receptor transphosphorylation (75).

22.6. SIGNALING FROM THE ErbB-2 RECEPTOR TYROSINE KINASE

Dimerization, in most cases, results in activation of the intrinsic kinase activity of those monomers within the receptor complex. This, in turn, causes the phosphorylation of tyrosine residues located in the cytoplasmic region of each receptor molecule. These phosphorylated tyrosine residues then serve as docking sites for a variety of cytoplasmic signaling molecules, which frequently contain SH2 or PTB/PI domains (76). A number of cytoplasmic signaling molecules are substrates of the Neu/ErbB-2 receptor, including proteins that possess enzymatic activities, as well as those that serve as adapter molecules. The identification of such molecules increases our understanding of how the Neu/ErbB-2 receptor induces cellular transformation.

The initial substrates of this receptor to be identified belong to a class of enzymes involved in phospholipid signaling. Phospholipase C-γ (PLC-γ) is an enzyme that converts phosphatidyl 4,5-bisphosphate into the second messengers inositol 1,4,5-trisphosphate and diacylglycerol (77). PLC-γ was first shown to associate with ErbB-2 by coimmunoprecipitation analysis in NIH3T3 cells (78,79), and the same interaction has subsequently been demonstrated in human mammary tumor-derived cell lines (80). Additional substrates of Neu/ErbB-2 that are likewise involved in generating lipid second messengers include the phosphatidyl inositol kinases. Activation of ErbB-2 with a monoclonal antibody stimulates both phosphatidylinositol 4′ kinase (PI4′K) and phosphatidylinositol 3′ kinase (PI3′K) activities in BT-474 cells (81). Additional experiments, utilizing a EGFR/Neu chimeric receptor have demonstrated that a small percentage of cellular PI3′K becomes associated with the chimera in response to ligand stimulation in NIH3T3 cells (82). Apart from its involvement in phospholipid signaling, Neu/ErbB-2 can also positively (see below) or negatively regulate the Ras pathway. The Ras-GTPase activating protein (Ras-GAP) is a target for ErbB-2-mediated phosphorylation (79), and the Ras-GAP SH2 domain has been shown to bind to activated Neu *in vitro* (83). Ras-GAP is capable of shuttling Ras into an inactive state by stimulating the intrinsic GTPase activity of Ras (84). Finally, a group of enzymes has recently been identified as Neu/ErbB-2 substrates that are themselves tyrosine kinases. Both c-Src and c-Yes are able to associate with the Neu/ErbB-2 receptor both *in vitro* and *in vivo* (83,85,86). Furthermore, the interaction of c-Src with Neu is direct and leads to an increase in the specific activity of c-Src (87). c-Yes also binds directly to Neu through the same phosphorylation site utilized by the c-Src tyrosine kinase (86). The relevance of these two cytoplasmic kinases in mammary tumorigenesis is substantiated by the fact that in murine mammary tumors expressing tyrosine phosphorylated Neu, the specific enzymatic activity of both proteins is greatly increased compared to that of adjacent tissue (83,86).

Adapter molecules involved in linking RTKs to the Ras pathway are also substrates of the Neu/ErbB-2 receptor. In particular, both Grb-2 and Shc proteins provide a means through which Neu/ErbB-2 can modulate Ras activity. Grb-2 exists in a complex with the guanine nucleotide exchange factor Son of Sevenless (Sos), which can activate Ras by catalyzing the exchange of bound GDP for GTP (88). The Grb-2–Sos interaction is mediated by the SH3 domain of Grb-2 and proline-rich regions within the C terminus of Sos, leaving the SH2 domain of Grb-2 free to bind either the Shc proteins or directly to RTKs (88). Interestingly, Grb-2 is overexpressed in a variety of breast cancer cell lines, many of which also overexpress ErbB-2 (89). Furthermore, physical complexes between ErbB-2 and Grb-2 have been demonstrated in several of these tumor-derived lines, and this association has been shown to be direct *in vitro* (90). Dominant-negative mutants of Grb2 lacking either the amino- or carboxy-terminal SH3 domain have been shown to suppress transformation from activated Neu in NIH3T3 cells (91). A role for Shc proteins as Neu/ErbB-2 substrates is suggested by the fact that they are tyrosine phosphorylated in response to ErbB-2 activation and form physical complexes with this receptor, as demonstrated by reciprocal immunoprecipitation experiments (92). Moreover, the presence of tyrosine phosphorylated Shc proteins and the presence of Shc/Grb2 complexes have been correlated with transformation mediated by the activated Neu receptor (93). A recent study has attempted to locate the binding sites on the ErbB-2 receptor for both the Shc and Grb2 proteins. A panel of ErbB-2 phosphorylation mutants and a combination of direct binding, *in vitro* binding, and peptide competition assays was used in this analysis (94). The binding sites for Grb2 and Shc have also been conclusively mapped on the Neu RTK (the rat homologue of ErbB-2). A series of autophosphorylation mutants combined with a panel of tyrosine add-back mutants was employed to identify the specific tyrosine residues that are both necessary and sufficient for mediating the binding of Grb2 and Shc to Neu (94a). Finally, several substrates that interact with the Neu/ErbB-2 RTK have been identified; their functions or importance remain unknown. These include the PTP1C, PTP1D (95), Grb7 (96), and Tob (97) proteins.

As indicated earlier, an important consideration when examining signal transduction from the Neu/ErbB-2 RTK is its ability to undergo heterodimerization with other members of the EGFR family. It is quite likely that the signal originating from a given receptor complex will vary greatly, depending on which family members are present in the dimer. This difference in signal output reflects a number of variables, not the least of which is the repertoire of signaling molecules that each monomer can recruit to the heterodimer complex (56,57). This concept is supported by recent observations reporting differential signaling between EGFR and EGFR/ErbB-3 receptor complexes. One study demonstrated that strong PI3′K activity was associated with an activated EGFR/ErbB-3 chimera (composed of the EGFR extracellular domain fused to the ErbB-3 cytoplasmic domain) compared to much weaker activity observed with the EGFR (98). The level of associated PI3′K activity reflected the ability of the p85 subunit to coimmunoprecipitate with each receptor (98). Similar findings were made using A431 cells in which increased PI3′K activity was detected in anti-ErbB-3, but not in anti-EGFR, immunoprecipitates in response to EGF treatment (65). This has led to the proposal that ErbB-3 mediates PI3′K activation induced by EGF stimulation. A similar situation has been described for ErbB-2, in which ErbB-3-mediated recruitment of PI3′K to ErbB-2/ErbB-3 heterodimers was shown to be important for transformation of NIH3T3 cells (71). It is interesting to note that, like the EGFR, Neu does not possess the Tyr-X-X-Met motif believed to be the docking site for PI3′K. This motif is, however, repeated seven times within the ErbB-3 receptor (99). One additional example for the selective recruitment of signaling molecules to certain EGFR family members has been presented for the c-Src tyrosine kinase. In this study, individual EGFR family members were immunoprecipitated from lysates obtained from Neu-induced murine mammary tumors or a tumor-derived cell line (NAFA). These immunoprecipitates were then separated by sodium dodecyl sulfate-polyacrylamid gel electrophoresis (SDS-PAGE) and probed with a radiolabeled c-Src SH2 domain. A direct interaction was observed only with the Neu RTK but not with the other three family members; however, it should be noted that the Neu RTK was the only predominantly phosphorylated receptor in these lysates (87). To ensure that the c-Src SH2 domain association was indeed specific, lysates from several cell lines were tested in which phosphorylated EGFR was obtained. In no case could an association between c-Src and activated EGFR be detected, either by direct binding assay or by coimmunoprecipitation (87). Conflicting observations have been reported in which an association between c-Src and the EGFR was demonstrated in breast carcinoma cell lines (85).

22.7. THE c-Src TYROSINE KINASE IN HUMAN MAMMARY CANCER

The Src family of protein tyrosine kinases, of which c-Src is the prototype, belongs to the cytoplasmic, nonreceptor class of tyrosine kinases. The members of this family display a common structural organization that includes an amino-terminal myristylation site followed by a unique region, a single SH3 and SH2 domain, the catalytic domain (SH1), and finally a regulatory carboxyl terminus (100,101). Like members of the EGFR family, c-Src has been implicated in the process of mammary tumorigenesis. Several studies have demonstrated elevated c-Src tyrosine kinase activity in human mammary tumors or tumor-derived cell lines (102–105). In one report, elevated specific activity of c-Src tyrosine kinase was observed in MCF-7, ZR75-1, and T47D but not in MDA-231 carcinoma cell lines (103). In this same study, analysis of a single matched set of normal human breast tissue and breast carcinoma revealed a significant increase in the specific kinase activity of c-Src within the tumor. A subsequent investigation examined the total tyrosine kinase activity associated with both membrane and cytosolic fractions for a number of malignant, benign, and normal breast tissues. The results indicated that a 25-fold increase in tyrosine kinase activity in the cytosolic fraction was observed between breast cancer samples and normal tissue and that high tyrosine kinase activities in this fraction were correlated with a strongly increased risk of relapse (104). Although the identity of the kinase was not determined, these observations were later extended by the same group; this time, approximately 70% of the cytosolic protein tyrosine kinase activity was attributed to c-Src (105). It is unclear why c-Src kinase activity is cytosolic in this study, given that this protein kinase is normally tethered to the inner face of the plasma membrane via a myristylated glycine residue in its amino terminus. These observations provide correlative evidence for involvement of the c-Src tyrosine kinase in the progression of this disease.

22.8. TRANSGENIC MOUSE MODELS INVOLVING THE c-Src PROTEIN TYROSINE KINASE

Another well-established model for mammary tumorigenesis has exploited the transforming properties of the polyoma virus middle T antigen (PyVmT). Animals expressing a MMTV/PyVmT transgene develop multifocal mammary adenocarcinomas that can be detected, by whole-mount analysis, as early as 3 weeks of age (106). Another striking phenotype exhibited by these mice was a high incidence of secondary lung metastasis, which ranged between 90% and 100% in female carriers from three independent lines (106). Because the PyV mT oncogene itself does not possess intrinsic enzymatic activity, it must rely on associations with various cellular signaling molecules. One group of proteins known to associate with mT antigen is the Src family of tyrosine kinases (107–111). As expected, the mT-associated kinase activity observed in mammary tumors from MMTV/PyVmT animals (106) was subsequently demonstrated to result from the activation of c-Src and c-Yes (112). A number of additional signaling proteins have also been shown to associate with middle T including the PI3′K (113,114), protein phosphatase 2A (PP2A) (115,116), Shc adaptor proteins (117,118), PLC-γ (119), and 14-3-3 proteins (120).

In an effort to understand the contributions made by both c-Src and c-Yes to PyV mT-mediated transformation of the mammary gland, the MMTV/PyV mT mice were mated to c-Src- (121) or c-Yes- (122) deficient mice. The results of these matings revealed that activation of c-Src was required for the induction of mammary tumors in the PyV mT strains (112). In contrast, the lack of c-Yes in these MMTV/PyV mT had no effect on the induction of mammary tumors (112). Moreover, the difference in the transforming activity of PyV mT in these crosses was not due to differences in the levels of PyV mT-associated kinase activity since tissues derived from either the c-Src- or c-Yes-deficient strains possessed comparable levels of activity. These data argue that c-Src, but not c-Yes, is required for the formation of mT-induced mammary tumors. Because the level of PyV mT-associated kinase activity was similar in each strain, these observations further suggest that c-Src and c-Yes have distinct substrates in mammary epithelial cells.

Although the above observations indicated that c-Src was required for the induction of mammary tumors by PyV mT antigen, it remained unclear whether c-Src activation was sufficient to induce the multifocal tumor phenotype. To address the question of whether the expression of an activated version of c-*src* was sufficient to induce mammary tumors *in vivo,* transgenic mice expressing an MMTV-activated c-*src* transgene in the mammary epithelium were generated (123). The version of c-Src that was used contained a tyrosine to phenylalanine substitution at position 527 in the c-Src regulatory domain, rendering the kinase constitutively active. The animals expressing this transgene developed mammary hyperplasias that could be detected as early as 8 weeks, which eventually developed into mammary tumors after a long latency period (T_{50} = 345 and 292 days for two independent lines) (123). The tumor phenotype displayed by these mice indicates that, although not sufficient for tumorigenesis, expression of c-Src is able to induce mammary epithelial hyperplasias that can eventually progress to mammary tumors. Together these results emphasize the importance of c-Src tyrosine kinase in the progression of mammary tumorigenesis in transgenic models of breast cancer.

A variety of molecules have now been demonstrated to act as substrates for both the viral and cellular forms of Src. However, one substrate is of particular interest due to its potential involvement in mammary tumorigenesis/metastasis. The focal adhesion kinase (FAK) has been shown to associate with both c-Src and v-Src (124,125), and the consequences of this interaction are now becoming more fully understood. The FAK protein kinase was originally cloned using a monoclonal antibody that recognized a 125-kDa phosphoprotein in v-Src-transformed chicken embryo cells (126). This protein was shown to possess intrinsic tyrosine kinase activity and to localize to sites of focal adhesion by immunofluorescence (126). A direct, stable interaction between Src (both cellular and viral) and FAK has been demonstrated that requires the SH2 domain of Src (124,125). Further work has demonstrated that Tyr-397, the major autophosphorylation site within FAK, provides the binding site for the c-Src SH2 domain (127). Mutation of this site reduces phosphorylation of FAK and abrogates binding of this kinase with c-Src. Formation of this complex has been proposed to facilitate subsequent interactions between FAK and other signaling molecules through c-Src-mediated phosphorylation of tyrosines within FAK (128). In this instance, the binding of Grb2 to tyrosine 925 of FAK was detected *in vitro* after phosphorylation of a FAK carboxy-terminal domain by c-Src (128). A subsequent study confirmed that the Grb2 binding site within FAK is a target for c-Src transphosphorylation rather than FAK autophosphorylation (129).

It is tempting to speculate that FAK may play a role in different aspects of mammary tumorigenesis. In particular, growing evidence for the involvement of integrin signaling, which utilizes FAK, in cell migration suggests a role for FAK in tumor metastasis (130). In fact, a study performed on a limited number of human mammary tumor samples revealed that FAK mRNA could be detected in 9/11 primary and 4/4 metastatic tumors (131). A subsequent report demonstrated increased FAK expression in 2 carcinoma-derived cell lines (BT474 and BT-20), as well as in 22 of 25 primary invasive and metastatic tumor samples (132). Further analyses utilizing transgenic models that display a metastatic phenotype may reveal the importance, if any, of the involvement of FAK in the metastatic progression of mammary tumors.

22.9. THE ROLE OF ADDITIONAL RTKs IN NORMAL MAMMARY GLAND DEVELOPMENT AND MAMMARY TUMORIGENESIS

It should be noted that protein tyrosine kinases are not solely involved in mammary gland transformation; they have also been implicated in certain aspects of normal mammary gland development. For example, mice carrying the naturally occurring *waved-2 (wa-2)* mutation express an altered form of the EGFR with impaired kinase activity (133,134). These animals display eye and skin abnormalities, as well as pronounced waviness of the hair and whiskers, and homozygous *waved-2* mothers exhibit impaired lactation resulting in a high mortality rate among their newborn pups (134). Moreover, ectopic expression of TGFα in the mammary epithelium of transgenic mice inhibits mammary gland involution following weaning (135). Both of these results suggest that signals emanating from the EGFR are important for the proper regulation of lactation. The importance of several other RTKs in various aspects of mammary gland development can be inferred from observations involving their cognate ligands. For example, the naturally occurring op/op mouse harbors a defect in the colony stimulating factor-1 (CSF-1) growth factor that results in a lactation deficiency. This inability to lactate properly reflects an inhibition of mammary ductal growth and differentiation (136). Still other growth factors can influence the process of involution in a fashion similar to that described above for TGFα. These include both int-2 (bFGF-3), a member of the fibroblast growth factor family, and the insulin-like growth factor (IGF) (137,138). It is apparent from these observations that a variety of growth factors, as well as their cognate receptors, play significant roles in normal mammary gland development.

Although most of this chapter has focused on the involvement of the EGFR family and c-Src in mammary tumorigenesis, a number of other protein tyrosine kinases have also been implicated in this process. For example, transgenic mice expressing an activated form of the *ret* gene under control of the MMTV LTR develop several tumor types including stochastic mammary tumors with latencies of 6–7 months (139). Similarly, widespread expression of an activated form of the Met receptor (Tpr-Met) under control of the metallothionein promoter leads to the development of focal mammary tumors in three independent lines (140). The importance of these observations, if any, to human breast cancer has not been determined. Additional RTKs, such as the CSF-1R, have been shown to be expressed in human breast carcinomas (141), and exposure of CSF-1R-expressing breast carcinoma cells (BT20) to CSF-1 resulted in increased invasiveness of these cells through an amnionic membrane (142). Finally, several reports have indicated that both the insulin receptor (IR) and the insulin-like growth factor receptor (IGFR) are overexpressed in primary human breast cancers compared to normal breast tissue and that many breast cancer cell lines also express these two receptors (143–146). The importance of IR and IGFR expression to the progression of the disease is unclear, although one report suggests that the presence of IGFR is a favorable prognostic indicator in breast cancer (146).

22.10. CONCLUSIONS

The establishment of transgenic animal models and examination of the process of mammary tumorigenesis in these systems is proving to be a rewarding venture. Information regarding the molecules involved and the signal transduction pathways that are exploited during the transformation process facilitates our understanding of the molecular biology underlying this disease. Continued analysis of existing transgenic models, the generation of novel or bigenic strains, and the combination of both overexpressed and disrupted genes within the same animal may help to extend our knowledge of mammary cancer in an *in vivo* setting.

ACKNOWLEDGMENTS

Work conducted in the laboratory of W.J. Muller was supported by grants received from the Medical Research Council of Canada, the Canadian Breast Cancer Initiative, and the U.S. Department of Defense. W.J.M. is a recipient of a Medical Research Council of Canada scientist award, and P.M.S. is supported by a studentship from the Medical Research Council of Canada.

REFERENCES

1. Ullrich A, Coussens L, Hayflick JS, Dull TJ, Gray A, Tam AW, Lee J, Yarden Y, Libermann TA, Schlessinger J, Downward J, Mayes ELV, Whittle N, Waterfield MD, Seeburg PH: Human epidermal growth factor receptor cDNA sequence and aberrant expression of the amplified gene in A431 epidermoid carcinoma cells. Nature 309: 418–425, 1984.
2. Bargmann CI, Hung M-C, Weinberg RA: The *neu* oncogene encodes an epidermal growth factor receptor-related protein. Nature 319:226–230, 1986.
3. Coussens L, Yang-Feng TL, Liao Y-C, Chen E, Gray A, McGrath J, Seeburg PH, Libermann TA, Schlessinger J, Francke U, Levinson A, Ullrich A: Tyrosine kinase receptor with extensive homology to EGF receptor shares chromosomal location with *neu* oncogene. Science 230:1132–1139, 1985.
4. Yamamoto T, Ikawa S, Akiyama T, Semba K, Nomura N, Miyajima N, Saito T, Toyoshima K: Similarity of protein encoded by the human c-*erb-B*-2 gene to epidermal growth factor receptor. Nature 319:230–234, 1986.
5. Kraus MH, Issing W, Miki T, Popescu NC, Aaronson SA: Isolation and characterization of *ERBB3,* a third member of the *ERBB*/epidermal growth factor family: Evidence for overexpression in a subset of human mammary tumors. Proc Natl Acad Sci USA 86:9193–9197, 1989.
6. Plowman GD, Whitney GS, Neubauer MG, Green JM, McDonald VL, Todaro GJ, Shoyab M: Molecular cloning and expression of an additional epidermal growth factor receptor-related gene. Proc Natl Acad Sci USA 87:4905–4909, 1990.
7. Plowman GD, Culouscou J-M, Whitney GS, Green JM, Carlton GW, Foy L, Neubauer MG, Shoyab M: Ligand-specific activation of HER4/p180^{erbB4}, a fourth member of the epidermal growth factor receptor family. Proc Natl Acad Sci USA 90:1746–1750, 1993.
8. Ullrich A, Schlessinger J: Signal transduction by receptors with tyrosine kinase activity. Cell 61:203–212, 1990.
9. Sainsbury JRC, Farndon JR, Needham GK, Malcolm AJ, Harris AL: Epidermal-growth-factor receptor status as predictor of early recurrence and death from breast cancer. Lancet 1:1398–1402, 1987.
10. Harris AL, Nicholson S, Sainsbury JRC, Farndon J, Wright C: Epidermal growth factor receptors in breast cancer: Association with early relapse and death, poor response to hormones and interactions with *neu*. J Steroid Biochem 34:123–131, 1989.
11. Klijn JGM, Berns PMJJ, Schmitz PIM, Foekens JA: The clinical significance of epidermal growth factor receptor (EGF-R) in human breast cancer: A review on 5232 patients. Endocrinol Rev 13:3–17, 1992.
12. Dittadi R, Donisi PM, Brazzale A, Cappellozza L, Bruscagnin G, Gion M: Epidermal growth factor receptor in breast cancer. Comparison with nonmalignant breast tissue. Br J Cancer 67:7–9, 1993.
13. Robertson KW, Reeves JR, Smith G, Keith WN, Ozanne BW, Cooke TG, Stanton PD: Quantitative estimation of epidermal growth factor receptor and c-*erb*B-2 in human breast cancer. Cancer Res 56:3823–3830, 1996.
14. Lemoine NR, Barnes DM, Hollywood DP, Hughes CM, Smith P, Dublin E, Prigent SA, Gullick WJ, Hurst HC: Expression of the *ERBB3* gene product in breast cancer. Br J Cancer 66:1116–1121, 1992.

15. Quinn CM, Ostrowski JL, Lane SA, Loney DP, Teasdale J, Benson EA: c-*erb*B-3 protein expression in human breast cancer: Comparison with other tumour variables and survival. Histopathology 25:247–252, 1994.

16. Gasparini G, Gullick WJ, Maluta S, Palma PD, Caffo O, Leonardi E, Boracchi P, Pozza F, Lemoine NR, Bevilacqua P: c-*erb*B-3 and c-*erb*B-2 protein expression in node-negative breast carcinoma–an immunocytochemical study. Eur J Cancer 30A:16–22, 1994.

17. Slamon DJ, Clark GM, Wong SG, Levin WJ, Ullrich A, McGuire WL: Human breast cancer: Correlation of relapse and survival with the amplification of the HER-2/*neu* oncogene. Science 235:177–182, 1987.

18. Slamon DJ, Godolphin W, Jones LA, Holt JA, Wong SG, Keith DE, Levin WJ, Stuart SG, Udove J, Ullrich A, Press MF: Studies of the HER-2/*neu* proto-oncogene in human breast and ovarian cancer. Science 244:707–712, 1989.

19. Press MF, Pike MC, Chazin VR, Hung G, Udove JA, Markowitz M, Danyluk J, Godolphin W, Sliwkowski M, Akita R, Paterson MC, Slamon DJ: Her-2/*neu* expression in node-negative breast cancer: Direct tissue quantitation by computerized image analysis and association of overexpression with increased risk of recurrent disease. Cancer Res 53:4960–4970, 1993.

20. Paterson MC, Dietrich KD, Danyluk J, Paterson AHG, Lees AW, Jamil N, Hanson J, Jenkins H, Krause BE, McBlain WA, Slamon DJ, Fourney RM: Correlation between c-*erb*B2 amplification and risk of recurrent disease in node-negative breast cancer. Cancer Res 51:556–567, 1991.

21. Gullick WJ, Love SB, Wright C, Barnes DM, Gusterson B, Harris AL, Altman DG: c-*erb*B-2 protein overexpression in breast cancer is a risk factor in patients with involved and uninvolved lymph nodes. Br J Cancer 63:434–438, 1991.

22. Ali IU, Campbell G, Lidereau R, Callahan R: Amplification of c-*erb*B-2 and aggressive human breast tumors? Science 240:1795–1796, 1988.

23. van de Vijver MJ, Peterse JL, Mooi WJ, Wisman P, Lomans J, Dalesio O, Nusse R: *Neu*-protein overexpression in breast cancer. N Engl J Med 319:1239–1245, 1988.

24. Parkes HC, Lillycrop K, Howell A, Craig RK: C-*erb*B2 mRNA expression in human breast tumours: Comparison with c-*erb*B2 DNA amplification and correlation with prognosis. Br J Cancer 61:39–45, 1990.

25. Clark GM, McGuire WL: Follow-up study of HER-2/*neu* amplification in primary breast cancer. Cancer Res 51:944–948, 1991.

26. Hynes NE, Stern DF: The biology of *erbB-2/neu/HER*-2 and its role in cancer. Biochim Biophys Acta 1198:165–184, 1994.

27. Kraus MH, Popescu NC, Amsbaugh SC, King CR: Overexpression of the EGF receptor-related proto-oncogene *erb*B-2 in human mammary tumor cell lines by different molecular mechanisms. EMBO J 6:605–610, 1987.

28. Bargmann CI, Hung M-C, Weinberg RA: Multiple independent activations of the *neu* oncogene by a point mutation altering the transmembrane domain of p185. Cell 45:649–657, 1986.

29. Bargmann CI, Weinberg RA: Increased tyrosine kinase activity associated with the protein encoded by the activated *neu* oncogene. Proc Natl Acad Sci USA 85:5394–5398, 1988.

30. Stern DF, Kamps MP, Cao H: Oncogenic activation of $p185^{neu}$ stimulates tyrosine phosphorylation in vivo. Mol Cell Biol 8:3969–3973, 1988.

31. Weiner DB, Kokai Y, Wada T, Cohen JA, Williams WV, Greene MI: Linkage of tyrosine kinase activity with transforming ability of the p185neu oncoprotein. Oncogene 4: 1175–1183, 1989.
32. Weiner DB, Liu J, Cohen JA, Williams WV, Greene MI: A point mutation in the neu oncogene mimics ligand induction of receptor aggregation. Nature 339:230–231, 1989.
33. Muller WJ, Sinn E, Pattengale PK, Wallace R, Leder P: Single-step induction of mammary adenocarcinoma in transgenic mice bearing the activated c-*neu* oncogene. Cell 54:105–115, 1988.
34. Bouchard L, Lamarre L, Tremblay PJ, Jolicoeur P: Stochastic appearance of mammary tumors in transgenic mice carrying the MMTV/c-*neu* oncogene. Cell 57:931–936, 1989.
35. Guy CT, Cardiff RD, Muller WJ: Activated *neu* induces rapid tumor progression. J Biol Chem 271:7673–7678, 1996.
36. Lemoine NR, Staddon S, Dickson C, Barnes DM, Gullick WJ: Absence of activating transmembrane mutations in the c-*erbB*-2 proto-oncogene in human breast cancer. Oncogene 5:237–239, 1990.
37. Zoll B, Kynast B, Corell B, Marx D, Fischer G, Schauer A: Alterations of the *c-erbB*2 gene in human breast cancer. J Cancer Res Clin Oncol 118:468–473, 1992.
38. Guy CT, Webster MA, Schaller M, Parsons TJ, Cardiff RD, Muller WJ: Expression of the *neu* protooncogene in the mammary epithelium of transgenic mice induces metastatic disease. Proc Natl Acad Sci USA 89:10578–10582, 1992.
39. Siegel PM, Dankort DL, Hardy WR, Muller WJ: Novel activating mutations in the *neu* proto-oncogene involved in induction of mammary tumors. Mol Cell Biol 14: 7068–7077, 1994.
40. Siegel PM, Muller WJ: Mutations affecting conserved cysteine residues within the extracellular domain of Neu promote receptor dimerization and activation. Proc Natl Acad Sci USA 93:8878–8883, 1996.
41. Normanno N, Ciardiello F, Brandt R, Salomon DS: Epidermal growth factor-related peptides in the pathogenesis of human breast cancer. Breast Cancer Res Treat 29:11–27, 1994.
42. Peles E, Bacus SS, Koski RA, Lu HS, Wen D, Ogden SG, Levy RB, Yarden Y: Isolation of the Neu/HER-2 stimulatory ligand: A 44 kd glycoprotein that induces differentiation of mammary tumor cells. Cell 69:205–216, 1992.
43. Wen D, Peles E, Cupples R, Suggs SV, Bacus SS, Luo Y, Trail G, Hu S, Sibiger SM, Ben Levy R, Koski RA, Lu HS, Yarden Y: Neu differentiation factor: A transmembrane glycoprotein containing an EGF domain and an immunoglobulin homology unit. Cell 69:559–572, 1992.
44. Wen D, Suggs SV, Karunagaran D, Liu N, Cupples RL, Luo Y, Janssen AM, Ben-Baruch N, Trollinger DB, Jacobsen VL, Meng S-Y, Lu HS, Hu S, Chang D, Yang W, Yanigahara D, Koski RA, Yarden Y: Structural and functional aspects of the multiplicity of Neu differentiation factors. Mol Cell Biol 14:1909–1919, 1994.
45. Carraway KL III, Sliwkowski MX, Akita R, Platko JV, Guy PM, Nuijens A, Diamonti AJ, Vandlen RL, Cantley LC, Cerione RA: The *erbB3* gene product is a receptor for heregulin. J Biol Chem 269:14303–14306, 1994.
46. Sliwkowski MX, Schaefer G, Akita RW, Lofgren JA, Fitzpatrick VD, Nuijens A, Fendly BM, Cerione RA, Vandlen RL, Caraway KL III: Coexpression of *erb*B2 and

*erb*B3 proteins reconstitutes a high affinity receptor for heregulin. J Biol Chem 269:14661–14665, 1994.

47. Plowman GD, Green JM, Culouscou J-M, Carlton GW, Rothwell VM, Buckley S: Heregulin induces tyrosine phosphorylation of HER4/p180^{erbB4}. Nature 366:473–475, 1993.
48. Dobashi K, Davis JG, Mikami Y, Freeman JK, Hamuro J, Greene MI: Characterization of a neu/c-erbB-2 protein-specific activating factor. Proc Natl Acad Sci USA 88:8582–8586, 1991.
49. Samanta A, LeVea CM, Dougall WC, Qian X, Greene MI: Ligand and p185$^{\text{c-neu}}$ density govern receptor interactions and tyrosine kinase activation. Proc Natl Acad Sci USA 91:1711–1715, 1994.
50. Sandgren EP, Luetteke NC, Palmiter RD, Brinster RL, Lee DC: Overexpression of TGFα in transgenic mice: Induction of epithelial hyperplasia, pancreatic metaplasia, and carcinoma of the breast. Cell 61:1121–1135, 1990.
51. Jhappan C, Stahle C, Harkins RN, Fausto N, Smith GH, Merlino GT: TGFα overexpression in transgenic mice induces liver neoplasia and abnormal development of the mammary gland and pancreas. Cell 61:1137–1146, 1990.
52. Matsui Y, Halter SA, Holt JT, Hogan BLM, Coffey RJ: Development of mammary hyperplasia and neoplasia in MMTV-TGFα transgenic mice. Cell 61:1147–1155, 1990.
53. Krane IM, Leder P: DNF/heregulin induces persistence of terminal end buds and adenocarcinomas in the mammary glands of transgenic mice. Oncogene 12:1781–1788, 1996.
54. Schlessinger J, Ullrich A: Growth factor signaling by receptor tyrosine kinases. Neuron 9:383–391, 1992.
55. Heldin C-H: Dimerization of cell surface receptors in signal transduction. Cell 80:213–223, 1995.
56. Carraway KL III, Cantley LC: A Neu acquaintance for ErbB3 and ErbB4: A role for receptor heterodimerization in growth signaling. Cell 78:5–8, 1994.
57. Earp HS, Dawson TL, Li X, Yu H: Heterodimerization and functional interaction between EGF receptor family members: A new signaling paradigm with implications for breast cancer research. Breast Cancer Res Treat 35:115–132, 1995.
58. Stern DF, Kamps MP: EGF-stimulated tyrosine phosphorylation of p185^{neu}: A potential model for receptor interactions. EMBO J 7:995–1001, 1988.
59. King CR, Borrello I, Bellot F, Comoglio P, Schlessinger J: EGF binding to its receptor triggers a rapid tyrosine phosphorylation of the *erb*B-2 protein in the mammary tumor cell line SK-BR-3. EMBO J 7:1647–1651, 1988.
60. Akiyama T, Saito T, Ogawara H, Toyoshima K, Yamamoto T: Tumor promoter and epidermal growth factor stimulate phosphorylation of the c-*erbB-2* gene product in MKN-7 human adenocarcinoma cells. Mol Cell Biol 8:1019–1026, 1988.
61. Wada T, Qian X, Greene MI: Intermolecular association of the p185^{neu} protein and EGF receptor modulates EGF receptor function. Cell 61:1339–1347, 1990.
62. Goldman R, Ben Levy R, Peles E, Yarden Y: Heterodimerization of the erbB-1 and erbB-2 receptors in human breast carcinoma cells: A mechanism for receptor transregulation. Biochemistry 29:11024–11028, 1990.
63. Kokai Y, Myers JN, Wada T, Brown VI, LeVea CM, Davis JG, Dobashi K, Greene MI: Synergistic interaction of p185c-*neu* and the EGF receptor leads to transformation of rodent fibroblasts. Cell 58:287–292, 1989.

64. Kim H-H, Sierke SL, Koland JG: Epidermal growth factor-dependent association of phosphatidylinositol 3-kinase with the *erb*B-3 gene product. J Biol Chem 269: 24747–24755, 1994.

65. Soltoff SP, Carraway KL III, Prigent SA, Gullick WJ, Cantley LC: ErbB3 is involved in activation of phosphatidylinositol 3-kinase by epidermal growth factor. Mol Cell Biol 14:3550–3558, 1994.

66. Tzahar E, Waterman H, Chen X, Levkowitz G, Karunagaran D, Lavi S, Ratzkin BJ, Yarden Y: A hierarchical network of interreceptor interactions determines signal transduction by Neu differentiation factor/neuregulin and epidermal growth factor. Mol Cell Biol 16:5276–5287, 1996.

67. Riese DJ II, van Raaij TM, Plowman GD, Andrews GC, Stern DF: The cellular response to neuregulins is governed by complex interactions of the erbB receptor family. Mol Cell Biol 15:5770–5776, 1995.

68. Cohen BD, Green JM, Foy L, Fell HP: Her-4-mediated biological and biochemical properties in NIH 3T3 cells. J Biol Chem 271:4813–4818, 1996.

69. Pinkas-Kramarski R, Soussan L, Waterman H, Levkowitz G, Alroy I, Klapper L, Lavi S, Seger R, Ratzkin BJ, Sela M, Yarden Y: Diversification of Neu differentiation factor and epidermal growth factor signaling by combinatorial receptor interactions. EMBO J 15:2452–2467, 1996.

70. Karunagaran D, Tzahar E, Beerli RR, Chen X, Graus-Porta D, Ratzkin BJ, Seger R, Hynes NE, Yarden Y: ErbB-2 is a common auxiliary subunit of NDF and EGF receptors: Implications for breast cancer. EMBO J 15:254–264, 1996.

71. Alimandi M, Romano A, Curia MC, Muraro R, Fedi P, Aaronson SA, Di Fiore PP, Kraus MH: Cooperative signaling of ErbB3 and ErbB4 in neoplastic transformation and human mammary carcinomas. Oncogene 10:1813–1821, 1995.

72. Wallasch C, Weisse FU, Niederfellner G, Jallal B, Issing W, Ullrich A: Heregulin-dependent regulation of HER2/neu oncogenic signaling by heterodimerization with HER3. EMBO J 14:4267–4275, 1995.

73. Graus-Porta D, Beerli RR, Hynes NE: Single-chain antibody-mediated intracellular retention of ErbB-2 impairs Neu differentiation factor and epidermal growth factor signaling. Mol Cell Biol 15:1182–1191, 1995.

74. Beerli RR, Graus-Porta D, Woods-Cook K, Chen X, Yarden Y, Hynes NE: Neu differentiation factor activation of ErbB-3 and ErbB-4 is cell specific and displays a differential requirement for ErbB-2. Mol Cell Biol 15:6496–6505, 1995.

75. Muller WJ, Arteaga CL, Muthuswamy SK, Siegel PM, Webster MA, Cardiff RD, Meise KS, Li F, Halter SA, Coffey RJ: Synergistic interaction of the Neu proto-oncogene product and transforming growth factor α in the mammary epithelium of transgenic mice. Mol Cell Biol 16:5726–5736, 1996.

76. Pawson T: Protein modules and signalling networks. Nature 373:573–580, 1995.

77. Berridge MJ: Inositol trisphosphate and calcium signaling. Nature 361:315–325, 1993.

78. Peles E, Ben Levy R, Or E, Ullrich A, Yarden Y: Oncogenic forms of the *neu*/HER2 tyrosine kinase are permanently coupled to phospholipase Cγ. EMBO J 10:2077–2086, 1991.

79. Fazioli F, Kim U-H, Rhee SG, Molloy CJ, Segatto O, Di Fiore PP: The *erb*B-2 mitogenic signaling pathway: Tyrosine phosphorylation of phopholipase C-γ and GTPase-activating protein does not correlate with erbB-2 mitogenic potency. Mol Cell Biol 11:2040–2048, 1991.

80. Jallal B, Schlessinger J, Ullrich A: Tyrosine phosphatase inhibition permits analysis of signal transduction complexes in p185$^{HER2/\mathit{neu}}$-overexpressing human tumor cells. J Biol Chem 267:4357–4363, 1992.

81. Scott GK, Dodson JM, Montgomery PA, Johnson RM, Sarup JC, Wong WL, Ullrich A, Shepard HM, Benz CC: p185^{HER2} signal transduction in breast cancer cells. J Biol Chem 266:14300–14305, 1991.

82. Peles E, Lamprecht R, Ben-Levy R, Tzahar E, Yarden Y: Regulated coupling of the *Neu* receptor to phosphatidylinositol 3′-kinase and its release by oncogenic activation. J Biol Chem 267:12266–12274, 1992.

83. Muthuswamy SK, Siegel PM, Dankort DL, Webster MA, Muller WJ: Mammary tumors expressing the *neu* proto-oncogene possess elevated c-Src tyrosine kinase activity. Mol Cell Biol 14:735–743, 1994.

84. Boguski MS, McCormick F: Proteins regulating Ras and its relatives. Nature 366: 643–654, 1993.

85. Luttrell DK, Lee A, Lansing TJ, Crosby RM, Jung KD, Willard D, Luther M, Rodriguez M, Berman J, Gilmer TM: Involvement of pp60$^{c\text{-}src}$ with two major signaling pathways in human breast cancer. Proc Natl Acad Sci USA 91:83–97, 1994.

86. Muthuswamy SK, Muller WJ: Activation of Src family kinases in Neu-induced mammary tumors correlates with their association with distinct sets of tyrosine phosphorylated proteins *in vivo*. Oncogene 11:1801–1810, 1995.

87. Muthuswamy SK, Muller WJ: Direct and specific interaction of c-Src with Neu is involved in signaling by the epidermal growth factor receptor. Oncogene 11:271–279, 1995.

88. McCormick F: Activators and effectors of *ras* p21 proteins. Curr Opin Gen Dev 4:71–76, 1994.

89. Daly RJ, Binder MD, Sutherland RL: Overexpression of the Grb2 gene in human breast cancer cell lines. Oncogene 9:9723–2727, 1994.

90. Janes PW, Daly RJ, deFazio A, Sutherland RL: Activation of the Ras signalling pathway in human breast cancer cells overexpressing *erb*B-2. Oncogene 9:3601–3608, 1994.

91. Xie Y, Pendergast AM, Hung M-C: Dominant-negative mutants of Grb-2 induced reversal of the transformed phenotypes caused by the point mutation-activated rat *HER-2/Neu**. J Biol Chem 270:30717–30724, 1995.

92. Segatto O, Pelicci G, Giuli S, Digiesi G, Di Fiore PP, McGlade J, Pawson T, Pelicci PG: Shc products are substrates of *erb*B-2 kinase. Oncogene 8:2105–2112, 1993.

93. Xie Y, Li K, Hung M-C: Tyrosine phosphorylation of Shc proteins and formation of Shc/Grb2 complex correlate to the transformation of NIH3T3 cells mediated by the point-mutation activated *neu*. Oncogene 10:2409–2413, 1995.

94. Ricci A, Lanfrancone L, Chiari R, Belardo G, Pertica C, Natali PG, Pelicci PG, Segatto O: Analysis of protein–protein interactions involved in the activation of the Shc/Grb-2 pathway by the ErbB-2 kinase. Oncogene 11:1519–1529, 1995.

94a. Dankort DL, Wang Z, Blackmore V, Moran MF, Muller WJ: Distinct tyrosine autophosphorylation sites negatively and positively modulate Meu-mediated transformation. Mol Cell Biol 17:5410–5425, 1997.

95. Vogel W, Lammers R, Huang J, Ullrich A: Activation of a phosphotyrosine phosphatase by tyrosine phosphorylation. Science 259:1611–1614, 1993.

96. Stein D, Wu J, Fuqua SAW, Roonprapunt C, Yajnik V, D'Eustachio P, Moskow JJ,

Buchberg AM, Osborne CK, Margolis B: The SH2 domain protein GRB-7 is co-amplified, and in a tight complex with HER2 in breast cancer. EMBO J 13:1331–1340, 1994.

97. Matsuda S, Kawamura-Tsuzuku J, Ohsugi M, Yoshida M, Emi M, Nakamura Y, Onda M, Yoshida Y, Nishiyama A, Yamamoto T: Tob, a novel protein that interacts with p185^{erbB2}, is associated with anti-proliferative activity. Oncogene 12:705–713, 1996.

98. Fedi P, Pierce JH, Di Fiore PP, Kraus MH: Efficient coupling with phosphatidylinositol 3-kinase, but not phospholipase Cγ or GTPase-activating protein, distinguishes ErbB-3 signaling from that of other ErbB/EGFR family members. Mol Cell Biol 14: 492–500, 1994.

99. Prigent SA, Gullick WJ: Identification of c-erbB-3 binding sites for phosphatidylinositol 3′-kinase and SHC using an EGF receptor/c-erbB-3 chimera. EMBO J 13:2831–2841, 1994.

100. Cooper JA, Howell B: The when and how of Src regulation. Cell 73:1051–1054, 1993.

101. Muthuswamy SK, Muller WJ: Activation of the Src family of tyrosine kinases in mammary tumorigenesis. Adv Cancer Res 64:111–123, 1994.

102. Jacobs C, Rubsamen H: Expression of pp60$^{c\text{-}src}$ protein kinase in adult and fetal human tissue: High activities in some sarcomas and mammary carcinomas. Cancer Res 43: 1696–1702, 1983.

103. Rosen N, Bolen JB, Schwartz AM, Cohen P, DeSeau V, Israel MA: Analysis of pp60$^{c\text{-}src}$ protein kinase activity in human tumor cell lines and tissues. J Biol Chem 261: 13754–13759, 1986.

104. Hennipman A, van Oirschot BA, Smits J, Rijksen G, Staal GEJ: Tyrosine kinase activity in breast cancer, benign breast disease, and normal breast tissue. Cancer Res 49:516–521, 1989.

105. Ottenhoff-Kalff AE, Rijksen G, van Beurden EACM, Hennipman A, Michels AA, Staal GEJ: Characterization of protein tyrosine kinases from human breast cancer: Involvement of the c-*src* oncogene product. Cancer Res 52:4773–4778, 1992.

106. Guy CT, Cardiff RD, Muller WJ: Induction of mammary tumors by expression of polyomavirus middle T oncogene: A transgenic mouse model for metastatic disease. Mol Cell Biol 12:954–961, 1992.

107. Courtneidge SA, Smith AE: Polyoma virus transforming protein associates with the product of the c-*src* cellular gene. Nature 303:435–439, 1983.

108. Bolen JB, Thiele CJ, Israel MA, Yonemoto W, Lipsich LA, Brugge JS: Enhancement of cellular *src* gene product associated tyrosyl kinase activity following polyoma virus infection and transformation. Cell 38:767–777, 1984.

109. Kornbluth S, Sudol M, Hanafusa H: Association of the polyomavirus middle-T antigen with c-*yes* protein. Nature 325:171–173, 1986.

110. Kypta RM, Hemming A, Courtneidge SA: Identification and characterization of p59^{fyn} (a *src*-like protein tyrosine kinase) in normal and polyoma virus transformed cells. EMBO J 7:3837–3844, 1988.

111. Cheng SH, Harvey R, Espino PC, Semba K, Yamamoto T, Toyoshima K, Smith AE: Peptide antibodies to the human c-*fyn* gene product demonstrate pp59$^{c\text{-}fyn}$ is capable of complex formation with the middle-T antigen of polyomavirus. EMBO J 7:3845–3855, 1988.

112. Guy CT, Muthuswamy SK, Cardiff RD, Soriano P, Muller WJ: Activation of the c-Src tyrosine kinase is required for the induction of mammary tumors in transgenic mice. Genes Dev 8:23–32, 1994.

113. Whitman M, Kaplan DR, Schaffhausen B, Cantley L, Roberts TM: Association of phosphatidylinositol kinase activity with polyoma middle-T competent for transformation. Nature 315:239–242, 1985.

114. Courtneidge SA, Heber A: An 81 kd protein complexed with middle T antigen and pp60$^{c\text{-}src}$: A possible phosphatidylinositol kinase. Cell 50:1031–1037, 1987.

115. Pallas DC, Shahrik LK, Martin BL, Jaspers S, Miller TB, Brautigan DL, Roberts TM: Polyoma small and middle T antigens and SV40 small t antigen form stable complexes with protein phosphatase 2A. Cell 60:167–176, 1990.

116. Walter G, Ruediger R, Slaughter C, Mumby M: Association of protein phosphatase 2A with polyoma virus medium tumor antigen. Proc Natl Acad Sci USA 87:2521–2525, 1990.

117. Dilworth SM, Brewster CEP, Jones MD, Lanfrancone L, Pelicci G, Pelicci PG: Transformation by polyoma virus middle T-antigen involves the binding and tyrosine phosphorylation of Shc. Nature 367:87–90, 1994.

118. Campbell KS, Ogris E, Burke B, Su W, Auger KR, Druker BJ, Schaffhausen BS, Roberts TM, Pallas DC: Polyoma middle tumor antigen interacts with SHC protein via the NPTY (Asn-Pro-Thr-Tyr) motif in middle tumor antigen. Proc Natl Acad Sci USA 91:6344–6348, 1994.

119. Su W, Liu W, Schaffhausen BS, Roberts TM: Association of polyomavirus middle tumor antigen with phospholipase C-γ1. J Biol Chem 270:12331–12334, 1995.

120. Pallas DC, Fu H, Haehnel LC, Weller W, Collier RJ, Roberts TM: Association of polyomavirus middle tumor antigen with 14-3-3 proteins. Science 265:535–537, 1994.

121. Soriano P, Montgomery C, Geske R, Bradley A: Targeted disruption of the c-*src* proto-oncogene leads to osteopetrosis in mice. Cell 64:693–702, 1991.

122. Stein PL, Vogel H, Soriano P: Combined deficiencies of Src, Fyn, and Yes tyrosine kinases in mutant mice. Genes Dev 8:1999–2007, 1994.

123. Webster MA, Cardiff RD, Muller WJ: Induction of mammary epithelial hyperplasias and mammary tumors in transgenic mice expressing a murine mammary tumor virus/activated c-*src* fusion gene. Proc Natl Acad Sci USA 92:7849–7853, 1995.

124. Cobb BS, Schaller MD, Leu T-H, Parsons JT: Stable association of pp60^{src} and pp59^{fyn} with the focal adhesion-associated protein tyrosine kinase, pp125^{FAK}. Mol Cell Biol 14:147–155, 1994.

125. Xing Z, Chen H-C, Nowlen JK, Taylor SJ, Shalloway D, Guan J-L: Direct interaction of v-Src with the focal adhesion kinase mediated by the src SH2 domain. Mol Cell Biol 5:413–421, 1994.

126. Schaller MD, Borgman CA, Cobb BS, Vines RR, Reynolds AB, Parsons JT: pp125^{FAK}, a structurally distinctive protein-tyrosine kinase associated with focal adhesions. Proc Natl Acad Sci USA 89:5192–5196, 1992.

127. Schaller MD, Hildebrand JD, Shannon JD, Fox JW, Vines RR, Parsons JT: Autophosphorylation of the focal adhesion kinase, pp125^{FAK}, directs SH2 dependent binding of pp60^{src}. Mol Cell Biol 14:1680–1688, 1994.

128. Schlaepfer DD, Hanks SK, Hunter T, van der Geer P: Integrin-mediated signal transduction linked to Ras pathway by GRB2 binding to focal adhesion kinase. Nature 372:786–791, 1994.

129. Schlaepfer DD, Hunter T: Evidence for in vivo phosphorylation of the Grb2 SH2-domain binding site on focal adhesion kinase by Src-family protein-tyrosine kinases. Mol Cell Biol 16:5623–5633, 1996.

130. Malik RK, Parsons JT: Integrin-mediated signaling in normal and malignant cells: A role of protein tyrosine kinases. Biochim Biophys Acta 1287:73–76, 1996.

131. Weiner TM, Liu ET, Craven RJ, Cance WG: Expression of focal adhesion kinase gene and invasive cancer. Lancet 342:1024–1025, 1993.

132. Owens LV, Xu L, Craven RJ, Dent GA, Weiner TM, Kornberg L, Liu ET, Cance WG: Overexpression of the focal adhesion kinase ($p125^{FAK}$) in invasive human tumors. Cancer Res 55:2752–2755, 1995.

133. Luetteke NC, Phillips HK, Qiu TH, Copeland NG, Earp HS, Jenkens NA, Lee DC: The mouse *waved-2* phenotype results from a point mutation in the EGF receptor tyrosine kinase. Genes Dev 8:399–413, 1994.

134. Fowler KJ, Walker F, Alexander W, Hibbs ML, Nice EC, Bohmer RM, Mann GB, Thumwood C, Maglitto R, Danks JA, Chetty R, Burgess AW, Dunn AR: A mutation in the epidermal growth factor receptor in waved-2 mice has a profound effect on receptor biochemistry that results in impaired lactation. Proc Natl Acad Sci USA 92:1465–1469, 1995.

135. Sandgren EP, Schroeder JA, Qui TH, Palmiter RD, Brinster RL, Lee DC: Inhibition of mammary gland involution is associated with transforming growth factor α but not c-myc-induced tumorigenesis in transgenic mice. Cancer Res 55:3915–3927, 1995.

136. Pollard JW, Hennighausen L: Colony stimulating factor 1 is required for mammary gland development during pregnancy. Proc Natl Acad Sci USA 91:9312–9316, 1994.

137. Muller WJ, Lee FS, Dickson C, Peters G, Pattengale P, Leder P: The int-2 gene product acts as an epithelial growth factor in transgenic mice. EMBO J 9:907–913, 1990.

138. Hadsell D, Greenberg N, Fligger J, Baumrucker C, Rosen J: Targeted expression of des(1–3) human insulin-like growth factor-1 in transgenic mice influences mammary gland development and IGF-binding protein expression. Endocrinology 137:321–330, 1996.

139. Iwamoto T, Takahashi M, Ito M, Hamaguichi M, Isobe K-I, Misawa N, Asai J-P, Yoshida T, Nakashima I: Oncogenicity of the *ret* transforming gene in MMTV/*ret* transgenic mice. Oncogene 5:535–542, 1990.

140. Liang TJ, Reid AE, Xavier R, Cardiff RD, Wang TC: Transgenic expression of tpr-met oncogene leads to development of mammary hyperplasia and tumors. J Clin Invest 97:2872–2877, 1996.

141. Kacinski BM, Scata KA, Carter D, Yee LD, Sapi E, King BL, Chambers SK, Jones MA, Pirro MH, Stanley ER, Rohrschneider LR: FMS (CSF-1 receptor) and CSF-1 transcripts and protein are expressed by human carcinomas *in vivo* and *in vitro*. Oncogene 6:941–952, 1991.

142. Filderman AE, Bruckner A, Kacinski BM, Deng N, Remold HG: Macrophage colony-stimulating factor (CSF-1) enhances invasiveness in CSF-1 receptor-positive carcinoma cell lines. Cancer Res 52:3661–3666, 1992.

143. Papa V, Pezzino V, Costantino A, Belfiore A, Giuffrida D, Frittitta L, Vannelli GB, Brand R, Goldfine ID, Vigneri R: Elevated insulin receptor content in human breast cancer. J Clin Invest 86:1503–1510, 1990.

144. Milazzo G, Giorgino F, Damante G, Sung C, Stampfer MR, Vigneri R, Goldfine ID, Belfiore A: Insulin receptor expression and function in human breast cancer cell lines. Cancer Res 52:3924–3930, 1992.

145. Cullen KJ, Yee D, Sly WS, Perdue J, Hampton B, Lippman ME, Rosen N: Insulin-like

growth factor receptor expression and function in human breast cancer. Cancer Res 50: 48–53, 1990.

146. Papa V, Gliozzo B, Clark GM, McGuire WL, Moore D, Fujita-Yamaguichi Y, Vigneri R, Golgfine ID, Pezzino V: Insulin-like growth factor-I receptors are overexpressed and predict a low risk in human breast cancer. Cancer Res 53:3736–3740, 1993.

23

THE *erb*B-2 GENE IN HUMAN CANCER: TRANSLATION FROM RESEARCH TO APPLICATION

FRANCISCO J. ESTEVA-LORENZO, LAKSHMI SASTRY, AND C. RICHTER KING

Vincent T. Lombardi Cancer Center, Georgetown University, Washington, DC 20007

23.1. INTRODUCTION

It is now clear that genetic changes occurring in somatic epithelial tissues cause human cancers. Among the first of these changes to be identified were abnormalities of the *erb*B-2 (or HER-2 or *neu*) gene. The *neu* gene was discovered in tumors of nitrosurea-treated rats, where it can be isolated as a dominant acting oncogene with a mutation in the transmembrane region (1,2). In human tumors the *erb*B-2 gene was identified by hybridization with v-*erb*B and epidermal growth factor (EGF) receptor probes and found to be amplified to 4–20 copies per cell (3). This amplification occurs frequently in different types of cancer and results in overexpression of an mRNA and protein of normal sequence. The human *erb*B-2 encodes $p185^{erbB\text{-}2}$, a transmembrane glycoprotein with intracellular tyrosine kinase activity and an extracellular domain that is very similar to the EGF binding domain of the EGF receptor (4). From the outset, the discoveries of *erb*B-2 gene amplification and its similarity to growth factor receptors were seen as opportunities for translational research. The consequences of this research have reached the clinic, with both therapeutic and diagnostic applications.

Hormones and Growth Factors in Development and Neoplasia, Edited by Robert B. Dickson and David S. Salomon
ISBN 0-471-16899-8

23.2. ROLE OF *erb*B-2 IN CANCER PATHOGENESIS

The available evidence indicates that abnormalities leading to *erb*B-2 overexpression occur very early in the pathogenesis of solid tumors. In experimental systems, overexpression of *erb*B-2 profoundly affects the cell phenotype. NIH 3T3 cells can be induced to the malignant phenotype by overexpression of a normal human *erb*B-2 coding sequence (5). Trangenic mice engineered to overexpress the normal (unactivated) *neu* gene under the mouse mammary tumor virus (MMTV) promoter develop mammary adenocarcinomas after a long latency period, with a tendency to metastasize (6). Introduction of the point mutated (activated) *neu* in mice embryos by homologous recombination results in the rapid development of breast carcinomas (7). In humans, the *erb*B-2 protein is not overexpressed in normal (8), hyperplastic, or dysplastic lesions of ductal breast epithelium (9). In contrast, *erb*B-2 is amplified and/or overexpressed in 60% of ductal carcinoma *in situ* (DCIS) (9–11), most of them being of the comedo type (10), and in almost 100% of cases of Paget's disease (12). *Erb*B-2 is also overexpressed in 20–30% of infiltrating ductal carcinomas of the breast (13–16), and this overexpression is functionally relevant to the conversion of normal cells to the malignant phenotype (9). Flow cytometry studies have shown that *erb*B-2 overexpression is an early step in the progression of breast cancer (17). Iglehart et al. (18) found that the DNA contents in the intraductal and invasive components of primary breast cancers were virtually identical, suggesting that the number of *erb*B-2 gene copies remains unchanged during progression from intraductal to invasive disease. In contrast to ductal neoplastic lesions, the involvement of *erb*B-2 overexpression is less clear in lobular breast cancers (19).

Similar to breast cancer, the *erb*B-2 gene is amplified in borderline ovarian tumors and invasive ovarian cancers but not in normal tissue (20), indicating a role in the initiation of the malignant process in ovarian carcinomas. In one study of colorectal cancer, *erb*B-2 overexpression was not observed in adenomas but was present in 50% of invasive carcinomas, suggesting that in this setting *erb*B-2 overexpression may be more prevalent in *de novo* carcinomas (21).

Our current understanding of human cancer indicates that development of the full malignant phenotype is part of a complex process involving different molecular events. There are clear interrelationships between *erb*B-2 overexpression and other changes leading to malignancy, such as activation of the Ras protein (22) and altered p53 expression (23). In one study (24), concomitant *erb*B-2 and p53 overexpression in breast cancer specimens was associated with high grade, large tumor size, involvement of axillary lymph nodes, negative progesterone receptors, and worse survival compared to either gene abnormality alone. *Erb*B-2 overexpression has been associated with the development of hormone-independent breast cancers (25). An inverse correlation between estrogen receptor and *erb*B-2 expression has been observed in human breast cancers (26).

23.3. *erb*B-2 OVEREXPRESSION, A COMMON CONTRIBUTOR TO CANCER

More than 70,000 patients with *erb*B-2-overexpressing malignancies are diagnosed every year in the United States (27) (Table 23.1). The *erb*B-2 protein is displayed

TABLE 23.1. Incidence of *erb*B-2-Overexpressing Human Tumors

Tumor	Incidence	Reference
Breast cancer	20–30%	13
Non-small cell lung cancer	30–50%	30
Gastric cancer	10–30%	34
Ovarian cancer	20–30%	162
Bladder cancer	8–39%	163, 164
Salivary gland cancer	21–38%	165, 166
Pancreatic cancer	20%	167
Cervical cancer	12%	168
Osteosarcoma	40%	169
Endometrial carcinoma	27%	32
Esophageal carcinoma	40%	170
Colorectal carcinoma	23–50%	21, 171, 172
Prostate adenocarcinoma	30–60%	173, 174

homogeneously on the surface of tumor cells, as disclosed by immunohistochemical methods (14), and there is a direct association between *erb*B-2 overexpression in the primary tumor and in metastatic lesions (28,29) (Fig. 23.1).

23.4. ROLE OF *erb*B-2 OVEREXPRESSION TO PREDICT PROGNOSIS

Amplification of the *erb*B-2 proto-oncogene and/or overexpression of the *erb*B-2 protein are associated with a poor prognosis in patients with breast (13), lung (30), ovarian (31), endometrial (32), nasopharyngeal (33), and gastric cancer (34). Women with *erb*B-2-overexpressing breast cancers are more likely to have involved axillary limph nodes and to present with rapidly proliferating tumors (11,35,36). To further define the clinical predictive value of *erb*B-2 overexpression in patients with node-negative breast cancer, the National Surgical Adjuvant Breast and Bowel Project (NSABP) (14) evaluated 292 cases and reported decreased survival for patients with *erb*B-2-overexpressing tumors in the absence of high nuclear grade. However, the correlation between *erb*B-2 overexpression and nuclear grade in node-negative patients is controversial (37). As with many prognostic factors, prospective clinical trials are needed to prove how best to utilize prognostic information provided by *erb*B-2 overexpression.

23.5. USING *erb*B-2 OVEREXPRESSION TO PREDICT THE RESPONSE TO THERAPY

One promising area of application of *erb*B-2 overexpression is in predicting the response to various local and systemic therapies. Evidence for interaction with cytotoxic agents was first obtained *in vitro*. Overexpression of the *erb*B-2 gene in fibroblasts confers resistance to tumor necrosis factor-α (TNFα) (38). Transfection of

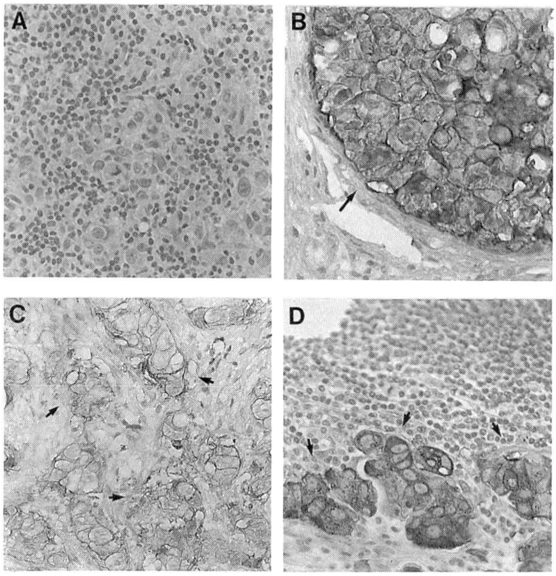

Fig. 23.1. Immunoperoxidase staining of breast cancer specimens with the anti-*erb*B-2 monoclonal antibody CB-11. Invasive ductal carcinoma showing no immunohistochemical staining of the *erb*B-2 protein (A). Intense membrane staining (arrows) is shown in a ductal carcinoma *in situ* (B), a primary invasive carcinoma (C), and a lymph node metastatic deposit (D). All cancer cells in panels B, C, and D overexpress the *erb*B-2 protein (ABC technique; $\times 400$).

the *erb*B-2 gene into breast cancer cells results in resistance to cisplatin (39). Anti-*erb*B-2 monoclonal antibodies can restore TNFα sensitivity to resistant *erb*B-2-overexpressing cells and enhance cisplatin- and doxorubicin-induced cytotoxicity (40). Recently, *erb*B-2 overexpression was found to predict ipsilateral local recurrence after conserving surgery in patients with breast cancer (41).

Overexpression of *erb*B-2 has now been shown to correlate with resistance to a variety of agents used in the treatment of breast, ovarian, and lung cancer. *Erb*B-2 overexpression in primary adenocarcinomas of the breast has been associated with resistance to adjuvant cyclophosphamide, methotrexate, and 5-fluorouracil (CMF) chemotherapy (11). Interestingly, *erb*B-2 overexpression was associated with longer survival in early-stage, node-positive breast cancer patients receiving increasing doses of doxorubicin. In this study (42), patients with stage II and stage III breast cancer were randomized to three doses of cyclophosphamide, doxorubicin, and 5-fluorouracil (CAF). The three regimens were identical except for the doxorubicin dose and schedule. A significant improvement in survival was observed in patients whose tumors overexpressed *erb*B-2 and who received higher doses of chemotherapy but not in patients with absent or minimal *erb*B-2 expression. This association between *erb*B-2 overexpression and drug sensitivity is intriguing. Either *erb*B-2-overexpressing tumors are more sensitive to doxorubicin or, alternatively, resistance is overcome by increasing the dose of doxorubicin. *Erb*B-2 overexpression has been associated with a poor response to hormonal therapy in breast cancer patients with estrogen and/or progesterone receptor-rich tumors (43). In patients with ovarian cancer, overexpression of *erb*B-2 has been associated with a poor response to carboplatin chemotherapy (44).

Recently, the *erb*B-2 extracellular domain (ECD) has been found to circulate in the serum of 25–30% of patients with infiltrating ductal carcinoma of the breast and in 40–50% of patients with metastatic disease (45–47). There is a good correlation between serum *erb*B-2 ECD levels and tissue overexpression in patients with breast cancer (48,49). The *erb*B-2 ECD can also be found in the serum of patients with ovarian cancer (48), stomach cancer (50,51), pancreatic cancer (52), and lung cancer (53,54). Elevated serum *erb*B-2 levels prior to hormonal therapy correlate with lower response rates and a shorter disease-free survival in patients with estrogen receptor- and/or progesterone receptor-positive metastatic breast cancer (55). Also, high serum levels of *erb*B-2 ECD before treatment correlate with partial resistance to adjuvant CMF chemotherapy (11). Recently, one of us (F.E.L.) (56) studied 188 patients with metastatic breast cancer treated with either CAF or CMF chemotherapy. Although there was no survival difference between high and low serum *erb*B-2 ECD, a significant improvement in disease-free-survival and overall survival was observed in patients with detectable *erb*B-2 ECD who had received doxorubicin-based chemotherapy compared to patients receiving CMF.

The molecular mechanisms that allow *erb*B-2 overexpression to effect drug resistance are unclear. *In vitro* data indicate that *erb*B-2-overexpressing chemoresistant lung cancer cells can be sensitized to cisplatin, doxorubicin, or etoposide by exposure to a specific tyrosine kinase inhibitor (57), suggesting that chemoresistance is mediated by *erb*B-2 tyrosine kinase activation. Arteaga et al. (58) showed that mechanisms of DNA repair important in reversing alkylator damage, such as DNA polymerase B, are increased in *erb*B-2-overexpressing breast cancer cells, making them more resistant to alkylators. However, resistance to cisplatin has been recently reported in an ovarian cell line with a low level of *erb*B-2 expression (59), indicating that other signaling pathways are implicated in the resistant phenotype.

23.6. SIGNALING FROM THE *erb*B-2 RECEPTOR

The *erb*B-2 tyrosine kinase is a cell surface receptor that consists of an extracellular cysteine-rich ligand binding domain, a single transmembrane domain, and an intracellular tyrosine kinase domain (4). The *erb*B-2 tyrosine kinase is a member of the class I type of receptors (60); other members of this family include the epidermal growth factor receptor (EGFR) (61), $p180^{erbB\text{-}3}$ (62), and $p180^{erbB\text{-}4}$ (63). It is proposed that *erb*B-2 overexpression leads to aberrant activation of the tyrosine kinase activity of the receptor and contributes to tumor progression (64). In fact, it has been demonstrated that tyrosine kinase activity is essential for the transforming ability of the *erb*B-2 oncoprotein (65). In addition, overexpression of *erb*B-2 in NIH 3T3 cells results in transformation of these cells, further suggesting that overexpression is sufficient for transformation (5). To understand how *erb*B-2 is involved in generating the malignant phenotype, it is essential to understand the signaling mechanisms originating from the *erb*B-2 tyrosine kinase.

The ligand for the *erb*B-2 protein has not yet been determined. However, heregulin (HRG), the ligand for *erb*B-3 and *erb*B-4, can activate *erb*B-2 via heterodimerization (62,66–69). Similarly, EGF can activate *erb*B-2 via heterodimerization with EGFR (70,71). The functional significance of heterodimerization is unclear, although it is proposed that such interreceptor activation and transphosphorylation may modulate growth regulatory signals by recruiting different intracellular signaling proteins.

The downstream signaling pathways from the *erb*B-2 tyrosine kinase are continuing to be defined. Autophosphorylation and dimerization of *erb*B-2 tyrosine kinase result in the generation of binding sites for a number of intracellular signaling molecules including PLC-γ, Ras-GAP, Grb2, Shc, and Grb7 (Fig. 23.2). Both PLC-γ and Ras-GAP are phosphorylated on tyrosines upon binding to p185 and form stable complexes (72). In experimental systems expressing EGFR/*erb*B-2 chimeras, the association and phosphorylation of PLC-γ and Ras-GAP are ligand (EGF) dependent (73,74). However, the importance of these two proteins in mediating the mitogenic signaling pathways from p185 is still unclear. Comparison of the phosphorylation levels of these proteins in NIH 3T3 cells transformed with point mutation-activated *neu* or normal *neu* shows no difference, suggesting that the phosphorylation of PLC-γ and Ras-GAP does not correlate with the cellular transformation mediated by transforming *neu* (75).

The adaptor protein Grb2 is composed of an src homology 2 domain (SH2) flanked by an amino-terminal SH3 domain (NSH3) and a carboxy-terminal SH3 domain (CSH3) (76). This adapter protein is expressed in all cell types, although increased expression of Grb2 has been observed in a number of *erb*B-2-overexpressing breast cancer cells (77,78). The SH2 domain of Grb2 binds phosphotyrosine sequences on activated p185. The major binding site of the SH2 domain has been localized to Y-1139 on the receptor (79). One target of the SH3 domains of Grb2 is the Ras exchange factor, Sos1. It is proposed that binding of Grb2 to the activated receptor results in the transport of the Grb2–Sos1 complex to the membrane and subsequent activation of Ras and the mitogen activated protein (MAP)-kinase signaling pathway. In support of this proposal, increased Grb2–Sos1 complex and increased activation of the Ras pathway have been observed in *erb*B-2-overexpressing breast cancer cells (77). However, the consequences of blockade of the Ras signaling pathway in these cells are yet to be determined, and therefore an absolute requirement of this pathway for transformation and malignancy of breast cancer cells is hard to assess. SH3 domain deletion mutants of Grb2 reverse the morphology of NIH 3T3 cells transformed with point mutation-activated *neu* and decrease Ras activation, lending support to the importance of the Ras signaling pathway in fibroblasts (81). We have recently demonstrated that the Grb2-SH3 domains can bind proteins (p228, p140, dynamin, p55, p28) other than Sos1 in a number of human cancer cells (82). These experiments strongly support the notion that Grb2 may have functions other than linking receptor tyrosine kinases to the Ras signaling pathway and that these functions may be mediated by the novel Grb2-SH3 domain binding proteins. Characterization of the novel Grb2 binding proteins may reveal still undiscovered signaling pathways from p185.

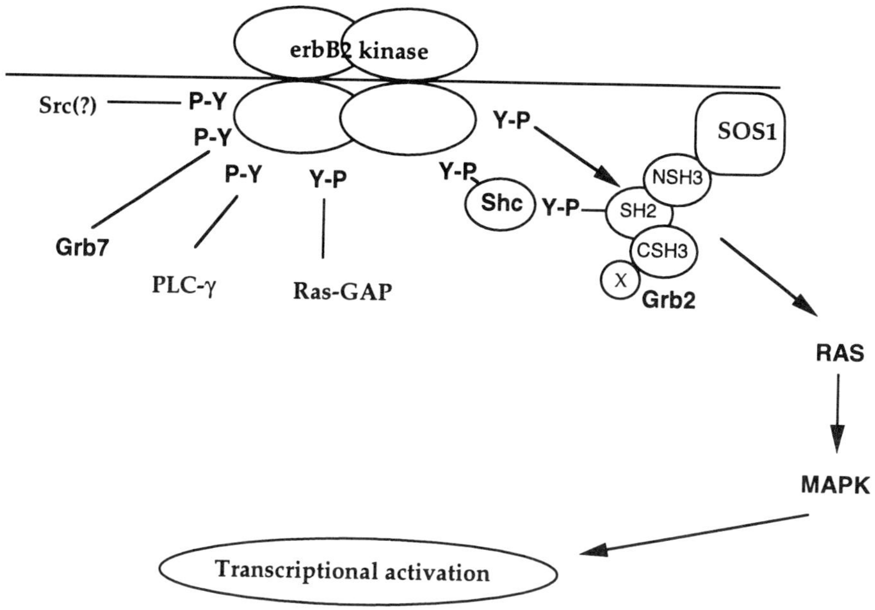

Fig. 23.2. Signaling pathways from *erb*B-2.

The Shc proteins are SH2 domain-containing proteins that bind various growth factor receptors including p185 and EGFR (83,84). Shc is expressed ubiquitously, and three isoforms of Shc (66,52,46) are detected in most cell types. The binding sites for Shc in *erb*B-2 have been localized to Y-1221 and Y-1248 *in vitro* (79). The Shc proteins may utilize a SH2 domain as well as a protein tyrosine binding domain (PTB) to interact with receptor tyrosine kinases (85). PTB domains have affinities similar to those of the SH2 domain, although they may have different kinetics (86). Interaction of Shc with receptor tyrosine kinases leads to the phosphorylation of the Shc proteins (84). The tyrosine phosphorylated proteins are substrates for the Grb2 SH2 domain (87). The trimeric Shc–Grb2–Sos1 complex has been determined to associate with p185 in NIH 3T3 cells and to play a role in the Ras signaling pathway. However, the contribution of the Shc pathway in p185-mediated transformation of breast cancer cells still needs to be determined. Shc is constitutively activated in a number of tumor cells, and overexpression of Shc in NIH 3T3 cells transforms these cells (88).

Another SH2 domain containing the protein Grb7 has been found to associate with p185 in some breast cancer cells (89). The Grb7 sequence also shows the presence of a pleckstrin homology domain (PH). Grb7 is amplified and overexpressed in some breast cancer cells, with overexpression correlating with *erb*B-2 overexpression. The Grb7 molecule has been mapped to chromosome 11 to a region that also contains the *erb*B-2 locus. Coimmunoprecipitation of Grb7 and p185 was observed in SK-BR-3 and BT 474 cells; association of Grb7 with Shc was also observed in EGF-stimulated SK-BR-3 cells. Although the downstream targets of Grb7 have not yet been identified and although the signaling pathways transduced by this molecule are still unclear, it is conceivable that Grb7 has an important function in the signaling pathways of activated p185.

Elevated activities of pp60c-src together with p185 have been reported in a number of primary human breast cancers (90). In *in vitro* assays, GST-Src protein binds p185 as well as EGFR in SK-BR-3 cells (91). Also, recent data suggest that EGFR is phosphorylated *in vivo* at nonautophosphorylation sites and that these sites can act as docking sites for Src, p85-a, and potentially other SH2 binding sites (92). These results suggest that Src and other nonreceptor tyrosine kinases may play an important role in the signals elicited from p185.

Discovery of novel intracellular proteins is likely to lead to new signaling pathways from receptor tyrosine kinases. Recent studies demonstrate the existence of a protein, sorting nexin (SNX1), responsible for internalization of EGFR, although this protein has no effect on p185 internalization (93). The mechanisms of internalization of p185 are still being investigated. The small G-proteins, Rac and Rho, are important for membrane ruffling and stress fiber formation (94). The involvement of these proteins in p185 signaling is not known. The connection to the STAT pathways and the cell cycle are other areas of research that would provide exciting mechanistic insights into p185 signaling. At the transcription factor level, *ets*, AP-1, and NF-kB have been shown to be activated by p185, and the activation appears to be through the Ras signaling pathway (95).

23.7. USING *erb*B-2 AS A THERAPEUTIC TARGET

The strong linkage to the malignant process and the developing information about its mode of action make *erb*B-2 a promising and popular target for the development of new cancer therapies. There have been many approaches, as outlined in Figure 23.3. All share the common strengths that originate with the characteristics of *erb*B-2-mediated malignancy. The level of *erb*B-2 expression found in human cancer cells where gene amplification occurs is much higher than that found in normal adult tissues. While expression in some normal tissues is found, it is at levels well below those identified in these cancers (8). This suggests that *erb*B-2 overexpression targeted therapies may have limited activity against the cells of normal tissues and thus manageable toxicities. A second attractive aspect of the *erb*B-2 target is that the target is present in a very high proportion of tumor cells. The great majority of primary breast cancers, where overexpression occurs, show uniform intense immunohistochemical staining in all cells of a tumor (14) (Fig. 23.1). This characteristic suggests that in a given patient an anti-*erb*B-2 therapy should be able to attack nearly all cancer cells. Finally, the *erb*B-2 overexpression phenotype is apparently shared by the primary tumor and metastatic sites (96). This is important in that it indicates that therapy for metastatic disease can be selected based on analysis of the primary tumor, and that an anti-*erb*B-2 therapy should be able to treat all sites of disease.

Experimental therapies designed to treat cancers that overexpress *erb*B-2 fall into four general classes. The first and most advanced strategies use the specificity

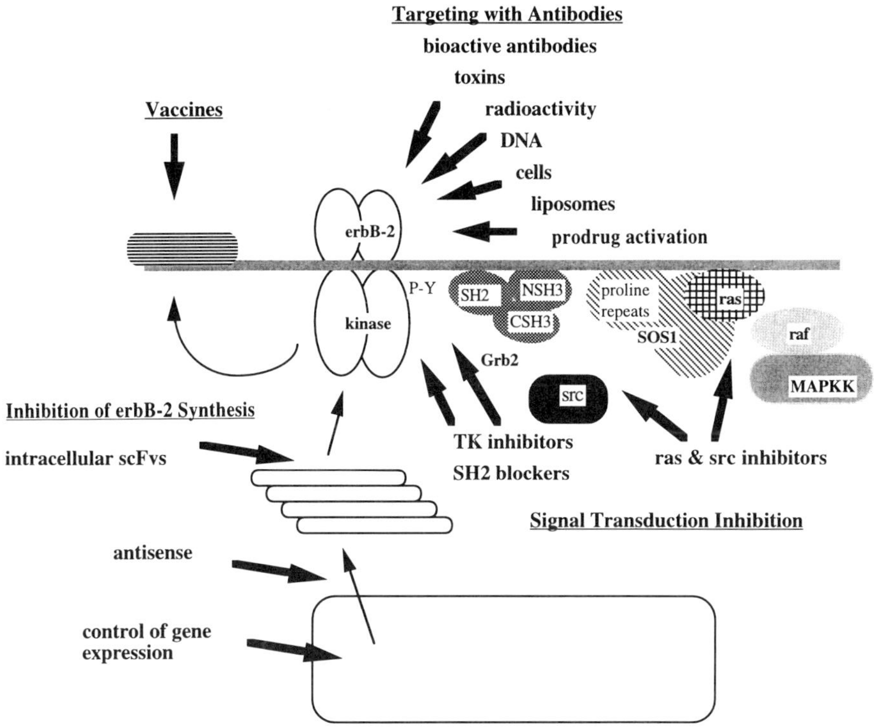

Fig. 23.3. Strategies of anti-*erb*B-2-directed therapies.

of an anti-*erb*B-2 monoclonal antibody to deliver biologic activity to cancer cells. A second approach seeks to inhibit the signals sent by the *erb*B-2 protein leading to cell malignancy. A third class of strategy interferes with the synthesis of mature $p185^{erbB\text{-}2}$ before it can impart its oncogenic action. Finally, there are attempts to generate vaccines that could recognize $p185^{erbB\text{-}2}$-derived peptides presented for recognition by cytolytic T-cells.

23.7.1. Targeting *erb*B-2 Using Antibody Specificity

The simplest approach to anti-*erb*B-2 therapy is to treat cells with monoclonal anti-$p185^{erbB\text{-}2}$ antibodies. A number of studies have shown that such antibodies could reduce the growth rate of human tumor cells (40,97–101) and sensitize them to chemotherapeutic agents (102,103). The underlying mechanism of action of these effects may require monoclonal antibody binding to modify the functional activity of $p185^{erbB\text{-}2}$ as a transforming protein in some way. There is evidence that internalization may be necessary for biologic activity (98,103). Whether this will be the dominant mechanism that causes anticancer activity in patients is not clear, however (104). The humanized versions are very effective in causing an antibody-directed cytotoxicity (ADCC) (105). Consequently, activity in human clinical trials may result from several biologic activities of antibodies. Monoclonal antibodies with biologic activity have been humanized and are undergoing clinical trial (106). Responses have been identified, and limited toxicity has been seen even with large doses of antibody (106). Clinical trials are currently being conducted in which antibody is given alone or in combination with chemotherapeutic drugs (107).

To increase the potency of antibody-directed therapy, the specificity of an antibody-combining site has been combined with a wide variety of effector agents. These approaches rely on the increased target density of $P185^{erbB\text{-}2}$ present on the tumor cell compared to normal tissue for their preferential antitumor effect. Toxins have been targeted to $p185^{erbB\text{-}2}$ using chemical conjugation with intact antibodies (108,109) and by generation of recombinant molecules in which an antibody-combining site is fused directly to the toxin (108,110). The antibody-combining site has been generated both as a single-chain structure and as a disulfide-linked Fv (111). These immunotoxins and oncotoxins are extremely potent molecules and show strong selectivity for $p185^{erbB\text{-}2}$ binding. Recombinant toxins show promise in that they can be safely delivered to experimental animals at effective doses (112). Recombinant toxins are also attractive in that they are relatively small proteins and may be able to penetrate tumors more effectively. This is indicated by pharmacokinetic measurements showing that recombinant toxins readily leave the circulation (112,113). One potential limitation facing the development of toxin targeting is the potential for an immune response to the protein.

A strategy that has some similarity to toxin targeting is the delivery of enzymes that can activate prodrugs to the environment of the tumor. This strategy has the theoretical advantages that not every cell must be bound and that the activated prodrug may be able to penetrate poorly vascularized tumors. Approaches of this kind have been reported for carboxypeptidase (114) and β-lactamase (115) attached to anti-$p185^{erbB\text{-}2}$ monoclonal antibodies. Another variation on the delivery strategy is to generate a vehicle for the transport of DNA to the tumor cell. In these constructs, an antibody combining site is fused to a DNA binding protein such as protamine (116) or Gal4 repressor (117).

Liposomes have the potential to deliver a variety of anticancer drugs to tumors. These have been generated with anti-$p185^{erbB\text{-}2}$ antibodies on their surface to increase delivery to the target tumor. It has been shown that such liposomes can increase the delivery of doxorubicin to mouse xenografts that overexpress $p185^{erbB\text{-}2}$ (118).

Radionuclides have also been attached to anti-$p185^{erbB\text{-}2}$ antibodies. These have shown the ability to image tumors in experimental animals and in humans (119–121). The therapeutic application of radiolabeled antibodies to this and other types of cancer may depend on the attachment of isotopes of sufficient specific activity and suitable decay pathlength.

Two approaches to immunotherapy have been developed that rely on targeting by anti-$p185^{erbB\text{-}2}$ antibodies. Both are designed to deliver immune effector cells to the tumor. In the first approach, a single chimeric protein molecule is designed to have two antibody binding specificities, one that binds a $p185^{erbB\text{-}2}$ and the other that binds an immune cell via either CD16, Fcγ receptor III (122), or CD3

(123,124). Phase I clinical studies have allowed toxicity to be assessed, and there is evidence that a biologically relevant concentration of the experimental therapy has been achieved (125).

Another immunotherapy approach rebuilds the recognition specificity of lymphocytes in such a way that they can bind to $p185^{erbB\text{-}2}$-expressing cells. To do this, a chimeric gene is generated that encodes an antibody combining site fused to the zeta chain of the T cell. The concept is to generate a cell capable of initiating an immune response following binding to $p185^{erbB\text{-}2}$-expressing tumor cells. When expressed at the cell surface, the chimeric protein serves as an analog of the natural T-cell receptor but binds tightly to $p185^{erbB\text{-}2}$. Cytolytic activity of such constructs for appropriate cell lines has been demonstrated *in vitro* (127–129). In animal models, such lymphocytes have been located in xenografts of human tumors with high levels of $p185^{erbB\text{-}2}$ and lytic activity identified (127).

23.7.2. Inhibition of $p185^{erbB\text{-}2}$-Initiated Signal Transduction

It is clear that $p185^{erbB\text{-}2}$ causes cell transformation because it initiates an inappropriate growth stimulus when overexpressed. The developing information on how cell surface growth factor receptors send their signals has created several efforts to develop agents that can block these events within tumor cells (130,131). Discovery efforts in this area have focused on the identification of small-molecule agents. This provides significant potential advantages for the development of a pharmaceutical. In particular, there may be advantages in the areas of tumor penetration, oral availability, and cost. Small-molecule drug development can also present some difficulties in the area of biochemical specificity of the drug and in its ability to cross cell membranes.

Inhibition of the $p185^{erbB\text{-}2}$ tyrosine kinase has been reported for different classes of compounds. Interestingly, some specificity for $p185^{erbB\text{-}2}$ has been reported compared to the very similar EGFR (131). Compounds can be very potent, with IC_{50} values of 1 μM or less (131). Some compounds have been tested in animal models and show antitumor activity (132). Sensitization to chemotherapeutic agents has also been shown (57,133). Tyrosine kinase inhibitors have also shown the ability to sensitize cells to chemotherapeutic agents. Agents that act to cause down-modulation of $p185^{erbB\text{-}2}$ have been reported (134,135). These compounds, the anasamycins, require a region of the extracellular domain of $p185^{erbB\text{-}2}$ for activity (134). They have been demonstrated to cause reduced activity of $p185^{erbB\text{-}2}$ in animal models of $p185^{erbB\text{-}2}$-overexpressing cancer (135).

Targets that are proximal to the tyrosine kinase are the SH2 and PTB domains used by Grb2 and SHC to bind to the receptor. Blockade at this point would limit communication to the ras/raf/MAP-kinase pathway. A antigrowth effect is the very likely consequence, although no direct experiments involving human cancer cells have been reported. In rodent fibroblasts transformed by mutationally activated *neu* gene dominant negative mutants of Grb2 can limit cell transformation (136). Construction of small-molecule SH2 domain blockers has been studied primarily with the intention of designing mimics of the natural phosphopeptide ligands. Very effective Grb2 SH2 domain antagonists have been generated that also contain the additional useful characteristic of resistance to phosphatases (137,138). Unfortunately, these compounds probably have limited activity, as they do not penetrate cells effectively due to the highly charged phosphate or phosphonate. An alternative may be possible based on the discovery of small, nonphosphorylated peptides that bind to the Grb2 SH2 domain with micromolar affinity (139). The design of non-peptide analogs of this structure is likely to be a significant advance in the testing of Grb2 SH2 domain inhibitors.

Farther downstream in the same pathway are farnesylation inhibitors (140–142). These compounds are designed to block the membrane localization of the Ras GTPase and thus prevent its function. The compounds have low toxicity in animal studies and can prevent the formation of experimental tumors that depend on an oncogenic form of Ras for their transformed character (143). Farnesylation inhibitors also act on a variety of other human tumor cell lines *in vitro* (144,145). In these experiments, tumor cell lines that overexpress p185$^{erb\text{B-2}}$ are inhibited in their ability to grow in anchorage-independent fashion. This suggests that inhibitors of Ras signaling could have valuable therapeutic effects against tumors of this kind.

Other signal transduction targets are under investigation and may have relevance to tumors that overexpress p185$^{erb\text{B-2}}$. These include pI3 kinase, which does not apparently bind directly to p185$^{erb\text{B-2}}$ but may be recruited by phosphorylation of *erb*B-3 (146). In addition, an important role for *src* is suggested in cells that overexpress p185$^{erb\text{B-2}}$, and therefore the development of *src* inhibitors is likely to be valuable (147).

23.7.3. Inhibition of p185$^{erb\text{B-2}}$ Synthesis

Since the primary biochemical event in inducing malignancy is the p185$^{erb\text{B-2}}$ tyrosine kinase, several investigators have developed strategies to prevent the synthesis of mature p185$^{erb\text{B-2}}$ at the cell membrane. The strategy most immediate to the underlying genetic defect derives from the finding that the *erb*B-2 gene can be repressed by the introduction of the adenovirus E1a gene (148). Delivery of E1a expression constructs into human tumor cell lines using liposomes has resulted in inhibition of p185$^{erb\text{B-2}}$ expression and loss of tumorigenicity (149). Antisense approaches to limiting *erb*B-2 expression have also been reported (150,151).

A novel method for blocking p185$^{erb\text{B-2}}$ has been developed that captures the p185$^{erb\text{B-2}}$ protein in the endoplasmic reticulum during synthesis (152–155). In this strategy, a gene is delivered to tumor cells that encodes a single chain antibody (SCFv) capable of binding p185$^{erb\text{B-2}}$. The SCFv-encoding gene has been directed for expression as a secretory protein by the inclusion of a leader sequence. The expression of SCFv in this location prevents the surface expression and thus the function of p185$^{erb\text{B-2}}$. Apoptotic cells have been observed as a result (156), and there is a dramatic reduction in tumorigenicity of *erb*B-2-overexpressing cell lines. The SCFv-encoding genes have been introduced into adenovirus vectors (154). Clinical trial in the ovarian cancer setting is anticipated.

23.7.4. Vaccines

As described, the level of expression of p185$^{erbB\text{-}2}$ seen in human tumors with gene amplification is unprecedented in normal tissues. Consequently, it has been proposed that peptides derived from the *erb*B-2 sequence could be presented by the human major histocompatibility complex (MHC) and recognized by the T-cell receptor on lymphocytes (157,158). Although the gene amplified in human tumors is apparently of a normal sequence, tolerance might be broadened by a quantitative increase in epitope display. Consistent with this concept is the finding that T-helper, cytotoxic, and antibody responses have been identified in patients whose tumor overexpresses *erb*B-2 (159). As a consequence, several peptide sequences within the *erb*B-2 have been identified that can mediate T-helper or cytotoxic responses (160,161). Such peptides, or potentially a larger expression construct containing more of the *erb*B-2 coding sequence, might be used as a vaccine. No evidence of autoimmune consequences has been found in these peptides or in tumor-bearing patients (157).

23.8. CONCLUSION

The attractiveness of the *erb*B-2 target has initiated a variety of therapeutic strategies based on the most current understanding of molecular biology. These strategies are being tested at different stages of development, with the most advanced ones showing utility in clinical trials. It seems very likely that more than one strategy will show effectiveness. It is also very likely that no one strategy will prove routinely curative. Consequently, such therapies will probably be most effective when combined with each other and with existing agents. The rapid advances in understanding of the role of *erb*B-2 in human cancer and in the development of new therapies brings hope for improved treatment in the future.

REFERENCES

1. Shih C, Padhy LC, Murray M, Weinberg RA: Transforming genes of carcinomas and neuroblastomas introduced into mouse fibroblasts. Nature 290:261–264, 1981.
2. Schechter AL, Stern DF, Vaidyanathan L, Decker SJ, Drebin JA, Greene MI, Weinberg RA: The *neu* oncogene: An *erb*B-related gene encoding a 185,000-Mr tumour antigen. Nature 312:513–516, 1984.
3. King CR, Kraus MH, Aaronson SA: Amplification of a novel v-*erb*B-related gene in a human mammary carcinoma. Science 229:974–976, 1985.
4. Yamamoto T, Ikawa S, Akiyama T, Semba K, Nomura N, Miyajima N, Saito T, Toyoshima K: Similarity of protein encoded by the human c-*erb*B-2 gene to epidermal growth factor receptor. Nature 319:230–234, 1986.
5. Di Fiore PP, Pierce JH, Kraus MH, Segatto O, King CR, Aaronson SA: *Erb*B-2 is a potent oncogene when overexpressed in NIH/3T3 cells. Science 237:178–182, 1987.
6. Guy CT, Webster MA, Schaller M, Parsons TJ, Cardiff RD, Muller WJ: Expression of

the *neu* protooncogene in the mammary epithelium of transgenic mice induces metastatic disease. Proc Natl Acad Sci USA 89:10578–10582, 1992.

7. Guy CT, Cardiff RD, Muller WJ: Activated *neu* induces rapid tumor progression. J Biol Chem 271:7673–7678, 1996.
8. Press MF, Cordon-Cardo C, Slamon DJ: Expression of the HER-2/*neu* proto-oncogene in normal human adult and fetal tissues. Oncogene 5:953–962, 1990.
9. Allred DC, Clark GM, Molina R, Tandon AK, Schnitt SJ, Gilchrist KW, Osborne CK, Tormey DC, McGuire WL: Overexpression of her-2/*neu* and its relationship with other prognostic factors change during the progression of *in situ* to invasive breast cancer. Hum Pathol 23:974–979, 1992.
10. van de Vijver MJ, Peterse JL, Mooi WJ, Wisman P, Lomans J, Dalesio O, Nusse R: *Neu*-protein overexpression in breast cancer. Association with comedo-type ductal carcinoma *in situ* and limited prognostic value in Stage II breast cancer. N Engl J Med 319:1239–1245, 1988.
11. Gusterson BA, Gelber RD, Goldhirsch A, et al: Prognostic importance of c-*erb*B-2 expression in breast cancer. International (Ludwig) breast cancer study group. J Clin Oncol 10:1049–1056, 1992.
12. Keatings L, Sinclair J, Wright C, Corbett IP, Watchorn C, Hennessy C, Angus B, Lennard T, Horne CH: C-*erb*B-2 oncoprotein expression in mammary and extramammary Paget's disease: An immunohistochemical study. Histopathology 17:243–247, 1990.
13. Slamon DJ, Clark GM, Wong SG, Levin WJ, Ullrich A, McGuire WL: Human breast cancer: Correlation of relapse and survival with amplification of the her-2/*neu* oncogene. Science 235:177–182, 1987.
14. Paik S, Hazan R, Fisher ER, Sass RE, Fisher B, Redmond C, Schlessinger J, Lippman ME, King CR: Pathologic findings from the National Surgical Adjuvant Breast and Bowel Project: Prognostic significance of *erb*B-2 protein overexpression in primary breast cancer. J Clin Oncol 8:103–112, 1990.
15. Klijn JG, Berns PM, Bontenbal M, Alexieva-Figusch J, Foekens JA: Clinical breast cancer, new developments in selection and endocrine treatment of patients. J Steroid Biochem Mol Biol 43:211–221, 1992.
16. Barnes DM, Bartkova J, Camplejohn RS, Gullick WJ, Smith PJ, Millis RR: Overexpression of the c-*erb*B-2 oncoprotein: Why does this occur more frequently in ductal carcinoma *in situ* than in invasive mammary carcinoma and is this of prognostic significance? Eur J Cancer 28:644–648, 1992.
17. Shackney SE, Smith CA, Pollice AA, Janocko LE, Singh SG, Groft DW, Brown KA, Hartsock RJ: Preferred genetic evolutionary sequences in human breast cancer: A case study. Cytometry 21:6–13, 1995.
18. Iglehart JD, Kerns BJ, Huper G, Marks JR: Maintenance of DNA content and *erb*B-2 alterations in intraductal and invasive phases of mammary cancer. Breast Cancer Res Treat 34:253–263, 1995.
19. Porter PL, Garcia R, Moe R, Corwin DJ, Gown AM: C-*erb*B-2 oncogene protein in *in situ* and invasive lobular breast neoplasia. Cancer 68:331–334, 1991.
20. Wong YF, Cheung TH, Lam SK, Lu HJ, Zhuang YL, Chan MY, Chung TK: Prevalence and significance of her-2/*neu* amplification in epithelial ovarian cancer. Gynecol Obstet Invest 40:209–212, 1995.
21. Shirai H, Ueno E, Osaki M, Tatebe S, Ito H, Kaibara N: Expression of growth factors

and their receptors in human early colorectal carcinomas: Immunohistochemical study. Anticancer Res 15:2889–2894, 1995.

22. Dati C, Muraca R, Tazartes O, Antoniotti S, Perroteau I, Giai M, Cortese P, Sismondi P, Saglio G, De Bortioli M: C-*erb*B-2 and ras expression levels in breast cancer are correlated and show a co-operative association with unfavorable clinical outcome. Int J Cancer 47:833–838, 1991.
23. Barbareschi M, Leonardi E, Mauri FA, Serio G, Palma PD: P53 and c-*erb*B-2 protein expression in breast carcinomas. An immunohistochemical study including correlations with receptor status, proliferation markers, and clinical stage in human breast cancer. Am J Clin Pathol 98:408–418, 1992.
24. Nakopoulou LL, Alexiadou A, Theodoropoulos GE, Lazaris AC, Tzonou A, Keramopoulos A: Prognostic significance of the co-expression of p53 and c-*erb*B-2 proteins in breast cancer. J Pathol 179:31–38, 1996.
25. Liu Y, el-Ashry D, Chen D, Ding IY, Kern FG: Mcf-7 breast cancer cells overexpressing transfected c-*erb*B-2 have an in vitro growth advantage in estrogen-depleted conditions and reduced estrogen-dependence and tamoxifen-sensitivity in vivo. Breast Cancer Res Treat 34:97–117, 1995.
26. Keshgegian AA: *Erb*B-2 oncoprotein overexpression in breast carcinoma: Inverse correlation with biochemically- and immunohistochemically-determined hormone receptors. Breast Cancer Res Treat 35:201–210, 1995.
27. Parker SL, Tong T, Bolden S, Wingo PA: Cancer statistics, 1996. CA Cancer J Clin 46:5–27, 1996.
28. Slamon DJ, Godolphin W, Jones LA, Holt JA, Wong SG, Keith DE, Levin WJ, Stuart SG, Udove J, Ullrich A, Press MF: Studies of the HER-2/*neu* proto-oncogene in human breast and ovarian cancer. Science 244:707–712, 1989.
29. Lovekin C, Ellis IO, Locker A, Robertson JF, Bell J, Nicholson R, Gullick WJ, Elston CW, Blamey RW: C-*erb*B-2 oncoprotein expression in primary and advanced breast cancer [published erratum appears in br j cancer 1991 jul; 64(1): 202]. Br J Cancer 63:439–443, 1991.
30. Kern JA, Schwartz DA, Nordberg JE, Weiner DB, Greene MI, Torney L, Robinson RA: p185*neu* expression in human lung adenocarcinomas predicts shortened survival. Cancer Res 50:5184–5187, 1990.
31. Felip E, Del Campo JM, Rubio D, Vidal MT, Colomer R, Bermejo B: Overexpression of c-*erb*B-2 in epithelial ovarian cancer. Prognostic value and relationship with response to chemotherapy. Cancer 75:2147–2152, 1995.
32. Berchuck A, Rodriguez G, Kinney RB, Soper JT, Dodge RK, Clarke-Pearson DL, Bast RC, Jr.: Overexpression of her-2/*neu* in endometrial cancer is associated with advanced stage disease. Am J Obstet Gynecol 164:15–21, 1991.
33. Roychowdhury DF, Tseng A Jr, Fu KK, Weinburg V, Weidner N: New prognostic factors in nasopharyngeal carcinoma tumor angiogenesis and c-*erb*B-2 expression. Cancer 77:1419–1426, 1996.
34. Lee HR, Kim JH, Uhm HD, Ahn JB, Rha SY, Cho JY, Lee JI, Lee KH, Chung HC, Roh JK, Min JS, Lee KS, Shin DH, Kim BS, Hong SW, Choi JH: Overexpression of c-*erb*B-2 protein in gastric cancer by immunohistochemical stain. Oncology 53:192–197, 1996.
35. Anbazhagan R, Gelber RD, Bettelheim R, Goldhirsch A, Gusterson BA: Association of c-*erb*B-2 expression and s-phase fraction in the prognosis of node positive breast cancer. Ann Oncol 2:47–53, 1991.

36. Gasparini G, Gullick WJ, Bevilacqua P, Sainsbury JR, Meli S, Boracchi P, Testolin A, La Malfa G, Pozza F: Human breast cancer: Prognostic significance of the c-*erb*B-2 oncoprotein compared with epidermal growth factor receptor, DNA ploidy, and conventional pathologic features. J Clin Oncol 10:686–695, 1992.

37. Schonborn I, Zschiesche W, Minguillon C, Spitzer E, Mohner M, Ebeling K, Grosse R: Prognostic value of proliferating cell nuclear antigen and c-*erb*B-2 compared with conventional histopathological factors in breast cancer. J Cancer Res Clin Oncol 121:115–122, 1995.

38. Hudziak RM, Schlessinger J, Ullrich A: Increased expression of the putative growth factor receptor p185HER2 causes transformation and tumorigenesis of NIH 3T3 cells. Proc Natl Acad Sci USA 84:7159–7163, 1987.

39. Benz CC, Scott GK, Sarup JC, Johnson RM, Tripathy D, Coronado E, Shepard HM, Osborne CK: Estrogen-dependent, tamoxifen-resistant tumorigenic growth of MCF-7 cells transfected with HER2/*neu*. Breast Cancer Res Treat 24:85–95, 1993.

40. Hudziak RM, Lewis GD, Winget M, Fendly BM, Shepard HM, Ullrich A: P185HER2 monoclonal antibody has antiproliferative effects in vitro and sensitizes human breast tumor cells to tumor necrosis factor. Mol Cell Biol 9:1165–1172, 1989.

41. Haffty BG, Brown F, Carter D, Flynn S: Evaluation of her-2 *neu* oncoprotein expression as a prognostic indicator of local recurrence in conservatively treated breast cancer: A case-control study. Int J Radiat Oncol Biol Phys 35:751–757, 1996.

42. Muss HB, Thor AD, Berry DA, Kute T, Liu ET, Koerner F, Cirrincione CT, Budman DR, Wood WC, Barcos M, Henderson IC: c-*erb*B-2 expression and response to adjuvant therapy in women with node-positive early breast cancer. N Engl J Med 330: 1260–1266, 1994.

43. Wright C, Nicholson S, Angus B, Sainsbury JR, Farndon J, Cairns J, Harris AL, Horne CH: Relationship between c-*erb*B-2 protein product expression and response to endocrine therapy in advanced breast cancer. Br J Cancer 65:118–121, 1992.

44. Berchuck A, Kamel A, Whitaker R, Kerns B, Olt G, Kinney R, Soper JT, Dodge R, Clarke-Pearson DL, Marks P, McKenzie S, Yin S, Bast RC, Jr: Overexpression of her-2/*neu* is associated with poor survival in advanced epithelial ovarian cancer. Cancer Res 50:4087–4091, 1990.

45. Isola JJ, Holli K, Oksa H, Teramoto Y, Kallioniemi OP: Elevated *erb*B-2 oncoprotein levels in preoperative and follow-up serum samples define an aggressive disease course in patients with breast cancer. Cancer 73:652–658, 1994.

46. Kandl H, Seymour L, Bezwoda WR: Soluble c-*erb*B-2 fragment in serum correlates with disease stage and predicts forshortened survival in patients with early-stage and advanced breast cancer. Br J Cancer 70:739–742, 1994.

47. Watanabe N, Miyamoto M, Tokuda Y, Kubota M, Ando Y, Tajima T, Mitomi T: Serum c-*erb*B-2 in breast cancer patients. Acta Oncol 33:901–904, 1994.

48. Wu JT, Astill ME, Gagon SD, Bryson L: Measurement of c-*erb*B-2 proteins in sera from patients with carcinomas and in breast tumor tissue cytosols: Correlation with serum tumor markers and membrane-bound oncoprotein. J Clin Lab Anal 9:151–165, 1995.

49. Andersen TI, Paus E, Nesland JM, McKenzie SJ, Borresen AL: Detection of c-*erb*B-2 related protein in sera from breast cancer patients. Relationship to *erb*B-2 gene amplification and c-*erb*B-2 protein overexpression in tumour. Acta Oncol 34:499–504, 1995.

50. Motoo Y, Sawabu N, Yamaguchi Y, Mouri I, Yamakawa O, Watanabe H, Ohta H, Okai T, Makino H: Serum levels of c-*erb*B-2 protein in digestive diseases. J Gastroenterol 29:616–620, 1994.
51. Vogel W, Kath R, Kosmehl H, Olschowsky E, Hoffken K: The *neu* oncogene product in serum and tissue of patients with metastatic gastrointestinal carcinomas. J Cancer Res Clin Oncol 122:118–121, 1996.
52. Okada N, Ohshio G, Yamaki K, Imamura T, Imamura M: Elevated serum c-*erb*B-2 protein levels in patients with pancreatic cancer: Correlation to metastasis and shorter survival. Oncology 52:392–396, 1995.
53. Osaki T, Mitsudomi T, Oyama T, Nakanishi R, Yasumoto K: Serum level and tissue expression of c-*erb*B-2 protein in lung adenocarcinoma. Chest 108:157–162, 1995.
54. Brandt-Rauf PW, Luo JC, Carney WP, Smith S, De Vivo I, Milling C, Hemminki K, Koskinen H, Vainio H, Neugut AI: Detection of increased amounts of the extracellular domain of the c-*erb*B-2 oncoprotein in serum during pulmonary carcinogenesis in humans. Int J Cancer 56:383–386, 1994.
55. Leitzel K, Teramoto Y, Konrad K, Chinchilli VM, Volas G, Grossberg H, Harvey H, Demers L, Lipton A: Elevated serum c-*erb*B-2 antigen levels and decreased response to hormone therapy of breast cancer. J Clin Oncol 13:1129–1135, 1995.
56. Harris LN, Trock B, Berris M, Esteva-Lorenzo F, Paik S: The role of *erb*B-2 extracellular domain in predicting response to chemotherapy in breast cancer patients. Proc Am Soc Clin Oncol 15:108, 1996.
57. Zhang L, Hung MC: Sensitization of HER-2/*neu*-overexpressing non-small cell lung cancer cells to chemotherapeutic drugs by tyrosine kinase inhibitor emodin. Oncogene 12:571–576, 1996.
58. Arteaga CL, Winnier AR, Poirier MC, Lopez-Larraza DM, Shawer LK, Hurd SD, Stewart SJ: p185c-*erb*B-2 signal enhances cisplatin-induced cytotoxicity in human breast carcinoma cells: Association between an oncogenic receptor tyrosine kinase and drug-induced DNA repair. Cancer Res 54:3758–3765, 1994.
59. Langton-Webster BC, Xuan JA, Brink JR, Salomon DS: Development of resistance to cisplatin is associated with decreased expression of the gp185c-*erb*B-2 protein and alterations in growth properties and responses to therapy in an ovarian tumor cell line. Cell Growth Differ 5:1367–1372, 1994.
60. Aaronson SA: Growth factors and cancer. Science 254:1146–1153, 1991.
61. Rajkumar T, Gullick WJ: The type I growth factor receptors in human breast cancer. Breast Cancer Res Treat 29:3–9, 1994.
62. Plowman GD, Whitney GS, Neubauer MG, Green JM, McDonald VL, Todaro GJ, Shoyab M: Molecular cloning and expression of an additional epidermal growth factor receptor-related gene. Proc Natl Acad Sci USA 87:4905–4909, 1990.
63. Plowman GD, Culouscou JM, Whitney GS, Green JM, Carlton GW, Foy L, Neubauer MG, Shoyab M: Ligand-specific activation of her4/p180*erb*B4, a fourth member of the epidermal growth factor receptor family. Proc Natl Acad Sci USA 90:1746–1750, 1993.
64. Pierce JH, Arnstein P, Di Marco E, Artrip J, Kraus MH, Lonardo F, Di Fiore PP, Aaronson SA: Oncogenic potential of *erb*B-2 in human mammary epithelial cells. Oncogene 6:1189–1194, 1991.
65. Weiner DB, Kokai Y, Wada T, Cohen JA, Williams WV, Greene MI: Linkage of tyrosine kinase activity with transforming ability of the p185*neu* oncoprotein. Oncogene 4:1175–1183, 1989.

66. Carraway KL, Cantley LC: A *neu* acquaintance for *erb*B3 and *erb*B4: A role for receptor heterodimerization in growth signaling. Cell 78:5–8, 1994.

67. Carraway KL, Sliwkowski MX, Akita R, Platko JV, Guy PM, Nuijens A, Diamonti AJ, Vandlen RL, Cantley LC, Cerione RA: The *erb*B3 gene product is a receptor for heregulin. J Biol Chem 269:14303–14306, 1994.

68. Sliwkowski MX, Schaefer G, Akita RW, Lofgren JA, Fizpatrick VD, Nuijens A, Fendly BM, Cerione RA, Vandlen RL, Carraway KL: Coexpression of *erb*B-2 and *erb*B3 proteins reconstitutes a high affinity receptor for heregulin. J Biol Chem 269: 14661–14665, 1994.

69. Plowman GD, Green JM, Culouscou JM, Carlton GW, Rothwell VM, Buckley S: Heregulin induces tyrosine phosphorylation of her4/p180*erb*B4. Nature 366:473–475, 1993.

70. King CR, Borrello I, Bellot F, Comoglio P, Schlessinger J: Egf binding to its receptor triggers a rapid tyrosine phosphorylation of the *erb*B-2 protein in the mammary tumor cell line sk-br-3. EMBO J 7:1647–1651, 1988.

71. Stern DF, Kamps MP: EGF-stimulated tyrosine phosphorylation of p185*neu*: A potential model for receptor interactions. EMBO J 7:995–1001, 1988.

72. Fazioli F, Kim UH, Rhee SG, Molloy CJ, Segatto O, Di Fiore PP: The *erb*B-2 mitogenic signaling pathway: Tyrosine phosphorylation of phospholipase C-gamma and GTPase-activating protein does not correlate with *erb*B-2 mitogenic potency. Mol Cell Biol 11:2040–2048, 1991.

73. Goldman R, Levy RB, Peles E, Yarden Y: Heterodimerization of the *erb*B-1 and *erb*B-2 receptors in human breast carcinoma cells: A mechanism for receptor transregulation. Biochemistry 29:11024–11028, 1990.

74. Jallal B, Schlessinger J, Ullrich A: Tyrosine phosphatase inhibition permits analysis of signal transduction complexes in p185HER2/*neu*-overexpressing human tumor cells. J Biol Chem 267:4357–4363, 1992.

75. Xie Y, Li K, Hung MC: Tyrosine phosphorylation of Shc proteins and formation of Shc/Grb2 complex correlate to the transformation of NIH3T3 cells mediated by the point-mutation activated *neu*. Oncogene 10:2409–2413, 1995.

76. Lowenstein EJ, Daly RJ, Batzer AG, Li W, Margolis B, Lammers R, Ullrich A, Skolnik EY, Bar-Sagi D, Schlessinger J: The SH2 and SH3 domain-containing protein GRB2 links receptor tyrosine kinases to ras signaling. Cell 70:431–442, 1992.

77. Janes PW, Daly RJ, de Fazio A, Sutherland RL: Activation of the Ras signalling pathway in human breast cancer cells overexpressing *erb*B-2. Oncogene 9:3601–3608, 1994.

78. Daly RJ, Binder MD, Sutherland RL: Overexpression of the Grb2 gene in human breast cancer cell lines. Oncogene 9:2723–2727, 1994.

79. Ricci A, Lanfrancone L, Chiari R, Belardo G, Pertica C, Natali PG, Segatto O: Analysis of protein–protein interactions involved in the activation of the Shc/Grb-2 pathway by the *Erb*B-2 kinase. Oncogene 11:1519–1529, 1995.

80. Schlessinger J, Mohammadi M, Margolis B, Ullrich A: Role of SH2-containing proteins in cellular signaling by receptor tyrosine kinases. Cold Spring Harb Symp Quant Biol 57:67–74, 1992

81. Xie Y, Hung MC: P66shc isoform down-regulated and not required for her-2/*neu* signaling pathway in human breast cancer cell lines with her-2/*neu* overexpression. Biochem Biophys Res Commun 221:140–145, 1996.

82. Sastry L, Cao T, King CR: Multiple Grb2-protein complexes in human cancer cells. Int J Cancer 70:208–213, 1997.

83. Pelicci G, Lanfrancone L, Grignani F, McGlade J, Cavallo F, Forni G, Nicoletti I, Pawson T, Pelicci PG: A novel transforming protein (SHC) with an SH2 domain is implicated in mitogenic signal transduction. Cell 70:93–104, 1992.

84. Segatto O, Pelicci G, Giuli S, Digiesi G, Di Fiore PP, McGlade J, Pawson T, Pelicci PG: Shc products are substrates of *erb*B-2 kinase. Oncogene 8:2105–2112, 1993.

85. Kavanaugh WM, Turck CW, Klippel A, Williams LT: Tyrosine 508 of the 85-kilodalton subunit of phosphatidylinositol 3-kinase is phosphorylated by the platelet-derived growth factor receptor. Biochemistry 33:11046–11050, 1994.

86. Laminet AA, Apell G, Conroy L, Kavanaugh WM: Affinity, specificity, and kinetics of the interaction of the shc phosphotyrosine binding domain with asparagine-x-x-phosphotyrosine motifs of growth factor receptors. J Biol Chem 271:264–269, 1996.

87. Rozakis-Adcock M, McGlade J, Mbamalu G, et al: Association of the Shc and Grb2/Sem5 SH2-containing proteins is implicated in activation of the Ras pathway by tyrosine kinases. Nature 360:689–692, 1992.

88. Pelicci G, Lanfrancone L, Salcini AE, Romano A, Mele S, Grazia Borrello M, Segatto O, Di Fiore PP, Pelicci PG: Constitutive phosphorylation of shc proteins in human tumors. Oncogene 11:899–907, 1995.

89. Stein D, Wu J, Fuqua SA, Roonprapunt C, Yajnik V, D'Eustachio P, Moskow JJ, Buchberg AM, Osborne CK, Margolis B: The SH2 domain protein GRB-7 is co-amplified, overexpressed and in a tight complex with HER2 in breast cancer. EMBO J 13:1331–1340, 1994.

90. Ottenhoff-Kalff AE, Rijksen G, van Beurden EA, Hennipman A, Michels AA, Staal GE: Characterization of protein tyrosine kinases from human breast cancer: Involvement of the c-src oncogene product. Cancer Res 52:4773–4778, 1992.

91. Luttrell DK, Lee A, Lansing TJ, Crosby RM, Jung KD, Willard D, Luther M, Rodriguez M, Berman J, Gilmer TM: Involvement of pp60c-src with two major signaling pathways in human breast cancer. Proc Natl Acad Sci USA 91:83–87, 1994.

92. Stover DR, Becker M, Liebetanz J, Lydon NB: Src phosphorylation of the epidermal growth factor receptor at novel sites mediates receptor interaction with src and p85 alpha. J Biol Chem 270:15591–15597, 1995.

93. Kurten RC, Cadena DL, Gill GN: Enhanced degradation of egf receptors by a sorting nexin, snx1. Science 272:1008–1010, 1996.

94. Hall A: Small GTP-binding proteins and the regulation of the actin cytoskeleton. Annu Rev Cell Biol 10:31–54, 1994.

95. Galang CK, Garcia-Ramirez J, Solski PA, Westwick JK, Der CJ, Neznanov NN, Oshima RG, Hauser CA: Oncogenic *neu*/*erb*B-2 increases ets, ap-1, and nf-kappab-dependent gene expression, and inhibiting ets activation blocks *neu*-mediated cellular transformation. J Biol Chem 271:7992–7998, 1996.

96. Webber KO, Reiter Y, Brinkmann U, Kreitman R, Pastan I: Preparation and characterization of a disulfide-stabilized Fv fragment of the anti-Tac antibody: Comparison with its single-chain analog. Mol Immunol 32:249–258, 1995.

97. Park JW, Stagg R, Lewis GD, Carter P, Maneval D, Slamon DJ, Jaffe H, Shepard HM: Anti-p185HER2 monoclonal antibodies: Biological properties and potential for immunotherapy. Cancer Treat Res 61:193–211, 1992.

98. Hurwitz E, Stancovski I, Sela M, Yarden Y: Suppression and promotion of tumor growth by monoclonal antibodies to *Erb*B-2 differentially correlate with cellular uptake. Proc Natl Acad Sci USA 92:3353–3357, 1995.

99. Kasprzyk PG, Song SU, Di Fiore PP, King CR: Therapy of an animal model of human gastric cancer using a combination of anti-*erb*B-2 monoclonal antibodies. Cancer Res 52:2771–2776, 1992.

100. Harwerth IM, Wels W, Schlegel J, Muller M, Hynes NE: Monoclonal antibodies directed to the *erb*B-2 receptor inhibit in vivo tumour cell growth. Br J Cancer 68: 1140–1145, 1993.

101. Xu F, Lupu R, Rodriguez GC, Whitaker RS, Boente MP, Berchuck A, Yu Y, De Sombre KA, Boyer CM, Bast RC, Jr: Antibody-induced growth inhibition is mediated through immunochemically and functionally distinct epitopes on the extracellular domain of the c-*erb*B-2 (HER-2/*neu*) gene product p185. Int J Cancer 53:401–408, 1993.

102. Hancock MC, Langton BC, Chan T, Toy P, Monahan JJ, Mischak RP, Shawver LK: A monoclonal antibody against the c-*erb*B-2 protein enhances the cytotoxicity of cis-diamminedichloroplatinum against human breast and ovarian tumor cell lines. Cancer Res 51:4575–4580, 1991.

103. Pietras RJ, Fendly BM, Chazin VR, Pegram MD, Howell SB, Slamon DJ: Antibody to HER-2/*neu* receptor blocks DNA repair after cisplatin in human breast and ovarian cancer cells. Oncogene 9:1829–1838, 1994.

104. Shawver LK, Mann E, Elliger SS, Dugger TC, Arteaga CL: Ligand-like effects induced by anti-c-*erb*B-2 antibodies do not correlate with and are not required for growth inhibition of human carcinoma cells. Cancer Res 54:1367–1373, 1994.

105. Tokuda Y, Ohnishi Y, Shimamura K, Iwasawa M, Yoshimura M, Ueyama Y, Tamaoki N, Tajima T, Mitomi T: In vitro and in vivo anti-tumour effects of a humanised monoclonal antibody against c-*erb*B-2 product. Br J Cancer 73:1362–1365, 1996.

106. Baselga J, Tripathy D, Mendelsohn J, Baughman S, Benz CC, Dantis L, Sklarin NT, Seidman AD, Hudis CA, Moore J, Rosen PP, Twaddell T, Henderson IC, Norton L: Phase II study of weekly intravenous recombinant humanized anti-p185HER2 monoclonal antibody in patients with HER2/*neu*-overexpressing metastatic breast cancer. J Clin Oncol 14:737–744, 1996.

107. Sliwkowski MX, Fox JA, Twaddell T: A humanized monoclonal antibody for the treatment of HER2 overexpressing breast cancer. Proc Am Assoc Cancer Res 37:625–626, 1996.

108. Batra JK, Kasprzyk PG, Bird RE, Pastan I, King CR: Recombinant anti-*erb*B-2 immunotoxins containing pseudomonas exotoxin. Proc Natl Acad Sci USA 89:5867–5871, 1992.

109. Rodriguez GC, Boente MP, Berchuck A, Whitaker RS, O'Briant KC, Xu F, Bast RC, Jr: The effect of antibodies and immunotoxins reactive with HER-2/*neu* on growth of ovarian and breast cancer cell lines. Am J Obstet Gynecol 168:228–232, 1993.

110. Wels W, Harwerth IM, Mueller M, Groner B, Hynes NE: Selective inhibition of tumor cell growth by a recombinant single-chain antibody-toxin specific for the *erb*B-2 receptor. Cancer Res 52:6310–6317, 1992.

111. Reiter Y, Brinkmann U, Jung SH, Lee B, Kasprzyk PG, King CR, Pastan I: Improved binding and antitumor activity of a recombinant anti-*erb*B-2 immunotoxin by disulfide stabilization of the fv fragment. J Biol Chem 269:18327–18331, 1994.

112. King CR, Fischer PH, Rando RF, Pastan I: The performance of e23(Fv)PE, recombinant toxins targeting the *erb*B-2 protein. Semin Cancer Biol 7:79–86, 1996.

113. Reiter Y, Pai LH, Brinkmann U, Wang QC, Pastan I: Antitumor activity and pharmacokinetics in mice of a recombinant immunotoxin containing a disulfide-stabilized Fv fragment. Cancer Res 54:2714–2718, 1994.

114. Eccles SA, Court WJ, Box GA, Dean CJ, Melton RG, Springer CJ: Regression of established breast carcinoma xenografts with antibody-directed enzyme prodrug therapy against c-*erb*B-2 p185. Cancer Res 54:5171–5177, 1994.

115. Rodrigues ML, Presta LG, Kotts CE, Wirth C, Mordenti J, Osaka G, Wong WL, Nuijens A, Blackburn B, Carter P: Development of a humanized disulfide-stabilized anti-p185HER2 Fv-beta-lactamase fusion protein for activation of a cephalosporin doxorubicin prodrug. Cancer Res 55:63–70, 1995.

116. Chen SY, Zani C, Khouri Y, Marasco WA: Design of a genetic immunotoxin to eliminate toxin immunogenicity. Gene Ther 2:116–123, 1995.

117. Fominaya J, Wels W: Target cell-specific DNA transfer mediated by a chimeric multidomain protein. Novel non-viral gene delivery system. J Biol Chem 271: 10560–10568, 1996.

118. Park JW, Hong K, Carter P, Asgari H, Guo LY, Keller GA, Wirth C, Shalaby R, Kotts C, Wood WI, Papahadjopoulos D, Benz CC: Development of anti-p185HER2 immunoliposomes for cancer therapy. Proc Natl Acad Sci USA 92:1327–1331, 1995.

119. Smellie WJ, Dean CJ, Sacks NP, Zalutsky MR, Garg PK, Carnochan P, Eccles SA: Radioimmunotherapy of breast cancer xenografts with monoclonal antibody icr12 against c-*erb*B-2 p185: Comparison of iodogen and n-succinimidyl 4-methyl-3-(tri-n-butylstannyl) benzoate radioiodination methods. Cancer Res 55:5842s–5846s, 1995.

120. Allan SM, Dean C, Fernando I, Eccles S, Styles J, McCready VR, Baum M, Sacks N: Radioimmunolocalisation in breast cancer using the gene product of c-*erb*B-2 as the target antigen. Br J Cancer 67:706–712, 1993.

121. Allan SM, Dean CJ, Eccles S, Sacks NP: Clinical radioimmunolocalization with a rat monoclonal antibody directed against c-*erb*B-2. Cell Biophys 24–25:93–98, 1994.

122. Weiner LM, Holmes M, Adams GP, La Creta F, Watts P, Garcia de Palazzo I: A human tumor xenograft model of therapy with a bispecific monoclonal antibody targeting c-*erb*B-2 and CD16. Cancer Res 53:94–100, 1993.

123. Shalaby MR, Carter P, Maneval D, Giltinan D, Kotts C: Bispecific HER2 × CD3 antibodies enhance T-cell cytotoxicity in vitro and localize to HER2-overexpressing xenografts in nude mice. Clin Immunol Immunopathol 74:185–192, 1995.

124. Zhu Z, Lewis GD, Carter P: Engineering high affinity humanized anti-p185HER2/anti-CD3 bispecific F(ab′)2 for efficient lysis of p185HER2 overexpressing tumor cells. Int J Cancer 62:319–324, 1995.

125. Weiner LM, Clark JI, Davey M, Li WS, Garcia de Palazzo I, Ring DB, Alpaugh RK: Phase I trial of 2b1, a bispecific monoclonal antibody targeting c-*erb*B-2 and fc gamma RIII.Cancer Res 55:4586–4593, 1995.

126. Weiner LM, Clark JI, Ring DB, Alpaugh RK: Clinical development of 2B1, a bispecific murine monoclonal antibody targeting c-erB-2 and Fc gamma RIII. J Hematother 4:453–456, 1995.

127. Moritz D, Wels W, Mattern J, Groner B: Cytotoxic T lymphocytes with a grafted recognition specificity for *ERB*B-2-expressing tumor cells. Proc Natl Acad Sci USA 91:4318–4322, 1994.

128. Moritz D, Groner B: A spacer region between the single chain antibody- and the CD3

zeta-chain domain of chimeric T cell receptor components is required for efficient ligand binding and signaling activity. Gene Ther 2:539–546, 1995.

129. Stancovski I, Schindler DG, Waks T, Yarden Y, Sela M, Eshhar Z: Targeting of T lymphocytes to *Neu*/HER2-expressing cells using chimeric single chain Fv receptors. J Immunol 151:6577–6582, 1993.

130. Levitzki A: Signal-transduction therapy. A novel approach to disease management. Eur J Biochem 226:1–13, 1994.

131. Levitzki A, Gazit A: Tyrosine kinase inhibition: An approach to drug development. Science 267:1782–1788, 1995.

132. Buchdunger E, Trinks U, Mett H, Regenass U, Muller M, Meyer T, McGlynn E, Pinna LA, Traxler P, Lydon NB: 4,5-Dianilinophthalimide: A protein-tyrosine kinase inhibitor with selectivity for the epidermal growth factor receptor signal transduction pathway and potent in vivo antitumor activity. Proc Natl Acad Sci USA 91:2334–2338, 1994.

133. Tsai CM, Levitzki A, Wu LH, Chang KT, Cheng CC, Gazit A, Perng RP: Enhancement of chemosensitivity by tyrphostin ag825 in high-p185(*neu*) expressing non-small cell lung cancer cells. Cancer Res 56:1068–1074, 1996.

134. Miller P, Di Orio C, Moyer M, Schnur RC, Bruskin A, Cullen W, Moyer JD: Depletion of the *erb*B-2 gene product p185 by benzoquinoid ansamycins. Cancer Res 54:2724–2730, 1994.

135. Schnur RC, Corman ML, Gallaschun RJ, Cooper BA, Dee MF, Doty JL, Foster BA, Pollack VA, Savage DM, Sloan DE, Pustilnik LR, Moyer MP: Inhibition of the oncogene product p185*erb*B-2 in vitro and in vivo by geldanamycin and dihydrogeldanamycin derivatives. J Med Chem 38:3806–3812, 1995.

136. Xie Y, Pendergast AM, Hung MC: Dominant-negative mutants of Grb2 induced reversal of the transformed phenotypes caused by the point mutation-activated rat HER-2/*Neu*. J Biol Chem 270:30717–30724, 1995.

137. Burke TR Jr, Barchi JJ Jr, George C, Wolf G, Shoelson SE, Yan X: Conformationally constrained phosphotyrosyl mimetics designed as monomeric Src homology 2 domain inhibitors. J Med Chem 38:1386–1396, 1995.

138. Ye B, Akamatsu M, Shoelson SE, Wolf G, Giorgetti-Peraldi S, Yan X, Roller PP, Burke TR, Jr. L-O-(2-malonyl)tyrosine: A new phosphotyrosyl mimetic for the preparation of Src homology 2 domain inhibitory peptides. J Med Chem 38:4270–4275, 1995.

139. Oligino L, Sastry LS, Lung FD, Bigelow J, Cao T, Curran M, Burke TR, Wang S, Krag D, Roller PP, King CR: Binding of a non-phosphorylated small peptide to the SH2 domain of Grb2. J Biol Chem In press: 1997.

140. Garcia AM, Rowell C, Ackermann K, Kowalczyk JJ, Lewis MD: Peptidomimetic inhibitors of Ras farnesylation and function in whole cells. J Biol Chem 268: 18415–18418, 1993.

141. James GL, Goldstein JL, Brown MS, Rawson TE, Somers TC, McDowell RS, Crowley CW, Lucas BK, Levinson AD, Marsters JC, Jr.: Benzodiazepine peptidomimetics: Potent inhibitors of Ras farnesylation in animal cells. Science 260:1937–1942, 1993.

142. Nigam M, Seong CM, Qian Y, Hamilton AD, Sebti SM: Potent inhibition of human tumor p21ras farnesyltransferase by IA2-lacking p21ras CIA2X peptidomimetics. J Biol Chem 268:20695–20698, 1993.

143. Kohl NE, Wilson FR, Mosser SD, et al: Protein farnesyltransferase inhibitors block the growth of ras-dependent tumors in nude mice. Proc Natl Acad Sci USA 91:9141–9145, 1994.

144. Nagasu T, Yoshimatsu K, Rowell C, Lewis MD, Garcia AM: Inhibition of human tumor xenograft growth by treatment with the farnesyl transferase inhibitor B956. Cancer Res 55:5310–5314, 1995.

145. Sepp-Lorenzino L, Ma Z, Rands E, Kohl NE, Gibbs JB, Oliff A, Rosen N: A peptidomimetic inhibitor of farnesyl:protein transferase blocks the anchorage-dependent and -independent growth of human tumor cell lines. Cancer Res 55:5302–5309, 1995.

146. Alimandi M, Romano A, Curia MC, Muraro R, Fedi P, Aaronson SA, Di Fiore PP, Kraus MH: Cooperative signaling of *Erb*B3 and *Erb*B-2 in neoplastic transformation and human mammary carcinomas. Oncogene 10:1813–1821, 1995.

147. Levizki A: Src as a target for anti-cancer drugs. Anticancer Drug Des 11:175–182, 1996.

148. Yu D, Shi D, Scanlon M, Hung MC: Reexpression of *neu*-encoded oncoprotein counteracts the tumor-suppressing but not the metastasis-suppressing function of E1A. Cancer Res 53:5784–5790, 1993.

149. Yu D, Matin A, Xia W, Sorgi F, Huang L, Hung MC: Liposome-mediated in vivo E1A gene transfer suppressed dissemination of ovarian cancer cells that overexpress HER-2/*neu*. Oncogene 11:1383–1388, 1995.

150. Bertram J, Killian M, Brysch W, Schlingensiepen KH, Kneba M: Reduction of *erb*B-2 gene product in mamma carcinoma cell lines by *erb*B-2 mRNA-specific and tyrosine kinase consensus phosphorothioate antisense oligonucleotides. Biochem Biophys Res Commun 200:661–667,1994.

151. Çolomer R, Lupu R, Bacus SS, Gelmann EP: *ERb*B-2 antisense olignucleotides inhibit the proliferation of breast carcinoma cells with *erb*B-2 oncogene amplification. Br J Cancer 70:819–825, 1994.

152. Richardson JH, Marasco WA: Intracellular antibodies: Development and therapeutic potential. Trends Biotechnol 13:306–310, 1995.

153. Deshane J, Loechel F, Conry RM, Siegal GP, King CR, Curiel DT: Intracellular single-chain antibody directed against *erb*B-2 down-regulates cell surface *erb*B-2 and exhibits a selective anti-proliferative effect in *erb*B-2 overexpressing cancer cell lines. Gene Ther 1:332–337, 1994.

154. Deshane J, Siegal GP, Alvarez RD, Wang MH, Feng M, Cabrera G, Liu T, Kay M, Curiel DT: Targeted tumor killing vĩa an intracellular antibody against *erb*B-2. J Clin Invest 96:2980–2989, 1995.

155. Graus-Porta D, Beerli RR, Hynes NE: Single-chain antibody-mediated intracellular retention of *Erb*B-2 impairs *Neu* differentiation factor and epidermal growth factor signaling. Mol Cell Biol 15:1182–1191, 1995.

156. Deshane J, Grim J, Loechel S, Siegal GP, Alvarez RD, Curiel DT: Intracellular antibody against *erb*B-2 mediates targeted tumor cell eradication by apoptosis. Cancer Gene Ther 3:89–98, 1996.

157. Disis ML, Bernhard H, Gralow JR, Hand SL, Emery SR, Calenoff E, Cheever MA: Immunity to the HER-2/*neu* oncogenic protein. Ciba Found Symp 187:198–211, 1994.

158. Disis ML, Smith JW, Murphy AE, Chen W, Cheever MA: In vitro generation of human cytolytic T-cells specific for peptides derived from the HER-2/*neu* protooncogene protein. Cancer Res 54:1071–1076, 1994.

159. Disis ML, Calenoff E, McLaughlin G, Murphy AE, Chen W, Groner B, Jeschke M, Lydon N, McGlynn E, Livingston RB, Moe R, Cheever MA: Existent T-cell antibody

immunity to HER-2/*neu* protein in patients with breast cancer. Cancer Res 54:16–20, 1994.

160. Yoshino I, Goedegebuure PS, Peoples GE, Parikh AS, Di Maio JM, Lyerly HK, Gazdar AF, Eberlein TJ: HER2/*neu*-derived peptides are shared antigens among human non-small cell lung cancer and ovarian cancer. Cancer Res 54:3387–3390, 1994.
161. Peoples GE, Smith RC, Linehan DC, Yoshino I, Goedegebuure PS, Eberlein TJ: Shared T cell epitopes in epithelial tumors. Cell Immunol 164:279–286, 1995.
162. Tanner B, Kreutz E, Weikel W, Meinert R, Oesch F, Knapstein PG, Becker R: Prognostic significance of c-*erb*B-2 mrna in ovarian carcinoma. Gynecol Oncol 62:268–277, 1996.
163. Lipponen P: Expression of c-*erb*B-2 oncoprotein in transitional cell bladder cancer. Eur J Cancer 29A:749–753, 1993.
164. Tetu B, Fradet Y, Allard P, Veilleux C, Roberge N, Bernard P: Prevalence and clinical significance of her/2*neu*, p53 and rb expression in primary superficial bladder cancer. J Urol 155:1784–1788, 1996.
165. Stenman G, Sandros J, Nordkvist A, Mark J, Sahlin P: Expression of the *erb*B-2 protein in benign and malignant salivary gland tumors. Genes Chromosom Cancer 3:128–135, 1991.
166. Semba K, Kamata N, Toyoshima K, Yamamoto T: A v-*erb*-2 related protooncogene, c-*erb*B-2, is distinct from the c-*erb*B-1/epidermal growth factor-receptor gene and is amplified in a human salivary gland adenocarcinoma. Proc Natl Acad Sci USA 82:6497–6501, 1985.
167. Hall PA, Hughes CM, Staddon SL, Richman PI, Gullick WJ, Lemoine NR: The c-*erb*B-2 proto-oncogene in human pancreatic cancer. J Pathol 161:195–200, 1990.
168. Kristensen GB, Holm R, Abeler VM, Trope CG: Evaluation of the prognostic significance of cathepsin d, epidermal growth factor receptor, and c-*erb*B-2 in early cervical squamous cell carcinoma. An immunohistochemical study. Cancer 78:433–440, 1996.
169. Onda M, Matsuda S, Higaki S, Iijima T, Fukushima J, Yokokura A, Kojima T, Horiuchi H, Kurokawa T, Yamamoto T: *Erb*B-2 expression is correlated with poor prognosis for patients with osteosarcoma. Cancer 77:71–78, 1996.
170. Duhaylongsod FG, Gottfried MR, Iglehart JD, Vaughn AL, Wolfe WG: The significance of c-*erb*B-2 and p53 immunoreactivity in patients with adenocarcinoma of the esophagus. Ann Surg 221:677–683, 1995; discussion 683–684.
171. Lazaris AC, Theodoropoulos GE, Anastassopoulos P, Nakopoulou L, Panoussopoulos D, Papadimitriou K: Prognostic significance of p53 and c-*erb*B-2 immunohistochemical evaluation in colorectal adenocarcinoma. Histol Histopathol 10:661–668, 1995.
172. Sun XF, Carstensen JM, Stal O, Zhang H, Nordenskjold B: C-*erb*B-2 oncoprotein in relation to DNA ploidy and prognosis in colorectal adenocarcinoma. APMIS 103:309–315, 1995.
173. Fox SB, Persad RA, Coleman N, Day CA, Silcocks PB, Collins CC: Prognostic value of c-*erb*B-2 and epidermal growth factor receptor in Stage I (T1a) prostatic adenocarcinoma. Br J Urol 74:214–220, 1994.
174. Zhau HE, Wan DS, Zhou J, Miller GJ, von Eschenbach AC: Expression of c-*erb*B-2/*neu* proto-oncogene in human prostatic cancer tissues and cell lines. Mol Carcinog 5:320–327, 1992.

INDEX

Acetylcholine receptor-inducing activity (ARIA), 148. *See also* Neuregulins
Achaetescute Complex (AS-C) proneural genes, 370–71
Achondroplasia, 295
Activin receptor
  in murine embryogenesis, 102t
  negative in *Xenopus,* 107
Activins
  BA, 100t
  BB, 100t
  in murine embryogenesis, 100t, 106–7
Acute myelogenous leukemia (AML), 292
Adenocarcinoma
  neural morphology in prostatic, 338
  of prostate, 260–61
  TGF-α in, 155
Aedes aegypti, 57–58
Aldosterone, mammary development and, 195
α-inhibin, in murine embryogenesis, 100t
Amniotic fluid prolactin, 199
Amphibians. *See* Invertebrates and amphibians; *Xenopus*
Amphiregulin, 96, 148, 217, 339. *See also* EGFs
  with ErbB protein, 147
  in murine embryogenesis, 101t, 103–4
  in prostate tumors, 337
Anchor cells, 4, 5
Androgen receptor, 339
Androgen receptors, *210,* 210–11
Androgens. *See also* Sex steroid hormones; Testosterone
  in prostate cancer progression, 338–39
  target organ stromal–epithelial interactions, 208–11
Andromedin (FGF-7), 337
Antibody-directed cytotoxicity, 429–31
apc gene, 369
Apoptosis. *See also* p53
  Gleason grade correlation with, 342
  IGF-I receptor in, 274
  MYC protein in, 331
  prolactin absence and, 195
  resistance to in prostate cancer metastasis, 343
  Schwann cell, 157
argos gene, 22, 147, 151
Argos protein, 151
ARIA (acetylcholine receptor-inducing activity), 148. *See also* Neuregulins
ARM protein, 367–69, 378. *See also wnt* genes
AS-C genes, 370-71
AT-20 cells, 369
Autoactivation, 29
Autocrine signaling
  model of, 12
  in wild-type animals, 11

Balb/c mouse model, 211
bax gene, 342
bcl-2 gene, 342
2Bc mutations, 63
β-catenin, 367–69
β-catenin gene, 378
Betacellulin, 147, 148–49
Betaglycan, in murine embryogenesis, 107
β-1 integrin, 97
β-lactamase, 430
β-lactoglobulin gene, 242–43
BMP-2, 23
BMP-4, 23, 25
BMP-4, 31
BMP-1 protein, 25
BMP-7 protein, 25
Bombesin/gastrin releasing peptide, 337
Bone morphogenetic proteins, *See* BMP
BRCA1/BRCA2, 253–54. *See also* Mammary tumors/breast cancer
BR-C gene, 55–56
Breast cancer. *See* Mammary tumors/breast cancer
br gene, 55, 138
br mutations, 63
Broad-Complex gene, 50
broad mutation, 55
BT-474 cell line, 404

e-cadherin, 97
Caenorhabditis elegans
  ErbB in, 150–52
  pattern formation in vulval development, 3–17, 152
    cell rate establishment, 11–13
    general sequence of, 4–6
    morphogen and sequential signaling models, 6–11
  point mutations and, 294
Calcitonin, 337
Cancer. *See* Tumorigenicity *and specific types*
Carboxypeptidase, 430
β-Catenin, 367–69
τ-Catenin, 369. *See also* Plakoglobin
β-Catenin. *See also wnt* genes
β-catenin gene, 378
CAT gene, 124
c-Cb1 proto-oncogenic protein, 148
cdc-2, 370
CD3 immunotherapy, 430–31
CD16 immunotherapy, 430–31
Cell fate, 4
  in *C. elegans* vulval induction, 11–13
  reproducibility of, 13
  rhomboid gene and, 27
Chemotherapeutic drugs, resistance to, 256, 424–25
Chicken
  embryogenesis
    activin in, 107
    FGF in, 107
  endoderm in patterning, 95
  v-*erbB* gene in, 299
  estrogen-withdrawn, 178
  pregastrulation embryogenesis in, 97
  TGF-β in, 335
Chironomus larvae chromosomal puffing, 60
Chironomus larval chromosome puffing, 46–50
chordin gene, 31
Chordin homolog of Soq, 25
Chromosome puffing, 46–50, 60. *See also* Ecdysone expression
Chronic myelomonocytic leukemia (CML), 292
ci gene, 379
Cisplatin resistance, 424–25
C57MG cell line, 369, 376
CNS development, *wnt* genes in, 372
Colon cancer, 369
  incremental multistep ("big bang") theory of progression, 328, *329*
Colony stimulating factor. *See* CSFs
Colony-stimulating factor. *See* CSFs
Compaction stage of embryogenesis, 95, 98
Competence factors, 315
Corkscrew mutation, 20
Costal-2, 370
COUP-TF, 65
Cox-2 growth factor, 96
Cranial nerve development, neuregulins in, 158
Craniosynostosis, 294
Crouzon syndrome, 294, 295
Crypto, in murine embryogenesis, 99t, 103
CSF-1, 97
  in mammary gland, 408–9
  in murine embryogenesis, 105
CSF-1/MCSF/fms receptor, 293, 298
CSF-1 receptor, 409
  in murine embryogenesis, 102t
CSFs, in murine embryogenesis, 100t
C-terminal truncation, of RTKs, 292
Cyclin D1, 334
Cyclin G, 334

Daughter of Sevenless mutation, 20
Decapentaplegic *Drosophila* signaling pathway. *See* Drop pathway; *Drosophila*
DER gene, 150
DER protein, 151–52
detached mutant, 32
Development. *See* Embryogenesis; Mammalian development; Pattern formation; Postnatal development
DHR23. See EcR gene
DHR78, 65
DHR96, 65
DHR3 gene, 59. *See also EcR gene*
DHR39 gene, 59
Diethylstilbestrol (DES), 180–81, 211–12, 316–20
dino mutant, 31
discs-at-large tumor suppressor gene, 367
dis-J mutants, 4
dl gene, 370
Dopamine, estrogen and, 173–74
dpp gene, 23, 370
Dpp pathway
  with EGF-R pathway, 28–33
    in adult wing development, 31–33
    in early neurogenesis, 28–31
  organization of, 23–25
  overview of developmental signaling, 27
  TGF-*β*-like signaling in pattern formation, 19–43. *See also under Drosophila*
DPP protein, 370
Drk adaptor protein, 5
drk gene, 21
Drosophila. See also specific genes and proteins
  branchless gene studies, 138
  ecdysone expression in, 45–78
    ecdysone receptor, 50–54
      EcR gene, 51–52
      nuclear hormone receptor superfamily, 51
      preliminary characterization, 50
      USP gene, 52–53
    ecdysone-responsive genes, 53–61, *See also specific genes and proteins*
      classical early, 54–58
      early late, 59–61
      other early, 58–59
    gene regulation, 46–68
      larval polytene chromosome puffing, 46–49
      molting and metamorphic response, 49
      reproductive response, 49–50
    physiologic functions of ecdysone, 46
  EGF-like ligands in, 150–51
  imaginal disc formation, 27
  Notch gene. *See INT3* gene
  Notch protein, 8, 9
  TFG-*β*-like Dpp signaling in pattern formation
    conclusions, 33
    Dpp pathway, 23–25
    early neurogenesis and adult wing development, 28–33
    EGF-R pathway, 20–23
    overview of EGF-R/Dpp signaling, 25–27
  torso RTK of, 293
  tyrosine kinase signaling pathways of, 21
  wg gene, 361. *See also wnt* genes
dsh gene, 366–67. *See also ent* genes
dsor gene, 21
DU 145 cell line, 332, 336
Dynamin, 426

E75A promoter, 64–65
e-cadherin, 97
EC cells, *FGF-4* transcriptional regulation in, 115–129. *See also FGF-4*
Ecdysiotropic hormone (PTTH), 68
Ecdysone expression, in *Drosophila,* 45–78
  ecdysone receptor, 50–54
    EcR gene, 51–52
    nuclear hormone receptor superfamily, 51
    preliminary characterization, 50
    USP gene, 52–53
  ecdysone-responsive genes, 53–61. *See also specific genes and proteins*
    classical early, 54–58
    early late, 59–61
    other early, 58–59
  gene regulation, 46–68
    larval polytene chromosome puffing, 46–49
    molting and metamorphic response, 49
    reproductive response, 49–50
  physiologic functions of ecdysone, 46
ecdysoneless-1 mutation, 50

Ecdysone receptor
  Drosophila, 50–54
    EcR gene, 51–52
    Ftz-F1 regulation of late prepupal response, 63–64
    nuclear hormone receptor interactions, 64–67
    DNA-binding competition model, 64–65
      endocrine integration model, 66–67
      heterodimer formation model, 65–66
    nuclear hormone receptor superfamily, 51
    perspectives on research, 67–68
      neuroendocrine control of production, 68
      reproduction and early development, 68
      steroid-regulated hierarchies, 67–68
    preliminary characterization, 50
    USP gene, 52–53
  early gene isoforms and, 62–63
  vertebrate versus *Drosophila,* 53
EcR gene, 51–52, 52–54
EcR protein, A, B1, and B2 isoforms of, 62–63
E63-1 gene, 58–59
E63-2 gene, 58–59
E74 gene, 55–58, 60
E75 gene, 55–58, 62
E78 gene, 59
E93 gene, 58–59
EGF motif, 148
EGF receptor (EGFR), 96–97, 97, 339
  as compared with *C. elegans let-3,* 5
  in *Drosophila*
    with Dpp signaling
      in adult wing development, 31–33
      in neurogenesis, 28–31
    organization of, 20–23
    overview of developmental signaling, 25–27
  with ErbB-2, 402–3
  with ErbB-3, 405
  in human breast cancer, 398–99
  in malignant versus normal prostate tissue, 337
  in murine embryrogenesis, 102t
  in mutational activation, 301
  mutational activation of, 298–99
  type III mutation, 292–93
Egf-r gene, 20, 22
EGFR-KO mouse, 218
EGFs (epidermal growth factors), 95. *See also* LET-23; *let-23;* Neuregulin
  ErbB-2 and, 149
  with ErbB protein, 147
  in estrogen induction, 312–16
  in GI tract development, 154–55
  HB, 217
  heparin-binding (HB-EGF), 96, 147–53
  in human cancer, 378–79
  with IGFs and IGF binding proteins in female reproductive tract, 235, *236*
  in lung development, 154
  in murine embryogenesis, 94, 99t, 100t
  in murine uterine development, 170–71
  in pregnancy, 155
  prolactin and, 195
  in prostate cancer progression, 337
  sex hormone target organ stromal–epithelial interactions, 217–18
  similarity of *C. elegans lin-3,* 5
EIP28/29 gene, 59
Embryogenesis
  compaction stage of, 95, 98
  ecdysone and, 50
  ecdysone in, 57–58
  indifferent stage of, 179–80. *See also* Progesterone receptor
  murine, *94*
  murine model of pre- and postimplantation, 93–114
    directionality of signaling, 96
    growth factor families, 97–108. *See also specific families*
      activins, 106–7
      colony-stimulating factor-1 (CSF-1), 105–6
      fibroblast growth factors (FGFs), 101–3
      growth factor receptor expression and function, 102–103t
      hepatocyte growth factors, 108
      insulin-like growth factors (IGFs), 97–101
      Interleukin-1 (IL-1), 107
      interleukin-6 (IL-6), 105
      leukemia-inhibiting factor (LIF), 105
      parathyroid hormone-related peptide (PhRO), 107
      placental lactogen-1 (PL-1), 107–8
      platelet-derived growth factors (PDGFs), 104–5

pre-gastrulation growth factor expression, 99–101t
stem cell factor (SCF), 106
transforming growth factor-β (TGF-β), 106
VG-1-related growth factor, 107
lower vertebrate comparisons, 97
null mutants, 96–97
significance and events, 93–96
Embryonal carcinoma cells. *See* EC cells
Endocrine integration model, of ecdysone response, 66–67
Endometrial cancer, 258–59
estrogen receptors in, 259
risk factors for, 258
tamoxifen in, 259
uterine biology and, 258
en-2 gene, 371
en genes, 370, 371
Enhancer of Split complex ([E(spl)-c], 386–87
EN protein, 371
v-*env* gene, 291
EPG-like repeats of *n* gene, 370
Epidermal development, ErbB-1 in, 154
Epidermal growth factors. *See* EGFs
Epidermal hyperkeratosis, 154
Epiregulin, 147, 148
Equivalence groups, 4
Erb-B-4, 148
ErbB1, 148, 152–55
in GI tract development, 154–55
in hair and skin development, 154
in lung development, 154
in placental and early embryonic development, 153
ErbB-2, 149–50
ErbB-3, 148
ErbB-3, in human breast cancer, 398
ErbB-2 coreceptor, 159–60
erbB-2 gene, 330
activated half-life of, 300
in human breast cancer, 398–99
in tumorigenesis
clinical application
antibody specificity in, 427–28
inhibition of $p185^{erbB-2}$-initiated signal transduction, 431–32
inhibition of $p185^{erbB-2}$ synthesis, 432
in prognosis, 423
of receptor signaling, 425–27
in therapeutic response prediction, 423–24
as therapeutic target, 428–33
vaccines, 433
overexpression in cancer, 422–23, 423t
pathogenetic role, 422
v-*erbB* gene, 291, 299
ErbB receptor
as compared with EGF receptor, 298–99
and ligands in mammalian development, 145–68
background of study, 145–46
developmental and physiological roles, 152–60
ErbB-1, 152–55
ErbB-2 coreceptor, 159–60
neuregulins and neuregulin receptors, 155–60
invertebrate modeling systems of, 160–62
perspectives on, 160–61
signaling network of, 149–50
structural homology of proteins and ligands, 146–49, *147*
ErbB2 receptor, 294
ErbB-2 receptor, with other Ern receptors, 402–3
c-*erb* receptors, with neuregulin, 136–37
ERKO mouse, 169–76, 212–14
Erythrocyte growth factor. *See* EGF
Erythrocyte growth factor receptor. *See* EGF receptor
Erythropoietin receptor, 294
Esophageal carcinoma, *wnt-2* in, 378
Estradiol. *See also* Estrogen receptors
EGF and EGF-R and, 173
Estrogen receptor
β form of, 180–81
ERKO mouse model of reproductive tract development, 169–76, 212–14
Estrogen receptors, 211
in breast cancer, 254–55
in endometrial cancer, 259
Estrogens
in reproductive tract carcinogenesis, *217, 318,* 318–20
target organ stromal-epithelial interactions, 211–16
ERKO mouse model, 212–14
Estromedins, 231
Evx-1, 97

FAK (focal adhesion kinase), 407–8
Fbp-1 gene, 59

FC*τ* receptor III delivery systems, 430–31
FDC-P1 cell line, 298
Feline sarcoma virus, 300
FGF-2, 96, 339
  in murine embryogenesis, 100t
  in prostate cancer, 336
FGF-3, 117
FGF-3. *See also* INT-2
  in murine embryogenesis, 100t, 101
  in prostate cancer, 336
FGF-4, 95, 96–97
FGF-5, 117
FGF-4, in murine embryogenesis, 94, 99t, 101
FGF-5, in murine embryogenesis, 100t
FGF-8, in prostate cancer, 337
FGF-4, in vitro model of murine transcriptional regulation, 115–29
  background of study, 115–16
  cis-regulatory elements in promoter region, 118–20
  future perspectives, 125–26
  of gene, 117–18, 118t, *119*
  growth factor expression in early murine embryos and EC cells, 116–17
  identification of transcription factors, 12–123, 120–21
  multiple cis-regulatory elements of distal enhancer, 121–22
FGF-7 (andromedin), 337
fgf gene, 375–78. *See also wnt* genes
FGF receptor-2, 294
  in murine embryogenesis, 102t
FGF receptor-3, 298
  in achondroplasia, 295
  in murine embryogenesis, 102t
FGF receptor-1, in murine embryogenesis, 102t
FGF receptor-4, in murine embryogenesis, 102t
FGFs (fibroblast growth factors)
  in epithelial development, 137–38
  keratinocyte growth factor, 218–20
  in mammary gland development and cancer, 375–78
  in murine embryogenesis, 101–3
  in prostate cancer progression, 336–37
  and *wnt* gene in development and tumorigenesis, 361, 375–78. *See also wnt* genes
Fibroblast growth factors. *See* FGFs
c-*fins,* mutational activation, 301
flb mutation, 20
FMA-3 cell line, 297–98
fma gene, 293
fms receptor, 298
Focal adhesion kinase (FAK), 407–8
Follistatin, 25
c-*fos* induction, EGF in, 172
c-*fos* proto-oncogene, 319–20
Ftz-F1 gene, 60–61
  in late prepupal response, 63–64
fu gene, 371
FU kinase, 370
fushi tarazu gene, 60
fzt gene, 366. *See also wnt* genes
FZT protein, 366. *See also wnt* genes

GADD45 gene, 334
τ-catenin, 369. *See also* Plakoglobin
gap-1 gene, 6, 21
Ga14 repressor, 430
Gastric carcinoma, *wnt-2* in, 378
Gastrointestinal development, ErbB-1 in, 154–55
G-CSF (granulocyte colony stimulating factor), in murine embryogenesis, 100t
GDP-RAS, 21
Genetic mosaic approach, 9–10
GFG-2, in murine embryogenesis, 101
GH, prolactin and, 195
GH receptor, inferred in murine embryogenesis, 102t
giant mutation, 57
Glial cell development, neuregulins in, 157
Glial growth factor (GGF), 148, 157. *See also* Neuregulin
gli gene, 371
glp-1 gene, 385. *See also INT3* gene; *n* gene
Glucocorticoid receptor, 53
GM-CSF (granulocyte macrophage colony stimulating factor), in murine embryogenesis, 100t
GNDF, 138–39
Granulocyte colony-stimulating factor. *See* G-CSF
Granulocyte macrophage colony-stimulating factor. *See* GM-CSF
Grb-2, 296
GRB2 adaptor protein, 5
Grb-2 protein, 404–5
Grb2 protein, 426–27
Grb7 protein, 425–26
Grb2 SH2 antagonists, 431–32
Grb-Sos1 complex, 426

Grk protein, 150–51
Growth factor families, in murien embryogenesis, 97–108, 99–101t, 102–103t. *See also* *specific families*
Growth hormone. *See* GH
gsb gene, 370
gsk-3 gene, 367. *See also wnt* genes
GTPase activating protein, 6
GTP-RAS, 21
Guanosine diphosphate. *See* GDP
Guanosine triphosphatase. *See* GTPase
gurken gene, 20
Gurken protein, 152

c-Ha-*ras* oncogene, 341
Hardy-Zuckerman-4 feline sarcoma virus, 300
HB-EGF, 96, 147, 217
  in murine embryogenesis, 97, 103–4
Hβ-58, 97
Heart development, neuregulins in, 156
Heat shock promoter experiment, 10–11
Henson's node, 3
Heparin-binding epidermal growth factor. *See* HB-EGF
Hepatocyte growth factor. *See* HGF; HGF receptor; Scatter factor (SF)
HER-2. *See erbB-2* gene
Heregulin, 148–49, 426. *See also* Neuregulins
Heterodimer formation model, of ecdysone response, 65–66
HGF, 108
HGF receptor (HGFR), in murine embryogenesis, 99t
HGF (scatter factor), in human cancer, 378–79
hh gene, 370, 371
HMG box, 123
HMG motifs, 123
HNF-4, 97
Hormone-resistant cancer
  of breast, 255–57
  of prostate, 261–62
hst-1 gene, 375
Hydrocortisone, mammary development and, 195
20-Hydroxyecdysone, 47. *See also* Ecdysone expression

ICM cells, 95, 98, 103
IGF-I, 97–98, 101t
  estrogen and, 173–74
  in estrogen induction, 312–16
  in murine embryogenesis, 101t
IGF-II, 97–98, 99t
  in murine embryogenesis, 101
IGF-II/mannose-6-phosphate receptor, 98
IGF (insulin-like growth factor) binding proteins, 216–17
  in female reproductive tract development, 229–38. *See also* IGFs
  in murine embryogenesis, 101, 101t
IGF-I receptor, 116, 296
  in apoptosis, 274
  in cell growth, 273
  functional domains of, 276–78, 277t, 278t
  in human tumors, 278
  in murine embryogenesis, 102t
  as target of abnormal growth, 279t, 279–80, 280t
  in transformation, 274–76
IGF receptors, 409
IGFs (insulin-like growth factors)
  in body growth and development, 272–73
  and IGF binding proteins in female reproductive tract, 229–38
    IGF system, 230
    ovarian regulation, 231
    uterine regulation, 231–34
      with EGF, 235, *236*
  in murine embryogenesis, 97–98, 99t, 101t
  in prostate cancer progression, 337
IL-1, 341
  in murine embryogenesis, 107
IL-1*β*, 99t, 341
IL-1*α*, 100t
IL-1 receptor, 102t
IL-2, 100t
IL-3, 100t
IL-4, 100t
IL-5, 100t
IL-6, 99t, 105
IL-6 receptor, 102t
IL-7, 99t
Immunosuppression, in prostate cancer, 341–42
Immunotherapy, 429–31
IMP genes, 60
Indifferent stage of embryogenesis, 179–80
α-Inhibin, in murine embryogenesis, 100t

Inner cell mass (ICM) cells. *See* ICM cells
InsR gene, 293
Insulin, 98
estrogen and, 173–74
mammary development and, 195
in murine embryogenesis, 97, 100t
Insulin-like growth factor receptors. *See* IGF-II receptor; IGF-I receptor
Insulin-like growth factors. *See* IGFs
Insulin receptor, 409
C-terminal deletions in, 300
in murine embryogenesis, 102t
mutational deletion and, 293
Insulin receptors, 116
Insulin-response substrate-1 (IRS-1), in murine embryogenesis, 101
INT-2, 336–37
int-2, 409
Integrins, 342
β-1, 97
Interferon, tau, in murine embryogenesis, 100t
Interleukins. *See* IL
int-1 gene, 361. *See also wnt* genes
int-2 gene, 375
int-3 gene
developmental role of, 390–93
discovery of, 384
involvement in mammary tumorigenesis, 387
as member of *n* (*notch*) gene family, 385–87
MMTV-LTR/INT3 mouse model, 388–89
WAP-INT3 mouse model, 389–90
Invertebrates and amphibians, 3–78
Aedes aegypti, 57–58
C. elegans vulval development, 3–17, 152
additional examples of pattern formation, 14
cell rate establishment, 11–13
general patterns of, 4–6
morphogen and sequential signaling models, 6–11
Drosophila
ecdysone expression in, 45–78
gene regulation in chromosome puffing, 46–68
physiologic functions of ecdysone, 46
TGF-β-like Dpp signaling in pattern formation, 19–43
conclusions, 33
Dpp pathway, 23–25
early neurogenesis and adult wing development, 28–33
EGF-R pathway, 20–23
overview of EGF-R/Dpp signaling, 25–27
IRS-1, 101

JAK/STAT pathway, 20, 247–48
JE, in murine embryogenesis, 99t
JH, 49, 62
JH receptor, 66–67
c-*jun* proto-oncogene, 319–20
Juvenile hormone, 49, 62
Juvenile hormone receptor, 66–67

KAII gene, 332
Keratinocyte differentiation, 159
Keratinocyte growth factor. *See* KGF
Keratin 14 promoter, 138
KGF, 137–38. *See also* FGFs
sex hormone target organ stromal–epithelial interactions, 218–20
Kidney development, 155
Kinase-containing receptors. *See* Receptor tyrosine kinases; Tyrosine kinases
c-*kit* proto-oncogene, 295, 297–98, 301
Kit receptor, 297–98
ksr-1, 6
ksr-1 mutations, 6

labial gene, 27
β-Lactamase, 430
Lactoferrin, in estrogen induction, 312–16
Lactogenesis, 240. *See also* Mammary gland development
β-Lactoglobulin gene, 242–43
LET-23. *See also* EGF receptor
let-23, 5, 6–11, 13. *See also* EGFs
let-60, 6
LET 23 EGFR/LET 60 Ras pathway, 5
let-23 (lf);lin-12(gf) double mutants, 9
Let-23 protein, 151–52
Let-60 protein, 151–52
LET-60 Ras, 5, 6
let-60 ras, 5
let-60 ras mutations, 6
LET-23 receptor, 5
LET 23 receptor, 5
LET-23 receptor/LET-60RAS/MPK-1 MAP kinase pathway, 6–8
let-23 RTK pathway, 12

Leukemia
  acute myelogenous (AML), 292
  chronic myelomonocytic (CML), 292
Leukemia inhibitory factor. *See* LIF
Leu926Pro substitution, 298
L71-6 gene, 60
LIF; 96–97
  in murine embryogenesis, 94, 105
LIF receptor, 96
  in murine embryogenesis, 102t
LIN-3, 5
lin-12, 8, 10
LIN-45, 5
LIN-12, as compared with Notch, 9
lin-2 gene, 385. *See also INT3* gene; *n* gene
lin-3 gene, 5, 150
lin-12(lf);let-60(fg) mutants, 9
lin-12(lf) mutants, 10
lin-3 mutants, 5
lin-15 mutants, 6–7
lin-12 pathway, 12
Lin-3 protein, 150–51
Lin-45 protein, 151–52
LIN-3 receptor, 11
LIN-23 receptor, 11–13
Liposome delivery systems, 430
LNCaP cell line, 336, 338, 339
Loss-of-function mutations, 8
Loss of function mutations, 5–6
Lung development, ErbB in, 154
Luteinizing hormone (LH), 199–200

mad gene, 23–24
MAD proteins, 23–24
Malignant progression
  incremental multistep ("big bang") model, 328, *329*
  three-compartment model for prostate cancer, 338–43, *339*
Mammalian development, 91–190
  ErbB receptors and ligands in, 145–68
    background of study, 145–46
    developmental and physiological roles, 152–60
      ErbB-1, 152–55
      ErbB-2 coreceptor, 159–60
      neuregulins and neuregulin receptors, 155–60
    invertebrate modeling systems of, 160–62
    perspectives on, 160–61
    signaling network of, 149–50
    structural homology of proteins and ligands, 146–49, *147*
  ERKO mouse model of estrogen receptor in, 169–76
  FGF-4 transcriptional regulation in, 115–29
    background of study, 115–16
    cis-regulatory elements in promoter region, 118–20
    future perspectives, 125–26
    of gene, 117–18, 118t, *119*
    growth factor expression in early murine embryos and EC cells, 116–17
    identification of transcription factors, 12–123, 120–21
    in vitro model for transcription and repression, 124–25
    multiple cis-regulatory elements of distal enhancer, 121–22
  murine model of embryogenesis, 93–114
    directionality of signaling, 96
    growth factor families, 97–108. *See also specific families*
      activins, 106–7
      colony-stimulating factor-1 (CSF-1), 105–6
      fibroblast growth factors (FGFs), 101–3
      growth factor receptor pregastrulation expression and function, 102–3t
      hepatocyte growth factors, 108
      insulin-like growth factors (IGFs), 97–101
      interleukin-1 (IL-1), 107
      interleukin-6 (IL-6), 105
      leukemia-inhibiting factor (LIF), 105
      parathyroid hormone-related peptide (PhRO), 107
      placental lactogen-1 (PL-1), 107–8
      platelet-derived growth factors (PDGFs), 104–5
      pre-gastrulation growth factor expression, 99–101t
      stem cell factor (SCF), 106
      transforming growth factor-β (TGF-β), 106
      VG-1-related growth factor, 107
    lower vertebrate comparisons, 97
    null mutants, 96–97
    significance and events, 93–96
  progesterone receptor and, 177–90

Mammalian development *(Continued)*
background of study, 177–79
mammary gland effects, 183–86
molecular basis for progesterone action, 178–79
uterine effects of progestin, 179–83
tyrosine kinase receptor interactions in, 131–43
background of study, 131–32
FGFs and receptors in epithelial development, 137–38
neuregulin and c-erb receptors, 136–37
c-ret receptor and ligand in vivo, 138–39
SF/HGF and c-*met* in vivo growth and motility, 133–36
SF/HGF and neuregulin in breast epithelia, 132–33
Mammary gland. *See also* Mammary tumors/breast cancer
anatomy of human, 183
development of, 209. *See also* Reproductive tract development
FGFs in, 375–76
INT3 gene in, 383–85. *See also INT3* gene
neuregulins in, 158–59
progesterone receptor in, 184
prolactin in postnatal adult
mammary-cell synthesis of PRL, 197–98
normal, 194–95
PRL receptors and, 194–97
signal networks in, 239–49
JAK-STAT pathway, 247–48
mammopoiesis and lactogenesis, 240
prolactin signaling, 240–47
TK receptor interactions in, 131–43
wnt genes in, 373–75
human cell lines and tissue, 374t, 374–75
mouse, 373t, 373–74
Mammary gland factor, 242–43
Mammary tumors/breast cancer, 253–57
ErbB-2 coreceptor in, 159
estrogen receptors in, 254–55
FGFs in, 376–78
hormone-resistant, 355–57
INT3 gene in, 383–95. *See also INT3* gene
mammary gland biology and, 253
risk factors for, 253–54
tamoxifen in, 255–57
tyrosine kinases and signal transduction in
additional RTKs, 408–9
human c-SRC TK, 406
mouse
EGF interreceptor interactions, 402–3
EGF receptor, 398–99
EGF receptor ligands, 401–3
Neu/ErbB-2 receptor expression, 399–401
signaling from ErbB-2 receptor TK, 403–6
transgenic mouse models of c-Src PTK, 406–8
Mammopoiesis, 240. *See also* Mammary gland: development of
Mandibular proliferation, 155
Manduca sexta, 60–61
MAP kinase, 5
MAPK pathway, 149, 151–52, 426
in prostate cancer, 330–31
Matrix metalloproteinases (MMPs), 341
MAX protein, 331
MCF-7 cell line, 403, 405
MDA-231 cell line, 405
Mdm-2/p53, 97
medea gene, 23
MEK, 5
MEK-s, 5
MEN (multiple endocrine neoplasia), 138
type 2A, 294
type 2B, 296–97
Menstrual cycle
as compared with wound healing, 315–16
progesterone and
in mammary gland, 183–84
in uterus, 181–82
Met receptor, 409
MGF, 148
with ErbB protein, 147
MHC I genes, 331
Milk protein gene expression, prolactin receptors and, 196–97
MMPs (matrix metalloproteinases), 341
MMTV-LTR/INT3 mouse, 388–89
MMTV/PyV mT mouse, 405–7
MMTV-*v*-Ha-*ras* mouse, 331
Molting, metamorphic ecdysone responsc and, 49
Morphogens, 30
Morphogen signaling model, 6–11

mos gene, in prostate cancer, 330–31
*mos*MAPK pathway, 330
mothers against Dpp gene. *See mad* gene
MPK-1/SUR-1, 5
msh gene, 28
Mullerian inhibiting substance, 169
Multiple endocrine neoplasia (MEN), 138, 294
Murine models
  Balb/c mouse, 211
  EGFR-KO mouse, 218
  ERKO mouse, 169–76, 212–14
  growth factors in pre- and postimplantation embryogenesis, 93–114. *See also under* Embryogenesis
  in vitro model of FGF-4 transcriptional regulation, 115–29. *See also* FGF-4
  in vivo studies of EGF receptor interactions, 403
  MMTV-LTR/INT3 mouse, 388–89
  MMTV/PyV mT mouse, 405–7
  MMTV-*v*-Ha-*ras* mouse, 331
  p53 knockout mouse, 333–34
  PRKO mouse, 177–86, 214–15
  Tfm mouse, 209, 216
  Tg.NK mouse, 399–401
  WAP-INT3 mouse, 389–90
  wt mouse, 216
MXII genes, 331
c*myc* oncogene, 314–15
c-*myc* oncogene, in prostate cancer, 331, 342–43
MYC protein, in prostate cancer, 331
Myelin-forming oligodendrocytes, 157
Myxoma virus factor. *See* MGF

n gene, 370. *See also INT3* gene
NAF (*neu*-activating factor), 401
National Surgical Adjuvant Breast and Bowel Project (NSABP), 433
NDF (Neu differentiation factor), 148–49. *See also* Neuregulins
Neiuwkoop's center, 107
neu-activating factor (NAF), 401
Neu differentiation factor, 148–49. *See also* Neuregulins
neu oncogene, 330. *See also erbB-2* gene
  in-frame mutations of, 400–1
  mutational activation of, 294–95
  rat, 295, 296
  RTK expression in mammary epithelium of transgenic mice, 399–401
Neural growth factor. *See* NGF
neu receptor. *See* erbB-2 receptor
Neuregulins, 20, 148–49
  in breast epithelial development, 132–33
  with c-erb receptors, 136–37
  in heart development, 156
  in mammalian development
    of cranial nerves, 158
    of glial cells, 157
    of heart, 156
    of mammary gland, 158–59
    of neuromuscular synapses, 156–57
    of Schwann cells, 157
  in neuromuscular synapse development, 156–57
  in Schwann and glial cell development, 157
Neuropeptide hormones, in prostate cancer progression, 337–38
Neurotensin, 337
NGFs
  in murine embryogenesis, 100t
  in prostate cancer progression, 338
NIH3T3 cell line, 402, 404, 405, 425, 426
nm23 genes, 332
Nodal growth factor, in murine embryogenesis, 100t
Noggin protein, 25
Notch gene. *See also INT3* gene
Notch protein, 8, 9
Notochord, 4–5
NRGs. *See* Neuregulins
NSCLC cell line, 292–93
Nuclear hormone receptor superfamily, 51
Null mutants, in murine embryogenesis, 96–97

Octamer motifs (OCT-1-OCT-6), 122–23
Oct-3 gene, 123, 124–25
Oogenesis, 152
  Drosophila, 58
  ecdysone in, 50, 68
Organizers, in pattern development, 4–5
Ovarian cancer, ErbB-2 coreceptor in, 159
Ovarian development, IGFs and IGF binding proteins in, 231
Ovarian-independent growth, 316

p28, 426
p53, 328, 332, 343–44
  Mdm-2, 97
  and prostate cancer, 330–33
p55, 426
p140, 426

p228, 426
Pancreatic development, 155
Paracrine embryo-to-maternal signaling, 96
Parathyroid hormone-related peptide. *See* PhRP
Pattern formation
  C. elegans vulval development, 3–17
    cell rate establishment, 11–13
    general sequence of, 4–6
    morphogen and sequential signaling models, 6–11
  Drosophila
    imaginal disc, 14
    TGF-β-like Dpp signaling in pattern formation, 19–43
      conclusions, 33
      Dpp pathway, 23–25
      early neurogenesis and adult wing development, 28–33
      EGF-R pathway, 20–23
      overview of EGF-R/Dpp signaling, 25–27
  endoderm in, 95
  Xenopus body axis, 14
Pax gene, 371
PC-3 cell line, 335, 336, 339
PC1 pheochromocytoma cell line, 158
PDGF receptors
  β, 292
  B in murine embryogenesis, 102t
  A in murine embryogenesis, 102t
  in mutational deletions, 293
PDGFs (plasma-derived growth factors), in estrogen induction, 312–16
PDGFs (platelet-derived growth factors)
  A, 104–5, 116
    in murine embryogenesis, 99t, 101
  B, 105
    in murine embryogenesis, 104–5
p21 gene, 334
Phospholipase C. *See* PLC
PhRP, 95
  in murine embrogenesis, 97, 99t, 107
PI-3 kinase, 20, 296
PI-3′ kinase, 404, 407
PI-4′ kinase, 404
p53 knockout mouse, 333–34
PL-1, in murine embryogenesis, 100t, 107–8
Plakoglobin, 367–69. *See also wnt* genes
Plasma lactogen. *See* PL; PL-1
Plasminogen activators, 341
Platelet-derived growth factors. *See* PDGF; PDGFs
PLCτ, 20
PLC-τ, 296, 404, 407, 426
p185[neu] receptor. *See* erbB-2 receptor
pointed gene, 21
Polytene chromosome puffs, 46–50. *See also* Ecdysone expression
Postnatal development, 91–249
  IGFs and IGF binding proteins in female reproductive tract, 229–238
    IGF system, 230
    ovarian regulation, 231
    uterine regulation, 231–34
      with EGF, 235, *236*
  mammary gland signal networks, 239–49
    JAK-STAT pathway, 247–48
    mammopoiesis and lactogenesis, 240
    prolactin signaling, 240–47
  prolactin and, 193–205
    in mammary gland, 194–98
    reproductive tract
      female, 198–99
      male, 199–201
  stromal–epithelial sex hormone interactions, 207–20
    with androgens, 208–11
    with estrogens, 211–12
      ERKO mouse model, 212–14
    historical perspective, 207–8
    IGFs in prostate, 216–17
    paracrine/autocrine mediation, 217–20
      by epidermal growth factor (EGF), 217–18
      by keratinocyte growth factor (KGF), 218–20
    with progesterone, 214–16
PR gene, 171
PRKO mouse, 177–86, 214–15
PRL. *See* Prolactin
Progesterone
  with prolactin, 195
  target organ stromal–epithelial interactions, 214–16
Progesterone receptor
  A, B, and C isoforms of, 178–79
  A and B isoforms of, 178–79
  in estrogen control, 314–15
  PRKO mouse model, 177–86, 214–15
  in reproductive development, 177–90
    mammary gland effects, 183–89
      anatomy of human mammary gland, 183

in utero, 184
in menstrual cycle, 183–84
molecular basis for action, 178–79
uterine effects
during embryogenesis, 179–81
in mature uterus, 181
menstrual cycle and, 181–82
progesterone carrier proteins and, 182–83
of progestin, 179–83
Progression factors, 315
Prolactin
amniotic fluid, 199
decidual, 199
with luteinizing hormone, 199–200
in mammary gland embryogenesis, 240–47
mitogenic activity of, 194–95
in postnatal development
mammary gland, 194–198
reproductive tract
female, 198–99
male, 199–201
with progesterone, 195
with testosterone, 200
Prolactin-receptor-related protein, 199
Prolactin receptors, 196–98
Prostate cancer, 259–62
androgen receptor in, 261
biology of prostate gland and, 260
hormone-resistant, 261–62
malignant progression, 327–59
amplification of malignant phenotype, 328
growth factors in, 335–38
epithelial growth factors (EGFs), 337
fibroblast growth factors (FEFs), 336–37
insulin-like growth factors (IGFs), 337
nerve growth factors (NGFs), 338
neuropeptide hormones with, 337–38
transforming growth factor-β (TGF-β), 335–36
oncogenes in, 328–35
c-*myc* oncogene, 331
Ras protein and Ras pathway, 330–31
tumor suppressor genes and, 332–35
steroid hormones in, 338–39
three-compartment model, 339–44
three-stage model of
local compartment: progression to extravasation, 341–43
metastasis: potential role of *p53*, 343–44
vascular compartment: apoptosis resistance, 343
risk factors for, 260
Prostate development, prolactin in postnatal, 200–1
Protamine, 430
Protein phosphatase 2A, 407
Psoriasis, 154
ptc gene, 379
PTTH hormone, 68
Puffing, 46–50
larval polytene chromosome. *See also* Ecdysone expression

Radionuclide delivery systems, 430
Raf, 5
raf gene, in prostate cancer, 330–31
Raf serine-threonine kinase, 21
Ras-GAP, 426
Ras-GAP pathway, 404
RAS/MAPK pathway, 20–23
Ras/MAPK pathway, 5
ras oncogene, in prostate cancer, 330–31
RAS protein, 330–31
Ras-Raf pathway, 151–52
Rb gene, 332
rbp gene, 55
rbp mutations, 63
Receptor tyrosine kinases. *See also* Tyrosine kinases
general function of, 145–46
mutational activation in tumorigenesis, 289–309
by genomic rearrangement, 292
by internal deletion or mutation within extracellular domain, 292–96
by truncation, 291
Recognition-of-pregnancy protein, 96
Reichardt's membrane, 95
Relaxin, 97–98
Reproduction, ecdysone response and, 49–50
Reproductive tract development
embryologic. *See also* Embryogenesis; Mammalian development
ErbB receptor in, 145–68
estrogen receptor in, 169–76
progesterone receptor in, 177–90

Reproductive tract development *(Continued)*
  ERKO mouse model of estrogen in, 169–76
  female, IGFs and IGF binding proteins, 229–38
  postnatal adult, prolactin in
    female, 198–99
    male, 199–201
Resistance, to chemotherapeutic agents, 424–25
ret gene, 294, 296–97, 409
  in MEN2A and MEN2B patients, 138
  point mutation in, 300
c-*ret* receptors, 136–39
Rhodnius larvae chromosomal puffing, 60
rhomboid gene, 20–21, 26–27, 28, 32
Rhomboid protein, 21, 151
Rieger syndrome, 155
rolled gene, 21
RXR gene, 53
RXR receptor, 54

sax gene, 22, 23
Scatter factor (hepatocyte growth factor). *See* SF (scatter factor)
SCF, in murine embryogenesis, 99t
SCF/c-kit, in murine embryogenesis, 102t
SCFτ encoding gene delivery, 432
Schwann cell development, neuregulins in, 157
Scw protein, 25
sea proto-oncogene, 291
Seed and soil concept, 343
Seminal vesical development, 209
Sem-5 protein, 5, 151–52, *See also* Grb-2
Sensory and motor neuron-derived factor (SMDF), 148–49
Sensory mother cells (SMCs), 371
Sequential signaling model, 6–11
ser gene, 370
Serines, of murine estrogen receptor, 172–73
Serine/threonine kinase (cdc-2), 370
Sevenless protein, 151–52
sev gene, 291
SFGF, 148
  with ErbB protein, 147
SF (scatter factor), 108
  in breast epithelial development, 132–33
  in human cancer, 378–79
  in murine embryogenesis, 105
  with neuregulin in growth and motility, 133–36
Shc adaptor proteins, 407
shc gene, 20
Shc proteins, 404, 425–26
SH2 domain antagonists, 431–32
Shope fibroma virus growth factor. *See* SFGF
SH-2 proteins, 146–48
Signaling
  in mammary gland development, 239–49
    JAK-STAT pathway, 247–48
    mammopoiesis and lactogenesis, 240
    prolactin signaling, 240–47
  paracrine embryo-to-maternal, 96
siml mutants, 28
skw gene, 24–25
SMC cells, 371
SMDF, 148–49
SNX1, 429
sog gene, 24–25, 28–31, 32
Sog protein, 25
Somatostatin, 337
Sorting nexin, 429
sos gene, 21
Sos protein, 151–52, 404
Sox genes, 123
SOX-2 transcription factor, 123
spalt gene, 27
Spemann's organizer, 3, 4–5, 107. *See also Xenopus*
Spi protein, 150–51
spitz gene, 20, 26, 28, 147
Sp1 transcription factor, 120–21
SP3 transcription factor, 123
v-*src* oncogene, 341
src oncogene, in prostate cancer, 330–31
c-Src protein, 404, 405, 407–8
v-Src protein, 407–8
sry gene, 180
Sry-related HMG box, 123
Stage specificity, of ecdysone response, 67–68
Star gene, 20–21, 26
Stat proteins, 241–48
Steroid sex hormones. *See also specific hormones and receptors*
  breast cancer and, 253–59
  cancer and, 253–67
  endometrial cancer and, 258–59
  prostate cancer and, 259–62
  stromal–epithelial growth factor interactions, 207–20
    of androgens, 208–11
    of estrogens, 211–12
      ERKO mouse model, 212–14

historical perspective, 207–8
paracrine/autocrine mediation, 217–20
by epidermal growth factor (EGF), 217–18
by keratinocyte growth factor (KGF), 218–20
of progesterone, 214–16
in prostate with IGFs, 216–17
Su(H) protein, 386–87
Surfactant promoter, 137, 154
SVP, 65
Syndecan, in murine embryogenesis, 102t

Tamoxifen
in breast cancer, 255–58
in endometrial cancer, 259
Tau interferon, in murine embryogenesis, 100t
T47D cell line, 405
tel gene, 292
Testicular feminization (Tfm) mouse, 209, 216
Testosterone. *See also* Androgens; Steroid sex hormones
with prolactin, 200
Tfm mouse, 209, 216
TGF-α, 148
TGF-α, 20, 116, 217–18, 339, 341. *See also* EGFs
in epidermal development, 154
with ErbB protein, 147
in estrogen induction, 312–16
in GI tract development, 154–55
in lung development, 154
in mammary tumorigenesis, 401–2
in murine embryogenesis, 99t, 101, 103–4
in murine embryos, 153
in prostate cancer, 337
spitz and, 26
transgenic models of, 155
wa-1 and, 153
TGF-β, 14, 25, 138, 370
in estrogen induction, 312–16
in murine embryogenesis, 106
in prostate cancer, 334
in prostate cancer progression, 335–36
TGF-β1, 116
immunosuppressive effects of, 342
in murine embryogenesis, 99t, 101
TGF-β2, 99t
TGF-β3, 99t
TGF-β receptor, in murine embryogenesis, 107
TGF-β1 receptor, in murine embryogenesis, 102t
TGF-β2 receptor, in murine embryogenesis, 102t
TGF-β3 receptor betaglycan, in murine embryogenesis, 102t
TGF-β-like Dpp signaling in pattern formation
Dpp pathway, 23–25
early neurogenesis and adult wing development, 28–33
EGF-R pathway, 20–23
overview, 25–27
Tg.NK mouse, 399–401
thick veins gene. *See tkv* gene
Thyroid hormone, mammary development and, 195
Thyroid-stimulating hormone (TSH), 337
Tissue specificity, of ecdysone response, 67–68
TK-CAT promoter/reporter gene, 124
tkv gene, 23
tld gene, 24–25, 28–29
TNF-α, 341
in murine embryogenesis, 99t
top mutation, 20
torso gene, 293
Torso protein, 151–52
Transcortin, 182
Transcription-factor-like *tel* gene, 292
Transforming growth factor. *See* TGF
Transforming growth factor-β. *See* TGF-β
Transforming growth factor-α. *See* TGF-α
tsg gene, 24–25, 28
Tumorigenesis, 253–443. *See also* Malignant progression and *specific lesions; specific lesions*
EGF-R and, 22
erbB-2 gene in
clinical application, 421–43
antibody specificity in, 427–28
inhibition of $p185^{erbB\text{-}2}$-initiated signal transduction, 431–32
inhibition of $p185^{erbB\text{-}2}$ synthesis, 432
in prognosis, 423
of receptor signaling, 425–27
in therapeutic response prediction, 423–24
as therapeutic target, 428–33
vaccines, 433
overexpression in cancer, 422–23, 423t

Tumorigenesis *(Continued)*
pathogenetic role, 422
estrogen and growth factors in
reproductive tract, 311–25. *See also* Diethylstilbestrol (DES)
background, 312–16
estrogen-induced carcinogenesis, 316–20, *317, 318*
IGF-I receptor in, 269–86
in apoptosis, 274
in cell growth, 273
functional domains of, 276–78, 277t, 278t
in human tumors, 278
IGF system, 269–71, *270*
receptor properties, 271–72
as target of abnormal growth, 279t, 279–80, 280t
in transformation, 274–76
INT3 gene in mammary, 383–95
developmental role of, 390–93
discovery of gene, 384
gene as member of *n* (*notch*) gene family, 385–87
MMTV-LTR/INT23 mouse model, 388–89
WAP-INT3 mouse model, 389-90
mutational activation of receptor tyrosine kinases in, 289–309
by genomic rearrangement, 292
by internal deletion or mutation within extracellular domain, 292–96
mutations inducing or stabilizing receptor dimerization, 290–96
by truncation, 291
prostate cancer, 327–59
malignant progression in
amplification of malignant phenotype, 328
growth factors in, 335–38
oncogenes in, 328–35
steroid hormones in, 338–39
three-compartment model, 339–44
reproductive tract carcinogenesis, 311–35
estrogen action, growth factors, lactoferrin, and wound healing, 312–16, *313*
estrogen-induced, 316–20, *317, 318*. *See also* Diethylstilbestrol (DES)
sex steroids and cancer, 253–67
breast cancer, 253–59
endometrial cancer, 258–59
prostate cancer, 259–62
wnt and FGFs in, 361–82
classification of *wnt* genes, 363–65
downstream alterations in human cancer, 378–79
Drosophila wnt family, 364
autocrine and paracrine pathways for WG signaling, 371
WG/*wnt* signaling, 366–71
FGF family, 375
fgf gene in mammary gland development, 375–76
interaction with *wnt* in development, 376
wnt family, 361–63, 362t
wnt family in normal development, 372–75
wnt/*fgf* expression in mammary carcinoma, 376–78
Tumor necrosis factor. *See* TNF
Tumor suppressor genes, in prostate cancer, 332–35
Tyrosine kinase receptors
for epidermal growth factors, 103
interactions in mammalian development, 131–43
background of study, 131–32
FGFs and receptors in epithelial development, 137–38
neuregulin and c-erb receptors, 136–37
c-ret receptor and ligand in vivo, 138–39
SF/HGF and c-*met* in vivo growth and motility, 133–36
SF/HGF and neuregulin in breast epithelia, 132–33
Tyrosine kinases. *See also* Receptor tyrosine kinase
in prostate cancer, 330
and signal transduction in mammary tumorigenesis
additional RTKs, 408–9
human c-SRC tyrosine kinase, 406
mouse, 397–419
transgenic mouse models of c-Src PTK, 406–8
and signal transduction mammary tumorigenesis, mouse
EGF interreceptor interactions, 402–3
EGF receptor, 398–99
EGF receptor, ligands, 401–3
Neu/ErbB-2 receptor expression, 399–401
signaling from ErbB-2 receptor TK, 403–6

Ubx gene, 27
ultraspiracle gene, 50
Urinary bladder, estrogen receptors in, 212
usp gene, 52–53, 54, 57–58
Uterine development. *See also* Reproductive tract development
IGFs and IGF binding proteins in, 231–34
with EGF, 235, *236*
Uterine (endometrial) cancer, sex steroids and, 258–59

Vaccines, 433
Vaccinia growth factor. *See* VGF
Vascular endothelial growth factor/vascular permeability factor, 104
Vein formation, 26–27, 31–33
vein gene, 20
Vein protein, 151
VGF, 148
with ErbB protein, 147
VG-1-related growth factor, in murine embryogenesis, 106
Vitellogenesis, ecdysones and, 50, 68
Vox/Xvent protein, 24
Vulval development. *See also* *Caenorhabditis elegans*
in *C. elegans*, 3–17

wa-1, 153
wa-2 mutation, 408–9
wap gene, 242–43, 389–90
WAP-INT3 mouse, 389–90
wg gene, 370–71, 371. *See also wnt* genes
wg genes, 351
wnt genes
autocrine and paracrine WG signaling in *Drosophila* embryos, 371
background and species occurrence, 361–63, 362t
classification of, 363–65
by function, 363–64
into subfamilies, 363
into transforming and nontransforming subgroups, 363
downstream alterations in human cancer, 378–79
Drosophila family of, 364–65
FGF family with, 375–78
mammary gland development and, 375–76
wnt/fgf expression in mammary carcinoma, 376–78
Xenopus studies of, 376
in normal development, 372–75
axis and mesoderm formation in Xenopus, 372
CNS development in mammals, 372
mammary gland development in mammals, 372–75
humans, 374t, 374–75
mice, 373t, 373–74
W/*wnt-1* signaling in *Drosophila* and mammalian cells, 366–71
Armadillo (ARM) (β-catenin, plakoglobin), 367–69
disheveled (*dsh*) gene and DSH receptors, 366–67
frizzled (*fz*) gene and FZ receptors, 366
other components of signaling, 369–71
shaggy/*zeste-white-3* (*zw-3*) and glycogen synthase kinase-3, 367
Wound healing, as compared with menstrual cycle changes, 315–16
wt mouse, 216

Xenopus
BMP-4 in, 31
body axis development, 3, 4–5
chordin gene, 31
Chordin homolog of Sog, 25
endoderm in patterning, 95
FGFs in embryogenesis, 101
negative activin receptor in, 107
Neiuwkoop's center in, 107
Noggin protein, 25
pregastrulation embryogenesis in, 97
Vox/Xvent protein, 24
wnt genes in, 363–64, 372, 376. *See also* *wnt* genes

c-Yes protein, 404
Yolk protein, ecdysone and, 50
Yolk sac endoderm, 97

Z isoforms of BR proteins, 55
Zebrafish, *dino* mutant, 31
zen gene, 28–29
Zen protein, 24
Zone of polarizing activity, 4–5
ZR75-1 cell line, 405
zw3 gene, 367. *See also wnt* genes
ZW-3 kinase, 371